Food Macromolecules and Colloids

Food Macromolecules and Colloids

Edited by

E. Dickinson
Procter Department of Food Science, University of Leeds, UK

D. Lorient
ENSBANA, Université de Bourgogne, Dijon, France

ROYAL SOCIETY OF CHEMISTRY

The Proceedings of a Conference organized by the Food Chemistry Group of The
Royal Society of Chemistry, held at ENSBANA, Université de Bourgogne, Dijon,
France on 23–25 March 1994

The cover diagram shows the microstructure of a commercial calcium caseinate. See
Figure 1b on page 183

Special Publication No. 156

ISBN 0-85404-700-X

A catalogue record for this book is available from the British Library.

Published by The Royal Society of Chemistry,
Thomas Graham House, Science Park, Milton Road,
Cambridge CB4 0WF, UK

Typeset in Great Britain by Computape (Pickering) Ltd, Pickering, North Yorkshire
Printed and bound in Great Britain by
Antony Rowe Ltd, Chippenham, Wiltshire

Preface

Food macromolecules play a crucial role in the formulation of a wide range of food products—beverages, bread, cheese, desserts, dressings, ice-cream, spreads and so on. The constituent phases in these food colloids may be liquid, crystalline, gaseous, glassy or gel-like. Product quality depends on interactions between the macromolecules and other food ingredients such as fats, sugars, surfactants, salts, flavours and aroma compounds. These interactions are sensitive to processing conditions during mixing, freezing, drying, baking, *etc.* An important functional property of proteins is in the stabilization of emulsions and foams by adsorption at oil–water and air–water interfaces. An important functional property of polysaccharides is the control of aqueous phase rheology through biopolymer association and gelation behaviour. Final food product structure and texture depends on a balance between different kinds of macromolecular interactions.

This volume records the proceedings of an International Symposium on 'Food Macromolecules and Colloids' held at ENSBANA (Ecole Nationale Supérieure de Biologie Appliquée à la Nutrition et à l'Alimentation), University of Burgundy, Dijon, France, on 23–25 March 1994. The conference was the fifth in a series of biennial Spring Symposia on the subject of food colloids to be organized by the Food Chemistry Group of The Royal Society of Chemistry. The main theme of the Dijon conference was the role of macromolecular interactions in determining the physical and biochemical properties of well-defined multi-phase, multi-component systems. The programme included invited overview lectures, contributed oral presentations, and a poster exhibition. The social highlight of the event was a memorable conference dinner at the Château du Clos de Vougeot. The meeting was attended by over 210 people from 22 different countries. This book collects together the lecture contributions and more than half the poster presentations.

We are very pleased to acknowledge here the important contributions of Dr R. D. Bee, Professor P. Richmond and Professor P. Walstra to the International Committee responsible for setting up the scientific programme of this meeting. For their enthusiasm and hard work, we are extremely grateful to all the members of the Local Organizing Committee (B. Colas, J.-L. Courthaudon, G. Lavoué, M. Le Meste, D. Simatos, A. Voilley), and to all those at ENSBANA, especially C. Compan, the secretary of the conference, who assisted in the smooth running of the local arrangements. Finally we acknowledge our gratitude to the many industrial companies who sponsored the conference.

E. Dickinson (Leeds)
D. Lorient (Dijon)
May 1994

Contents

INTRODUCTORY LECTURE   1
Recent Trends in Food Colloids Research
E. Dickinson

Adsorbed Layers

INVITED LECTURE   23
Structure and Properties of Adsorbed Layers in Emulsions
Containing Milk Proteins
D. G. Dalgleish

INVITED LECTURE   34
Structure of Proteins Adsorbed at an Emulsified Oil Surface
M. Shimizu

A Phenomenological Model for the Dynamic Interfacial   43
Behaviour of Adsorbed Protein Layers
G. A. van Aken

Association of Chymosin with Adsorbed Caseins   50
A. L. de Roos, P. Walstra, and T. J. Geurts

Surface Activity and Competitive Adsorption of Milk Component 3   58
and Porcine Pancreatic Lipase at the Dodecane–Water Interface
J.-L. Courthaudon, J.-M. Girardet, C. Chapal, D. Lorient, and G. Linden

Application of Polymer Scaling Concepts to Purified Gliadins   71
at the Air–Water Interface
J. Hargreaves, R. Douillard, and Y. Popineau

A Neutron Reflectivity Study of the Adsorption of β-Casein at   77
the Air–Water Interface
P. J. Atkinson, E. Dickinson, D. S. Horne, and R. M. Richardson

Effect of Temperature on Lipid–Protein Interactions at the   81
Oil–Water Interface
*J.-L. Gelin, P. Tainturier, L. Poyen, J.-L. Courthaudon, M. Le Meste,
and D. Lorient*

Modification of the Interfacial Properties of Whey by Enzymic   85
Hydrolysis of the Residual Fat
C. Blecker, V. Cerne, M. Paquot, G. Lognay, and A. Sensidoni

Influence of Charge on the Adsorption of Proteins to Surfaces   90
J. Leaver, D. S. Horne, C. M. Davidson, and D. V. Brooksbank

Surface Properties of the Milk Fat Globule Membrane: Competition   95
between Casein and Membrane Material
S. Chazelas, H. Razafindralambo, Q. Dumont de Chassart, and M. Paquot

Surface-active Properties of Mixed Protein Films Containing   99
Caseinate + Gelatin
V. B. Galazka, B. T. O'Kennedy, and M. K. Keogh

Protein Adsorption and Protein–Monoglyceride Interactions at   103
Fluid–Fluid Interfaces
J. M. Rodríguez Patino and M. R. Rodríguez Niño

Destabilization of Monoglyceride Monolayers at the Air–Water   109
Interface: Structure and Stability Relationships
J. M. Rodríguez Patino and J. de la Fuente Feria

Competitive Adsorption of Spherical Particles of Different Sizes   114
by Molecular Dynamics
E. G. Pelan and E. Dickinson

Protein Interactions and Functionality

INVITED LECTURE   123
Protein–Aroma Interactions
S. Langourieux and J. Crouzet

Some Changes to the Properties of Milk Protein Caused by High-   134
Pressure Treatment
D. E. Johnston and R. J. Murphy

Surface Energy at the Ice–Solution Interface for Systems Containing   141
Antifreeze Biopolymers
D. S. Reid, W. L. Kerr, J. Zhao, and Y. Wada

Studies of Interactions between Casein and Phospholipid Vesicles 146
Y. Fang and D. G. Dalgleish

Effect of Protein on the Retention and Transfer of Aroma Compounds 154
at the Lipid–Water Interface
B. A. Harvey, C. Druaux, and A. Voilley

Emulsifying and Oil-binding Properties of the Enzymic Hydrolysate 164
of Bovine Serum Albumin
M. Saito

Conformational Stability of Globular Proteins: A Differential Scanning 167
Calorimetry Study of Whey Proteins
P. Relkin, A. Muller, and B. Launay

Thermal Denaturation and Aggregation of β-Lactoglobulin Studied by 171
Differential Scanning Calorimetry
M. A. M. Hoffmann, P. J. J. M. van Mil, and C. G. de Kruif

Changes in Molecular Structure and Functionality during Purification 178
and Denaturation of Faba Bean Proteins
H. M. Rawel and G. Muschiolik

Microstructural, Physico-chemical, and Functional Properties of 182
Commercial Caseinates
P. Bastier, E. Dumay, and J.-C. Cheftel

Biochemical and Physico-chemical Characteristics of the Protein 189
Constituents of Crab Analogues Prepared by Thermal Gelation or
Extrusion Cooking
M. Thiebaud, E. Dumay, and J.-C. Cheftel

Interactions between Fat Crystals and Proteins at the Oil–Water 194
Interface
L. G. Ogden and A. J. Rosenthal

Emulsions

INVITED LECTURE 201
Surface Structures and Surface-active Components in Food Emulsions
B. Bergenståhl, P. Fäldt, and M. Malmsten

On the Stability of Milk Protein-Stabilized Concentrated Oil-in-Water 215
Food Emulsions
B. van Dam, K. Watts, I. Campbell, and A. Lips

Ultrasonic Studies of the Creaming of Concentrated Emulsions   223
E. Dickinson, J. G. Ma, V. J. Pinfield, and M. J. W. Povey

Formulation and Properties of Protein-Stabilized Water-in-Oil-   235
in-Water Multiple Emulsions
J. Evison, E. Dickinson, R. K. Owusu Apenten, and A. Williams

Effect of Non-Starch Polysaccharide on the Stability of Model   244
Physiological Emulsions
A. Fillery-Travis, L. Foster, S. Moulson, M. Garrood, S. Clark, and M. Robins

Investigation of the Function of Whey Protein Preparations in   248
Oil-in-Water Emulsions
G. Muschiolik, S. Dräger, H. M. Rawel, P. Gunning, and D. C. Clark

Shear Induced Instability of Oil-in-Water Emulsions   252
A. Williams and E. Dickinson

Surfactant–Protein Competitive Adsorption and Electrophoretic   256
Mobility of Oil-in-Water Emulsions
J. Chen, J. Evison, and E. Dickinson

Osmotic Pressure of Emulsions Containing Polysaccharide + Non-ionic   261
or Anionic Surfactants
E. Dickinson, M. I. Goller, and D. J. Wedlock

Interfacial and Stability Properties of Emulsions: Influence of   269
Protein Heat Treatment and Emulsifiers
E. Dickinson and S.-T. Hong

Foams

Surface and Bulk Properties in Relation to Bubble Stability in Bread   277
Dough
J. J. Kokelaar, T. van Vliet, and A. Prins

Comparison of the Foaming and Interfacial Properties of Two   285
Related Lipid-binding Proteins from Wheat in the Presence of a
Competing Surfactant
F. Husband, P. J. Wilde, D. Marion, and D. C. Clark

Reflectance Studies on Ice-Cream Models   297
R. D. Bee and R. J. Birkett

Disproportionation in Aerosol Whipped Cream   309
M. E. Wijnen and A. Prins

Determination of Protein Foam Stability in the Presence of 312
Polysaccharide
E. Izgi and E. Dickinson

Bubble Growth on an Active Site: Effect of the Cavity Volume 316
A. F. Zuidberg and A. Prins

Mixed Biopolymer Systems

INVITED LECTURE 321
Thermal Behaviour of Kappa-Carrageenan + Galactomannan Mixed
Systems
P. B. Fernandes, M. P. Gonçalves, and J.-L. Doublier

Whey Protein + Polysaccharide Mixtures: Polymer Incompatibility 328
and Its Application
A. Syrbe, P. B. Fernandes, F. Dannenberg, W. Bauer, and H. Klostermeyer

Effect of Sodium Caseinate on Pasting and Gelation Properties of 340
Wheat Starch
C. Marzin, J.-L. Doublier, and J. Lefebvre

Colloidal Stability and Sedimentation of Pectin-Stabilized Acid Milk 349
Drinks
T. P. Kravtchenko, A. Parker, and A. Trespoey

Decrease of *In Vitro* Hydrolysis of Soybean Protein by Sodium 356
Carrageenan
*J. Mouécoucou, C. Villaume, H. M. Bau, A. Schwertz, J. P. Nicolas,
and L. Méjean*

Gels and Networks

INVITED LECTURE 363
The Importance of Biopolymers in Structure Engineering
A.-M. Hermansson

INVITED LECTURE 376
Physical Chemistry of Heterogeneous and Mixed Gels
V. J. Morris and G. J. Brownsey

Investigation of Sol–Gel Transitions of β-Lactoglobulin by Rheological 390
and Small-angle Neutron Scattering Measurements
D. Renard, M. A. V. Axelos, and J. Lefebvre

High Pressure Gelation of Fish Myofibrillar Proteins   400
A. Carlez, J. Borderias, E. Dumay, and J.-C. Cheftel

Gelation of Protein Solutions and Emulsions by Transglutaminase   410
Y. Matsumura, Y. Chanyongvorakul, T. Mori, and M. Motoki

Sintering of Fat Crystal Networks in Oils   418
D. Johansson, B. Bergenståhl, and E. Lundgren

Thermal Gelation of Sunflower Proteins   426
A. C. Sánchez and J. Burgos

Binding of Calcium Ions by Pectins and Relationship to Gelation   431
C. Garnier, M. A. V. Axelos, and J.-F. Thibault

Heat-induced Denaturation and Aggregation of β-Lactoglobulin: 437
Influence of Sodium Chloride
M. Verheul, S. P. F. M. Roefs, and C. G. de Kruif

Rheological and Mechanical Properties

INVITED LECTURE   447
Mechanical Properties of Concentrated Food Gels
T. van Vliet

Scaling Behaviour of Shear Moduli during the Formation of Rennet   456
Milk Gels
D. S. Horne

Sol–Gel Transition of ι-Carrageenan and Gelatin Systems: 462
Dynamic Visco-elastic Characterization
C. Michon, G. Cuvelier, B. Launay, and A. Parker

Effect of Retrogradation on the Structure and Mechanics of   472
Concentrated Starch Gels
C. J. A. M. Keetels, T. van Vliet, and H. Luyten

Mechanical Properties of Thermo-reversible Gels in Relation to their   480
Structure and the Conformations of their Macromolecules
E. E. Braudo and I. G. Plashchina

Effect of Hydrocolloid Concentration on Mechanical Behaviour of   488
Orange Gels
S. M. Fiszman, M. C. Trujillo, and L. Durán

Effect of Starter Culture on Rheology of Yoghurt   492
H. Rohm

Rheology of Mixed Carrageenan Gels: Opposing Effects of Potassium   495
and Iodide Ions
A. Parker

Rheology of Semi-sweet Biscuit Doughs   499
G. Oliver and S. S. Sahi

Bulk and Surface Rheological Properties of Wafer Batters   503
G. Oliver and S. S. Sahi

Effect of Dry Ultra-fine Size Reduction on Physico-chemical Properties   507
of Pea Starch
S. Jacqmin and M. Paquot

Influence of Fat Globule Size on the Rheological Properties of a   512
Model Acid Fresh Cheese
C. Sanchez, K. Maurer, and J. Hardy

Glasses

INVITED LECTURE   519
Influence of Macromolecules on the Glass Transition in Frozen
Systems
D. Simatos, G. Blond, and F. Martin

Phenomenon of Enthalpy Relaxation at the Glass Transition   534
Temperature in Granular Starches
C. C. Seow

Kinetic Processes in Highly Viscous, Aqueous Carbohydrate Liquids   543
T. R. Noel, R. Parker, and S. G. Ring

Calculation of Glass Transition Temperature of Food Proteins and   552
Plasticizer Effects of Different Ingredients
Yu. I. Matveev

Water Adsorption and Plasticization of Amylopectin Glasses   556
K. Jouppila, T. Ahonen, and Y. Roos

Influence of Moisture Content on Glass Transition Temperature of the   560
Amorphous Matrix in 'Xixona Turrón'
N. Martínez, M. P. Betrán, and A. Chiralt

Phase Transitions of Tapioca Starch   566
V. Garcia, A. Buleon, P. Colonna, G. Della Valle, and D. Lourdin

CONCLUDING REMARKS   572
D. Lorient

Subject Index   **575**

Recent Trends in Food Colloids Research

By Eric Dickinson

PROCTER DEPARTMENT OF FOOD SCIENCE, UNIVERSITY OF LEEDS, LEEDS LS2 9JT, UK

1 Introduction

At the RSC Food Chemistry Group meeting on *Food Emulsions and Foams* in Leeds in 1986, Dr Don Darling gave an introductory lecture entitled 'Food colloids in practice' in which he stated his perception of the current level of understanding of the various factors affecting the formulation and stability of food colloids. To a mixed audience of physical chemists and food scientists the lecturer made some forthright suggestions about where useful new research should be done to apply the principles of colloid science in real food systems. As the published version[1] of the Darling lecture has been widely cited, its content is presumably fairly well known to most of the participants at the present symposium. It therefore offers a convenient arbitrary benchmark against which to measure progress in food colloids research in recent years. In the context of the present article, then, the term 'recent' describes research reported or carried out during the past eight years.

As in most fields of scientific endeavour, the main spurs to progress in the field of food colloids are, firstly, the increased availability of new experimental techniques and advanced instrumentation, and, secondly, the emergence of influencial new concepts and theoretical ideas. As demonstrated clearly by the contributions to this volume, scientists working on food colloids and macromolecules are now making extensive use of a wide range of advanced physico-chemical techniques based on principles of spectroscopy, microscopy, calorimetry, scattering, ultrasonics, rheology, and so on. The increased availability of small powerful computers has led to substantial enhancements in instrument sensitivity and data handling capacity, as well as to much greater opportunities for testing experiments against theories and for developing realistic computer simulation models. Some new theoretical terms that have found their way into the language of food colloids are 'fractal dimension', 'glass transition', 'depletion flocculation', 'thermodynamic incompatibility'— to name just four.

The direction and pace of developments is also influenced by various non-technical considerations such as consumer preference, commercial interests and

1

the priorities of research funding organizations. Two particular issues which are a continuing stimulus to fundamental research are the development of low-fat products and the enhancement of the functionality of 'natural' food ingredients. Removal of fat from traditional products without loss of texture or flavour characteristics is a formidable challenge for the food technologist. The attempted solution normally involves a greatly increased usage of food macromolecules—proteins and/or polysaccharides—which focuses increased attention on the interactions of these biopolymers with other ingredients (*i.e.* lipids, water, other macromolecules, *etc.*). This in turn provides commercial justification for the tendency in recent years for researchers to study in some detail the properties of well-characterized *mixed* systems, *e.g.* the rheology of mixed biopolymer gels, the competitive adsorption of mixed emulsifiers, *etc.* Another noticeable trend in recent years has been the move away from research on *chemical* modification of biopolymers. More emphasis is now placed by food researchers on the implications of *physical* modification or physical processing—heating, drying, high pressure, extrusion, or just simple mixing.

In this article, the author describes a few areas in which there have been some interesting developments during the past few years. Three of these topics are preceded by quotations from the lecture by Don Darling[1] as a way of showing the extent (or lack) of progress during the intervening period. The view expressed here is necessarily highly subjective because it is based on work either done in Leeds or closely related to work done in Leeds. Where other topics are not mentioned, it is certainly not because the author considers them to be any less important—it is just that there is not space to cover all aspects of food colloids in this short introductory talk, and so it seems appropriate to concentrate on what the author knows best.

2 Properties of Adsorbed Protein Layers

> ". . . thickness of the stabilizing layer around a droplet is a practical indicator of the strength of the repulsive force between two colloidal particles. While characterization techniques exist, it [the thickness] is rarely measured in practice."

Protein-stabilized emulsions are the most important class of food colloids.[2] A pure protein film adsorbed at the oil–water interface is a thin dense layer of strongly interacting macromolecules. It is useful to distinguish between an adsorbed layer formed from a disordered flexible protein like β-casein and that from a compact globular protein like β-lactoglobulin. The former is known[3] to be considerably thicker, less dense and more mobile than the latter. The monolayer surface coverage of β-casein at hydrophobic surfaces lies in the range 2–3 mg m^{-2} whereas that for a globular protein is typically 1–2 mg m^{-2}. The surface shear viscosity of β-casein at the oil–water interface is 2 or 3 orders of magnitude smaller than that of a globular protein under similar conditions.[4,5]

A sensitive new technique for studying adsorbed layer structure at fluid interfaces is specular neutron reflectance.[6,7] The essence of the experiment is to measure the neutron reflectivity R as a function of the wave-vector transfer Q perpendicular to the surface. Wavelength λ is varied at constant incident angle θ by scanning the time of flight of the polychromatic pulsed neutron beam. In the absence of any adsorbed material, the so-called Fresnel reflectivity from a clean sharp interface is

$$R_F = 16\pi^2 \Delta\rho^2/Q^4 \qquad (1)$$

where $Q = (4\pi/\lambda)\sin\theta$ and $\Delta\rho$ is the difference in scattering length density between the incident medium and the subphase. At the air–water interface, when the aqueous phase consists of a 92:8 mixture (by volume) of H_2O to D_2O, the subphase is contrast matched to the air, and $\Delta\rho$ is zero. In the presence of an adsorbed layer at the interface between air and air-contrast-matched water, the reflectivity at low Q is given by

$$R = (16\pi^2/Q^2)m^2\exp(Q^2\sigma^2) \qquad (2)$$

where σ is the second moment of the adsorbate distribution,

$$\sigma = (<z^2> - <z>^2)^{\frac{1}{2}} \qquad (3)$$

and m is the scattering length density integrated over the adsorbed layer:

$$m = \int_{-\infty}^{+\infty} \rho(z)\mathrm{d}z \qquad (4)$$

The thickness σ of the adsorbed layer is obtained from a Guinier plot of $\ln RQ^2$ against Q^2; the plot is a straight line of slope $-\sigma^2$ and intercept $\ln[(16\pi^2)m^2]$. The adsorbed amount Γ (mass per unit area) is related to m by

$$\Gamma = (Mm/N_A)\left[\sum_i b_i\right]^{-1} = Mm/N_A b_m \qquad (5)$$

where M is the adsorbate molecular weight, N_A is Avogadro's number, and b_i is the scattering length of the ith atom in the adsorbate molecule. The total molecular scattering length for β-casein is $b_m = (5.43 \pm 0.13) \times 10^{-12}$ m, where the quoted error reflects uncertainty about the extent of H/D exchange when the protein is dissolved in contrast-matched water.[8] Table 1 records values of the quantities m, σ, and Γ calculated from Guinier plots for β-casein adsorbed at the surface of air-contrast-matched water at pH 5.4, pH 6.0 and pH 7.0. The bulk protein concentration is 5×10^{-3} wt% in each case. The results show a substantially increased layer thickness and adsorbed amount as the pH is lowered towards the protein's isoelectric point. Using a non-linear least-squares fitting routine, data for β-casein at air–water and oil–water interfaces can be fitted[9,10] by a two-layer model. The inner layer, which is directly at the fluid interface, has a thickness of ca. 2 nm and a high protein volume fraction ($>90\%$). Extending beyond this into the aqueous phase is an outer layer of thickness 5–7 nm and a much lower volume fraction (ca. 15%).

Table 1   *Properties of adsorbed layer of β-casein at the air–water surface as a function of pH from neutron reflectance. The tabulated quantities are defined as follows: m is the total scattering length density from eqn (4); Γ is the surface concentration from eqn (5); σ is the second moment of the adsorbate distribution; σ$_{tot}$ is the total layer thickness obtained by fitting the reflectivity data by a two-layer model*

pH	$m/10^{-5}$ Å^{-1}	Γ/mg m^{-2}	σ/nm	σ_{tot}/nm
7.0	2.82	2.05 ± 0.1	1.65 ± 0.07	5.6 ± 0.3
6.0	3.90	2.85 ± 0.1	2.39 ± 0.08	7.2 ± 0.4
5.4	5.35	3.90 ± 0.15	2.57 ± 0.08	8.5 ± 0.5

Taken together with measurements from ellipsometry,[11] X-ray scattering[12] and dynamic light-scattering[13] the picture which emerges for β-casein is one of a rather thick adsorbed layer with about 70% of the total integrated polymer density located in trains or very small loops very close to the Gibbs surface (within 2 nm) and the rest present as longer loops and long tails extending up to 10 nm into the aqueous phase.[3] Such a configuration is consistent with that simulated by a simple Monte Carlo model of the adsorption of a β-casein-like polymer chain.[14]

When a globular protein adsorbs at a fluid interface, there is a loss of the native tertiary structure as hydrophobic side-chains distribute themselves towards the non-aqueous side of the interface. The degree of molecular extension, flattening and unfolding depends on the strength of the protein–surface interaction, the strength of the intramolecular bonds, and the space available at the interface. Strong intramolecular bonds (disulfide linkages) are unaffected by initial adsorption, although for some globular proteins (*e.g.* β-lactoglobulin) there is slow polymerization via sulfhydryl–disulfide interchange in the ageing adsorbed layer.[15] Evidence exists[16,17] to suggest that secondary structures of adsorbed and unadsorbed proteins are not very different.

Under certain conditions for some globular proteins in solution, a so-called 'molten globule' state has been described[18] having well-defined structural and dynamic properties between the native and the completely unfolded states. The molten globule state is defined[19,20] as a protein configurational state with native-like ordered secondary structure but no tertiary structure. One may speculate[21] that, for some globular proteins, the native protein structure changes to the molten globule state on adsorption. Furthermore, for adsorption from the molten globule state in solution, it might reasonably be supposed that adsorption and the associated interfacial conformational rearrangements could occur more rapidly than for adsorption from the native state. So, in addition to its role in gelation,[22] the molten globule state concept may provide new useful insight into the relationship between protein structure and functionality at liquid interfaces.[21]

An adsorbed layer of globular protein molecules may be represented as a dense quasi-two-dimensional assembly of interacting deformable particles.[23] A convenient deformable particle model suitable for investigation by computer

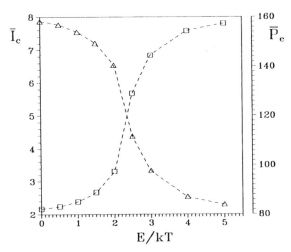

Figure 1   *Structure of 30-subunit deformable particle simulated by Brownian dynamics as a function of the inter-subunit interaction energy E. Plotted on the same graph are the average subunit coordination number $\bar{I}_c$ (□) and the average particle perimeter area $\bar{P}_e$ (△). (See reference 26 for full details of how these quantities are defined and calculated)*

simulation[24] consists of a large number of loosely linked subunits, each of which is free to move relative to all the others, subject only to maintenance of the integrity of the whole connected structure. (Notice that this deformable particle model with freely interchangeable subunits is different from normal models of linear or branched polymers which have *fixed* inter-subunit bonds.) When the subunits are identical rigid spheres, the deformable particle is analogous to a floc of colloidal spheres[25] which is not allowed to fall apart, and may adopt a compact close-packed structure or an open fractal-type structure depending on the strength of the subunit–subunit interaction.[26] Figure 1 shows how the average configuration of a multi-subunit three-dimensional deformable particle depends on the inter-subunit interaction energy E in an isolated particle of 30 spherical subunits. The results were obtained by Brownian dynamics simulation.[26] The quantities plotted are the average subunit coordination number $\bar{I}_c$ and the average perimeter area $\bar{P}_e$. Both functions, $\bar{I}_c(E)$ and $\bar{P}_e(E)$, have a sigmoidal shape with the largest change in property occurring at $E = 2–3\ kT$. As the E value is lowered, there is a transition from a compact structure to a stringy structure. This is somewhat analogous to the transition from collapsed coil to random walk chain on increasing the solvent quality in a solution of an excluded-volume linear polymer.[27] When adsorbed at a plane surface, the stringy 30-subunit deformable particle gives quite a different spatial distribution of subunits from that for the equivalent compact globular particle[26] or the equivalent flexible linear chain.[28] Exploration of the behaviour of this simple model may provide useful insight into the role of the molten globule state in protein adsorption.

oil

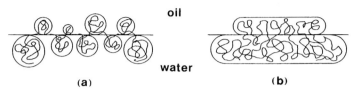

water

(a)   (b)

Figure 2   *(a) Schematic representation of adsorbed protein viewed as a series of loops segregated between oil and water phases. Each circle denotes the excluded volume of an individual loop. (b) Representation of general protein conformation as two pancakes after taking account of the entropic elasticity. The envelope describes the overall excluded volume of the whole molecule. (Adapted from reference 31)*

An interesting scaling approach to the thermodynamics of globular proteins at fluid interfaces has recently been introduced.[29–32] In the simplest version of the theory,[29] which is more applicable to the air–water interface, each adsorbed molecule is represented as a single 'pancake' (*une 'galette'*) of radius R and thickness D. In the refined version,[31] more applicable to the oil–water interface, each adsorbed molecule is represented by two pancakes (*deux galettes*) with radii R_w and R_o and thicknesses D_w and D_o, where the subscripts refer to the aqueous and oil sides of the interface, respectively. As illustrated in Figure 2(a), the adsorbed protein molecule at the oil–water interface is viewed as a series of loops located alternately in the oil and aqueous phases. It is argued that each loop region experiences 'good' solvent conditions—otherwise the loop would not choose to form in that phase. Train regions (*i.e.* flattened loops) exist where the solvent quality is poor for both phases. All loops floating within the phase (oil or water) within the pancake are assumed to be replaced by a single chain block as illustrated in Figure 2(b). This hypothetical block, which has the same affinity for the solvent as the loops, is trapped between the Gibbs surface and another parallel plane at a distance D_w or D_o from the Gibbs surface. The pancake radius R_w is calculated as the radius of a single chain of fN monomers where f is the fraction of monomers in the aqueous phase. From scaling arguments[33] for a polymer trapped in a slit in a good solvent, the pancake radius is given by

$$R_w \sim (fN)^{\frac{3}{4}} D_w^{-\frac{1}{4}} \tag{6}$$

and similarly for R_o. In the dilute regime, the surface pressure Π is directly proportional to the surface concentration Γ. Above a certain critical overlap concentration $\Gamma^* = (2R)^{-2}$, the surface pressure varies as

$$\Pi \sim \Gamma^y \tag{7}$$

where y is a parameter (>1) which depends on the solvent quality and the dimensionality of the local behaviour. According to scaling theory,[34] the behaviour is two-dimensional when the correlation length ξ exceeds the slit thickness D_w, and three dimensional for $\xi < D_w$. The total surface pressure in the refined theory[31,32] is taken as the sum of the pressures in the aqueous and oil phases:

$$\Pi = \Pi_w + \Pi_o \qquad (8)$$

It is possible for one layer (water or oil) to be in the semi-dilute regime ($\Gamma > \Gamma^*$) whereas the other is in the dilute regime ($\Gamma < \Gamma^*$). As proteins are predominantly hydrophilic, it seems reasonable to assume that the semi-dilute regime will normally be reached at a lower protein surface coverage for pancakes in the aqueous phase than for those in the oil phase. Under good solvent conditions, scaling theory gives

$$\Pi_w \sim \begin{cases} \Gamma^3 D_w^{-1} & (\xi > D_w) \\ \Gamma^{9/4} D_w^{-5/4} & (\xi < D_w) \end{cases} \qquad (9)$$

and equivalently for Π_o. In practice, the effective experimental exponent y in eqn (7) will reflect the behaviour of the layer (probably aqueous) with the strongest functional dependence on Γ; the other (oil) layer may still be in the dilute regime ($y = 1$) or in the semi-dilute regime with a lower value of y. A comparison has been made between this theory and thermodynamic data for β-lactoglobulin at the air–water interface[29] and bovine serum albumin at the oil–water interface.[32] The correlation between the measured exponent y and the inferred solvent quality of the aqueous phase appears reasonably satisfactory, considering the substantial simplifications made in the theoretical analysis and the substantial experimental error in the data (30–50 years old). Further progress with this promising approach will probably require more sensitive and reliable measurements of the surface equation of state of pure globular proteins over a wide range of conditions (temperature, pH and ionic strength).

3 Displacement of Proteins from Interfaces

"... emulsifiers destabilize sterically stabilized emulsions. Initial results suggest that the ability of emulsifiers to displace macromolecules is highly correlated with their surface pressures. So, can a 'league table' of specific surface pressures at equivalent concentrations be established which will define the equilibrium interfacial composition in foams and emulsions?"

Considerable progress has been made during the past few years in the area of the competitive adsorption of milk proteins and small-molecule emulsifiers. By studying model emulsion systems containing a single pure protein (β-casein or β-lactoglobulin) and a single emulsifier, considerable information has been obtained[34–45] concerning the various physical and chemical factors affecting the composition of adsorbed layers in oil-in-water emulsions. Further insight

Table 2 *Competitive adsorption of β-lactoglobulin + sucrose monoesters at neutral pH. The tabulated quantities are defined as follows: n is the fatty acid chain length of the emulsifier; R(50%) is the surfactant/ protein molar ratio for 50% displacement of protein from the oil–water interface in emulsions (0.4 wt% protein, 20 wt% oil); CMC is the surfactant critical micelle concentration at 30 °C; c_s is the surfactant concentration at which the surface tension at 30 °C is the same in the presence or absence of protein (0.1 wt%)*

Emulsifier[a]	n	R(50%)	CMC/mmol l^{-1}	c_s/mmol l^{-1}
Sucrose monolaurate	12	13	0.2	0.2
Sucrose monocaprate	10	16	2.0	1.0
Sucrose monocaprylate	8	60	8.5	2.5

[a] Supplied by Mitsubishi–Kasei Foods Corporation (Tokyo, Japan)

has been gained from complementary surface shear viscosity measurements[36,43,46] and comparisons of experimental data with theories and computer simulation.[46-50] The competitive displacement of β-casein from the air–water interface by non-ionic surfactant $C_{12}E_6$ has recently been demonstrated[8] by neutron reflectance. The relevance of these fundamental studies to realistic food emulsion product formulations has also been demonstrated.[51-55]

In a typical competitive adsorption experiment a fine oil-in-water emulsion is prepared from known amounts of protein and purified oil (e.g. *n*-tetradecane, soybean oil, silicone oil). If the emulsifier under investigation is oil-soluble, it is dissolved in the oil phase prior to homogenization. If the emulsifier is water-soluble, it may be present during emulsification or added afterwards. Analysis of the serum phase after centrifugation yields the amount of unadsorbed protein. From the known emulsion droplet-size distribution can then be calculated the protein surface coverage (in mg m^{-2}). By varying the emulsifier content at constant protein (and oil) content, the molar ratio *R* of surfactant to protein corresponding to partial or complete protein removal from the surface can be estimated. Table 2 gives experimental values[56] of *R* for 50% displacement of β-lactoglobulin from the oil–water interface by three pure sucrose esters (>99.9%) in *n*-tetradecane oil-in-water emulsions (0.4 wt% protein, 20 wt% oil, pH 7). The water-soluble surfactant was added after emulsification to the protein-stabilized emulsion with an initial protein surface coverage of 1.2 mg m^{-2}. We can see from the results in Table 2 that the emulsifier with the longest fatty acid chain (sucrose monolaurate) is the most effective at displacing protein from the interface. Also shown in the Table are values of the critical micelle concentration (CMC) for the three sucrose esters estimated from surface tension data at the air–water interface (pH 7), as well as the surfactant concentrations at which the surface tension is the same (within experimental error) in the presence or absence of β-lactoglobulin (0.1 wt% protein, pH 7). The sucrose ester which is the least effective at displacing β-lactoglobulin from the interface (sucrose monocaprylate) is the one with the highest CMC and the lowest hydrophobicity, and it is also the one which requires the highest surfactant concentration to produce equal tensions (*i.e.* equal surface free energies) in the presence or absence of protein. These results suggest that surface tension data do have a useful predictive role in determining the relative effectiveness of emulsifiers for protein displacement.

Some conclusions drawn from various recent experimental studies of competitive adsorption of milk proteins + surfactants in emulsions are listed below.

(1) Water-soluble non-ionic emulsifiers (Tweens, sucrose esters) are generally more effective at displacing proteins from liquid oil–water interfaces than oil-soluble non-ionic emulsifiers (Spans, monoglycerides) or zwitterionics (lecithins).

(2) Whether the water-soluble surfactant is added before or after emulsification has relatively little influence on its competitive adsorption behaviour.

(3) Protein surface coverage is usually lower at the triglyceride–water interface than at the less polar hydrocarbon–water interface, and the amount of surfactant required for complete protein displacement also tends to be lower.

(4) For emulsion droplets stabilized by β-lactoglobulin, the amount of water-soluble emulsifier required for complete displacement increases with the age of the protein adsorbed layer.

(5) For *n*-hexadecane droplets stabilized by β-casein, the amount of water-soluble emulsifier required for complete protein displacement is less at $0\,^{\circ}C$ when the droplets are solid than at $20\,^{\circ}C$ when they are liquid.

(6) The presence of hydrophobic emulsifier (*e.g.* glycerol monostearate) in the oil phase can lead to a reduction in the amount of water-soluble surfactant required for partial or complete protein displacement.

The ease of displacement of milk proteins from the oil–water interface by small-molecule surfactants contrasts sharply with the difficulty of displacing one milk protein by another different milk protein.[57–60] Extensive rapid exchange of proteins after emulsification seems to be possible only for a mixture of disordered proteins (*e.g.* α_{s1}-casein + β-casein) in the absence of aggregation (low calcium ion content, pH away from isoelectric point). As a general rule, the first protein component to arrive at the surface remains adsorbed and cannot readily be displaced by another protein component (even β-casein).

4 Prediction of Emulsion Stability

" . . . . precise measurements of changes in particle size, either under quiescent or shear conditions, could be used to predict long-term colloid stability. Modern particle-size analysis methods, particularly automated particle counters and light-scattering techniques, make this a practical possibility."

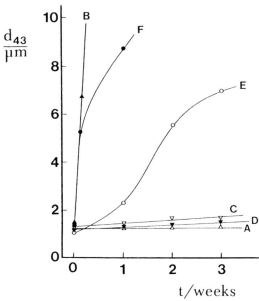

Figure 3   *Coalescence stability under quiescent conditions for oil-in-water emulsions (0.5 wt% protein, 10 wt% oil, pH 7) made with a 1:3 mixture of BSA + polysaccharide or a 1:3 complex produced by dry-heating. The average droplet diameter d_{43} is plotted against the storage time t at 25 °C: (A) BSA–dextran conjugate; (B) BSA + dextran mixture; (C) BSA–dextran sulfate conjugate; (D) BSA + dextran sulfate mixture; (E) BSA–amylopectin conjugate; (F) BSA + amylopectin mixture. (Taken from reference 62)*

One such convenient and reliable instrument for determining changes in emulsion droplet-size distribution is the Malvern Mastersizer,[61] which is a multi-angle static laser light-scattering apparatus. An example of its use for studying the coalescence stability of oil-in-water emulsions is provided by the data in Figure 3 taken from a recent investigation[62] of stabilization by protein–polysaccharide complexes. The average droplet size d_{43} is plotted against the storage time t for emulsions made at pH 7 with 10 wt% n-hexadecane, 0.5 wt% bovine serum albumin (BSA) and 1.5 wt% polysaccharide (dextran, dextran sulfate or amylopectin). The equivalent emulsion made without polysaccharide present (not shown) has poor stability: d_{43} changes from 1.5 to 4 μm in the first 24 h. Figure 3 shows that emulsions made with simple mixtures of BSA + dextran (5×10^5 daltons) or BSA + amylopectin also have poor stability. The best stability occurs for the emulsion made with the BSA–dextran complex produced by dry-heating a mixture of BSA + dextran (3 weeks at 60 °C and 40% relative humidity): after 3 weeks storage of the emulsion, the value of $d_{43} = 1.26 \pm 0.02$ μm is identical to that for the freshly made emulsion. Similar excellent stability has also been found for emulsions made with dry-heated complexes of dextran with other proteins.[63–65] Also showing good stability with respect to coalescence in

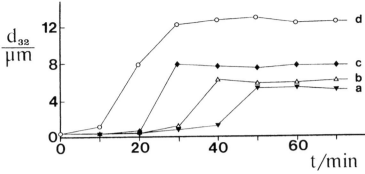

Figure 4 *Influence of heat treatment on shear-induced coalescence stability of oil-in-water emulsions (0.3 wt% protein, 20 wt% oil, pH 7). The average droplet diameter* d_{32} *is plotted against the shearing time* t *for emulsion samples pre-stored for 2* h *at various temperatures: (a) 20°C, (b) 40°C, (c) 60°C, and (d) 80°C*

Figure 3 is the emulsion made with a simple mixture of BSA + dextran sulfate: d_{43} changes from 1.19 to just 1.50 μm over a period of 3 weeks. This greatly improved stability compared with the pure BSA emulsion is explained[62] in terms of an electrostatic complex between protein and polysaccharide at the emulsion droplet surface.

The results described above refer to changes in average droplet size under quiescent conditions (perikinetic stability). The use of the Mastersizer to study emulsion breakdown under shear conditions (orthokinetic stability) has also been reported recently.[44,66,67] Figure 4 shows some recent results[68] for heat-treated β-lactoglobulin-stabilized emulsions (20 wt% *n*-tetradecane, 0.3 wt% protein, pH 7). After preparation at room temperature, the emulsion samples were stored for 2 h at a set temperature (20, 40, 60 or 80 °C), cooled to room temperature for 30 min, and then stirred in a high-speed mixer (9000 r.p.m.) with samples analysed for droplet-size distribution every 10 min. The average droplet size d_{32} for all the emulsions at the start of shearing is 0.40 ± 0.05 μm. After about 1 h of shearing, d_{32} reaches a steady-state value more than an order of magnitude larger than the initial value. The higher the emulsion heating temperature prior to shearing, the larger is the steady-state particle size after extended shearing. The time at which there is the maximum rate of increase in average particle size, $d(d_{32})/dt$, is denoted as the destabilization time. The reduction in the orthokinetic stabilization time with increasing pretreatment storage temperature may be due to the increasing presence of flocculated droplets (possibly caused by increased formation of disulfide cross-links between droplets in the heat-treated emulsions).

There is a limit, however, to what can be achieved from measurements of droplet-size distributions using techniques like light-scattering that require extensive dilution prior to analysis. Weak flocculation by non-adsorbing polymers, which is often associated with poor creaming stability, cannot be detected by light-scattering techniques because the flocs are rapidly disrupted by dilution and/or mild stirring. The way forward here seems to be to place more emphasis on experimental techniques such as low-stress rheology and ultrasonics which can probe the structure of the concentrated emulsion *in situ*,[69-73] and also to make greater use of recent advances in the theory of depletion flocculation.[74-77]

5 Biopolymer Gelation

There has been considerable progress in recent years in understanding the structure and rheology of biopolymer gels.[78-80] In particular, substantial advances have been made in understanding the properties of mixed poly-saccharide gels[81-87] and heat-set globular protein gels.[88-91]

Some important food colloids are protein particle gels formed by the strong aggregation of proteinaceous (or protein-coated) colloidal particles. In most cases the colloidal particles are roughly spherical in shape, but their sizes may range over several orders of magnitude depending on whether they are single globular protein molecules, proteinaceous particles (*e.g.* casein micelles), or protein-stabilized oil droplets. Gelation may be induced by heating, by addition of acid or salts, or by treatment with enzymes. The usefulness of the concept of fractal dimensionality in describing the structure of real and simulated aggregates formed by irreversible Brownian coagulation has prompted renewed theoretical interest in the structure of particle gels.[92-94] In any particular case, however, the gel structure and rheology are dependent on the nature and strength of the interactions between the particles both during and after gelation.

Computer simulation provides a way of understanding the link between the particle interactions and the gel structure. This is illustrated in Figure 5 by the two contrasting pictures of two-dimensional particle gels formed by slow irreversible cross-linking of monodisperse particles with (a) non-bonded repulsive interactions or (b) non-bonded attractive interactions. Starting from a pseudo-random distribution in a square cell with periodic boundary conditions, aggregation was simulated[95] by the technique of Brownian dynamics[96] until a network structure was produced with all (or nearly all) the particles in a single self-connected aggregate. The structure of the simulated particle gel is controlled by two parameters f and P. The parameter f measures the strength of the medium-range interparticle force which is typically of the same order of magnitude as the Brownian force and may be attractive ($f > 0$) or repulsive ($f < 0$). The parameter P is the probability that a bond will form during a specified time interval Δt when the surface-to-surface separation of a pair of particles is less than a certain specified distance. The aggregates formed from bonded particles are flexible but they cannot fall apart. (Details of the simulation model may be found elsewhere.[95]) The picture shown in Figure 5(a) was generated from a simulation of 400 particles after 30 000 time-steps each of length Δt with $f = -4$ and $P = 0.01$. A particular feature of the connected structure is the wiggly character of the chains of particles between the junction zones. This arises from the strong chain–chain repulsion ($f \ll 0$) which favours a fine maze-like pore structure with characteristic spacing of the order of the

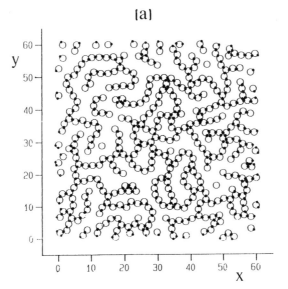

(a)

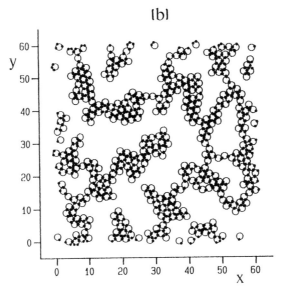

(b)

Figure 5   *Snapshots of two-dimensional particle gels generated by Brownian dynamics simulation with (a) non-bonded repulsive interparticle force (f = −4 and P = 0.01) and (b) non-bonded attractive interparticle force (f = 2 and P = 0.01). The simulation cell is periodic in the x and y directions. Each open circle represents a particle. Each small dot denotes a bond. (See reference 95 for full computational details)*

range of the interparticle repulsion (set at twice the particle diameter). With an average of just 2.2 bonds per particle, the structure in Figure 5(a) is characteristic of a highly cross-linked polymer gel having a substantial entropic contribution to its elastic properties. This contrasts sharply with the picture in Figure 5(b) from the corresponding simulation with $f = 2$ and $P = 0.01$. The particle–particle interaction is sufficiently attractive here to cause clustering of particles. In the absence of bonding, this would lead to phase separation analogous to the gas–liquid transition of a molecular fluid.[97] However, when as here the particles have a substantial reactivity, the drive towards phase separation only produces domains of a finite average size before they get 'pinned' due to gelation.[98] Compared with picture (a), the structure has a coarse pore structure and a high proportion of close-packed multi-bonded particles. These features become even more strongly enhanced in simulations with larger f and/or lower P. They are characteristic of phase-separated particle gels with elasticity determined by energetic factors.[93]

The simple simulation model described above has features that can be identified with experimental results for protein-containing particle gels. For instance, the dependence of the structure and rheology of β-lactoglobulin gels on pH and heating conditions[90,91] can be interpreted, at least in part, as a competition between (i) the tendency towards cross-linking of the thermally denatured globular protein molecules and (ii) the tendency towards flocculation or phase separation. Another important example of this type is the influence of pH, enzyme concentration, *etc.*, on the properties of particle gels formed by coagulation of casein micelles.[94]

6 Dispersions of Gas Microcells

In principle, one of the simplest ways of reducing fat content in many food colloids would be to replace the emulsion oil droplets by gas cells of the same size. Such dispersed microcells would have the potential for imparting a creamy texture and visual appearance like that of normal dairy-type oil-in-water systems. Typically, however, gas cells falling into the colloidal size range are very unstable with respect to disproportionation due to the high Laplace pressure.[2]

Recently, it has been demonstrated by Bee and co-workers[99] that stable gas microcells in the size range 1–10 μm can be prepared using stearate esters of sucrose. A striking feature of these stable gas microcells is a surface structure consisting of polyhedral domains as illustrated in Figure 6. The individual domains, principally pentagons and especially hexagons, also have a dome-like curvature, and they are considered to be solid-like since the hydrocarbon chains of the surfactant are crystalline. As each domain is convex outwards, it follows that the boundary between adjacent domains must have a negative radius of curvature relative to that of the gas microcell. It has been postulated[100] that such an arrangement can lead to the Laplace pressure diminishing to zero, thereby eliminating the thermodynamic driving force towards disproportionation. Furthermore, it has been postulated[100] that the

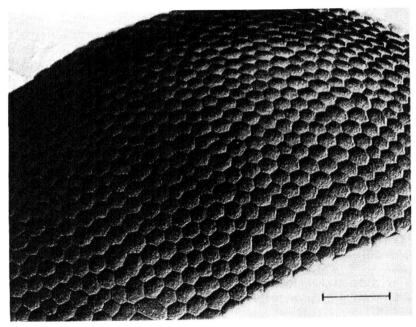

Figure 6   *Surface of single gas microcell as studied by transmission electron microscopy. The foam was prepared from an aqueous solution containing 70 wt%
maltodextrin 63DE and 2 wt% sucrose monostearate ester by whipping with
air for 1 h using a Kenwood Chef Mixer operating at speed setting 5. The bar
represents a distance of 0.2 μm.*
(Photograph supplied by Dr R. D. Bee)

average radius of the polyhedral domain is determined by the area occupied by
the surfactant head-group (sucrose) relative to its hydrophobic tail (stearate).
Introducing a co-surfactant with a smaller head-group (*e.g.* sodium stearate)
was found to lead to an increase of the domain curvature in agreement with
the postulate.

7 Concluding Remarks

This introductory article has described just a few of the areas where progress
has been made over the past few years. Many other exciting advances are
reported in the rest of this conference volume. Much of the work has its
scientific origins outside the field of food science and technology, and in many
cases the relevance to actual food products is unproven or distinctly tenuous.
Nevertheless, it does seem likely that only through incorporation of ideas and
techniques from new basic science will genuine advances in the field of food
colloids continue to emerge.

   In his lecture in 1986, Don Darling expressed the view[1] that "too much time is spent burying our heads in the proverbial 'model system' rather than facing up to the complex world of real food systems". In one important respect, of course, he is quite correct: the detailed study of model systems for their own sake is extremely unlikely to lead to the development of new product formulations in "the dirty world of food colloids in practice". That having been said, however, I am happy to report that the model system does still seem to be alive and well some 8 years on. And I suspect that it will continue to play a crucial role in the field for quite a long time to come!

References

1. D. F. Darling and R. J. Birkett, in 'Food Emulsions and Foams', ed. E. Dickinson, Special Publication No. 58, The Royal Society of Chemistry, London, 1987, p. 1.
2. E. Dickinson, 'An Introduction to Food Colloids', Oxford University Press, 1992.
3. E. Dickinson, *J. Chem. Soc., Faraday Trans.*, 1992, **88**, 2973.
4. J. Castle, E. Dickinson, B. S. Murray, and G. Stainsby, *ACS Symp. Ser.*, 1987, **343**, 118.
5. E. Dickinson, S. E. Rolfe, and D. G. Dalgleish, *Int. J. Biol. Macromol.*, 1990, **12**, 189.
6. R. W. Richards and J. Penfold, *Trends Polym. Sci.*, 1994, **2**, 5.
7. T. Cosgrove, J. S. Phipps, and R. M. Richardson, *Colloids Surf.*, 1992, **62**, 199.
8. E. Dickinson, D. S. Horne, and R. M. Richardson, *Food Hydrocolloids*, 1993, **7**, 497.
9. E. Dickinson, D. S. Horne, J. S. Phipps, and R. M. Richardson, *Langmuir*, 1993, **9**, 242.
10. E. Dickinson, D. S. Horne, J. S. Phipps, and R. M. Richardson, in 'Food Colloids and Polymers: Stability and Mechanical Properties', ed. E. Dickinson and P. Walstra, Special Publication No. 113, The Royal Society of Chemistry, Cambridge, UK, 1993, p. 396.
11. B. W. Morrissey and C. C. Han, *J. Colloid Interface Sci.*, 1978, **65**, 423.
12. A. R. Mackie, J. Mingins, and A. N. North, *J. Chem. Soc., Faraday Trans.*, 1991, **87**, 3043.
13. D. G. Dalgleish, *Colloids Surf.*, 1990, **46**, 141.
14. E. Dickinson and S. R. Euston, *Adv. Colloid Interface Sci.*, 1992, **42**, 89.
15. E. Dickinson and Y. Matsumura, *Int. J. Biol. Macromol.*, 1991, **13**, 26.
16. D. C. Clark, L. J. Smith, and D. R. Wilson, *J. Colloid Interface Sci.*, 1988, **121**, 136.
17. D. G. Cornell and D. L. Patterson, *J. Agric. Food Chem.*, 1989, **37**, 1455.
18. M. Ohgushi and A. Wada, *FEBS Lett.*, 1983, **164**, 21.
19. K. Kuwajima, *Proteins: Struct. Funct. Genet.*, 1989, **6**, 87.
20. O. B. Ptitsyn, in 'Protein Folding', ed. T. E. Creighton, Freeman, New York, 1992, p. 243.
21. E. Dickinson and Y. Matsumura, *Colloids Surf. B*, 1994, **3**, 1.
22. H. Hirose, *Trends Food Sci. Technol.*, 1993, **4**, 48.
23. E. Dickinson and S. R. Euston, *Mol. Phys.*, 1989, **66**, 865.
24. E. Dickinson and S. R. Euston, *J. Colloid Interface Sci.*, 1992, **152**, 562.
25. G. C. Ansell and E. Dickinson, *J. Colloid Interface Sci.*, 1986, **110**, 73.
26. M. C. Buján-Núñez and E. Dickinson, *Mol. Phys.*, 1993, **80**, 431.
27. M. Bishop and J. H. R. Clarke, *J. Chem. Phys.*, 1990, **93**, 1455.

28. M. C. Buján-Núñez and E. Dickinson, *J. Chem. Soc., Faraday Trans.*, 1993, **89**, 573.
29. R. Douillard, *Colloids Surf. B*, 1993, **1**, 333.
30. R. Douillard, M. Daoud, J. Lefebvre, C. Minier, G. Lecannu, and J. Coutret, *J. Colloid Interface Sci.*, 1994, **163**, 277.
31. R. Douillard, 'Proceedings of First World Congress on Emulsions', Paris, October 1993, vol. 2, paper 2–21–202.
32. R. Douillard, *Colloids Surf. A*, 1994, **91**, 113.
33. M. Daoud and P. G. de Gennes, *J. Phys. (Paris)*, 1977, **38**, 85.
34. P. G. de Gennes, 'Scaling Concepts in Polymer Physics', Cornell University Press, Ithaca, NY, 1979.
35. J.-L. Courthaudon, E. Dickinson, and D. G. Dalgleish, *J. Colloid Interface Sci.*, 1991, **145**, 390.
36. J.-L. Courthaudon, E. Dickinson, Y. Matsumura, and D. C. Clark, *Colloids Surf.*, 1991, **56**, 293.
37. J.-L. Courthaudon, E. Dickinson, Y. Matsumura, and A. Williams, *Food Struct.*, 1991, **10**, 109.
38. J.-L. Courthaudon, E. Dickinson, and W. W. Christie, *J. Agric. Food Chem.*, 1991, **39**, 1365.
39. E. Dickinson and S. Tanai, *J. Agric. Food Chem.*, 1992, **40**, 179.
40. E. Dickinson and S. Tanai, *Food Hydrocolloids*, 1992, **6**, 163.
41. E. Dickinson and J.-L. Gelin, *Colloids Surf.*, 1992, **63**, 329.
42. E. Dickinson, R. K. Owusu, S. Tan, and A. Williams, *J. Food Sci.*, 1993, **58**, 295.
43. E. Dickinson and G. Iveson, *Food Hydrocolloids*, 1993, **6**, 533.
44. J. Chen, E. Dickinson, and G. Iveson, *Food Struct.*, 1993, **12**, 135.
45. J. Chen and E. Dickinson, *J. Sci. Food Agric.*, 1993, **62**, 283.
46. E. Dickinson, S. R. Euston, and C. M. Woskett, *Prog. Colloid Polym. Sci.*, 1990, **82**, 65.
47. E. Dickinson, *Mol. Phys.*, 1988, **65**, 895.
48. E. Dickinson and C. M. Woskett, in 'Food Colloids', ed. R. D. Bee, P. Richmond, and J. Mingins, Special Publication No. 75, The Royal Society of Chemistry, Cambridge, UK, 1989, p. 74.
49. E. Dickinson and S. R. Euston, *Mol. Phys.*, 1989, **68**, 407.
50. E. Dickinson, in 'Interactions of Surfactants with Polymers and Proteins', ed. E. D. Goddard and K. P. Ananthapadmanabhan, CRC Press, Boca Raton, FL, 1993, p. 295.
51. H. D. Goff and W. K. Jordan, *J. Dairy Sci.*, 1989, **72**, 18.
52. P. Paquin and E. Dickinson, 'Proceedings of XXIII International Dairy Congress', Montreal, October 1993, vol. 2, p. 1492.
53. N. M. Barfod, N. Krog, G. Larsen, and W. Buchheim, *Fat Sci. Technol.*, 1991, **93**, 24.
54. O. Robin, N. Remillard, and P. Paquin, *Colloids Surf. A*, 1993, **80**, 211.
55. A. Thomas, J.-L. Courthaudon, D. Paquet, and D. Lorient, *Food Hydrocolloids*, 1994.
56. C. Bireau and E. Dickinson, unpublished results.
57. E. Dickinson, S. E. Rolfe, and D. G. Dalgleish, *Food Hydrocolloids*, 1988, **2**, 397.
58. E. Dickinson, S. E. Rolfe, and D. G. Dalgleish, *Food Hydrocolloids*, 1989, **3**, 193.
59. D. G. Dalgleish, S. E. Euston, J. A. Hunt, and E. Dickinson, in 'Food Polymers, Gels and Colloids', ed. E. Dickinson, Special Publication No. 82, The Royal Society of Chemistry, Cambridge, UK, 1991, p. 485.

60. T. Nylander and N. M. Wahlgren, *J. Colloid Interface Sci.*, 1994, **162**, 151.
61. A. Lips, P. M. Hart, I. D. Evans, and M. Debet, in 'Gums and Stabilisers for the Food Industry', ed. G. O. Phillips, D. J. Wedlock, and P. A. Williams, IRL Press, Oxford, 1992, vol. 6, p. 335.
62. E. Dickinson and V. B. Galazka, in 'Gums and Stabilisers for the Food Industry', ed. G. O. Phillips, D. J. Wedlock, and P. A. Williams, IRL Press, Oxford, 1992, vol. 6, p. 351.
63. E. Dickinson and V. B. Galazka, *Food Hydrocolloids*, 1991, **5**, 281.
64. E. Dickinson and M. G. Semenova, *Colloids Surf.*, 1992, **64**, 299.
65. A. Kato, R. Mifuru, N. Matsudomi, and K. Kobayashi, *Biosci. Biotech. Biochem.*, 1992, **56**, 567.
66. A. Lips, T. Westbury, P. M. Hart, I. D. Evans, and I. J. Campbell, in 'Food Colloids and Polymers: Stability and Mechanical Properties', ed. E. Dickinson and P. Walstra, Special Publication No. 113, The Royal Society of Chemistry, Cambridge, UK, 1993, p. 31.
67. E. Dickinson, R. K. Owusu, and A. Williams, *J. Chem. Soc., Faraday Trans.*, 1993, **89**, 865.
68. E. Dickinson and N. I. Wilson, unpublished results.
69. D. J. Hibberd, A. M. Howe, and M. M. Robins, *Colloids Surf.*, 1988, **31**, 347.
70. S. J. Gouldby, P. A. Gunning, D. J. Hibberd, and M. M. Robins, in 'Food Polymers, Gels and Colloids', ed. E. Dickinson, Special Publication No. 82, The Royal Society of Chemistry, Cambridge, UK, 1991, p. 244.
71. Y. Cao, E. Dickinson, and D. J. Wedlock, *Food Hydrocolloids*, 1991, **5**, 443.
72. H. Luyten, M. Jonkman, W. Kloek, and T. van Vliet, in 'Food Colloids and Polymers: Stability and Mechanical Properties', ed. E. Dickinson and P. Walstra, Special Publication No. 113, The Royal Society of Chemistry, Cambridge, UK, 1993, p. 224.
73. E. Dickinson, M. I. Goller, and D. J. Wedlock, *Colloids Surf. A*, 1993, **75**, 195.
74. B. Vincent, J. Edwards, S. Emmett, and R. Croot, *Colloids Surf.*, 1988, **31**, 267.
75. M. J. Snowden, S. M. Clegg, P. A. Williams, and I. D. Robb, *J. Chem. Soc., Faraday Trans.*, 1991, **87**, 2201.
76. T. Biben and J.-P. Hansen, *Phys. Rev. Lett.*, 1991, **66**, 2215.
77. H. N. W. Lekkerkerker and A. Stroobants, *Physica A*, 1993, **195**, 387.
78. P. Harris, ed., 'Food Gels', Elsevier Applied Science, London, 1990.
79. W. Burchard and S. B. Ross-Murphy, ed., 'Physical Networks', Elsevier Applied Science, London, 1990.
80. A. H. Clark, in 'Physical Chemistry of Foods', ed. H. G. Schwartzberg and R. W. Hartel, Marcel Dekker, New York, 1992, p. 263.
81. V. J. Morris, in 'Gums and Stabilisers for the Food Industry', ed. G. O. Phillips, D. J. Wedlock, and P. A. Williams, Elsevier Applied Science, London, 1986, vol. 3, p. 87.
82. P. Cairns, M. J. Miles, and V. J. Morris, *Carbohydr. Polym.*, 1988, **8**, 99.
83. K. P. Shatwell, I. W. Sutherland, S. B. Ross-Murphy, and I. C. M. Dea, *Carbohydr. Polym.*, 1991, **14**, 131.
84. P. A. Williams, S. M. Clegg, D. H. Day, G. O. Phillips, and K. Nishinari, in 'Food Polymers, Gels and Colloids', ed. E. Dickinson, Special Publication No. 82, The Royal Society of Chemistry, Cambridge, UK, 1991, p. 339.
85. E. Costell, M. H. Damasio, L. Izquierdo, and L. Durán, *Food Hydrocolloids*, 1992, **6**, 275.
86. J.-L. Doublier and G. Llamas, in 'Food Colloids and Polymers: Stability and

Mechanical Properites', ed. E. Dickinson and P. Walstra, Special Publication No. 113, The Royal Society of Chemistry, Cambridge, UK, 1993, p. 138.

87. K. Kohyama, H. Iida, and K. Nishinari, *Food Hydrocolloids*, 1993, **7**, 213.
88. D. M. Mulvihill, D. Rector, and J. E. Kinsella, *Food Hydrocolloids*, 1990, **4**, 267.
89. R. Vreeker, L. L. Hoekstra, D. C. den Boer, and W. G. M. Agterof, *Food Hydrocolloids*, 1992, **6**, 423.
90. M. Stading, M. Langton, and A.-M. Hermansson, *Food Hydrocolloids*, 1992, **6**, 455.
91. M. Stading, M. Langton, and A.-M. Hermansson, *Food Hydrocolloids*, 1993, **7**, 195.
92. E. Dickinson, *J. Colloid Interface Sci.*, 1987, **118**, 286.
93. L. G. B. Bremer, B. H. Bijsterbosch, R. Schrijvers, T. van Vliet, and P. Walstra, *Colloids Surf.*, 1990, **51**, 159.
94. P. Walstra, T. van Vliet, and L. G. B. Bremer, in 'Food Polymers, Gels and Colloids', ed. E. Dickinson, Special Publication No. 82, The Royal Society of Chemistry, Cambridge, UK, 1991, p. 369.
95. E. Dickinson, *J. Chem. Soc., Faraday Trans.*, 1994, **90**, 173.
96. E. Dickinson, *Chem. Soc. Rev.*, 1985, **14**, 421.
97. E. Dickinson, C. Elvingson, and S. R. Euston, *J. Chem. Soc., Faraday Trans. 2*, 1989, **85**, 891.
98. F. Sciortino, R. Bansil, H. E. Stanley, and P. Alstrøm, *Phys. Rev. E*, 1993, **47**, 4615.
99. R. D. Bee, International Patent Application No. PCT/EP92/01326, published as WO 92/21255, December 1992.
100. R. D. Bee, *Colworth News*, 1994, January, p. 4.

Adsorbed Layers

Structures and Properties of Adsorbed Layers in Emulsions Containing Milk Proteins

By Douglas G. Dalgleish

DEPARTMENT OF FOOD SCIENCE, UNIVERSITY OF GUELPH, GUELPH, ONTARIO N1G 2W1, CANADA

1 Introduction

It is probably no exaggeration to say that there are more papers written on the adsorption and surfactant properties of milk proteins than on the same properties of all other food proteins put together. As a result, we understand more of the behaviour of the milk proteins than of most other groups of food proteins. For example, descriptions have been given of systems which contain not only protein and oil but other components such as small molecule surfactants (Tweens, lecithins, glycerol monoesters). Most of this research has been performed in the course of the last 15 years, and has coincided with an increase in the industrial uses of dairy proteins in food emulsions, and a more general interest in the emulsifying properties of proteins.

For the researcher, milk provides a number of proteins of different structures and properties which can be fairly readily separated from one another (in the laboratory, at least). The major proteins in milk, the caseins (α_s, β and κ) generally lack large amounts of regular structure; they are considered to be rather flexible molecules, a property which is likely to enhance their surfactant properties. In contrast to the caseins, there are the whey, or serum, proteins (α-lactalbumin, β-lactoglobulin, bovine serum albumin and immunoglobulins) which are characterized by well defined three-dimensional structures held together by disulfide bridges; these proteins are much more rigid than the caseins. The use of individual proteins, and mixtures of them, to prepare emulsions may potentially allow for the formation of particles having a range of structures and potential functionalities, although this has still to be established in detail.

Most of the research has so far been performed using single purified proteins, or on simple mixtures of them; however, it should be remembered that in milk the caseins naturally form aggregated particles (the casein

23

micelles), and that these can also be used to emulsify fats and oils, as in homogenized milks. Although the latter have received comparatively little research attention relative to other model emulsions, they may become of increased importance, since it is now possible to use microfiltration techniques to isolate a slightly modified micellar fraction from milk; the functional properties, including emulsification, of this fraction therefore need to be established.

This review deals with the properties of the different milk proteins adsorbed to oil–water interfaces. In particular, it contrasts the behaviour of the whey proteins and the caseins in terms of adsorption, and it describes how this may affect the structures of the adsorbed layers of protein at the oil–water interface.

2 Protein Adsorption and Particle Sizes in Emulsions

Both of the major whey proteins, β-lactoglobulin and α-lactalbumin, adsorb to oil–water interfaces and are capable of giving stable emulsions.[1] Indeed, singly or in combination, these proteins are excellent emulsifying agents, and their emulsions are only a little less stable than those produced using caseins. For emulsions prepared under the same conditions[2] (concentrations of oil and protein, pH, homogenization pressure), the droplet sizes in the whey protein emulsions are somewhat greater than those in the casein-stabilized emulsions when the concentration of protein is low ($<1\%$), but identical at higher concentrations (Figure 1). This Figure also emphasizes the effect of the concentration of protein on the sizes of the emulsion droplets. It has been known in practice that such a relationship exists,[3] but it has been recently established that the protein concentration affects not only the droplet sizes, but also the protein load[4] (amount of protein per unit area of the oil–water interface); it therefore may also potentially affect the properties of the emulsion droplets.

Figure 2 shows that in emulsions prepared with 20 wt% oil and whey protein in the concentration range 0–3 wt%, the surface concentration (Γ) increases in a series of steps, from 1.5 to 3 mg m^{-2} as the concentration of protein in the mixture is increased from 0.3 to 3 wt%. It is possible that the last of these steps may arise from multilayer formation, but the others appear to arise from conformational rearrangement of the protein, or different spatial distribution of the protein molecules at the interface.[2] In casein-stabilized emulsions, the surface concentration of protein depends even more strongly on the total concentration of protein in the emulsion.[4] It is possible to make a stable emulsion containing casein when the surface concentration is as little as 1 mg m^{-2} (Figure 2). That is, if droplets are coated with less than that amount of casein, they will coalesce until the surface area has decreased sufficiently for there to be enough casein to cover it at about this minimum level of Γ. For an emulsion containing 20 wt% oil, the lower limit of the casein concentration is about 0.3%. However, as the amount of casein increases, so does the surface concentration, until an upper limit is reached at about 3 mg m^{-2}, when the casein concentration is about 2%. Unlike the

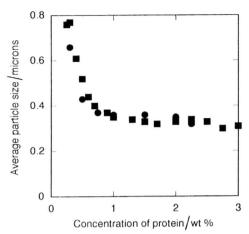

Figure 1   *Comparison of the values of average particle size d₃₂ of emulsion droplets formed in emulsions containing 20 wt% soya oil, and varying concentrations of caseinate (●) and whey protein isolate (■). Emulsions were made using a microfluidizer, and particle sizes were measured by light scattering. Details of the experiments are given in reference 2*

results found for whey proteins, Γ increases smoothly with the concentration of casein, and there are no steps in the function to suggest multilayer formation. The results from both sets of proteins show that, by careful manipulation of the homogenization conditions, and the concentrations of oil and protein, it is possible to produce a range of emulsions with different protein surface concentrations.

3 Competition between Proteins during and after Emulsion Formation

Once the emulsion has been formed, and the proteins have adsorbed, it is generally accepted that the adsorption process is largely irreversible, and that, in the absence of other agents, adsorbed proteins will not wash off the interface.[5] It is possible that β-casein is an exception to this rule:[5] exchange of labelled β-casein between the solution and the interface has been demonstrated. Also,[6] extremely thorough washing of latex particles to which β-casein has adsorbed, by passing them through a long column, reduces the surface concentration from 3 mg m^{-2} to 1 mg m^{-2}. It is not established whether a similar washing phenomenon occurs when the protein is adsorbed to oil–water interfaces.

If mixtures of proteins are present when the emulsion is formed, there seems to be little evidence of competition between the different proteins for sites on the surface of the oil droplets. For example, in mixtures of α-lactalbumin and β-lactoglobulin, the proteins adsorb in proportion to their concentrations, especially when emulsions are prepared at neutral pH.[2,7] Only when casein is

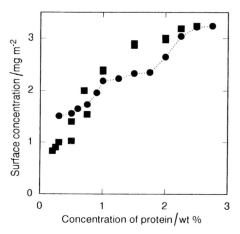

Figure 2   *Surface concentrations Γ in emulsions containing 20 wt% soya oil and varying concentrations of caseinate (■) and whey protein isolate (●). Surface concentrations were measured using an electrophoretic method as described in reference 2. The smooth curve of the caseinate results is to be contrasted with the stepwise function found using whey protein. Results for the caseinate emulsion are a composite of results in references 2 and 4*

present, and the total amount of protein is greatly in excess of that needed to cover the oil–water interface, is there any evidence of competition between the whey proteins and caseins as the emulsion is formed. Although the caseins have been less thoroughly studied, similar lack of competition during homogenization occurs, except at high concentrations of casein where κ-casein is less readily adsorbed than the α_s or β-caseins.[2] All of these results confirm the fact that during homogenization the controlling factor is not the different affinities of the proteins for the interface, but simply their overall concentration.

In emulsions containing whey proteins, there is little or no exchange between adsorbed and soluble protein after the emulsion is formed. However, in the presence of competing surfactants, such as Tweens, some competition does appear to be established, not only between the proteins and the surfactant, but also between the individual proteins.[8] The small molecule surfactant competes with the proteins, but in doing so it mobilizes the proteins, so that a more 'thermodynamic' scale of competitive adsorption can be established in the presence of moderate quantities of surfactant. Excess quantities of surfactant, however, remove the protein completely.[9,10] In contrast to the whey proteins, individual caseins can displace one another from the interface.[11] Thus, β-casein can displace α_s-casein and *vice versa* if the concentrations of the proteins are adjusted after the emulsion is formed. Similarly, β-casein can displace phosvitin from the interface but only when the concentration of calcium is low.[12] In general, it seems that β-casein (being apparently the most flexible as well as the most hydrophobic of the caseins) can also to some extent displace whey proteins from an interface,[13] although the displacement may only be partial, or may lead to the formation of multilayers. It seemed at one time possible that caseins could be used in this way to establish some kind of thermodynamic scale for adsorption, but it is probable that they are the only proteins to show truly competitive displacement.

Exchange between adsorbed and soluble proteins after emulsion formation may be enhanced by changing temperture or by altering the solution conditions, *e.g.* pH.[14] Since the strength of the adsorption of a protein depends on its conformation and conformational mobility, and possibly its quaternary structure, and since these may be affected by the pH of the solution,[15] especially if the protein is near to its isoelectric point, it can easily be seen that solution conditions may be important, so that the adsorption behaviour may be pH-dependent. For example, β-lactoglobulin adsorbs less strongly, compared with α-lactalbumin,[16] when emulsions are made at pH values below 7. Furthermore, there is a further effect of buffer in this system; at pH 6.0, when imidazole is used, less β-lactoglobulin is replaced by α-lactalbumin than when citrate is used. The origin of this effect is not known, unless it is possible for the citrate to remove the calcium from the α-lactalbumin, which effectively changes its conformation.[17] This effect, however, is only found when the pH is adjusted before the emulsion is made: there is only a very small exchange between adsorbed and non-adsorbed protein once the initial adsorption has occurred, even if the pH is subsequently adjusted.[16]

4 The Effect of Surfactants on Adsorption of Milk Proteins

It is perhaps rare for a true food emulsion to be made only using protein as the surfactant: in most formulations, other surfactants are added deliberately, or are naturally present, as for example lecithin is present when egg yolks are used in the preparation of mayonnaise. It is worthwhile, therefore, in model systems to consider the effect of combining other surfactants with proteins in forming emulsions.

Studies have been made on a variety of surfactant molecules and their effect on the adsorption of milk proteins. Soluble surfactants, such as Tweens and sodium dodecyl sulfate, compete effectively with caseins even if they are added after the emulsion has been formed.[18] Similarly, hydrophilic polyoxyethylene surfactants added to an emulsion can displace the adsorbed protein.[19] However, more hydrophobic detergents need to be present at the time the emulsions are formed for displacement of protein to occur, since they have such low solubility that they cannot be present in the solution in sufficient concentration to displace the protein.[18,19]

The interactions of phospholipids in emulsions are more complex. Since they are not soluble, they need to be present at the time of emulsion formation, but even then there is some measure of disagreement over the extent of action, which may depend on the nature of the non-aqueous phase.[20] They do not displace casein, for example, as well as do other surfactants.[20–22] However, they seem to be able to stabilize emulsions where only low concentrations of protein are present, so that there is a synergy between protein and phospholipid in stabilizing the emulsion.[21]

5 States of the Adsorbed Layers of Protein

It has been established that the interfacial viscosities of mixed α-lactalbumin/β-lactoglobulin interfaces are defined by the composition of the protein mixtures, and also by the order of addition of the two proteins,[23] suggesting that interchange was possible between adsorbed and soluble proteins, rather in contradiction to the results where the overall composition of the interface was measured directly in aged emulsions. However, it was subsequently demonstrated that adsorbed β-lactoglobulin was capable of forming intermolecular disulfide bonds,[24] suggesting that the protein was denatured at least partially in the adsorbed state, and that the increase in intefacial viscosity with time could be caused by the progressive formation of a covalently linked monolayer of protein. This kind of cross-linking will be impossible if there are other proteins present on the interface which cannot participate in disulfide bond formation.

Other experiments have confirmed that β-lactoglobulin is denatured at the oil–water interface, independently of whether the protein is the only one present. We have demonstrated, using differential scanning calorimetry, that β-lactoglobulin adsorbed to a polystyrene latex had a much reduced thermal transition during denaturation than did the protein in solution.[6] We have recently been able to repeat this experiment in more detail, using emulsions rather than latex, where β-lactoglobulin, or mixtures of whey proteins, are used as the surfactants (Figure 3). Although this work is at an early stage, it is possible to conclude from these results that (a) for β-lactoglobulin adsorbed to an oil–water interface, there is no intake of heat in the normal temperature range of denaturation; (b) the effect on α-lactalbumin is similar but not so pronounced, so that it appears that this protein when adsorbed is somewhat less denatured than β-lactoglobulin; (c) when mixtures of whey proteins are adsorbed, the β-lactoglobulin contribution to the thermogram is much reduced. These make it clear that adsorption profoundly alters the denaturation characteristics of the proteins. The simplest interpretation is that the protein is denatured by its adsorption to the oil–water interface, but it is also possible to suppose that the proteins are stabilized because of multipoint attachment to the interface, and that complete denaturation only occurs at a much higher temperature. It has been impossible to test this hypothesis within the temperature range of existing equipment, but since it is known that the emulsion particles can be gelled when they are heated, it is probable that at least partial denaturation acompanies the adsorption.

6 Dimensions of Adsorbed Protein Layers

Attempts have been made to measure the thicknesses of adsorbed layers of whey proteins, and it has been shown that the layers are not thick. If the

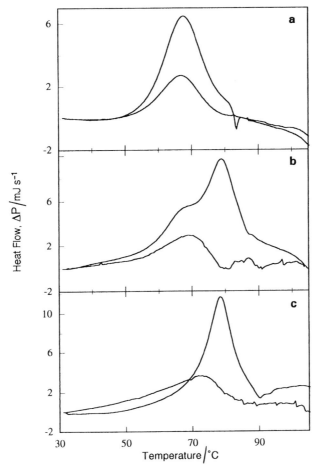

Figure 3 *Differential scanning calorimetry thermograms for the denaturation of whey proteins, in emulsions and in solution. (a) Denaturation of α-lactalbumin (2 wt%) in solution (large peak) and in emulsion containing 2 wt% protein and 20 wt% soya oil; (b) denaturation of whey protein isolate (2 wt%) in solution (large peak) and in emulsion containing 2% protein and 20% soya oil; (c) denaturation of β-lactoglobulin (2%) in solution (large peak) and in emulsion containing 2 wt% protein and 20 wt% soya oil*

adsorbed proteins maintain a relatively compact structure, the adsorbed layer would have about the same dimension as the protein, and this has been confirmed by scattering studies on latex particles to which β-lactoglobulin had been adsorbed.[25] In emulsions, there seems to be little decrease in diameter when the adsorbed layer of whey protein is broken down by proteolytic attack, and therefore it seems that the thickness of the layer can only be in the region of 1–2 nm. Since such thin layers are formed, it has not possible to determine whether there are time-dependent changes in thickness, following the denaturation and the subsequent formation of intermolecular disulfide bonds. However, the fact that the layers are only thin demonstrates that the proteins, although denatured, do not gain sufficient flexibility to protrude far from the interface into the solution.

Caseins form much thicker adsorbed layers than do whey proteins, as a result of their extended and flexible conformations:[26] this has been confirmed by a number of studies, using a variety of techniques.[27,28] The four different caseins appear to form layers of different thickness,[29] which may reflect their different strutures, but this may also be a result of experimental conditions. What is clearly apparent is that the thickness of the adsorbed layers depends strongly on the surface concentration.[4] In emulsions prepared at the low limit of Γ (about 1 mg m^{-2}), the adsorbed layer is thinner (about 5 nm thick) than is the layer at the high limit of adsorption (about 10 nm) when Γ is of the order of 3 mg m^{-2} (Figure 4). Since it is believed that the casein forms only a monolayer, it is probable that the different layer thicknesses represent different conformations of the adsorbed protein molecules. At low Γ, the casein has to spread as much as possible to cover the maximum interfacial area, otherwise the emulsion cannot be stable and the droplets will coalesce when they collide. Conversely, at high Γ, the protein molecules are crowded together and do not need to spread, so that they are forced to change conformation and to project more from the surface. It has already been described that it is not possible to modify the surface layers of whey protein-stabilized emulsion droplets after they have been formed, by adding more protein; however, in emulsions prepared using casein, with low Γ, this is not the case. When casein is added after the emulsion has been made, it adsorbs to the emulsion droplets, giving increased values of Γ together with an increase in layer thickness.[4] This observation does suggest that the adsorption of casein is not completely irreversible (or at least that the protein is mobile on the interface), and may help to explain the observations where β-casein was found to be washed from the surface of a latex particle.

These extended layers of caseins must play a large part in stabilizing the emulsions, and it is relevant to enquire how they are affected by other factors. For example, ionic strength, particularly for ions such as calcium, is likely to affect the layers,[12] consisting as they do of charged polymers. Also, addition of small molecule surfactants plays a part. It has been shown that the presence of lecithin does not greatly diminish the amount of casein which is adsorbed, especially at low concentrations of protein.[21] However, the presence of lecithin does seem to force the adsorbed caseins to adopt an extended conformation and project into solution at a lower concentration than normal (Figure 4): presumably, by adsorbing in the gaps between the protein molecules, it allows them to extend as if they were surrounded by other protein molecules. Therefore it seems that the conformation of adsorbed casein may depend on the overall concentration of surfactant, not simply that of casein. Although it may not be relevant to the emulsions, it is evident that casein and phospholipid also interact together when the homogenized, and the structures formed by these are described in a separate paper in this volume.[30]

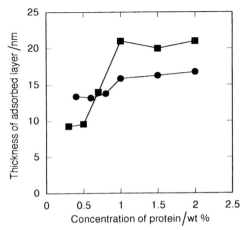

Figure 4   *Thicknesses of adsorbed layers of casein in emulsions containing 20 wt%
soya oil and different concentrations of casein (■). The effect of incorpor-
ating 0.5 wt% phospholipid (egg phosphatidylcholine) on the layer thickness
is also shown (●) for comparison*

7 Emulsions Formed using Casein Micelles

Semi-intact casein micelles are present in such foods as homogenized milk
and ice-cream, and the presence of these particles adds an additional degree
of complexity to the structures of the emulsions. Casein micelles are particles
containing thousands of individual protein molecules, which are held
together by hydrophobic bonds and by calcium phosphate; the sizes of the
particles are in the range 50–300 nm.[31] In addition to natural micelles in
milk, it is now possible to obtain micelle-like particles by microfiltration and
diafiltration of milk.

It is not certain that casein micelles adsorb naturally to an oil–water
interface, since the particles are sterically stabilized by a surface layer of κ-
casein, which is hydrophilic and unlikely to facilitate direct adsorption.
However, when milk, or mixtures of casein micelles and fat or oil are
homogenized, the stresses of homogenization cause the micelles to become
asociated with the oil.[32] Whether the micelles are broken apart by the
homogenization is not certain: the binding to the fat may simply be the result
of violent collisions between fat globules and casein micelles in the homogenizer valve. Whatever the cause, the micelles, or fragments of them, are
found at the oil–water interface, and they form a stabilizing layer around the
oil droplets. The thickness of this layer is less than the average dimensions of a
casein micelle,[33] especially if the homogenization has been vigorous, but the
layer is equally clearly not simply a monolayer of casein. By dissociating the
adsorbed micelles, the layer thickness can be found to be in the range of 50–
20 nm, depending on the homogenization pressure and the number of times

the milk is passed through the homogenizer. Since the casein micelles do not lose their calcium phosphate,[34] dissociation to monomeric proteins before or after adsorption would not be possible. Therefore, protein loads are much higher than in emulsions prepared using either simple caseinate or whey proteins,[32] and as a result the emulsifying power of micellar casein is less than that of an equivalent weight of caseinate.

Individual caseins appear to adsorb in the order β-casein $>$ α_s-caseins $>$ κ-casein, but if micellar casein is used to make emulsions, a different situation applies. When the adsorbed micelles are dissociated using EDTA and urea, the casein which remains associated with the fat globule is composed mainly of the α_{s2} and κ-caseins.[35] This may suggest, surprisingly, that these caseins form the part of the micelle which is responsible for the adsorption, *i.e.* they are nearest to the oil–water interface. However, other experiments lead to the conclusion that the κ-casein is to be found on the outside of the particles since (a) the globules are colloidally stable and (b) rennet, which specifically destroys κ-casein, destabilizes the particles. Thus the κ-casein seems to be in two places at once. The problem remains to be resolved, but presumably the answer is to be found in the suggestion that the dissociation of the adsorbed micellar moieties is different from that of the micelles in free solution. For example, the α_{s2} and κ-caseins are capable of forming disulfide linked polymers, and possibly these polymers tend to adsorb well to the oil–water interface; and there have been suggestions that such macromolecular arrays might adsorb more than the smaller individual proteins.[36]

8 Conclusions

The foregoing description of the formation of emulsions involving milk proteins demonstrates the combined versatility of the whole group, *i.e.* the caseins, the whey proteins, and casein micelles, in producing emulsions of different structures. These structures arise because of the different proteins which form them, and the conditions under which they are formed. What is less understood, and what needs to be established, is the relation which these different structures have to the properties of the emulsions. For example, do the emulsions with different coverages of casein, and, it is believed, different conformations of the adsorbed protein, possess significantly different properties? Research is needed to answer these queries. Moreover, since so much is known of the simple structures of the emulsions, we may now be on the threshold of understanding what happens to these types of systems when they are subjected to processing conditions, *e.g.* when they are heated or the pH or other solution conditions are altered. So far, what have been studied have been mainly physical processes, such as adsorption, but with heating will come a need to understand the effect of chemical changes on the adsorbed protein layers. These will be the challenges for the future.

References

1. K. Yamauchi, M. Shimizu, and T. Kamiya, *J. Food Sci.*, 1980, **45**, 1237.
2. J. A. Hunt and D. G. Dalgleish, *Food Hydrocolloids*, 1994, **8**, 175.
3. O. Robin, V. Blanchot, J. C. Vuillemard, and P. Paquin, *Lait*, 1992, **72**, 511.
4. Y. Fang and D. G. Dalgleish, *J. Colloid Interface Sci.*, 1993, **156**, 329.
5. J. R. Hunter, R. G. Carbonell, and P. K. Kilpatrick, *J. Colloid Interface Sci.*, 1991, **143**, 37.
6. K. D. Caldwell, J. Li, J.-T. Li, and D. G. Dalgleish, *J. Chromatogr.*, 1992, **604**, 63.
7. E. Dickinson, S. E. Rolfe, and D. G. Dalgleish, *Food Hydrocolloids*, 1989, **3**, 193.
8. J.-L. Courthaudon, E. Dickinson, Y. Matsumura, and A. Williams, *Food Struct.*, 1991, **10**, 109.
9. J. A. de Feijter, J. Benjamins, and M. Tamboer, *Colloids Surf.*, 1987, **27**, 243.
10. P. J. Wilde and D. C. Clark, *J. Colloid Interface Sci.*, 1993, **155**, 48.
11. E. Dickinson, S. E. Rolfe, and D. G. Dalgleish, *Food Hydrocolloids*, 1988, **2**, 397.
12. J. A. Hunt, E. Dickinson, and D. S. Horne, *Colloids Surf. A*, 1993, **71**, 197.
13. D. G. Dalgleish, S. E. Euston, J. A. Hunt, and E. Dickinson, in 'Food Polymers, Gels and Colloids', ed. E. Dickinson, Special Publication No. 82, The Royal Society of Chemistry, Cambridge, UK, 1991, p. 485.
14. M. Shimizu, M. Saito, and K. Yamauchi, *Agric. Biol. Chem.*, 1985, **49**, 189.
15. H. Pessen, J. M. Purcell, and H. M. Farrell, Jr., *Biochim. Biophys. Acta*, 1985, **828**, 1.
16. J. A. Hunt and D. G. Dalgleish, *J. Agric. Food Chem.*, 1994, **42**, 2131.
17. S. J. Prestrelski, D. M. Byler, and M. P. Thompson, *Biochemistry*, 1991, **30**, 8797.
18. J. Chen and E. Dickinson, *J. Sci. Food Agric.*, 1993, **62**, 283.
19. E. Dickinson and S. Tanai, *J. Agric. Food Chem.*, 1992, **40**, 179.
20. J.-L. Courthaudon, E. Dickinson, and W. W. Christie, *J. Agric. Food Chem.*, 1991, **39**, 1355.
21. Y. Fang and D. G. Dalgleish, *Colloids Surf. B*, 1993, **1**, 357.
22. E. Dickinson and G. Iveson, *Food Hydrocolloids*, 1993, **6**, 533.
23. E. Dickinson, S. E. Euston, and D. G. Dalgleish, *Int. J. Biol. Macromol.*, 1990, **12**, 189.
24. E. Dickinson and Y. Matsumura, *Int. J. Biol. Macromol.*, 1991, **13**, 26.
25. A. R. Mackie, J. Mingins, R. Dann, and A. N. North, in 'Food Polymers, Gels and Colloids', ed. E. Dickinson, Special Publication No. 82, The Royal Society of Chemistry, Cambridge, UK, 1991, p. 96.
26. D. G. Dalgleish, *Colloids Surf.*, 1990, **46**, 141.
27. A. R. Mackie, J. Mingins, and A. N. North, *J. Chem. Soc., Faraday Trans.*, 1991, **87**, 3043.
28. E. Dickinson, D. S. Horne, J. S. Phipps, and R. M. Richardson, *Langmuir*, 1993, **9**, 242.
29. D. G. Dalgleish, *Colloids Surf. B*, 1993, **1**, 1.
30. Y. Fang and D. G. Dalgleish, this volume, p. 146.
31. C. Holt, *Adv. Protein Chem.*, 1992, **43**, 63.
32. H. Oortwijn and P. Walstra, *Neth. Milk Dairy J.*, 1979, **33**, 134.
33. S. Tosh and D. G. Dalgleish, unpublished observations.
34. D. G. Dalgleish and E. W. Robson, *J. Dairy Res.*, 1985, **52**, 539.
35. S. K. Sharma and D. G. Dalgleish, *J. Agric. Food Chem.*, 1993, **41**, 1407.
36. D. E. Brooks and R. G. Greig, *J. Colloid Interface Sci.*, 1981, **83**, 661.

Structure of Proteins Adsorbed at an Emulsified Oil Surface

By Makoto Shimizu

DEPARTMENT OF AGRICULTURAL CHEMISTRY, UNIVERSITY OF TOKYO,
BUNKYO-KU, TOKYO 113, JAPAN

1 Protein Adsorption—Its Complicated Features

When oil is homogenized in a protein solution, the protein molecules are adsorbed at the surface of the oil droplets thereby stabilizing the emulsion. Proteins with better interfacial properties and adsorptivity are generally thought to produce finer and more stable emulsions. However, protein adsorption is a highly complex phenomenon; in some cases, a protein is denatured at the surface during emulsification. In other cases, proteins retain their conformation and biochemical functions even after being adsorbed at the oil surface. The amount of protein adsorbed at the oil–water interface during emulsification is also quite variable for different proteins.

Figure 1 shows the quite different surface behaviour of proteins during emulsification.[1] In this experiment, aqueous solutions of five different bioactive proteins (*i.e.* enzymes or antibodies) were homogenized with oil, and the proteins remaining in the aqueous phase after emulsification were determined by measuring their enzyme or antibody activity (Figure 1a). Chicken immunoglobulin G (IgG) and catalase were almost completely lost from the aqueous phase by emulsification, suggesting that they were rapidly adsorbed to the emulsified oil droplet surface, whereas adsorption of α-amylase to the oil phase was minimal. The adsorbed proteins were then extracted from the oil droplet surface by a detergent or organic solvent treatment, and recovery of the activity in the aqueous phase was determined (Figure 1b). Although the amount of protein recovered was almost 100% for the five proteins, the recovery of activity varied considerably among the proteins. In the case of chicken IgG, only 40% of the antibody activity was recovered, suggesting that the adsorption of this protein had been, at least partly, accompanied by irreversible conformational changes. In contrast, catalase, which had also been rapidly adsorbed to the oil phase by emulsification (Figure 1a), had its activity

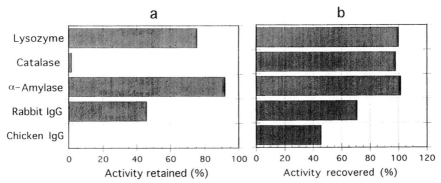

Figure 1   *Changes in the activity of proteins during adsorption at and desorption from
an emulsified oil surface. A protein solution (2 ml, 0.5%) in a 50 mM
phosphate buffer at pH 7.0 was homogenized with 0.5 g of soybean oil
containing 2% polyglyceryl condensed ricinolate. After being emulsified, the
protein-coated oil droplets were separated by membrane filtration, and the
protein remaining in the aqueous phase was determined by measuring the
activity (a). To the emulsion was added a detergent solution or an organic
solvent, and the adsorbed proteins were removed from the oil phase. The
proteins recovered to the aqueous phase were determined by measuring the
activity (b)*

completely restored after desorption. The results demonstrate the highly
complicated features of protein adsorption at an oil–water interface during
emulsification.

2 Factors Determining the Adsorptivity of Proteins

Although protein adsorption is a complicated phenomenon, there must be
some general principles which determine the relative adsorptivity of different
proteins. Structural factors which determine the emulsifying properties of
proteins have been proposed by many researchers. Nakai and his co-workers[2,3]
have demonstrated the surface hydrophobicity of a protein molecule to be an
important factor, strongly correlating with the emulsifying and surface properties. The conformational flexibility of a protein molecule has also been
considered to be important for protein emulsification.[4,5] It seems obvious that
these factors are prerequisites for a protein to express good emulsifying
properties, because a hydrophobic surface is essential for interacting with an
oil surface, and a flexible molecular structure, which would induce a rapid
conformational change in a protein at the oil–water interface, will tend to
accelerate lipid–protein interaction. However, it is still difficult to describe the
structure of an adsorbed protein at the molecular level. For example, the
following questions need to be answered: Are there any specific hydrophobic
sites for oil binding in each protein molecule? Does the quality of hydrophobicity (*e.g.* aromatic or aliphatic) of those sites affect the interaction? How

numerous and how large are the hydrophobic sites required for permanent adsorption of a protein molecule to an oil surface? What type of conformation is preferable for a protein to interact with an oil surface? To answer all these questions, more detailed studies using various methodologies are needed.

A variety of studies have been made during the last 2 or 3 decades to analyse protein properties at an oil surface. Measurement of the interfacial tension and surface shear viscosity of protein layers, using such model proteins as caseins, bovine serum albumin and egg lysozyme, has been one of the most common approaches, as reviewed by many researchers.[6-8] The thickness and density of the adsorbed protein layer can be evaluated, which gives useful information on the structure and orientation of protein molecules at the surface, although the information is, in most cases, restricted to the structure of the 'whole molecule'. In order to reveal more detailed (or more local) structural information, the application of other methods is needed. Physical approaches, using small-angle X-ray scattering or neutron reflectance, to the conformational analysis of adsorbed proteins seem promising as described by Dickinson in his recent review.[9] These techniques, as well as nuclear magnetic resonance, will provide a wealth of valuable information on the structure of adsorbed proteins in the near future. Analysis of protein–lipid interaction from the viewpoint of lipid-phase mobility using electron spin resonance may also be useful for understanding protein adsorption.[10]

In addition to these physical or physicochemical approaches, we have proposed biochemical approaches which could be valuable for estimating the local protein structure at an emulsified oil surface.

3 Biochemical Approaches

The Enzymic Method

The accessibility of proteinases to peptide bonds reflects the state of those peptide bonds in the local environment. To distinguish the external moieties from the internal moieties of a polypeptide, studying the susceptibility of the polypeptide to proteinases is quite useful. Furthermore, by using proteinases which recognize specific amino acid residues in a polypeptide chain, more detailed information on a certain portion of a protein molecule can be obtained.

We have applied this proteinase-digestion method to elucidate the structure of proteins adsorbed to an emulsified oil surface.[11,12] As a model protein, we use α_{s1}-casein, because this protein has little conformational rigidity and most of the peptide bonds are thought to be accessible to proteinases in an aqueous solution. If some of the peptide bonds in α_{s1}-casein become inaccessible to proteinases after emulsification, these peptide portions can be considered to be involved in direct interaction with the oil surface.

By digesting with trypsin and chymotrypsin, α_{s1}-casein in aqueous solution could be cleaved at 47 peptide bonds. When α_{s1}-casein adsorbed to emulsified oil droplets was digested by the same proteinases, however, some 13 of the 47

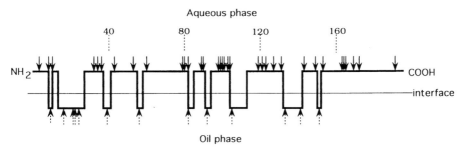

Figure 2 *Schematic representation of an α_{s1}-casein polypeptide chain adsorbed to an emulsified oil surface.[11] The arrows in the aqueous phase (downward) indicate the positions of peptide bonds cleavable by proteinases. The arrows in the oil phase (upward) indicate the positions which were cleaved in a solution but not in an emulsion*

peptide bonds were found not to be cleaved, suggesting that these peptide bonds were in parts of the adsorbed layer inaccessible to proteinases. From these results, we have presented a schematic structure of the α_{s1}-casein chain adsorbed at an emulsified oil surface (Figure 2). Although the proteinase resistance may be partly due to steric hindrance by the newly formed protein conformation, the adsorption of α_{s1}-casein to the oil–water interface is likely to occur through multiple binding sites. These binding sites are mostly rich in hydrophobic amino acid residues, and their distribution coincides well with that of the hydrophobic regions located in the α_{s1}-casein molecule. One of the presumed binding sites, located around residues 16–25, is particularly interesting. The peptide bond Phe[23] – Phe[24] is known to be cleaved very rapidly by such enzymes as pepsin and chymosin. The peptide bonds Leu[21] – Arg[22], Arg[22] – Phe[23] and Phe[24] – Val[25] are also readily cleaved by trypsin or chymotrypsin when soluble α_{s1}-casein is digested. However, when the adsorbed α_{s1}-casein is treated with these enzymes, none of these peptide bonds is cleaved. The susceptibility of the Leu[16] – Asn[17] peptide bond is also eliminated by adsorption. These results strongly suggest that the polypeptide portion around residues 16–25 plays an important role in the protein binding with the oil surface. The contribution of this portion to oil binding is supported by the fact that the adsorption of α_{s1}-casein to emulsified oil droplets is markedly reduced by removing the N-terminal 1–23 residues by pepsin digestion.[13]

Observation of the cleavability of peptide bonds is thus quite useful for evaluating the structure of a protein adsorbed to an emulsified oil surface, particularly for proteins with low conformational rigidity and good digestibility. Leaver and Dalgleish[14] have applied a similar method for the analysis of β-casein adsorbed to an emulsified oil surface. Exposure of the N-terminal hydrophilic portion to the aqueous phase as a single mobile loop was suggested by a kinetic approach, using trypsin-catalysed hydrolysis.

Immunochemical Method

The enzymic method just described uses proteinases as the tools. We have also attempted to use antibodies as the tools to analyse the protein structure. Like proteinases, antibodies gain access to their antigen proteins from the aqueous phase. By measuring the binding activity of certain antibodies with antigen proteins, we can deduce whether the epitopes recognized by the antibodies are exposed to the aqueous phase or not.

We have first applied this method to analyse α_{s1}-casein adsorbed to emulsified oil droplets.[15] An anti-α_{s1}-casein antiserum produced after immunizing mice was mixed with α_{s1}-casein-coated oil droplets, and binding of the antibodies to the droplets was measured. By increasing the concentration of the oil droplets, all of the antibodies were observed to bind to the α_{s1}-casein-coated droplets, indicating that the epitopes of α_{s1}-casein at the emulsified oil surface were all accessible to the antibodies. The epitopes were then mapped[16] and the structure of α_{s1}-casein at the oil–water interface was deduced from these data. The results supported our model structure deduced from the enzymic method (Figure 2).

Monitoring the structure of a globular protein such as β-lactoglobulin (β-LG) adsorbed to an emulsified oil surface can also be performed by the immunochemical method. We have recently succeeded in raising some monoclonal antibodies (MAbs) against β-LG.[17,18] Since these MAbs were found to recognize the conformations of different regions of the β-LG molecule (Table 1), they are thought to be useful for monitoring local structural changes in the β-LG molecule. The local structural changes occurring in a β-LG molecule during heating[18] and denaturant treatment[19] have been successfully monitored by using these antibodies. We also applied these MAbs to the structural analysis of β-LG at an emulsified oil surface.[20]

Four MAbs (21B3, 31A4, 61B4 and 62A6) were used for the analysis. By incubating these with emulsified oil droplets coated with β-LG, all of the MAbs were absorbed, suggesting that the epitopes for these MAbs were exposed to the aqueous phase. The binding affinity of the MAbs with the β-LG molecule adsorbed to the emulsified oil surface was then estimated by inhibition ELISA (enzyme-linked immunosorbent assay) and compared with that against the soluble β-LG molecule. The results are shown in Table 2. The binding affinity (K_{as}) of MAbs 61B4 and 62A6 with the adsorbed β-LG was similar to or only slightly lower than that with the soluble β-LG, suggesting that the structure of the epitopes for MAbs 61B4 and 62A6 (the region around Thr^{125}–Lys^{135}) was scarcely changed by emulsification. However, the value of K_{as} of MAbs 21B3 and 31A4 with the adsorbed β-LG was about 100 times higher than that with the soluble β-LG, suggesting that the structure of the epitopes for MAbs 21B3 and 31A4 (the region around Lys^8–Ile^{29}) was greatly changed by emulsification. The K_{as} values measured in this experiment may not have been very accurate because the competing antigen molecule was not in a soluble form, but in a form bound with the dispersed oil droplets. However, it is obvious from these data that the β-LG molecule was not

Table 1 *Properties of anti-β-lactoglobulin monoclonal antibodies*

MAb	Epitopes	Binding affinity
61B4	$Thr^{125} - Lys^{135}$	Native β-LG $\gg$ Denatured β-LG[a]
62A6	Close to the epitope for 61B4	Native β-LG $\gg$ Denatured β-LG
21B3	$Val^{15} - Ile^{29}$	Denatured β-LG $\gg$ Native β-LG
31A4	$Lys^8 - Trp^{19}$	Denatured β-LG $\gg$ Native β-LG

[a] Denatured β-LG was prepared by reduced carboxymethylation.

Table 2 *Changes in the binding affinity (K_{as}) of MAbs with β-LG after adsorption to an emulsified oil surface*

MAb	Native β-LG	Adsorbed β-LG	Structure of the epitope
61B4	1.5×10^8	1.2×10^8	Unchanged (native)
62A6	2.0×10^8	4.7×10^7	Slightly changed
21B3	1.2×10^6	1.6×10^8	Changed (unfolded)
31A4	6.0×10^5	5.5×10^7	Changed (unfolded)

completely unfolded at the emulsified oil surface, a certain portion of the molecule (*e.g.* the region around $Thr^{125} - Lys^{135}$) still maintaining its native conformation. Thus, it appears that the immunochemical approach could provide highly specific information on the local structural changes in an adsorbed protein molecule.

4 Possible Role of the Amphiphilic α-Helix Structure in Protein Emulsification

Various types of lipid–protein interaction are apparent in biological systems such as cell membranes and plasma lipoproteins. The penetration of polypeptides into lipid bilayers by forming hydrophobic α-helices is a typical form of lipid–protein interaction in cell membranes.[21] Integral membrane proteins such as receptors and nutrient transporters generally have several transmembrane domains with hydrophobic α-helices. The interaction between apolipoproteins and plasma lipoprotein particles is another type of lipid–protein interaction.[22] For example, Apo–CIII of the plasma lipoproteins is known to have an amphiphilic α-helix that has one polar and one non-polar face, by which this protein is strongly adsorbed to hydrophobic lipoprotein surfaces. This type of interaction possibly plays a role in protein adsorption at an emulsified oil surface.

Residues 16–25 of α_{s1}-casein were estimated to comprise the major oil-binding site of α_{s1}-casein, as described above (Figure 2). Although this region is rich in hydrophobic amino acids, it contains two charged amino acids (Glu and Arg) and two Asn residues, which would tend to reduce the hydrophobicity of this region. However, when the region forms an α-helix, the

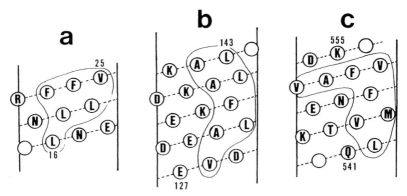

Figure 3   *Cylindrical plots of the α-helix formed by a segment of (a) α_{s1}-casein,*
residues 16–25, (b) β-lactoglobulin, residues 127–143, and (c) bovine serum
albumin, residues 541–555. Clusters of non-polar residues are encircled

hydrophobic amino acid residues are more concentrated on one side of the
helix (Figure 3a). The lipid-binding activity may depend on this amphiphilic
helix structure.

As described above (Tables 1 and 2), the region containing residues 125–
135 of β-LG are assumed to retain their native conformation after adsorption
to an emulsified oil surface. This region is partly involved in the α-helix
portion of β-LG, residues 130–140,[23] which also shows a typical amphiphilic
structure (Figure 3b). The amphiphilic structure may enable this region to
interact with an oil surface without changing its conformation at the oil–
water interface.

Bovine serum albumin (BSA) is known to have good emulsifying properties.
We have recently found that the third domain of BSA plays an important role
in its adsorption to an emulsified oil surface[24] although the lipid-binding
portion in this domain has not yet been determined in detail. Interestingly, this
domain also contains a sequence which could provide a typical amphiphilic α-
helix structure as shown in Figure 3(c).

Although there is still no direct evidence to support the importance of
amphiphilic α-helices, these findings suggest that the amphiphilic helix
could be one of the basic conformations in proteins adsorbed to an
emulsified oil surface. To investigate the importance of the amphiphilic
helix in protein adsorption, we recently started an analysis using synthetic
peptides.[25] Peptides of 16 residues with the same amino acid composition
(eight Leu and eight Glu), but with different sequences, have been designed
and synthesized. We observe that the peptide which can form an
amphiphilic helix shows higher emulsifying activity than the peptide whose
helix structure cannot be amphiphilic. These synthetic studies will be
particularly useful for evaluating the structural factors which determine
protein functionality.

5 Other Aspects of Protein Adsorption

As already described above, biochemical approaches are providing a lct of interesting information on the lipid–protein interaction at an emulsified oil surface. Accumulation of these data will be helpful for deducing the oil-binding properties of protein emulsifiers. Oil binding is, however, only one aspect of protein adsorption at an emulsified oil surface. Protein adsorption is accompanied by more complicated phenomena such as competitive and co-operative adsorption[9] and time-dependent rearrangement or polymerizaticn of the proteins adsorbed to the surface.[26,27] Further studies by various approaches will therefore be needed to obtain definitive information on the structure of proteins adsorbed at an emulsified oil surface.

References

1. M. Shimizu and Y. Nakane, *Biosci. Biotech. Biochem.*, 1995, **59**, in the press.
2. S. Nakai, *J. Agric. Food Chem.*, 1983, **31**, 676.
3. S. Nakai and E. Li-Chan, 'Hydrophobic Interactions in Food Systems', CRC Press, Boca Raton, FL, 1988, p. 44.
4. A. Kato and K. Yutani, *Protein Eng.*, 1988, **2**, 153.
5. M. Shimizu, M. Saito, and K. Yamauchi, *Agric. Biol. Chem.*, 1985, **49**, 189.
6. E. Tornberg, A. Olsson, and K. Persson, in 'Food Emulsions', ed. K. Larsson and S. E. Friberg, Marcel Dekker, New York, 2nd edn, 1990, p. 247.
7. M. C. Phillips, *Food Technol.*, 1981, **35**, 50.
8. E. Dickinson and G. Stainsby, 'Colloids in Food', Applied Science, London, 1982, p. 285.
9. E. Dickinson, *J. Chem. Soc., Faraday Trans.*, 1992, **88**, 2973.
10. S. Aynie, M. Le Meste, B. Colas, and D. Lorient, *J. Food Sci.*, 1992, **57**, 883.
11. M. Shimizu, A. Ametani, S. Kaminogawa, and K. Yamauchi, *Biochim. Biophys. Acta*, 1986, **869**, 259.
12. S. Kaminogawa, M. Shimizu, A. Ametani, S. W. Lee, and K. Yamauchi, *J. Am. Oil Chem. Soc.*, 1987, **64**, 1688.
13. M. Shimizu, T. Takahashi, S. Kaminogawa, and K. Yamauchi, *J. Agric. Food Chem.*, 1983, **31**, 1214.
14. J. Leaver and D. G. Dalgleish, *Biochim. Biophys. Acta*, 1990, **1041**, 217.
15. A. Ametani, M. Shimizu, S. Kaminogawa, and K. Yamauchi, *Agric. Biol. Chem.*, 1989, **53**, 1279.
16. A. Ametani, S. Kaminogawa, M. Shimizu, and K. Yamauchi, *J. Biochem.*, 1987, **102**, 421.
17. S. Kaminogawa, M. Hattori, O. Ando, J. Kurisaki, and K. Yamauchi, *Agric. Biol. Chem.*, 1987, **51**, 797.
18. S. Kaminogawa, M. Shimizu, A. Ametani, H. Hattori, O. Ando, S. Hachimura, Y. Nakamura, M. Totsuka, and K. Yamauchi, *Biochim. Biophys. Acta*, 1989, **998**, 50.
19. M. Hattori, A. Ametani, Y. Katakura, M. Shimizu, and S. Kaminogawa, *J. Biol. Chem.*, 1993, **268**, 22414.
20. M. Shimizu, A. Ametani, S. Kaminogawa, and K. Yamauchi, 'Protein and Fat Globule Modifications by Heat Treatment, Homogenization & Other Technological

Means for High Quality Dairy Products', International Dairy Federation, Brussels, 1993, p. 362.

21. D. C. Rees, L. De Antonio, and D. Eisenberg, *Science*, 1989, **245**, 510.
22. C. Edelstein, F. J. Kezdy, A. M. Scanu, and B. W. Shen, *J. Lipid Res.*, 1979, **20**, 143.
23. M. Z. Papiz, L. Sawyer, E. E. Eliopoulos, A. C. T. North, J. B. C. Findlay, R. Sivaprasadarao, T. A. Jones, M. E. Newcomer, and P. J. Kraulis, *Nature (London)*, 1986, **324**, 383.
24. M. Saito, M. Monna, K. Chikuni, and M. Shimizu, *Biosci. Biotech. Biochem.*, 1993, **57**, 952.
25. M. Saito, M. Ogasawara, K. Chikuni, and M. Shimizu, *Biosci. Biotech. Biochem.*, 1995, **59**, in the press.
26. R. D. Waniska and J. E. Kinsella, *J. Agric. Food Chem.*, 1985, **33**, 1143.
27. E. Dickinson and Y. Matsumura, *Int. J. Biol. Macromol.*, 1991, **13**, 26.

A Phenomenological Model for the Dynamic Interfacial Behaviour of Adsorbed Protein Layers

By George A. van Aken

DEPARTMENT OF BIOPHYSICAL CHEMISTRY, NETHERLANDS INSTITUTE
FOR DAIRY RESEARCH (NIZO), PO BOX 20, 6710 BA EDE, THE NETHERLANDS

1 Introduction

Water-soluble proteins, such as the major milk proteins, have an excellent capability for stabilizing foams and emulsions. Recently, increasing interest has emerged for the application of modified proteins such as protein hydrolysates for this purpose. To improve our knowledge in this field, a research programme was set up to investigate the relationship between the molecular structure of milk proteins and peptides and their emulsifying and foaming capacity.

Part of this research programme focusses on the dynamic aspects of the adsorption behaviour of proteins, and more specifically on the adsorption kinetics and surface-dilational properties, because these are thought to be related to the formation and breakdown of foams and emulsions. It is very difficult to quantify the dynamic interfacial properties with relevance to foams and emulsions, because these properties are strongly dependent on the experimental conditions and especially on the time-scale of the measurements. Therefore we have developed a phenomenological theory that enables us to characterize the dynamic behaviour of a layer of protein adsorbed at an air–water interface in terms of a small set of experimentally accessible parameters.

2 Reversibility of Adsorbed Protein Layers with Respect to Surface Dilation

It is well known that for most proteins desorption from the air–water interface at surface pressures below approximately 18 mN m^{-1} occurs at such a slow rate that the adsorbed layer may be considered irreversibly adsorbed on a time-scale of a few hours.[1,2] Hence the properties of an adsorbed protein layer can be studied experimentally by spreading the protein on the surface of a

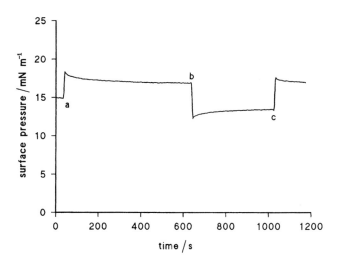

Figure 1   *Stress–relaxation curves for an adsorbed layer of β-lactoglobulin. Relative area variations: (a) 1 → 0.667; (b) 0.667 → 1; (c) 1 → 0.667. Solvent: aqueous imidazole solution adjusted to pH 7.0, ionic strength 0.075*

buffered solution and expanding, compressing or shearing the adsorbed layer without substantial release of protein into the bulk liquid. We note that for many proteins the adsorbed layer properties depend on the way the protein has been transferred to the surface.[3,4] The differences are probably due to differences in the surface pressure (or surface load) when new protein molecules reach the surface.

Adsorbed protein layers are usually viscoelastic, as can be established by stress–relaxation studies of the adsorbed layer.[5–8] In these experiments a step-wise area variation is imposed on the adsorbed layer, inducing an instantaneous change in the surface pressure (an elastic effect), followed by a relaxation of this change (a viscous effect). A typical stress–relaxation curve for β-lactoglobulin is shown in Figure 1. The relaxation process reflects a slow reorganization in the adsorbed layer, during which time the molecules adapt to the altered surface pressure. Another way to study the visco-elastic dilational behaviour is by imposing sinusoidal area dilations with various frequencies on the adsorbed layer, and monitoring the 'storage' and 'loss' contributions of the resulting sinusoidal response of the surface pressure.[6,9] In principle, the frequency-dependent surface-dilational modulus is related to a stress–relaxation curve by a Laplace transformation,[10,11] and this mathematical relationship has been verified experimentally for an adsorbed layer of bovine serum albumin (BSA).[7]

A very interesting feature, that can be seen in Figure 1 for β-lactoglobulin, is that the surface pressure returns to the original value if the area is set at the original value.[8] This reversibility has been observed for most proteins that we have studied so far (β-casein, BSA, triacylglycerol lipase (*Candida cylindracea*), soybean glycinin and soybean β-conglycinin); however, reversibility was not found for hen-egg lysozyme. The experimental reversibility indicates that the molecular reorganizations in the adsorbed layer are reversible. Consequently, in these cases a functional relationship should exist between the surface pressure and the surface load.

3 Surface Equation of State of Adsorbed Protein Layer

De Feijter and Benjamins[12] have used the equation of Helfand *et al.*[13] for the osmotic pressure of a two-dimensional fluid of hard disks to describe the surface pressure of an adsorbed protein layer,

$$\Pi = \frac{kT\Gamma}{(1-\theta)^2} \tag{1}$$

where Γ is the surface load, and θ is the fraction of the area that is covered by the protein,

$$\theta = \Gamma\Omega \tag{2}$$

where Ω is the molecular area. In practice, the application of eqn (1) to adsorbed protein layers reveals that, for adsorbed protein layers at surface pressures above approximately 1 mN m^{-1}, the value of θ must be close to unity, which means that already at this low surface pressure the surface is almost completely covered. Higher surface pressures are obtained at higher surface loads. Higher surface loads are, however, only possible if the adsorbed molecules are compressible. On this basis De Feijter and Benjamins[12] calculated the compression of a few types of protein molecules as a function of the surface load. We extend their approach by separating the variation of the area per molecule into an elastic contribution and a viscous contribution. The elastic contribution is governed by the surface pressure Π. We will assume that the viscous contribution can be described by a single parameter α, which can be viewed as an 'unfolding' parameter. Because of the reversibility of the area variations, infinitesimal variations of Ω can be described by the total differential

$$d\Omega = \left(\frac{\partial\Omega}{\partial\Pi}\right)_\alpha d\Pi + \left(\frac{\partial\Omega}{\partial\alpha}\right)_\pi d\alpha \tag{3}$$

If we define the elastic compressibility of the adsorbed molecules by

$$\beta = -\frac{1}{\Omega}\left(\frac{\partial\Omega}{\partial\Pi}\right)_\alpha \qquad (4)$$

Then the first partial derivative in eqn (3) equals $-\beta\Omega$. At high surface pressures, where $\theta \to 1$, $1/\beta$ equals the high-frequency limit of the surface-dilational modulus of the adsorbed layer. A practical definition for α is the area per molecule divided by the area of the fully unfolded molecule $\Omega_u(\pi)$ at the same surface pressure. In that case the second partial differential of eqn (3) equals $\Omega_u(\pi)$. If we assume β to be a constant, it follows for a completely unfolded molecule ($\alpha = 1$) from eqn (4) that we have

$$\Omega_u(\pi) = \Omega_0 \exp(-\beta\pi) \qquad (5)$$

and, for other values of α we have

$$\Omega = \alpha\Omega_0 \exp(-\beta\pi) \qquad (6)$$

where Ω_0 is the area of the completely unfolded molecules at zero surface pressure.

MacRitchie[2] performed surface balance experiments in which adsorbed protein layers at low initial surface pressure (< 1 mN m^{-1}) were quickly compressed to an area A_0 at a surface pressure of π_e, after which the surface pressure was maintained constant at π_e. After compression the adsorbed layer

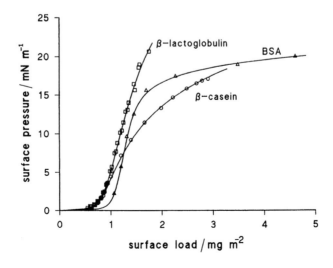

Figure 2   *Surface pressure versus surface load for β-casein, BSA and β-lactoglobulin A. Fitting parameters are listed in Table 1*

Table 1   *Experimental fitting parameters for the surface pressure versus surface load curves. Symbols: Ω_0 is the area of completely unfolded adsorbed molecules at zero surface pressure; β is the elastic compressibility of the adsorbed molecules; π^* and Ω^* are the fitting parameters for the viscous compressibility of the adsorbed molecules*

Protein	Molecular mass	Ω_0	β	π^*	Ω^*
	g mol^{-1}	m^2 mg^{-1}	m N^{-1}	mN m^{-1}	nm^2
β-Casein	23983	1.136	25.6	14.8	0.59
BSA	65400	0.847	18.9	18.7	2.15
β-Lactoglobulin A	17500	0.962	26.3	33.0	0.80

was seen to slowly relax towards a new equilibrium value of the area, A_e. The area variations appeared to obey the relationship

$$E = \frac{A_e}{A_0 - A_e} = \exp\left(-\frac{(\pi_e - \pi^*)\Omega^*}{kT}\right) \qquad (7)$$

where π^* and Ω^* are constants. According to MacRitchie[2] this equation reflects the equilibrium between adsorbed and desorbed segments of the protein molecule. In this interpretation, the parameter E equals the area ration of (adsorbed segments)/(desorbed segments), Ω^* is the area of one such segment and $-\pi^*\Omega^*/kT$ is the free energy of adsorption of such a segment when $\pi_e = 0$. If E is large, the molecules are in a highly unfolded state and occupy a large area in the interface; and the opposite is the case for small E values. The parameter E can therefore be related to α_e, the equilibrium value of α, by

$$\alpha_e = \frac{A_e}{A_0} = \frac{E}{1 + E} \qquad (8)$$

The set of eqns (1, 2, 6–8) was used to fit π_e *versus* Γ curves by adjusting the parameters Ω_0, β, π^* and Ω^*. Figure 2 shows best fits through the experimental data for the proteins β-casein,[14] BSA[1] and β-lactoglobulin[4] taken from the literature; the fitting parameters are given in Table 1.

4 Relaxation Processes in the Adsorbed Protein Layer

In order to describe the process of relaxation, we need to specify the speed at which α approaches its equilibrium value. The assumption of a first-order process would lead to the linear differential equation

$$\frac{d\alpha}{dt} = -k_r(\alpha - \alpha_e) \qquad (9)$$

where k_r is the relaxation constant. Integration of eqn (9) then leads to an approximately exponential decay of the surface pressure. Experimentally,

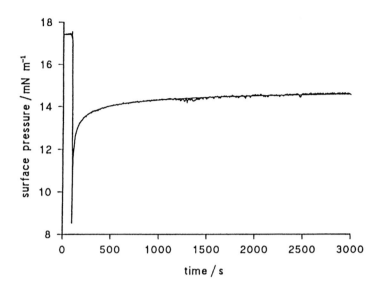

Figure 3   *Stress–relaxation curve for BSA together with a best-fit obtained with the theoretical equations. Fitting parameters: values for Ω_0, β, π^* and Ω^* taken from Table 1; additional parameters: $k_r = 10^{-4}$ s^{-1}, $\kappa = 35$. Solvent as in Figure 1*

however, the decay curves often deviate from an exponential shape (Figures 1 and 3), indicating the inadequacy of eqn (9). Due to the complex nature of protein molecules, it is most likely that a multitude of reorganizations take place in the adsorbed layer during the decay process. Such behaviour can be modelled as a sum of several exponential decay functions. However, this requires many additional fitting parameters, which is not desirable for the present purpose. An empirical equation that needs only two adjustable parameters, and still renders a fairly accurate description of the stress–relaxation of the surface pressure of a pre-equilibrated adsorbed protein layer is the equation

$$\frac{d\alpha}{dt} = -\frac{k_r}{\kappa} \cdot \sinh[\kappa \cdot (\alpha - \alpha_e)] \tag{10}$$

where κ (> 0) accounts for the deviation from linearity (in the limit of $\kappa \to 0$ eqn (9) is retrieved). Figure 3 shows the best fit according to eqn (10) of a stress–relaxation curve for an adsorbed layer of BSA. The two curves match almost exactly.

5 Discussion and Conclusion

It has been shown that it is possible to give an accurate description of curves of surface pressure *versus* surface load for an adsorbed protein layer using an expression with four parameters, and that at least two additional parameters are needed to include a fairly accurate description of the kinetics of the relaxation processes. In principle, these six parameters cover all dilational properties of an adsorbed layer. For example, the frequency-dependent surface-dilational moduli, for either small and large surface dilations and at any surface pressure, can be calculated by simulating a stress–relaxation experiment and a Laplace transformation of the relaxation curve. This approach allows us to compare the surface-dilational properties of proteins and peptides on the basis of these six parameters.

References

1. F. MacRitchie and L. Ter-Minassian-Saraga, *Colloids Surf.*, 1984, **10**, 53.
2. F. MacRitchie, *J. Colloid Interface Sci.*, 1981, **79**, 461.
3. H. J. Trurnit, *J. Colloid Sci.*, 1960, **15**, 1.
4. J. Mitchell, L. Irons, and G. J. Palmer, *Biochim. Biophys. Acta*, 1970, **200**, 138.
5. R. Maksymiw and W. Nitsch, *J. Colloid Interface Sci.*, 1991, **147**, 67.
6. G. Serrien, G. Geeraerts, L. Ghosh, and P. Joos, *Colloids Surf.*, 1992, **68**, 219.
7. G. A. van Aken and M. T. E. Merks, in 'Food Colloids and Polymers: Stability and Mechanical Properties', ed. E. Dickinson and P. Walstra, Special Publication No. 113, The Royal Society of Chemistry, UK, 1993, p. 402.
8. G. A. van Aken and M. T. E. Merks, *Prog. Colloid Polym. Sci.*, 1994, **97**, 281.
9. E. H. Lucassen-Reynders and D. T. Wasan, *Food Struct.*, 1993, **12**, 1
10. G. Loglio, U. Tesei, and R. Cini, *J. Colloid Interface Sci.*, 1979, **71**, 316.
11. R. Miller, G. Loglio, U. Tesei, and K.-H. Schano, *Adv. Colloid Interface Sci.*, 1991, **37**, 73.
12. J. A. de Feijter and J. Benjamins, *J. Colloid Interface Sci.*, 1982, **90**, 289.
13. E. Helfand, H. L. Frisch, and J. L. Lebowitz, *J. Chem. Phys.*, 1961, **34**, 1037.
14. J. V. Boyd, J. R. Mitchell, L. Irons, P. R. Mussellwhite, and P. Sherman, *J. Colloid Interface Sci.*, 1973, **45**, 478.

Association of Chymosin with Adsorbed Caseins

By André L. de Roos, Pieter Walstra, and Tom J. Geurts

DEPARTMENT OF FOOD SCIENCE, WAGENINGEN AGRICULTURAL UNIVERSITY, PO BOX 8129, 6700 EV WAGENINGEN, THE NETHERLANDS

1 Introduction

In studies aimed at the acceleration of cheese ripening by means of adding exo- and endoproteases, isolated from starter bacteria, to the cheese milk, it is often observed that, during the cheese making process, almost all of the extra added enzymes go to the whey fraction and very little goes to the curd. It was reasoned that these enzymes could in principle be transported into the curd by immobilizing them onto milk fat globules assuming that the enzyme activity is retained.

Recently we have used the renneting enzyme chymosin as the sole emulsifying agent for stabilizing a soya oil-in-water emulsion. In this study (to be published) it was found that the adsorbed enzyme had lost its activity after having been displaced from the oil–water interface by a small-molecule surfactant.

It is well established that chymosin is transported partly (depending on production conditions) into the curd during cheese-making, and that it plays a key role in cheese ripening. The mechanism of transport of chymosin into the curd is believed to involve adsorption onto (or association with) casein micelles. The mechanism by which chymosin becomes associated with caseins is, however, not well understood, nor is it known which individual casein is primarily involved. In this paper we use casein as the emulsifier with which chymosin could be associated. Various aspects of the mechanism of chymosin association with adsorbed caseins are described. The influences on adsorption isotherms of pH, ionic strength, temperature and substrate concentration are investigated. We have to keep in mind, however, that the presented adsorption isotherms reflect the protein–protein interaction between the adsorbed casein and chymosin and not the adsorption of proteins onto a conventional solid–liquid or fluid–liquid interface.

50

2 Materials and Methods

Emulsions were made with a Condi lab homogenizer using purified soya oil (volume fraction $\varphi = 0.02$) and various caseins (2 wt%, Sigma) as the emulsifying agents. The emulsions were made in 25 mM phosphate buffer, pH 6.7, repeating the homogenization ten times. Commercial soya oil (Reddy) was made monoglyceride free by stirring it for 1 hour with pre-dried Kieselgel 60 (Merck) followed by centrifugation. The soya oil volume fraction was determined by the Gerber method[1] (a correction factor of 1.11 was used for the soya oil density).

In a preliminary experiment, meant to establish which casein the chymosin mainly associates with, the cream layers of emulsions made with α_s-, β- or κ-casein (2 g l^{-1}) were, after having washed away the unbound caseins twice by centrifugation (20 min, 12000 g) and resuspending with 25 mM phosphate buffer solutions of pH 5.0, 5.4, 5.8, 6.2 or 6.8, suspended in phosphate buffers containing 0.7 μM chymosin.

In a subsequent experiment, the cream layers of a κ-casein emulsion were washed twice with phosphate buffers of different concentration, pH, temperature and ionic strength (by adding KCl), and finally resuspended with the same buffers with chymosin at concentrations of 10, 5, 4, 3, 2, 1 or 0.5 μM. The chymosin (Chymax, Pfizer) was lyophilized, after having been made salt-free by means of dialysis against water, in order to minimize the influence of salts on the adsorption conditions due to addition of the enzyme. The emulsion was incubated with the chymosin for 20 min and then centrifuged in 1 ml vials (Eppendorf) for 1 min at 10000 r.p.m. The residual chymosin activity in the supernatant was determined by the Berridge flocculation test.[2] The percentage of adsorbed chymosin could be calculated according to the relation of Storch and Segelcke, which states that the flocculation time is inversely proportional to the chymosin concentration.[3] Chymosin solutions corresponding to the experimental conditions, at a concentration to give a flocculation time of about 300 s (on 4000 times dilution), were used as the blank.

Emulsion droplet size analysis was performed by means of laser diffraction (Coulter LS130) yielding the volume/surface average diameter d_{vs}. The specific surface area A of the emulsion was calculated from the equation $A = 6\,\varphi/d_{vs}$.

The casein concentrations in the starting solution and the emulsion supernatant were determined by the BCA protein assay (Pierce). From the difference between these two concentrations, the casein surface load (Γ in mg m^{-2}) could be calculated.

The chymosin surface load on adsorbed casein (mmol of chymosin/mol of casein) was determined assuming that a 13.3 μM chymosin concentration leads to a rennet clotting time of about 300 s (4000-fold dilution).

Chymosin activity *in situ*, *i.e.* with the enzyme associated with the casein-coated droplets, was determined using a small synthetic hexapeptide substrate HLeu-Ser-Phe(NO$_2$)-Nle-Ala-Leu-Ome (Bachem) which is split at the Phe(NO$_2$)–NLe linkage.[4] The emulsion droplet layer was washed three times in

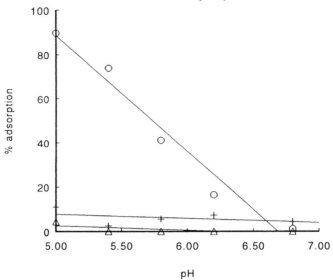

Figure 1   *The association of chymosin, expressed as a percentage of the initial amount of chymosin (0.7 μM), with adsorbed α_s-casein (+), β-casein (△) and κ-casein (○) as a function of pH, after 20 min of incubation at room temperature in 25 mM phosphate buffer*

a 0.05 M acetate buffer of pH 4.7 and resuspended in a 0.5 mM hexapeptide solution in the same buffer. After 20 min of incubation the emulsion was filtered (pore size 0.2 μm) and the absorbance of the filtrate measured spectrophotometrically at 310 nm. Alternatively, chymosin activity *in situ* was measured by following caseinomacropeptide release by means of HPLC[5] after various times of incubation (0, 0.5, 1, 2, 5, 15 and 30 min) of washed emulsion droplets with a κ-casein solution (2 g l^{-1} at room temperature).

3 Results and Discussion

In Figure 1 the association of chymosin with the emulsions made with the three different individual caseins is shown. Chymosin was found to be associated strongly and exclusively with adsorbed κ-casein; the association was higher at lower pH, probably as a result of electrostatic interaction between enzyme and substrate. Association with α_s- or β-casein was weak or absent. Fast creaming of the κ-casein emulsion was observed after addition of the chymosin. This must be due to flocculation of the emulsion droplets, caused by loss of steric repulsion after splitting off the caseinomacropeptide of the κ-casein molecule. This would imply that the caseinomacropeptide moiety is not attached to the soya oil–water interface. The association of chymosin probably takes place at the remaining positively charged *para*-κ-casein part, most likely after splitting off the caseinomacropeptide. Emulsions made with α_s- and β-casein were far more stable after adding chymosin. In the supernatant of these

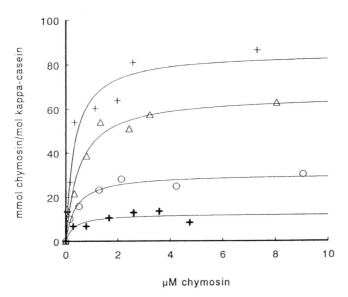

Figure 2 *The influence of pH on the association of chymosin with κ-casein, adsorbed onto emulsion droplets, after 20 min of incubation in 25 μM phosphate buffer at room temperature: pH 5.0 (+); pH 5.4 (△); pH 5.8 (○); and pH 6.2 (bold +)*

emulsions with added chymosin, the presence of various casein fragments were indicated by means of HPLC gel permeation measurements. Identification of these fragments was not carried out.

Chymosin adsorption isotherms under various conditions were determined starting from a single κ-casein/soya oil emulsion with the following characteristics: $\varphi = 0.0185$; volume/surface average weighed diameter $d_{vs} = 1.63 \times 10^{-6}$ m; specific surface area $A = 0.068$ m^2 m^{-3}; κ-casein surface load $\Gamma = 8.97$ mg m^{-2} corresponding to a κ-casein concentration of 31 μM for the emulsion. The chymosin surface load on this emulsion is expressed as mmol of chymosin/mol of κ-casein. The results were fitted to a Langmuir-type equation.

Figure 2 shows the influence of the pH on the chymosin association. It was found that the chymosin association increased with decreasing pH. This pH dependence points to electrostatic interactions being involved. At pH 5.0 the chymosin molecule will have a small net negative charge (pI = 4.7) that will increase with increasing pH. In this pH range *para*-κ-casein will be positively charged. (The pI is unknown but electrophoretic mobility towards the cathode is observed at pH 7.[7]) As we do not know the mechanism of enzyme–substrate complex formation and the pK values of the amino acids involved, it is difficult to interpret this result. The three histidine groups (amino acids 98, 100, 102) are believed to play an essential role in the formation of the enzyme–substrate complex that preceeds the cleavage of the Phe–Met bond of κ-casein.[8] At pH

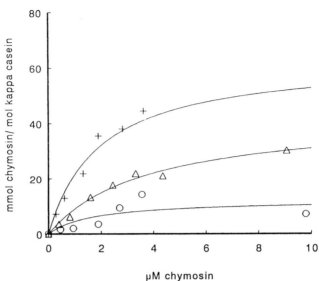

Figure 3   *The influence of ionic strength on the association of chymosin with κ-casein, adsorbed onto emulsion droplets, after 20 min of incubation in 25 mM phosphate buffer pH 5.5 at room temperature: [KCl]=0 mM (+); [KCl] = 75 mM (△); [KCl] = 150 mM (○)*

values above pH 6.0, the histidine groups become less positive, and more and more the negatively charged chymosin becomes less able to associate. This speculation on the mechanism of association can only be valid if it would be coupled to the active centre of the enzyme.

If the protein/protein interaction was of an electrostatic nature, ionic strength should also play a part. We varied the ionic strength at constant pH by adding KCl in various amounts. As can be seen in Figure 3, chymosin association becomes weaker at greater ionic strength. Although the Debye length is expected to become shorter[7] (for 0 mM KCl ≈ 0.95 nm and for 150 mM ≈ 0.63 nm) and closer approach of the two proteins would thus be possible, the degree of association becomes less. This may be due to direct association of the ions with the charged groups, which play an essential role in the electrostatic protein interaction. Other salts, like NaCl, LiCl and varying concentrations of the potassium phosphate buffer, have shown comparable effects.

In cheese-making the transfer of chymosin into the curd is greatly affected by temperature.[9] A similar effect is found for the system examined here (Figure 4). Chymosin association becomes much stronger with decreasing temperature. For this reason the interaction of chymosin and *para*-κ-casein is not believed to be a hydrophobic one.

When chymosin was added to dilutions of the emulsion with adsorbed κ-casein, the absolute percentage of association was reduced. At first sight this looked quite obvious. However, if we expressed the association in values of surface load, *i.e.* mmol of chymosin associated per mol of casein, we found,

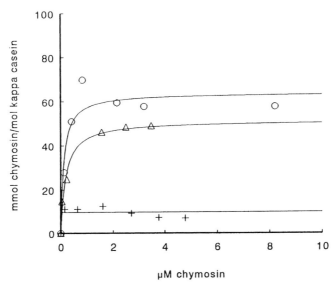

Figure 4   *The influence of temperature on the association of chymosin with κ-casein adsorbed onto emulsion droplets, after 20 min of incubation in 25 mM phosphate buffer at pH 5.5; 4°C (○); 20°C (△); and 40°C (+)*

surprisingly, at comparable chymosin equilibrium concentration a higher surface load for the diluted system (Figure 5). Dilution of the emulsion resulted in a decreased number of fat globules and hence an increased water/adsorbed protein ratio. In other words, at a higher water to protein ratio of the system, chymosin association became higher.

In a very concentrated system like cheese, a surface load of 0.1 mmol of chymosin/mol of κ-casein was found, and for milk (pH 5.0, $I = 120$ mM, [κ-casein] = 100 μM), the diluted form, 1.4 mmol of chymosin/mol of κ-casein.[10] By comparing the surface load for the emulsion system (under the same experimental conditions as in milk and at a chymosin equilibrium concentration of 0.1 μM), we have found the same surface load of 1.4 mmol of chymosin/mol of κ-casein. An explanation for the chymosin surface load being dependent on the κ-casein concentration has not been worked out completely yet, but we feel it must be due to a shift in the association equilibrium with competing associating molecules, presumably other caseins, upon dilution. These other caseins can be regarded, in fact, as contaminants in the κ-casein material. The effect of temperature on the chymosin surface load may also be the consequence of the shift in the association equilibrium of the competing, associating molecules. Hydrophobic interactions may play a role here. The presence of the competing caseins may also account for the low surface load of chymosin.

As the association of chymosin is so delicately dependent on environmental conditions like pH and ionic strength, the reverse process of release of already

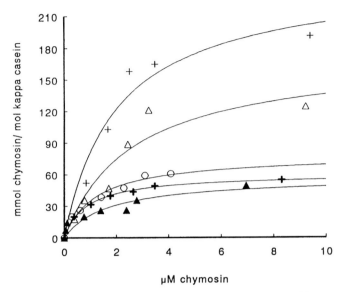

Figure 5   *The influence of the κ-casein concentration, in various dilutions of the soya oil/κ-casein emulsions, on the association with chymosin after 20 min of incubation in 25 mM phosphate buffer, pH 5.5, at room temperature. Emulsion dilution: 10 × (+); 5 × (△); 2 × (○); 0 × (bold +); 0.5 × (▲)*

associated chymosin would be expected to be induced by altering these conditions. If emulsion droplets loaded with κ-casein and chymosin (made, for instance, at pH 5.5) were added to milk in the Berridge flocculation test (experimental pH 6.3), a chymosin activity would be found, corresponding to the differences in association between pH 5.5 and 6.3. When emulsion droplets, made at pH 6.3 and at an ionic strength comparable with that in the flocculation of milk,[11] were used in the flocculation test, no chymosin activity was found, presumably because the associated chymosin had not been released. This means that the association of chymosin with adsorbed κ-casein did not inactivate the enzyme. It also means that the concentration of associated chymosin *in situ* could not be determined by the flocculation assay because diffusion limitation and steric hindrance may have affected their affinity. No further quantification of this finding was made, but it was shown that chymosin associated with κ-casein remained active. In an experiment where κ-casein was used as the substrate, complete conversion of the substrate was found.

If the association interaction is restricted to the active centre of the enzyme, as has been speculated previously, it is hard to imagine that the enzyme can be associated to *para*-κ-casein and be able to be active at the same time. The amino acids involved in the association interaction are thus most likely to be different from those involved in the enzyme–substrate reaction.

It is known that other biologically active proteins, like lysozyme, lipase, plasmin, bacteriocines and nisine also have an associating affinity for caseins. All these proteins can be transferred into the curd during the cheese-making process. Although we have to keep in mind that the process of emulsification may well cause a conformational change of the casein molecules and thereby lead to exposure of different functional groups for association, there is still a fair chance that adsorbed caseins will also associate with the proteins mentioned above. Moreover other proteins may possess binding affinity for caseins. In this way emulsion droplets can serve as a liquid carrier for immobilizing proteins with caseins as the intermediate ligand.

Acknowledgement

The authors acknowledge the skilled technical assistance of Miss Jennifer van Dijk.

References

1. L. Radema and H. Mulder, *Neth. Milk Dairy J.*, 1951, **5**, 104.
2. N. J. Berridge, *Analyst*, 1952, **77**, 57.
3. P. Walstra and R. Jenness, 'Dairy Chemistry and Physics', Wiley, New York, 1984, p. 243.
4. H. N. Raymond, E. Bricas, R. Salese, J. Garnier, P. Garnot, and B. Ribadeau-Dumas, *J. Dairy Sci.*, 1973, **56**, 419.
5. A. C. M. van Hooydonk and C. Olieman, *Neth. Milk Dairy J.*, 1982, **36**, 153.
6. A. Andrén, K. Larsson, H. Andersson, and L. Björck, Brief communications and abstracts of posters of the XXIII International Dairy Congress, Montreal, vol. 1, 1990, p. 248.
7. D. G. Dalgleish, in 'Developments in Dairy Chemistry—1 Proteins', ed. P. F. Fox, Applied Science, London, 1982, p. 329.
8. N. M. C. Kay and P. Jollès, *Biochim. Biophys. Acta*, 1978, **536**, 329.
9. J. Stadhouders and G. Hup, *Neth. Milk Dairy J.*, 1975, **29**, 335.
10. T. J. Geurts, unpublished results.
11. R. Jenness and J. Koops, *Neth. Milk Dairy J.*, 1962, **16**, 185.

Surface Activity and Competitive Adsorption of Milk Component 3 and Porcine Pancreatic Lipase at the Dodecane– Water Interface

By Jean-Luc Courthaudon, Jean-Michel Girardet,[1] Claire Chapal, Denis Lorient, and Guy Linden[1]

LABORATOIRE DE BIOCHIMIE ET TOXICOLOGIE ALIMENTAIRES, ECOLE NATIONALE SUPÉRIEURE DE BIOLOGIE APPLIQUÉE À LA NUTRITION ET À L'ALIMENTATION (ENSBANA), 1 ESPLANADE ERASME, 21000 DIJON, FRANCE
[1] LABORATOIRE DE BIOCHIMIE APPLIQUÉE, ASSOCIÉ À L'INSTITUT NATIONAL DE LA RECHERCHE AGRONOMIQUE, UNIVERSITÉ DE NANCY, BP 239, 54506 VANDŒUVRE-LÈS-NANCY, FRANCE

1 Introduction

During the time which elapses between the milking of cows on the farm and milk pasteurization at the dairy plant, some spontaneous lipolysis of milk fat occurs under the influence of lipoprotein lipase.[1] This hydrolysis of milk fat, say triglycerides, leads to a bad flavour: rancid taste and bitterness. Therefore there is much interest, in terms of product quality, in studying the mechanism responsible for this spontaneous lipolysis occurring in milk upon storage at $+4\,^{\circ}C$ for several days.

From previous work it is known[2–5] that milk lipoprotein lipase activity can be reversibly inhibited by a minor fraction of milk proteose-peptone: milk component 3. However, the mechanism involved has not been clearly established yet. It is also known that, in milk, the enzyme is largely bound to casein micelles; this reduces the concentration of free enzyme and hence its activity.[1] Many studies have been carried out using the porcine pancreatic lipase instead of milk protein lipase since these two enzymes are similar in that they belong to the same family.[6,7] In addition, each of these two lipases is activated by a polypeptide; lipoprotein lipase is activated by apolipoprotein C-II and porcine pancreatic lipase by pancreatic colipase.[8–10]

In an oil-in-water emulsion, the activity of porcine pancreatic lipase is reversibly inhibited by some proteins, e.g. β-lactoglobulin, ovalbumin,

58

serum albumin, and, in particular, milk component 3.[6,7] The activity of porcine pancreatic lipase is restored by the addition of colipase and bile salts (*e.g.* taurodeoxycholate, glycocholate or taurocholate).[7,10] However, sodium glycocholate or taurocholate do not reactivate lipase activity if it was inhibited by milk component 3.[11] The amount of sodium taurodeoxycholate required for lipase reactivation is four times higher than that required if lipase was inactivated by β-lactoglobulin. This means lipase activity is more difficult to restore in the presence of milk component 3 than in the presence of β-lactoglobulin.[11] It has been shown, by gel filtration chromatography, that proteins do not interact with porcine pancreatic lipase in solution.[6] Fluorescence measurements of 8-anilino-naphthalene-1-sulfonate bound to the surface hydrophobic sites of porcine pancreatic lipase in the absence or in the presence of milk component 3 have not shown any interaction of the last two molecules through hydrophobic interactions.[11] A full recovery of the lipolytic activity of porcine pancreatic lipase is obtained on addition of taurodeoxycholate and colipase.[11] Thus the reversible inactivation of porcine pancreatic lipase by milk component 3 is not due to the association of milk component 3 and porcine pancreatic lipase by hydrophobic interactions.[11]

The objective of the present study, carried out in a simplified system, is to determine whether the reversible inactivation of porcine pancreatic lipase by milk component 3 is due to one or both of the two following phenomena: a preferential adsorption of milk component 3 over porcine pancreatic lipase at the oil–water interface, or an association of these two proteins by hydrophilic interactions.

Surface tension measurements and oil-in-water emulsions were made under pH and temperature conditions close to those of milk storage, *i.e.* pH 7 and +4 °C. The oil phase was *n*-dodecane since this oil is not a substrate for the enzyme. Otherwise lipase, which acts at the lipid–water interface, would hydrolyse triglycerides and consequently generate free fatty acids which are surface active. The protein was component 3, porcine pancreatic lipase, or a mixture of these two protein substrates. Porcine pancreatic lipase was used instead of expensive lioprotein lipase because these two enzymes are very similar in terms of structure and activity.[9,10] Additional experiments were made with the two proteins introduced successively in the system.

2 Materials and Methods

Materials and Reagents

The porcine pancreatic lipase (EC 3.1.1.3) was from Boehringer Mannheim (Mannheim, Germany). Tween 20 (1224 g mol^{-1}; CMC $= 3.5 \times 10^{-5}$ M; estimated HLB $= 16.7$) was a high purity sample (Surfact-Amp 20) obtained from Pierce Chemicals (Rockford, USA). The *n*-dodecane was from Sigma. Other reagents were AnalaR grade.

Isolation of Milk Component 3

The proteose-peptone fraction was prepared from bulk skim milk after heat treatment (95 °C, 30 min) and isoelectric precipitation at pH 4.6 (by 1 M HCl).[12] The proteose-peptone fraction was precipitated using ammonium sulfate (380 g l^{-1}) and freeze-dried after dialysis. The hydrophobic fraction of proteose peptone (HFPP) was isolated from the proteose-peptone fraction by hydrophobic interaction chromatography using a TSK-Phenyl-5PW column (21 × 150 mm) connected to a FPLC system (Pharmacia Fine Chemicals, Uppsala, Sweden).[13] The HFPP was shown to be mainly constituted of milk component 3 (see electrophoresis results below). Therefore, it is equally referred to as HFPP or milk component 3 in the rest of the text. The component 3 fraction was dialysed and freeze-dried.

Surface Tension Measurements

Surface tension measurements were made with a platinum Wilhelmy plate at the n-dodecane–water interface using a tensiometer (Tensimat no. 3, Prolabo, Paris, France) equipped with a recorder.[14] The interfacial tension was registered as a function of time. Temperature was maintained at $+4 \pm 0.2$ °C using a double-envelope jacket dish. The aqueous phase was 50 ml of citrate/phosphate/HCl buffer (0.1 M, pH 7.0) and the oil phase was 65 ml of n-dodecane. A sample of 25 or 50 μl of protein solution was injected in the aqueous phase, under the oil–water interface, to obtain a final protein concentration of 20 mg l^{-1}.[11]

Emulsion Formation

Oil-in-water emulsions (oil phase 20 g in 100 g; aqueous phase 80 g in 100 g) were prepared using a jet homogenizer operating at a pressure of 300 bar.[15] The aqueous phase of the emulsion was citrate/phosphate/borate/HCl buffer (0.1 M, pH 7.0) containing 0.5 g of protein in 100 g of solution. The oil phase was n-dodecane. Protein was porcine pancreatic lipase, milk component 3, or a mixture (50 wt% of each protein) of both proteins.

Protein/Protein and Protein/Tween 20 Displacement Experiments

Immediately after emulsification, to an aliquot of emulsion stabilized by one protein, component 3 or porcine pancreatic lipase, was added the right amount of the other protein (freeze-dried sample), say porcine pancreatic lipase or component 3, respectively, in order that the emulsion aliquot contained the same amount of each protein. Part of the freshly made emulsion stabilized with milk component 3 was divided into aliquots to which appropriate amounts of Tween 20 were added to produce a range of values of surfactant-to-protein molar ratio (R) from 0 to 90.

Determination of Droplet Size and Aggregation

The droplet size distribution of the emulsions was determined using a light scattering apparatus (Malvern Mastersizer MS 20; Malvern Instruments, Malvern, UK). Droplet size measurements were performed on the freshly made emulsion. Another measurement was made after the emulsion had been left in the presence of a dissociating medium (1 wt% sodium dodecyl sulfate (SDS), 30 °C, 30 min) to disperse the aggregates formed during emulsification.[16] This procedure allowed us to obtain the so-called apparent and real mean droplet diameters.[16] The apparent droplet diameter is actually the average diameter obtained when the measurement medium is water. The real droplet diameter refers to the average droplet diameter obtained when the measurements are made in the dissociating medium, that is, the aqueous SDS solution.[16]

Protein Analysis

One hour after emulsification or emulsifier addition, the emulsion samples were centrifuged (Jouan, Saint-Herblain, France) at 4×10^3 *g* for 20 min at room temperature (20 °C) to separate the oil droplets from the aqueous serum phase. The protein concentration was determined in the aqueous phase before and after emulsion formation.[17] The fraction of protein adsorbed was calculated from the difference between the protein concentration in the aqueous phase after centrifugation and the protein concentration in the aqueous phase used for making the emulsions.

In the presence of 0.1 wt% SDS and 5 wt% 2-mercaptoethanol, SDS/polyacrylamide gel electrophoresis (PAGE) analysis was performed with a 4.9%T 2.7%C concentration gel in 0.125 M Tris/HCl buffer (pH 6.8) and with a 15.4%T 2.7%C separation gel in 0.38 M Tris/HCl buffer (pH 8.8).[18] Gels were stained with Comassie Brilliant Blue.[19]

Determination of Enzymic Activity

The method used was adapted from that of Quinn *et al.*[20] The porcine pancreatic lipase-catalysed hydrolysis of *p*-nitrophenyl butyrate (PNPB) was followed by monitoring production of *p*-nitrophenol at 400 nm in a 1 cm path length cell, using a Shimadzu UV 2100 spectrophotometer (Kyoto, Japan). The esterasic activity was determined at 37 °C in 1.5 ml of a 0.08 M dimethylmalonylurea buffer, pH 7.6, containing 15 μl of 62 mM PNPB (solubilized in acetonitrile) and 150 μl of the aqueous phase of the emulsion. Initial velocity (V_i) was calculated from the slope of the time course ($\Delta A_{400}/\text{min}^{-1}$) using the equation

$$V_i/\text{nmol min}^{-1} = 10^6 (\Delta A_{400}/\text{min}^{-1})\, v/\epsilon_{PNP} \tag{1}$$

where v is the assay volume (ml) and ϵ_{PNP} the absorptivity constant, in absorbance units ($\text{M}^{-1}\,\text{cm}^{-1}$), determined from the Beer–Lambert law. Under

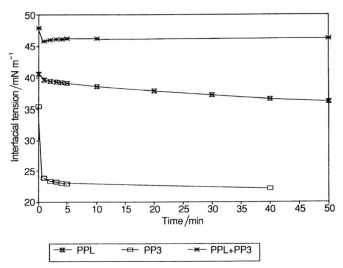

Figure 1   *Surface activity of component 3 (PP3), porcine pancreatic lipase (PPL) and a PP3 + PPL mixture (50/50(wt%)) at 4 °C at the n-dodecane–water interface. The interfacial tension is plotted against time. (Aqueous phase: citrate/phosphate/borate/HCl buffer, 0.1 M, pH 7.0; protein concentration 20 mg l^{-1})*

our experimental conditions, ϵ_{PNP} is 15 740 M^{-1} cm^{-1}. A titration value of 0.05 nmol min^{-1} was obtained in a blank experiment with citrate/phosphate/borate/HCl buffer (0.1 M, pH 7.0) added to the reaction system instead of the aqueous phase and substracted from each value of activity measured.

3 Results and Discussion

Surface Tension Measurements

Figure 1 shows the surface tension values obtained at +4 °C at the *n*-dodecane/water interface in the presence of milk component 3, porcine pancreatic lipase or a 50/50 wt% mixture of both proteins.

Surface tension values are much lower in the presence of milk component 3 than in the presence of lipase. Similar results have been obtained recently at +20 °C.[11] Milk component 3 is hence much more surface active than lipase. The mixture of both proteins is not efficient in decreasing the surface tension (Figure 1). This very low surface activity of the protein mixture suggests the occurrence of one or both of the following possibilities: (i) competition of both proteins for the oil–water interface, and (ii) formation of a complex between these two proteins.

Droplet Size Measurements

Values of average droplet size are given in Figure 2 as a function of the protein substrate used to make the emulsion, *i.e.* component 3, porcine pancreatic

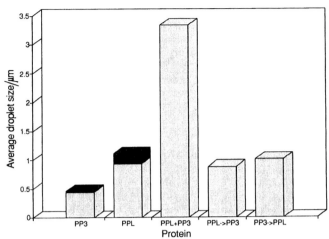

Figure 2 *Effect of protein substrate on the average droplet size d_{32} of oil-in-water emulsions (0.4 g of protein/100 g in citrate/phosphate/borate/HCl buffer, 0.1 M, pH 7.0; 20 g/100 g n-dodecane) made at 4 °C. Protein substrate was component 3 (PP3), porcine pancreatic lipase (PPL) or a PP3 + PPL mixture (50:50). Also given are the results for emulsions made with one protein substrate (0.4 g/100 g PPL or PP3) to the freshly made emulsion. Dark shading shows change in droplet size on dilution in a dissociating medium (1 g of SDS l^{-1})*

lipase, or a mixture (50/50 wt%) of both proteins. Also given are the results for emulsions made with one protein substrate to which a second protein substrate is added to the freshly made emulsion. Component 3 gives the finest emulsions ($d_{32} = 0.44 \pm 0.02$ μm) with a monomodal size distribution (results not shown). No droplet aggregation is present since the average droplet size is the same irrespective of whether or not measurements are made in a dissociating medium. Droplets are a bit larger when the protein is lipase; the droplet size distribution also shows two peaks. Some aggregates are clearly present since the d_{32} value obtained in water ($d_{32} = 1.23 \pm 0.02$ μm) is significantly higher than that obtained in the dissociating medium ($d_{32} = 0.93 \pm 0.02$ μm). The mixture of lipase and component 3 gives much coarser emulsions ($d_{32} = 3.33 \pm 0.04$ μm) than either protein alone. These data show that component 3 is a better emulsifier than lipase, which in turn is better than the mixture of the two proteins. These results are consistent with the surface tension data.

Addition of component 3 protein to the emulsions made with lipase ($d_{32} = 1.22 \pm 0.02$ μm) reduces the d_{32} value to a value similar to that obtained in the dissociating medium ($d_{32} = 0.88 \pm 0.02$ μm). Component 3 presumably causes a disruption of the aggregates present. In contrast, adding lipase to a component 3-stabilized emulsion ($d_{32} = 0.44 \pm 0.02$ μm) increases the average droplet size ($d_{32} = 1.01 \pm 0.02$ μm), which means that it causes aggregation.

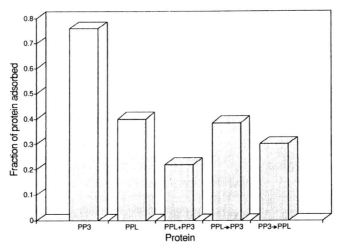

Figure 3   *Effect of protein substrate on the fraction of protein adsorbed at the oil–water interface in oil-in-water emulsions (0.4 g of protein/100 g in citrate/ phosphate/borate/HCl buffer 0.1 M, pH 7.0; 20 g/100 g n-dodecane) made at 4 °C. Protein substrate was component 3 (PP3), porcine pancreatic lipase (PPL) or a PP3 + PPL mixture (50:50). Also given are the results for emulsions made with one protein substrate (0.4 g/100 g PP3 or PPL) to which a second protein substrate was added (0.4 g/100 g PPL or PP3) to the freshly made emulsion*

Fraction of Protein Adsorbed

The fraction of protein adsorbed at the oil–water interface has been determined for each emulsion and the values are reported in Figure 3. As far as we can see, most of the component 3 is adsorbed ($F_{ADS} = 0.76 \pm 0.04$) at the oil–water interface whereas the fraction of protein adsorbed is much lower with lipase ($F_{ADS} = 0.40 \pm 0.02$). It is shown that lipase adsorption requires the presence of colipase, especially in the 20–30 mN m^{-1} zone.[7] When the mixture of both proteins is used to make the emulsion, only a small part of the total protein adsorbs ($F_{ADS} = 0.22 \pm 0.02$). If we compare these results with those obtained for surface activity (Figure 1) and emulsion droplet size (Figure 2), we note that, among the three protein samples tested (component 3, lipase, component 3 + lipase), component 3 is the most surface active, and it gives the finest emulsions and the highest fraction of protein adsorbed. However, compared with component 3 and lipase, the mixture of component 3 + lipase is the least surface-active sample, giving coarser emulsions with a low fraction of protein adsorbed. This could result from competition of both proteins for the oil–water interface or from the formation of a complex which has a low surface activity. Values of the fraction of protein adsorbed are close to that of β-casein for component 3 and that of β-lactoglobulin for lipase, in similar systems.[21,22]

After addition of component 3 to the emulsions made with lipase ($F_{ADS} = 0.40 \pm 0.02$), the fraction of protein adsorbed is not changed ($F_{ADS} = 0.38 \pm 0.02$) although twice as much protein is present. This implies that the protein surface coverage and the protein content in the aqueous phase are multiplied by a factor of two. The addition of lipase to a component 3-stabilized emulsion reduces the fraction of protein adsorbed from 0.76 ± 0.04 to 0.30 ± 0.02, which means that the protein surface coverage is slightly lowered; the protein content in the aqueous phase is increased.

Esterasic Activity in Aqueous Phases of Emulsions

In the aqueous phases of the emulsions was also measured the esterasic activity, whose value is directly proportional to the lipase concentration, *i.e.* the amount of lipase which is not adsorbed at the oil–water interface. Figure 4 shows the values of esterasic activity in the aqueous phases of the different emulsions. Of course, there is no esterasic activity (no lipase) in the aqueous phase of a component 3-stabilized emulsion. The value of esterasic activity in the aqueous phase is found to be higher if the emulsion is made with the component 3 + lipase mixture than if it is made only with lipase (2.8 ± 0.5 nmol min^{-1}), although the overall amount of lipase is lower in the first emulsion than in the second. This demonstrates that the presence of component 3, jointly with lipase, during homogenization prevents some of the

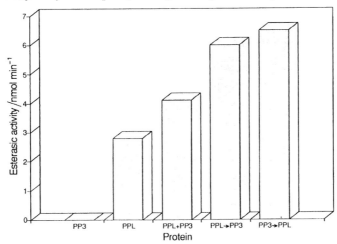

Figure 4   *Esterasic activity in aqueous phases of oil-in-water emulsions (0.4 g of protein/100 g in citrate/phosphate/borate/HCl buffer, 0.1 M, pH 7.0; 20 g/ 100 g n-dodecane) made at 4 °C with component 3 (PP3), porcine pancreatic lipase (PPL) or a PP3 + PPL mixture (50:50). Also given are the results for emulsions made with one protein substrate (0.4 g/100 g PP3 or PPL) to which a second protein substrate was added (0.4 g/100 g PPL or PP3) to the freshly made emulsion. Activity is given in nmol of* p-nitrophenol *released per minute in 0.08 M dimethylmalonylurea buffer, pH 7.6, containing 15 μl of emulsion aqueous phase*

lipase adsorbing at the interface. This is consistent with the result given in Figure 2. The lowest fraction of protein adsorbed at the oil–water interface is obtained with the lipase component 3 mixture (0.22 ± 0.02); most of the protein remains in the aqueous phase.

The highest values of esterasic activity are observed in aqueous phases of emulsions to which the second protein is added after the emulsion is made, and these values are close to the value of esterasic activity in the initial lipase solution. The addition of component 3 to a lipase-stabilized emulsion increases its esterasic activity from 2.8 ± 0.5 nmol min^{-1} to 6.0 ± 0.5 nmol min^{-1}. From that, we can infer that most of the component 3 added to the lipase-stabilized emulsion does come to the interface, whereas most of the adsorbed lipase molecules are released in the aqueous phase. In the aqueous phase of a component 3-stabilized emulsion, to which has been added lipase, the esterasic activity is found to be 6.5 ± 0.5 nmol min^{-1}, *i.e.* very similar to that of the lipase solution at the same protein concentration. This means that the added lipase mainly remains in the aqueous phase. This explains why the protein content in the aqueous phase is much increased upon addition of lipase to the emulsion.

Competitive Displacement of Milk Component 3 by Tween 20 from the Oil–Water Interface

Figure 5 shows protein surface coverage, Γ, and fraction of protein adsorbed, F_{ADS}, at the oil–water interface as a function of the amount of Tween 20 added

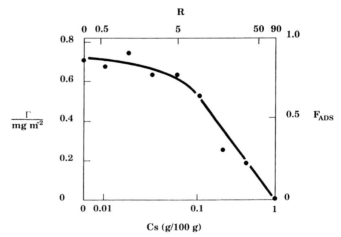

Figure 5   *Competitive displacement of HFPP from the n-dodecane–water interface by Tween 20 added after formation of the emulsion (0.4 g of protein/100 g in citrate/phosphate/borate/HCl buffer, 0.1 M, pH 7.0; 20 g/100 g n-dodecane) at 4 °C. Protein surface coverage Γ (left) and the fraction of protein adsorbed at the oil–water interface F_{ADS} (right) are plotted against Tween 20 content C_s (bottom) and surfactant-to-protein molar ratio R (top)*

to aliquots of an emulsion made with the HFPP. The amount of this water-soluble surfactant added is expressed both in terms of overall content of Tween 20 in the emulsion and in terms of the surfactant-to-protein molar ratio. Considering the fact that the protein used for making the emulsions is HFPP—mainly made of component 3—the molar mass used for calculations is 29 000 g mol^{-1}.

Under the homogenization conditions employed, the average droplet size is 0.43 ± 0.02 μm and 83% of the protein is adsorbed at the oil–water interface in the emulsion made with component 3 in the absence of any added surfactant. The calculated protein surface coverage in the absence of surfactant is $\Gamma = 0.70 \pm 0.04$ mg m^{-2}. While at low Tween 20 content ($R < 5$), there is no significant change in protein surface coverage, the protein is completely displaced from the droplet surface for a Tween content of 1 g/100 g of emulsion, *i.e.* for a surfactant-to-protein molar ratio of 90. This value is much higher than that required ($\sim$0.25 g/100 g) for a full displacement of β-lactoglobulin from the oil–water interface by Tween 20, in *n*-tetradecane-in-water emulsions.[22] It is interesting to note that, if the activity of porcine pancreatic lipase is inhibited by component 3, the amount of sodium taurodeoxycholate required for lipase reactivation, in the

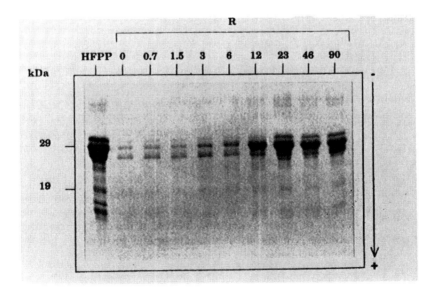

Figure 6   *Competitive displacement of component 3 from the n-dodecane–water interface by Tween 20 added after formation of the emulsion (0.4 g of protein/ 100 g in citrate/phosphate/borate/HCl buffer, 0.1 M, pH 7.0; 20 g/100 g n-dodecane) at 4°C. SDS/polyacrylamide gel electrophoresis (SDS/PAGE) patterns of proteins from aqueous phases are shown as a function of surfactant-to-protein molar ratio, R. (HFPP = hydrophobic fraction of proteose-peptone = component 3)*

presence of colipase, is four times higher than that required if lipase is inactivated by β-lactoglobulin.[11] Component 3 seems therefore to be more strongly bound to the oil–water interface. This makes lipase activity more difficult to restore in the presence of component 3 than in the presence of β-lactoglobulin. This might also explain the fact that reactivation of lipase, inhibited by some proteins such as β-lactoglobulin, ovalbumin or serum albumin, can be achieved by the addition of colipase and bile salts (*e.g.* sodium taurodeoxycholate, glycocholate or taurocholate), whereas the reactivation of lipase inhibited by milk component 3 can only be achieved by colipase and sodium taurodeoxycholate; sodium taurocholate or glycocholate are inefficient.[11]

For each surfactant-to-protein molar ratio tested, there was also carried out, in addition to the protein content measurement, an electrophoretic analysis of the aqueous phases of the component 3-stabilized emulsion. Results are shown in Figure 6. It gives the electrophoretic pattern of the initial aqueous phase of component 3 (HFPP) used to make the emulsion and the electrophoretic pattern of the aqueous phase of emulsion for each surfactant-to-protein molar ratio tested. We can see that, beyond $R = 12$, the 29 kDa band[23,24] (*i.e.* the component 3 band) is markedly predominant relative to other bands. This suggests that desorption of component 3 in the free state or associated to other minor proteins requires a higher Tween 20 content than other protein fractions of HFPP.

4 Conclusions

At the *n*-dodecane–water interface, at a planar surface, or in an emulsion, milk component 3 is more surface active than porcine pancreatic lipase, which in turn is more surface active than a mixture of milk component 3 + porcine pancreatic lipase. The displacement studies of one protein by the other show that component 3 added to an emulsion made with lipase is able to displace most of the lipase adsorbed at the interface. They also show that the presence of component 3 at the oil–water interface in an emulsion made with component 3 prevents some lipase added to the emulsion from adsorbing at the interface. Moreover, displacement experiments of protein from the hydrophobic fraction of proteose-peptone show that the 29 kDa fraction, *i.e.* the proper component 3, is the most strongly bound protein fraction to the oil–water interface. Previous studies have shown that, in well-defined systems, there is no complex formation between milk component 3 and porcine pancreatic lipase *via* hydrophobic interactions.[6,11] Consequently, when both proteins are introduced successively into our system, the reversible inactivation of porcine pancreatic lipase by milk component 3 is due to a preferential adsorption of milk component 3 over porcine pancreatic lipase at the *n*-dodecane–water interface. When they are introduced together, the inactivation could be due to the occurrence of a complex between the two proteins. However, we have to be aware that the system and the conditions used in the present study are still far away from those which exist in milk. Although pH and temperature are similar to those of milk storage, the natural milk emulsion differs from our system in the type of oil phase (milk triglycerides instead of *n*-dodecane), the nature of the enzyme (lipoprotein lipase instead of porcine pancreatic lipase), the presence of many other components, and of course the presence of the membrane around of the milk fat globules. Milk triglycerides are obviously less hydrophobic than a pure alkane and are mainly crystallized at + 4 °C since their melting point is higher (final melting point $\sim$37 °C) than that of *n*-dodecane (-9.6 °C). The natural fat-globule membrane makes the interfacial tension so low (<2.5 mN m^{-1}) that it prevents lipase adsorption at the fat surface. Therefore, much work has still to be undertaken to understand the real mechanism of lipase-reversible inactivation in milk by milk component 3.

An additional point of interest of the present study is that it could help us to understand the way lipases act *in vivo* during digestion.

References

1. P. Walstra and R. Jenness, 'Dairy Chemistry and Physics', Wiley, New York, 1984.
2. M. Anderson, *J. Dairy Res.*, 1981, **48**, 247.
3. P. Cartier and Y. Chilliard, *J. Dairy Sci.*, 1986, **69**, 951.
4. P. Cartier, Y. Chilliard, and D. Paquet, *J. Dairy Sci.*, 1990, **73**, 1173.
5. Y. Chilliard and G. Lamberet, *Lait*, 1984, **64**, 5544.
6. Y. Gargouri, R. Julien, A. Sugihara, R. Verger, and L. Sarda, *Biochim. Biophys. Acta*, 1984, **795**, 326.
7. Y. Gargouri, G. Pieroni, C. Riviere, A. Sugihara, L. Sarda, and R. Verger, *J. Biol. Chem.*, 1985, **260**, 2268.
8. Q. Cheng, P. Blackett, K. W. Jackson, W. J. McConathy, and C. S. Wang, *Biochem. J.*, 1990, **269**, 403.
9. B. Persson, G. Bengtsson-Olivecrona, S. Enerback, T. Olivecrona, and H. Jörnvall, *Eur. J. Biochem.*, 1989, **179**, 39.
10. J. Rathelot, R. Julien, P. Canioni, C. Coeroli, and L. Sarda, *Biochimie (Paris)*, 1975, **57**, 1117.
11. J.-M. Girardet, G. Linden, S. Loye, J.-L. Courthaudon, and D. Lorient, *J. Dairy Sci.*, 1993, **76**, 2156.
12. D. Paquet, Y. Nejjar, and G. Linden, *J. Dairy Sci.*, 1988, **71**, 1464.
13. J.-M. Girardet, A. Mati, T. Sanogo, L. Etienne, and G. Linden, *J. Dairy Res.*, 1991, **58**, 85.
14. A. G. Gaonkar, *J. Am. Oil Chem. Soc.*, 1989, **66**, 1090.
15. I. Burgaud, E. Dickinson, and P. V. Nelson, *Int. J. Food Sci. Technol.*, 1990, **25**, 39.
16. A. Tomas, D. Paquet, J.-L. Courthaudon, and D. Lorient, *J. Dairy Sci.*, 1994, **77**, 413.
17. O. H. Lowry, N. J. Rosebrough, A. L. Farr, and R. J. Randall, *J. Biol. Chem*, 1951, **193**, 265.
18. U. K. Laemmli and M. Favre, *J. Mol. Biol.*, 1973, **80**, 575.
19. A. T. Andrews, in 'Methods of Enzymatic Analysis', ed. H. U. Bergmeyer, Academic Press, New York, 1981, vol. 5, p. 277.
20. D. M. Quinn, K. Shirai, R. L. Jackson, and J. A. K. Harmony, *Biochemistry*, 1982, **21**, 6872.

21. J.-L. Courthaudon, E. Dickinson, and D. G. Dalgleish, *J. Colloid Interface Sci.*, 1991, **145**, 390.
22. J.-L. Courthaudon, E. Dickinson, Y. Matsumura, and A. Williams, *Food Struct.*, 1991, **10**, 109.
23. E. S. Sorensen and T. E. Petersen, *J. Dairy Res.*, 1993, **60**, 535.
24. J.-M. Girardet, F. Saulnier, A. Driou, G. Linden, B. Coddeville, and G. Spik, *J. Diary Sci.*, 1994, **77**, 1205.

Application of Polymer Scaling Concepts to Purified Gliadins at the Air–Water Interface

By Jeremy Hargreaves, Roger Douillard,[1] and Yves Popineau[1]

DEPARTEMENT DE BIOLOGIE PHYSICO-CHIMIQUE, ENSBANA,
1 ESPLANADE ERASME, 21000 DIJON, FRANCE
[1] LABORATOIRE DE BIOCHIMIE ET TECHNOLOGIE DES PROTÉINES, INRA,
BP 527, 44026 NANTES, FRANCE

1 Introduction

In order to get a better understanding of protein adsorption at fluid interfaces, an analogy has often been made with polymers.[1,2] Recently, proteins in the semi-dilute regime at interfaces have been shown[3,4] to follow scaling laws similar to those used in polymer physics.[5,6] These laws enable us to relate the interfacial concentration and surface pressure of polymers, taking into account their conformation in the interfacial layer.

Gliadins are monomeric wheat storage proteins representing about 40% of gluten. They are often defined as the water or salt solution insoluble, ethanol soluble fraction of gluten (the Osborne definition). The low levels of lysine, arginine and histidine, along with the low levels of free carboxylic groups due to almost total amidation of glutamic acid, place the gliadins as among the least charged proteins known.[7] The sulfur-rich gliadins (α, β and γ) possess cysteine residues that are all implicated in intramolecular disulfide bridges, whereas the sulfur-poor gliadins (ω) contain no cysteine. Gliadins contribute to the viscosity of doughs, and they can be involved at interfaces in these types of systems where numerous air bubbles are trapped.

2 Materials and Methods

Gliadins were purified from gluten by chromatographic techniques as described elsewhere.[8-10] The proteins were solubilized in 70% ethanol and the resulting solution was filtered (0.22 μm pores). Protein concentration was determined spectrophotometrically ($E_{\alpha,\beta} = 0.475$, $E_\gamma = 0.497$, $E_\omega = 0.305$ ml cm^{-1} mg^{-1}).

A film balance (system 5000, KSV Instruments, Helsinki) was filled with purified water and the interface was cleaned. The protein solutions were

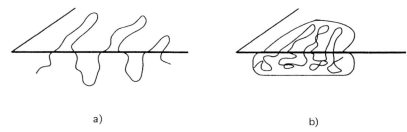

a)   b)

Figure 1   *Polypeptide at an air–liquid interface: (a) according to the 'tail–train–loop'*
theory, hydrophobic sequences form three-dimensional tails and loops in the
water phase, and hydrophilic sequences form two-dimensional trains at the
interface; (b) in the 'maximum entropy' conformation the loops and tails
gather in two 'pancakes' according to their solvent affinity

dropped onto the interface using a microsyringe. After an equilibration
period of about 30 min, the surface was swept with a barrier at the speed of
15 mm min^{-1}. The interfacial pressure was measured by the Wilhelmy plate
method.

3 Theoretical Background

Polypeptides tend to adopt a 'tail–train–loop' conformation at interfaces;
'loops' are hydrophilic sequences with three-dimensional character and
'trains' are hydrophobic, flattened loops. To be under maximum entropy
conditions, trains and loops on both sides of the surface will be in a
minimum excluded volume, forming two 'pancakes' (Figure 1). The
boundary between the two types of pancakes is defined by the behaviour of
the sequences with respect to the medium; at this level there is a sharp
variation in solvent concentration rather than a physical air–liquid barrier.
Several different regimes can be defined for polymers at interfaces;[3,4] we
shall only consider two of these here.

In the *dilute regime*, the molecules behave as individual objects. The
interfacial pressure Π in this regime is described by the ideal gas law

$$\Pi = k_b T \Gamma \tag{1}$$

where K_b is the Boltzmann constant, T is the absolute temperature, and Γ is
the concentration. At the critical overlap concentration Γ_1, characterizing the
border between dilute and semi-dilute regimes, the Flory radius of the largest
pancake (R_{F2}) is defined by:

$$R_{F2} = (4\Gamma_1)^{-0.5} \tag{2}$$

In the *constant thickness semi-dilute regime*, when the molecules begin to

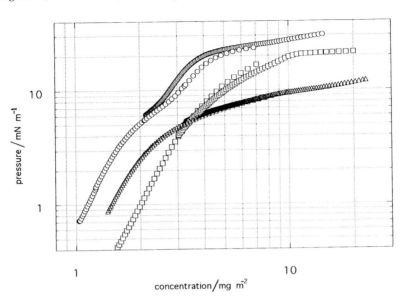

Figure 2 *Isotherms of purified water-insoluble gliadins. Surface pressure measured with a film balance is plotted against surface concentration. Two isotherms are shown for each protein so as to cover a wide surface concentration range. The subphase is water;* $\bigcirc$, *γ-gliadin;* $\square$, *β-gliadin;* $\triangle$, *ω-gliadin*

overlap, at least one loop layer enters the semi-dilute regime where the surface properties are dominated by interactions between polypeptide segments. In the semi-dilute regime, the interfacial pressure of the loop layer follows the power law[5]

$$\pi = k_n k_b T \Gamma^y \tag{3}$$

where k_n is a numerical constant. The exponent y depends on the affinity of the polymer for the solvent and on the spatial dimension d of the system; its value is related to the Flory exponent v (Table 1).[5] The surface pressure is dominated by the loop layer having the highest y value. While in the dilute regime the thermodynamic units are individual molecules, in the semi-dilute regime these units are polymer 'blobs' whose characteristic size is the correlation length. When the correlation length is larger than the layer thickness, the regime is purely two-dimensional. At the concentration of critical cross-over to the semi-dilute regime (Γ_2), the blobs are spherical. The correlation length is equal to the thickness of the layer. Further increase in interfacial concentration results in a change of regime.

Table 1 *Characteristic values of the exponents y and v in relation to solvent quality and dimensionality d*

Conditions	$v(d=2)$	$v(d=3)$	$y(d=2)$	$y(d=3)$
Extended chain	1	1	2	$\frac{3}{2}$
Good Solvent	$\frac{3}{4}$	$\frac{3}{5}$	3	$\frac{9}{4}$
Θ conditions	$\frac{4}{7}$	$\frac{1}{2}$	8	3
Poor solvent	$\frac{1}{2}$	$\frac{1}{3}$	∞	∞

Table 2 *Numerical results for three gliadins at the air–water interface*

Type of gliadin	MW^a g mol^{-1}	Number of residues^b	y	Γ_1/mg m$^{-2\,c}$	Γ_2/mg m^{-2}	$R_{F2}(at\ \Gamma_1)$ Å
β	29 000	252	3.43	0.14	3	91
γ	30 700	267	3.73	0.33	1.5	62
ω	52 000	452	3.24	0.03	2	267

[a] Deduced from size exclusion HPLC data of reference 10.
[b] Calculated from the molecular weight MW (115 g mol^{-1} per residue).
[c] Obtained by equating equations (1) and (3).

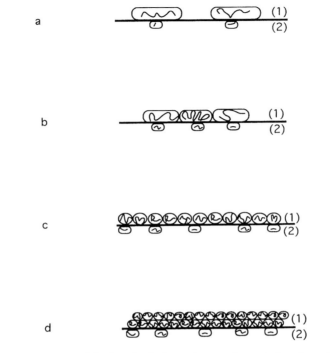

Figure 3 *Speculative model of γ-gliadin in various regimes at the air–water interface (see Figure 2). The thick horizontal line represents the boundary between polypeptide chains of different solvent affinities: (1) towards the air side of the interface, (2) towards the bulk water phase. (a) In the dilute regime, the polypeptide is represented by two pancakes; (b) region (1) is in the first semi-dilute regime, and the blobs represent uncorrelated parts of the chain; (c) region (1) is at the concentration of the first critical cross-over (Γ_2), and the blobs are spherical; (d) region (1) in a second semi-dilute regime, so that the size of the blobs is decreased but the thickness of the layer is increased*

4 Results and Discussion

A power law dependence of the surface pressure is observed for surface concentrations between 1 and 3 mg m^{-2} (Figure 2). According to the theory, this should be indicative of a semi-dilute regime in at least one loop layer. The values of the exponent y are above 3 (Table 2). This may correspond to two possibilities—the polypeptides are in nearly θ conditions with a local three-dimensional behaviour, or in quite good solvent conditions in two dimensions (Table 1). The first possibility seems less likely, as the molecules occupy 'flat' pancakes at the beginning of this regime. The two-dimensional condition is consistent with the insolubility of gliadins in water. The concentrations of critical overlap found (Table 2) show that ω-gliadin molecules are largely spread out at the surface in the dilute regime, while the effective radius of the other gliadins is smaller.

The values of Γ_2 (Table 2) can be compared with the value of theoretical maximum coverage of the interface by a monolayer of amino acids[3,11] (1.3 mg m^{-2}). This shows that the β-gliadin molecules have at least half the polypeptide chain in the water phase, the ω-gliadins only have a minimum of 25% of amino acids immersed, and the γ-gliadin residues may be almost totally at the surface. In the case of γ-gliadin, an inflexion in the isotherm at around 2 mg m^{-2} may be interpreted as a transition to a second semi-dilute regime with a value of y also close to 3. Together with the high value of the surface pressure reached at the end of this second regime, these facts may be indicative of the formation of a 'double layer' of similar blobs as shown in Figure 3. Because of the small number of charges and the insolubility of γ-gliadins, it is speculated that the polypeptides are mainly in the superficial, water-depleted layer labelled (1) in Figure 3.

References

1. S. J. Singer, *J. Chem. Phys.*, 1948, **16**, 872.
2. H. L. Frisch and R. Simha, *J. Chem. Phys.*, 1975, **27**, 702.
3. R. Douillard, *Colloids Surf. B*, 1993, **1**, 333.
4. R. Douillard, M. Daoud, J. Lefebvre, C. Minier, G. Lecannu, and J. Coutret, *J. Colloid Interface Sci.*, 1994, **163**, 277.
5. M. Daoud and P.G. de Gennes, *J. Phys. (Paris)*, 1978, **38**, 85.
6. P. G. de Gennes, 'Scalling Concepts in Polymer Physics', Cornell University Press, Ithaca, NY, 1979.
7. R. Lasztity, 'The Chemistry of Cereal Proteins', CRC Press, Boca Raton, FL, 1984.
8. Y. Popineau and F. Pineau, *J. Cereal Sci.*, 1985, **3**, 363.
9. Y. Popineau, J.-L. le Guerroué, and F. Pineau, *Lebensm.-Wiss. Technol.*, 1986, **19**, 226.
10. Y. Popineau and F. Pineau, *Lebensm.-Wiss. Technol.*, 1988, **21**, 113.
11. K. S. Birdi, 'Lipid and Biopolymer Monolayers at Interfaces', Plenum, New York, 1989.

A Neutron Reflectivity Study of the Adsorption of β-Casein at the Air–Water Interface

By Peter J. Atkinson, Eric Dickinson, David S. Horne,[1] and Robert M. Richardson[2]

PROCTER DEPARTMENT OF FOOD SCIENCE, UNIVERSITY OF LEEDS, LEEDS LS2 9JT, UK
[1] HANNAH RESEARCH INSTITUTE, AYR KA6 5HL, UK
[2] SCHOOL OF CHEMISTRY, UNIVERSITY OF BRISTOL, BRISTOL BS8 1TS, UK

1 Introduction

The technique of neutron reflectance has been used to determine directly the segment density profile of the milk protein β-casein[1] at the air–water interface as a function of solution pH in the range 5.5–7.0. By fitting a simple model of the adsorbed protein segment density profile to the reflectivity data, a picture of the changes occurring in the protein layer as the solution pH is lowered towards the casein isoelectric point (pI = 5.3) is obtained. These results are interpreted in terms of the formation of multiple layers of casein molecules at the air–water interface on approaching the pI.

Slow neutrons can interact with matter in a similar fashion to light, and their behaviour may be interpreted using similar theory.[2] Reflectivity of neutrons from a surface therefore depends on the neutron refractive index profile perpendicular to the interface, and consequently on the chemical composition of the interface. Because of the short neutron wavelength used ($0.5 \text{Å} \leqslant \lambda \leqslant 10 \text{ Å}$), experimental resolution is on the atomic scale and small molecules are easily 'seen'. Contrast between molecules and phases of interest is achieved by exploiting the large difference in neutron scattering cross-section between hydrogen and deuterium atoms. By using a suitable mixture of H_2O and D_2O, the water solvent can be contrast-matched to air so that only the protein layer contributes to the reflectivity from the air–water interface.

2 Experimental and Data Analysis

Protein solutions containing 5×10^{-3} wt% β-casein were made up in 20 mM imidazole/HCl buffer using air-contrast-matched water as solvent. Reflectivity

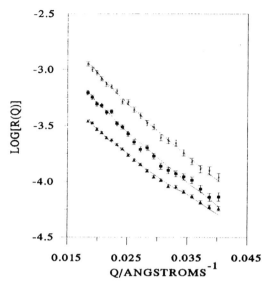

Figure 1   *Reflectivity data for 5×10^{-3}% β-casein solution as a function of solution pH:* ▲, *pH 7;* ■, *pH 6;* ×, *pH 5.5. Solid lines are fits using a two-layer model of the adsorbed protein film as described in the text*

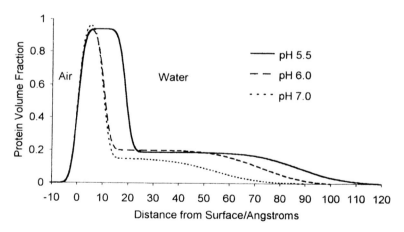

Figure 2   *Protein volume fraction profiles* versus *distance normal to the air–water interface as a function of solution pH for a 5×10^{-3}% β-casein solution*

profiles were measured using the CRISP instrument of the ISIS facility at the Rutherford–Appleton Laboratory, Chilton, Didcot, UK. A polychromatic beam of neutrons was made to impinge on the sample surface at fixed angle of incidence $\theta = 0.5°$. Reflected neutron intensity R was measured as a function of wavelength λ by time-of-flight analysis, and plotted as $\log[R(Q)]$ *versus* Q, where $Q = (4\pi/\lambda)\sin \theta$.

Reflectivity data were analysed using a two-layer model of the β-casein molecules adsorbed at the air–water interface: an inner layer representing hydrophobic train segments adsorbed closest to the air, and an outer layer mainly consisting of the more hydrophilic tail segments extenting into the aqueous phase.[3] Layer thickness, scattering length density (and hence protein volume fractions) of the layers, and a 'roughness' parameter were all adjusted to achieve a best fit to the data using a non-linear least-squares fitting routine. The fitted scattering length density distribution of the adsorbed layers normal to the interface, $\rho(z)$, is directly related to the protein volume fraction profile $\phi(z)$ by

$$\rho(z) = \phi(z)\rho_p + [1 - \phi(z)]\rho_s$$

where ρ_p and ρ_s are the pure protein and solvent scattering length densities.

3 Results and Discussion

As shown in Figure 1, a two-layer model profile of β-casein adsorbing at the air–water interface fits the observed reflectivity data very well. Figure 2 shows plots of the fitted protein volume fraction profile *versus* the distance normal to the air–water interface as a function of substrate pH. The protein volume fraction in the layer closest to the air is high (0.95) and remains approximately constant with changing pH. At pH 5.5, however, the inner layer is nearly twice as thick as at higher pH (18.5 Å and 10 Å, respectively). The outer layer thickness and volume fraction both increase on reducing the substrate pH, the former from 48 Å to 68 Å and the latter from 0.14 to 0.20. The amount of protein adsorbed at the interface almost doubles over the same pH range, increasing from 2.13 to 4.05 ± 0.15 mg m^{-2}.

These observations can be interpreted in terms of a tendency towards the formation of multiple protein layers at the air–water interface as the substrate pH is reduced. It is well known that β-casein molecules begin to aggregate more extensively close to the isoelectric point, since electrostatic repulsion between neighbouring molecules becomes small.[4] The fitted data at pH 6 suggest that the effect of aggregation here is weak: only the outer layer protein volume fraction and thickness are changed, increasing as compared with pH 7. At pH 5.5, however, adsorption appears to be much stronger and the structure of the whole interface is affected.

Adding a second layer of β-casein onto the first layer would approximately double the inner layer thickness, but would not noticeably affect the protein volume fraction, which is already very high. The outer layer thickness and protein volume fraction should increase by at least the pH 7 value of the inner layer thickness. In fact, the amount of protein adsorbed almost doubles, lending support for this interpretation. A more extended tail segment conformation could also account for the thicker outer layer at pH 5.5. Figure 3 shows a possible simplistic model of the changes in the structure of the protein film at the air–water interface as the solution conditions are changed over the pH range studied.

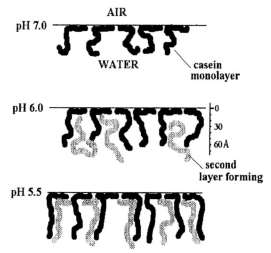

Figure 3   *Schematic diagram of the changes occurring in a β-casein film at the air–water interface as the solution pH is reduced from 7.0 to 5.5, just above the protein pI*

4 Conclusion

Neutron reflectance can provide detailed information on the structure of adsorbed protein films. It has been shown that, at pH 7, a monolayer of the milk protein β-casein adsorbed at the air–water interface consists of a dense inner layer of approx. 10 Å thickness closest to the interface, and a more diffuse outer layer of approx. 47 Å thickness, extending into the aqueous phase. On reducing the solution pH towards the isoelectric point of the protein, the layers thicken and the protein coverage at the surface increases almost two-fold. This observation is interpreted as being due to adsorption of further β-casein molecules onto the existing monolayer to form an extensively adsorbed second layer.

References

1.   E. Dickinson, 'An Introduction to Food Colloids', Oxford University Press, 1992.
2.   J. Penfold and R. K. Thomas, *J. Phys. Condens. Matter*, 1990, **2**, 1369.
3   E. Dickinson, D. S. Horne, J. S. Phipps, and R. M. Richardson, *Langmuir*, 1993, **9**, 242.
4.   E. Dickinson, B. S. Murray, and G. Stainsby in 'Advances in Food Emulsions and Foams', ed. E. Dickinson and G. Stainsby, Elsevier Applied Science, London, 1988, p. 123.

Effect of Temperature on Lipid–Protein Interactions at the Oil–Water Interface

By J.-L. Gelin, P. Tainturier, L. Poyen,[1] J.-L. Courthaudon, M. Le Meste, and D. Lorient

ENSBANA, 1 ESPLANADE ERASME, 21000 DIJON, FRANCE
[1] SANOFI BIO-INDUSTRIES, CDA, 50500 CARENTAN, FRANCE

1 Introduction

Dairy foods emulsions are commonly stabilized by surface-active species (*e.g.* milk proteins, monoglycerides, Tweens) which facilitate the formation of small oil droplets and slow down destabilization processes.[1] Both monoglycerides and polyoxyethylene (20) sorbitan monooleate have been observed to displace proteins from the surface of oil droplets in oil-in-water emulsions.[2-4] The aim of the present work is to investigate the effect of temperature on lipid–protein interactions using tensiometry and electron spin resonance (ESR) spectroscopy.

2 Materials and Methods

Interfacial and surface tensions were measured using the Wilhelmy plate method (tensiometer Prolabo N3). Triolein (Sigma, T7752; melting point = 6 °C) was percolated through a Florisil column to remove the impurities present in the oil.[5] Butter oil (fat content > 99.9 wt%, melting range −40 °C/ +30 °C) was provided by France-Beurre. Glycerol monostearate (Sanofi Bio-Industries) was added to triolein and heated (0.5 wt%, 75 °C, 30 min) to ensure the melting of the crystalline form of monoglycerides. The oil phase was then cooled to 40 °C before the experiment. When used alone, triolein was heated at a lower temperature (40 °C, 30 min). The water phase was either distilled water or a suspension of skim milk powder dispersed in distilled water (8 wt%, 40 °C, 1 h) (the protein content of the latter was 2.65 wt%). In some experiments, polyoxyethylene (20) sorbitan monooleate (polysorbate 80, Sanofi Bio-Industries) was added (0.65 wt%) before heating the aqueous phase. Initially the tension was measured at 40 °C for approx. 1 h to reach a constant value. A step-by-step decreasing temperature gradient from 40 to 6 °C was applied and tension was measured for approximately 1 h at each temperature to allow the system to reach a steady state. For each system the values of air–oil ($\gamma_{A/O}$), air–water ($\gamma_{A/W}$) and oil–water ($\gamma_{O/W}$) tensions were measured to calculate the work of adhesion, W_A, according to the equation

$$W_A = \gamma_{A/O} + \gamma_{A/W} - \gamma_{O/W} \tag{1}$$

Paramagnetic homologues of stearic acid were used as spin probes to study the interactions between lipids and proteins. Probes with a nitroxide group covalently linked to the fifth (5SA) carbon atom of the aliphatic chain of stearic acid were used. Changes in the mobility of the 5SA probes due to the presence of interacting species reflected the behaviour of the polar end of 5SA molecules. Rotational correlation times τ_c were either estimated according to the equation

$$\tau_c = (6.5 \times 10^{-10})\Delta H_{+1}[\sqrt{(I_0/I_{-1})} - 1] \tag{2}$$

(ΔH_{+1} is the width of the low field line, and I_0 and I_{-1} are the amplitudes of the central and the high field lines, respectively) or by comparison with calculated spectra.

Skim milk suspension and triolein were prepared as described above. The molar ratio of spin probe to protein was fixed at 0.3. Spin probes were dispersed in oil before emulsification (40 °C, 30 min). Emulsions (8 wt% fat, 2.44 wt% protein) were prepared by high-speed stirring (polytron Kinematica, 16 500 r.p.m., 40 °C, 4 min). ESR experiments were carried out using a controlled temperature spectrometer (Brücker 300E).

3 Results and Discussion

Figure 1 shows the low temperature dependence of the interfacial tension measurements for milk proteins or polysorbate 80. It also shows the temperature dependence of interfacial tension when glycerol monostearate is present. The present results are in accordance with previous work, which has led to the idea that monoglycerides exhibit an increased surface activity below a certain critical temperature which depends on the nature and the concentration of the monoglycerides.[6]

Figure 2 shows that hydrophilic surface-active species generally reduce the interactions between oil and water. On the contrary, the presence of saturated monoglycerides increases the work of adhesion: the curve exhibits a negative slope above 20 °C (-0.49 mN m^{-1} °C^{-1}) and a break point at 20 °C, which is interpreted as a physical change in the state of the monoglyceride molecules adsorbed at the interface. The increase in the work of adhesion is interpreted as an increase in the interactions between the monoglyceride-covered surface and water molecules.

ESR studies have been performed on bulk fats and emulsions in the temperature range 4–40 °C. For the two fats studied, similar spectral features were obtained for the bulk fat and for the emulsions. These results suggest that the mobility of spin probes is not affected by the change in the physical state

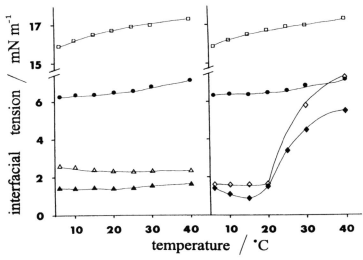

Figure 1   *Interfacial tensions* versus *temperature.* □ *Triolein/water,* ● *triolein/milk;* △ *triolein/water + polysorbate 80 (0.65 wt%);* ◇ *triolein + GMS (0.5 wt%)/water;* ▲ *triolein/milk + polysorbate 80 (0.65 wt%);* ◆ *triolein + GMS (0.5 wt%)/milk*

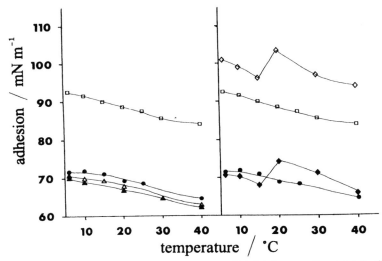

Figure 2   *Work of adhesion* versus *temperature.* □ *Triolein/water;* ● *triolein/milk;* △ *triolein/water + polysorbate 80 (0.65 wt%);* ◇ *triolein + GMS (0.5 wt%)/water;* ▲ *triolein/milk + polysorbate 80 (0.65 wt%);* ◆ *triolein + GMS (0.5 wt%)/milk*

of bulk lipids. Previous ESR studies on solid and liquid fractions of lard had shown similar values of τ_c.[7] The present results show that spin probes are preferentially located in the liquid fraction of bulk fats. Figure 3 shows ESR

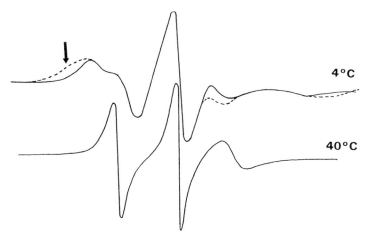

Figure 3   *ESR spectra of 5-doxyl stearic acid in bulk triolein or butter oil for two temperatures (————) and in triolein/milk emulsion at 4 °C (- - -). Arrow denotes the detection of protein/spin probe interactions*

spectra of both bulk triolein and triolein/milk emulsions containing 5SA probes for two temperatures. Decreasing temperature from 40 to 4 °C led to a reduced value of τ_c from 2×10^{-9} s to 2×10^{-10} s for bulk triolein. In the emulsion, a population of spin probes with a reduced mobility (broken line in Figure 3) was found: this reduced mobility could be explained by lipid–protein interactions mainly occuring at the oil droplet surface. The similarity of the spectra observed for both emulsified butter oil and triolein indicated that the interactions between lipids and proteins are not significantly affected by the change in the physical state of the emulsified fat.

References

1. E. Dickinson, 'An Introduction to Food Colloids', ed. E. Dickinson, Oxford University Press, 1992.
2. H. Oortwijn and P. Walstra, *Neth. Milk Dairy J.*, 1979, **33**, 134.
3. H. D. Goff and W. K. Jordan, *J. Dairy Sci.*, 1989, **72**, 18.
4. H. D. Goff, M. Liboff, W. K. Jordan, and J. E. Kinsella, *Food Microstruct.*, 1987, **1**, 23.
5. A. G. Gaonkar, *J. Am. Oil Chem. Soc.*, 1989, **66**, 1090.
6. N. J. Krog, in 'Microemulsions and Emulsions in Food', ed. M. El-Nokaly and D. J. Cornell, ACS Symp. Ser. No. 448, American Chemical Society, Washington, DC, 1991, p. 138.
7. M. Le Meste, G. Cornilly, and D. Simatos, *Lipids*, 1985, **20**, 296.

Modification of the Interfacial Properties of Whey by Enzymic Hydrolysis of the Residual Fat

By Christophe Blecker, Virna Cerne,[1] Michel Paquot, Georges Lognay,[2] and Alessandro Sensidoni[1]

FACULTÉ DES SCIENCES AGRONOMIQUES, UER DE TECHNOLOGIE DES INDUSTRIES AGRO-ALIMENTAIRES, 2 PASSAGE DES DÉPORTÉS, 5030 GEMBLOUX, BELGIUM
[1]UNIVERSITA DEGLI STUDI DI UDINE, FACOLTA DI AGRARIA, INSTITUTO DI TECNOLOGIE ALIMENTARI, 33100 UDINE, ITALY
[2] UER DE CHIMIE GÉNÉRALE ET ORGANIQUE, GEMBLOUX, BELGIUM

1 Introduction

European legislation authorizes the use of sweet whey as a skimmed milk substitute to improve the texture and flavour of food systems. Indeed, whey contains a large proportion of proteins with outstanding nutritional quality.[1] Food products for which whey is often mentioned as having potential utility are emulsions and foams. The ability of whey proteins to stabilize emulsions and foams is, however, rather poor,[2] and is known to be considerably reduced by the presence of residual milk lipids.[3] Usually, demineralization treatment (electrodialysis, ion exchange) followed by acidification leads to a significant removal of the residual fat.[4]

Instead of eliminating the residual fat in whey by separation, we are proposing in this study to modify it by enzymic hydrolysis with a specific lipase in order to produce mono- and diglycerides which may enhance the emulsifying properties of the proteins.

2 Materials and Methods

Commercial sweet whey powder was obtained from SIAB (Chateaubourg, France). It was produced by the Emmental cheese-making process. The 1,3-specific lipase of *Mucor miehei* (lipozyme 10 000L) was purchased from NOVO Nordisk (Bagsvaerd, Denmark).

Sweet whey powder (5 wt%) was suspended in distilled water, and 6 ml of this solution were incubated with 0.4 ml of the enzyme solutions (50 wt% in distilled water) at 30 °C with constant agitation (180 min^{-1}) for 24 h. Blanks were examined by replacing the enzyme solution with distilled water and heat-deactivated (100 °C for 5 min) enzyme solution.

The total fat contents of whey and modified whey were determined by the method of Folch *et al.*[5] with analysis by thin layer chromatography on silica gel plates with eluent petroleum ether (b.p. 60–80 °C)/ethyl ether/formic acid (60 : 40 : 1.5 by volume) and gas liquid chromatography on a Carlo Erba Mega 5160 chromatograph equipped with a Chrompack CPSIL 5 CB (100% methylpolysiloxane) column (5 m × 0.32 mm i.d., d.f. = 0.2 μm) and a FID detector at 350 °C.

The surface tension was measured continuously at 25 °C for 30 min with a Wilhelmy plate tensiometer (Tensimat N3, Prolabo) connected to a recorder (Servogor 120). For this purpose, 50 μl of whey solution were injected in 50 ml of water with a Hamilton microsyringe. Immediately after the injection, the solution was stirred with an integrated magnetic stirrer for 10 s and the recording of surface tension was started. Surface pressure *Π versus* surface area *A* isotherms were measured on a FW-2 film balance (Lauda GmbH Germany) at 25 ± 0.1 °C and constant compression rate of 185.4 cm^2 min^{-1}. Whey solution (50 μl) was spread using a Hamilton microsyringe with its needle point positioned in the surface.

3 Results and Discussion

Thin layer chromatography of the fat fraction extracted from non-modified and modified whey shows that almost all the triglycerides are converted into mono- and diglycerides by the action of *M. miehei* lipase. This is confirmed by gas liquid chromatography (Figure 1).

The interfacial properties of whey are significantly affected by the enzymic lipolysis. Firstly, adsorption kinetics of whey at the air–water interface indicate that the equilibrium surface tension value is lower and more rapidly reached after hydrolysis of the triglycerides (Figure 2). Indeed, the rate of adsorption of whey (taken as the slope of the tangent at 5 mN m^{-1} of surface tension decrease[6]) increases from 0.9 mN m^{-1} min^{-1} before lipolysis to 24 mN m^{-1} min^{-1} after lipolysis. Such an increase in adsorption rate would be expected to improve the emulsifying capacity of the protein.[6] However, it is deduced from the compression isotherms that the modified whey films spread at the air–water interface exhibit different properties from the films of the untreated whey (Figure 3). After fat hydrolysis, a transition zone is observed in the *Π*–*A* diagram for areas from 0.1 to 0.035 m^2 mg^{-1}. Then, in the condensed zone (for surfaces below 0.01 m^2 mg^{-1}), the modulus of compressibility (defined as $C_s^{-1} = -A(\partial\Pi/\partial A)$[7]) is higher than before lipolysis (82.6 mN m^{-1} against 73.7 mN m^{-1}). So, the modified whey forms more elastic films which therefore possess a better resistance to mechanical disturbance. This property is required for good foam stability.[8]

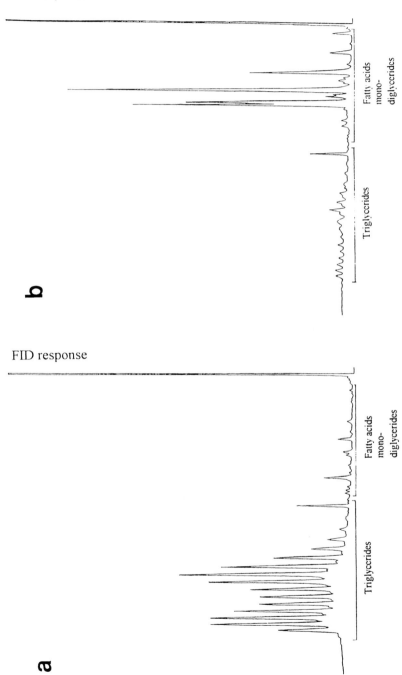

Figure 1   *Gas liquid chromatograms of the fat fraction extracted from (a) unmodified whey and (b) modified whay*

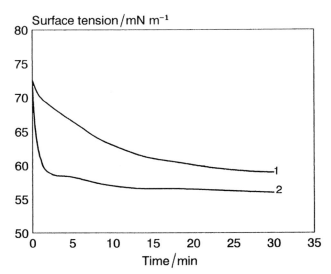

Figure 2   *Adsorption kinetics at pH 7 at the air–water interface. Surface tension is plotted against adsorption time: (1) untreated whey, (2) modified whey*

By comparison with adsorption kinetics and compression isotherms obtained with the blanks, it is clear that fat hydrolysis is the only process responsible for the appearance of these new properties. It is, however, difficult to go further in the interpretation because the lipolysis modifies the whey properties not only by eliminating the triglycerides, but also by creating fatty

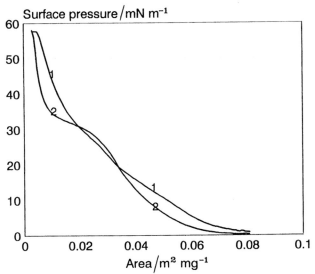

Figure 3   *Compression isotherms at 25 °C of unmodified (1) and modified (2) whey. Surface pressure is plotted against specific surface area*

acids, mono- and diglycerides, which may interact with the whey proteins However, the flavour of the short, free fatty acids from milk (such as C_6, C_8 or C_{10} fatty acids) may be undesirable in food emulsions and foams.

4 Conclusion

The enzymic hydrolysis of the fat in whey could be considered as a method of enhancing the value of this by-product of the cheese industry—producing surface-active mono- and diglycerides. The study of the interfacial properties of the modified whey suggests that this new product would be more suitable for emulsification and foaming than the unmodified whey. Therefore, it is possible that such transformed whey could be used advantageously to improve the texture and structure of some food products such as ice-cream.

References

1. J.-L. Maubois and G. Brulé, *Lait*, 1982, **62**, 484.
2. C. V. Morr, *J. Dairy Res.*, 1979, **46**, 369.
3. R. Peltonen-Shalaby and M. E. Mangino, *J. Food Sci.*, 1986, **51**, 91.
4. J. N. De Wit, *Neth. Milk Dairy J.*, 1986, **40**, 41.
5. J. Folch, M. Lees, and G. H. Sloane Stanley, *J. Biol. Chem.*, 1957, **226**, 497.
6. S. L. Turgeon, S. F. Gauthier, and P. Paquin, *J. Agric. Food Chem.*, 1991, **39**, 673.
7. G. L. Gaines, 'Insoluble Monolayers at Liquid–Gas Interfaces', Interscience, New York, 1966, p. 136.
8. J. Poré, 'Dispersions Aqueuses, Suspensions, Emulsions, Mousses', Le Cuir, 1976, p. 189.

Influence of Charge on the Adsorption of Proteins to Surfaces

By Jeffrey Leaver, David S. Horne, Celia M. Davidson, and
Dawn V. Brooksbank[1]

HANNAH RESEARCH INSTITUTE, AYR KA6 5HL, UK
[1] ENSBANA, 1 ESPLANADE ERASME, 21000 DIJON, FRANCE

1 Introduction

The polyelectrolyte nature of proteins means that inter- and intramolecular
Coulombic interactions are important with respect to both their structure
and interfacial behaviour. Protein-stabilized oil-in-water emulsions, in which
protein molecules adsorbed at the interface form a steric stabilizing layer
which serves to inhibit the aggregation and subsequent coalescence of
individual droplets, constitute the most important class of food colloids.
Due to its excellent emulsifying properties, sodium caseinate, prepared by
the isoelectric precipitation of the major protein fraction of milk, is
frequently used in the preparation of these emulsions. The caseins are a
family of proteins whose amino acid sequences and hence charge distributions are well characterized. Furthermore, they are relatively lacking in
secondary structure, and so the predicted distribution of charged groups
along the polypeptide chain may be similar to the actual distribution on the
adsorbed protein molecules (κ-casein excepted, which has a less random
structure).

Changes in the surface properties of initially uncoated emulsion droplets
resulting from the progressive adsorption of protein molecules cannot be
measured. Therefore, imitation emulsion droplets have been used in model
studies. Amongst these are polystyrene latices (the most widely studied), which
can be prepared with a variety of charged groups on the surface, and
unilamellar liposomes, where, by varying the phospholipid class, small
particles with different net surface charges can again be prepared. We have
been investigating the influence of the charge on the surface, and of the net
charge on the protein (as modified by phosphorylation, glycosylation, and the
presence of salts) on the adsorption of a number of individual, purified caseins
in these model systems. Some of the results are reported here.

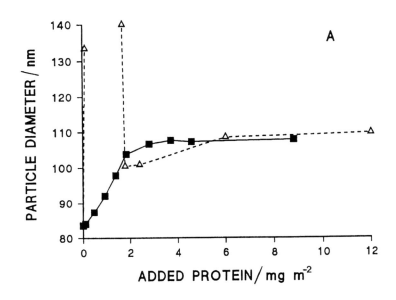

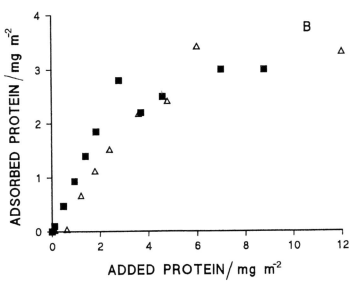

Figure 1   *Adsorption of native, phosphorylated β-casein to a negatively charged polystyrene latex in the presence (△) and absence (■) of 50 mM NaCl at pH 7.0. (A) Mean particle diameter is plotted against added protein concentration; (B) surface protein concentration is plotted against added protein concentration*

2 Experimental

Caseins were purified by ion exchange chromatography.[1] Liposomes were prepared by sonication[2] and the thickness of adsorbed protein layers on liposomes and polystyrene latices, and the surface coverage of individual proteins on the latices, were determined by photon correlation spectroscopy and solution depletion techniques respectively.[2-4]

3 Results and Discussion

Upon addition of protein to latex suspensions, the mean diameter of the particles and the amount of protein adsorbed to the surface were found to increase until plateau values were attained. The effect of the interaction between native β-casein and a negatively charged latex in the presence and absence of NaCl is shown in Figure 1. When small amounts of any of the caseins were added to the positively charged latex, or to a negatively charged latex in the presence of 50 mM NaCl, the mean size was observed to increase dramatically as the latex particles were flocculated (Figure 1A). This is due to protein molecules forming bridges between the latex particles. Further additions of protein fail to disrupt these aggregates, presumably because the additional protein molecules cannot penetrate into the flocs. When fresh latex suspensions are used at each concentration of added protein, it is observed that above a critical surface loading, flocculation no longer occurs and a plateau layer thickness, similar to that on the negatively charged latex in the absence of NaCl, is attained. This critical value is approx. 1 mg m^{-2} for native β-casein on both the negatively charged latex in the presence of 50 mM NaCl and on the positively charged latex in the absence of NaCl.

Bovine β-casein contains a cluster of five phosphoserine residues in the hydrophilic N-terminal region of the molecule. Studies of the kinetics of the proteolysis of β-casein-stabilized oil-in-water emulsions suggests that this region forms a loop or tail which projects from the interface into the aqueous phase.[5] The net negative charge on this tail region of the β-casein molecule can be reduced by enzymically dephosphorylating the protein. Similarly, κ-casein possesses a highly negatively charged C-terminal region (net charge at pH 6.6 of -10 for the B variant) which projects into the aqueous phase from the surface of the casein micelles in milk, thus preventing aggregation of the individual micelles. κ-Casein can exist in either glycosylated or non-glycosylated forms, the glycosylated residues all being located in this negatively charged tail region. In addition to increasing the hydrophilicity of this κ-casein tail, each sialic acid residue contributes an additional negative charge. Hence, three tetrasaccharide chains would give an overall net negative charge in this region of -16. We have therefore dephosphorylated β-casein and purified non-glycosylated and highly-glycosylated κ-casein fractions in order to investigate the effects of these modifications on the behaviour of the proteins. The influence of the net charge of the β- and κ-casein molecules on

Table 1   *Influence of polystyrene latex surface charge on the thickness* d *of adsorbed protein layers and the protein surface coverage* Γ *at pH 7*

Surface Charge		Native β-casein		Dephos. β-casein		Non-glyc. κ-casein B		Glyc. κ-casein B	
		d/nm	Γ/mg^{-2}	d/nm	Γ/mg^{-2}	d/nm	Γ/mg^{-2}	d/nm	Γ/mg^{-2}
Positive	−ME[a]	14	5.2	13	2.8				
	+ME[b]					10	3	12	1.7
Negative	−ME	15	3.0	11	3.4	11		14	
	+ME					13	5.0	15	3.0

Dephos. = dephosphorylated; glyc. = glycosylated.
[a] Without mercaptoethanol.
[b] With mercaptoethanol.

the thickness and surface loading on positively and negatively charged latices is shown in Table 1. Decreasing the net negative charge of both proteins results in a reduction in the thickness of the adsorbed layer on the negatively charged surface, presumably as a result of the reduction in charge repulsion between these tail regions of the proteins and the surface. On the contrary, decreasing the protein net negative charge does not change significantly the thickness of the adsorbed layer on a positively charged surface, but decreases the protein layer density as a result of less charge attraction. Table 1 also shows that addition of mercaptoethanol (ME) to the κ-casein causes a thickening of the protein layer presumably as a result of relaxation of the protein structure resulting from the reduction of disulfide bridges.

Table 2   *Decrease in the thicknesses of preadsorbed protein layers on negatively charged latex as a result of adding salts at pH 7*

Protein	Decrease in layer thickness (nm)	
	+ 50 mM NaCl	+ 2 mM CaCl$_2$
Phosphorylated β-casein	2	5
Dephosphorylated β-casein	0	2
Glycosylated κ-casein	1	3
Non-glycosylated κ-casein	1	3

Changes in the thickness of preadsorbed protein layers on the negatively charged latex as a result of adding either NaCl or CaCl$_2$ are shown in Table 2. Protein layers are clearly considerably more sensitive to CaCl$_2$ although the maximum decrease in the thickness of preadsorbed phosphorylated β-casein layers was similar (about 5 to 6 nm) in both cases but was not achieved until the concentration of NaCl reached 100 mM as opposed to 0.5 mM CaCl$_2$. This suggests that, unlike the reduction in the presence of NaCl, which results from a simple ionic strength effect, the layer thinning resulting from the addition of Ca^{2+} is due to specific binding of the Ca^{2+} to highly negatively charged

Table 3   *Influence of net surface charge on the thickness (nm) of protein layers adsorbed to lipsomes made from phosphatidylcholine (PC) and phosphatidylglycerol (PG) at pH 7*

Lipid class	Native β-casein	Dephos. β-casein	κ-casein	α_{s1}-casein	β-Lacto-globulin
PC (neutral)	6.0	6.0	6.0	7.0	4.0
PG (negative)	11.0	9.5	14.0	50[*]	8.5

* Probably due to bridging flocculation, or to very strong repulsion between PG and highly phosphorylated protein.

regions of the adsorbed proteins which reduces repulsion and allows the proteins to lie closer to the latex surface due to ionic bridging.

In addition to measuring binding of proteins to latices, we have also investigated their adsorption to unilamellar liposomes. This area is of interest to the pharmaceutical and cosmetic industries where the use of liposomes is becoming increasingly common. The surface charge of the liposomes was varied by using different classes of phospholipids. Liposomes prepared using phosphatidylcholine (PC), which is zwitterionic, possess a surface which has no net charge at the pH values which we investigated, whereas those prepared from acidic phosphatidylglycerol (PG) have a negative charge. With all of the proteins which were tested, the maximum layer thickness on the neutral PC liposomes was significantly less than that on the negatively charged PG surface (Table 3). Again, this is due to charge repulsion between the negatively charged surface and proteins. In the case of the interaction between α_{s1}-casein and the PG liposomes, the large increase in particle size was again probably a result of limited bridging between liposomes or of very strong ionic repulsion between PG liposomes and very highly phosphorylated casein (9 serine P per mole).

Acknowledgement

This research was funded by The Scottish Office Agriculture and Fisheries Department.

References

1. J. Leaver and A. J. R. Law, *J. Dairy Res.*, 1992, **59**, 557.
2. J. Leaver, D. V. Brooksbank, and D. S. Horne, *J. Colloid Interface Sci.*, 1994, **162**, 463.
3. D. V. Brooksbank, J. Leaver, and D. S. Horne, *J. Colloid Interface Sci.*, 1993, **161**, 38.
4. D. V. Brooksbank, C. M. Davidson, D. S. Horne, and J. Leaver, *J. Chem. Soc., Faraday Trans.*, 1993, **89**, 3419.
5. J. Leaver and D. G. Dalgleish, *Biochim. Biophys. Acta*, 1990, **1041**, 217.

Surface Properties of the Milk Fat Globule Membrane: Competition between Casein and Membrane Material

By S. Chazelas, H. Razafindralambo,[1] Q. Dumont de Chassart,[1] and M. Paquot[1]

ENSBANA, 1 ESPLANADE ERASME, 21000 DIJON, FRANCE
[1]FACULTÉ DES SCIENCES AGRONOMIQUES DE GEMBLOUX, UER DE TECHNOLOGIE DES INDUSTRIES AGRO-ALIMENTAIRES, 2 PASSAGE DES DÉPORTÉS, 5030 GEMBLOUX, BELGIUM

1 Introduction

Heat and mechanical treatments in the dairy industry lead to modifications in fat globule membrane composition and structure. Because of the large increase of interfacial area, the original milk fat globule membrane is almost completely replaced by skim milk constituents. Casein appears by far to be the major constituent adsorbed to the oil–water interface.

Replacement of the native milk fat globule membrane by casein causes important losses of fat globule functionality in milk and milk products. The whipping properties of cream and the interaction of globules with the casein network in fermented milk products are two examples in which modification of the fat globule membrane plays an important role. In this study, the surface coverages of casein and membrane material at the air–water interface are compared to those developed in oil-in-water emulsion, with a view to explaining their competitive adsorption in milk processing.

2 Material and Methods

Sodium caseinate was made from raw milk.[1] Milk fat globule membrane (MFGM) was purified by washing raw cream three times with phosphate buffer (NaH$_2$PO$_4$/Na$_2$HPO$_4$, pH 7.2, 0.01 M) and NaCl (0.015 M).[2] Emulsions were made with a constant ratio of emulsifier:fat (2 wt%) and 10% of sunflower oil using a Microfluidizer (Microfluidics model 110 T). Emulsifiers were dispersed in phosphate buffer. Fat globule size was determined with a Malvern Zeta Sizer III (Malvern UK) after dilution in 8 M urea.[3] Oil–water

95

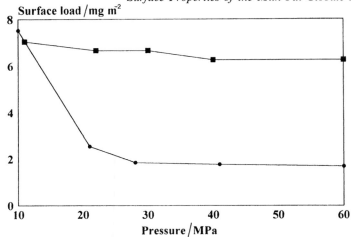

Figure 1   *Change of surface load Γ with homogenization pressure in 10% oil-in-water emulsion: (■) 2% MFGM; (●) 2% sodium caseinate*

interfacial areas were calculated for each emulsion assuming fat globules to be spheres. Surface loads Γ were estimated for oil–water emulsions from a determination of protein in the aqueous phase.[4,5] The relationship between surface pressure Π and surface load Γ was determined with a Langmuir film balance (Lauda FW2) using the phosphate buffer described above as subphase.

3 Results

Figure 1 shows the change in surface load Γ as the homogenization pressure increases for emulsions made with sodium caseinate and MFGM. In emulsions made with MFGM alone, the average fat globule size varies little, and therefore the quantity of emulsifier adsorbed remains constant. In contrast, caseinate gives increasingly smaller fat globules as the pressure increases, and consequently the surface load is gradually reduced.

Figure 2 presents plots of film pressure Π *versus* surface load Γ curves for sodium caseinate and MFGM; monolayer saturation limits are pointed out with arrows.[6]

4 Discussion

For a fixed quantity of emulsifier, an increase in homogenization pressure increases the oil–water interfacial area. This new interface must be covered by surface-active molecules to ensure fat globule stability. If the interface is not totally covered, coalescence reduces the interfacial area to a new steady-state value.

The Π/Γ curves show clearly the spreading capacity of the casein which is driven by the unfolding of β-casein.[7] This behaviour allows us to obtain finer dispersions by simply increasing homogenization energy. The newly formed

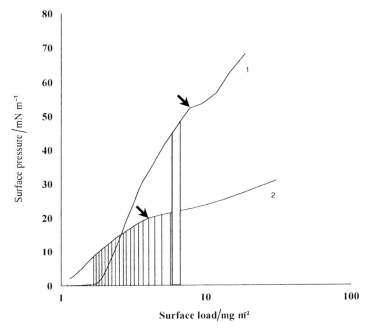

Figure 2   *Surface pressure Π versus surface load Γ for (1) MFGM and (2) sodium caseinate at the air–water interface. Γ values corresponding to the oil-in-water emulsion are hatched. Monolayer saturation limits are indicated with arrows*

interface is covered by adsorbing casein. In contrast, for MFGM, the fat globule size reduction is limited by the emulsifier availability. Its film pressure increases strongly above a definite surface load; the Π/Γ curve is quite similar to that obtained for globular proteins or polar lipids, which are less compressible than casein films.

Above a homogenization pressure of 20 MPa, the surface load obtained with the microfluidizer in the oil–water emulsion could be compared with that of the monolayer obtained with the film balance. Shear rates provided by this homogenizer are sufficient to maximize the area occupied by molecules at the oil–water interface, and minimize the surface load. From the Π/Γ curves we see that it is possible to optimize the amount of emulsifier required to stabilize an emulsion and to maintain the properties of membrane material which are needed at the interface for controlling viscosity and whipping time. If membrane isolates are used alone, the droplet size does not depend on homogenization pressure; and so, from the Π/Γ curves, it is possible to calculate the interfacial area and the fat globule size for a given quantity of emulsifier. If membrane material and caseins are in competition a high homogenization pressure improves casein adsorption to the fat globules. In principle, the ratio of MFGM to casein at the interface could be estimated by particle sizing. In milk or cream, stability improvement could be achieved by addition of MFGM isolates (*i.e.* as obtained by buttermilk fractionation) instead of by increasing homogenization pressure.

References

1. D. M. Mulvihill and P. C. Murphy, *Int. Dairy J.*, 1991, **1**, 13.
2. J. J. Basch, R. Greenberg, and H. M. Farrel, Jr, *Biochim. Biophys. Acta*, 1985, **830**, 127.
3. O. Robin and P. Paquin, *J. Dairy Sci.*, 1991, **74**, 2440.
4. S. Chazelas, Mémoire de fin d'études, ENSBANA, Dijon, 1992.
5. O. H. Lowry, N. J. Rosenbrough, A. L. Farr, and R. J. Randall, *J. Biol. Chem.*, 1951, **193**, 265.
6. G. L. Gaines, Jr, 'Insoluble Monolayer at Liquid/gas Interface', Interscience, New York, 1966.
7. P. Walstra and A. L. de Roos, *Food Rev. Int.*, 1993, **9**, 503.

Surface-active Properties of Mixed Protein Films Containing Caseinate + Gelatin

By V. B. Galazka,* B. T. O'Kennedy, and M. K. Keogh

NATIONAL DAIRY PRODUCTS RESEARCH CENTRE, TEAGASC, MOOREPARK, FERMOY, CO. CORK, REPUBLIC OF IRELAND

1 Introduction

In the formulation of food colloid systems, the caseins are commonly used as emulsifiers and stabilizers.[1,2] These biopolymers have the ability to adsorb rapidly at a newly created interface to form a coherent viscoelastic layer which protects freshly formed droplets against aggregation and coalescence. By way of contrast, gelatin is generally added as a gelling agent due to its water-binding properties both in the bulk aqueous phase and at the oil–water interface.[3]

As far as proteins go, caseins are exceptionally surface-active because of a unique combination of high molecular flexibility and fairly high hydrophobicity. Gelatin is also disordered but is less hydrophobic, and therefore less surface-active than casein.[1,4] Early observations have shown that sodium caseinate will displace gelatin from a planar oil–water interface,[5,6] and will inhibit gelatin adsorption when both are present during emulsification.[3,7] Further studies have demonstrated[4] that interfacial gelatin in a freshly made emulsion is readily replaced by the more surface-active caseinate when added to the continuous phase, although the ability to exchange diminishes considerably as the film is aged.

The present study looks into time-dependent changes in surface tension for individual and mixed protein films containing a pure milk protein (α_{s1}-, β-, κ-casein) or sodium caseinate and gelatin under various temperature conditions (5, 15, 25 and 40 °C).

2 Materials and Methods

Gelatin obtained from Extraco (Sweden) was of special grade (SG 720 - N, BS 275) with an isoionic point at pH = 8–9. The mean molecular weight was 71×10^3 g mol^{-1}. Sodium caseinate was obtained from Kerry Ingredients

---

* Present address: Department of Food Science and Technology, University of Reading, Whiteknights, PO Box 226, Reading RG2 2AP, UK

(Ireland); α_{s1}-casein, β-casein, and κ-casein were purchased from Sigma. Buffer solutions were prepared with AnalaR-grade reagents (Sigma) and single-distilled, deionized, Millipore water.

Surface tensions were measured at the planar air–water interface using the K12 Processor Tensiometer (Krüss) Wilhelmy plate technique. Measurements were made over a period of 24–60 h for pure and mixed (casein:gelatin = 1:9) protein solutions (usually 10^{-3} wt%) in imidazole buffer (pH 7.0, ionic strength 0.05 M).

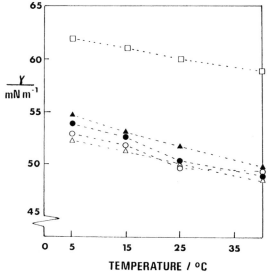

Figure 1   *Surface activity curves for individual protein (10^{-3} wt%, 0.05 M, pH 7.0) films at the air–water interface. The surface tension γ at 10 h is plotted against temperature (5, 15, 25 or 40 °C): gelatin ($\square$); α_{s1}-casein ($\blacktriangle$); β-casein ($\bullet$); sodium caseinate ($\bigcirc$); κ-casein ($\triangle$)*

3 Results and Discussion

Time-dependent surface tension measurements (not shown) at the planar air–water interface (pH 7.0, 0.05 mol dm^{-3}, 25 °C) for individual proteins show that the caseins give a much more rapid lowering of the tension as well as a lower steady-state tension than gelatin as found by Castle *et al.*[4] and Murray.[8] Figures 1 and 2 present values of surface tension γ at 10 h for individual and mixed protein films at various temperatures (5, 15, 25, and 40 °C). The tension for the 1:9 mixtures is very close to that for the individual casein component (10^{-3} wt%) with the exception of κ-casein which may become polymerized through S–S linkages. Duplicate consistent results were difficult to obtain with this last combination, and the steady-state tension was achieved only after extraordinarily long periods of time. It can be inferred from these measurements that caseines predominate over gelatin at the air–water interface,[5,6] and

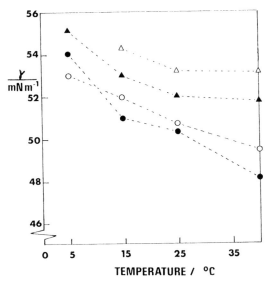

Figure 2   *Surface tension data for mixed (casein:gelatin = 1:9) protein films at the air–water interface (10⁻³ wt%, 0.05 M, pH 7.0). The surface tension γ at 10 h is plotted against temperature (5, 15, 25 or 40 °C): κ-casein + gelatin (△); αₛ₁-casein + gelatin (▲); β-casein + gelatin, (●); sodium caseinate + gelatin (○)*

that the proteins become more surface-active with an increase in temperature, possibly due to a greater number of hydrophobic interactions at the interface.

Table 1   *Time taken (hours) to reach constant values* [a] *(±0.3 mN m⁻¹) of surface tension γ for individual and mixed (casein:gelatin = 1:9) protein films at the air–water interface (10⁻³ wt%, 0.05 M, pH 7.0) at various temperatures*

Sample	5 °C	15 °C	25 °C	40 °C
α_{s1}-Casein	28.5	21.0	16.0	2.5
β-Casein	20.5	15.5	9.0	1.5
κ-Casein	31.5	21.5	15.5	4.0
Sodium caseinate	22.0	17.5	9.0	1.0
Gelatin	44.5	30.0	21.5	9.5
α_{s1}-Casein + gelatin	49.5	31.0	16.5	6.5
β-Casein + gelatin	29.5	21.5	11.5	4.5
κ-Casein + gelatin	–	59.0	19.5	8.5
Sodium caseinate + gelatin	40.0	21.0	12.5	8.0

[a] Estimated error ± 5%.

The time-dependent results for individual proteins at the air–water interface (Table 1) show that β-casein reaches its final value (± 0.3 mN m⁻¹) faster than the other caseins at all temperatures. It was also found that there is a general trend of faster diffusion to the interface with increasing temperature. Similar trends were observed for the 1:9 mixtures. As well as the increase in hydrophobicity at higher temperatures, the decrease in viscosity and increase in thermal movement should further increase the rate of adsorption.

In conclusion, the results presented in this paper show that κ-casein is marginally the most surface-active of the caseins in the absence of gelatin. The caseins dominate the interface in the presence of gelatin, with β-casein being the most efficient at displacing gelatin. Finally, temperature has a significant effect on surface activity and the rate of diffusion to the interface.

Acknowledgement

V. B. G. and B. T. O'K. are grateful to the N.D.P.R.C. for providing the K12 Krüss Processor Tensiometer. This research was supported by the EC AIR1 programme.

References

1. E. Dickinson, A. Murray, B. S. Murray, and G. Stainsby, in 'Food Emulsions and Foams', ed. E. Dickinson, Special Publication No. 58, The Royal Society of Chemistry, London, 1987, p. 86.
2. E. Dickinson, in 'Food Colloids and Polymers: Stability and Mechanical Properties', ed. E. Dickinson and P. Walstra, Special Publication No. 113, The Royal Society of Chemistry, Cambridge, UK, 1993, p. 77.
3. S. M. Chesworth, E. Dickinson, A. Searle, and G. Stainsby, *Lebensm.-Wiss. Technol.*, 1985, **18**, 230.
4. J. Castle, E. Dickinson, B. S. Murray, and G. Stainsby, *ACS Symp. Ser.*, 1987, **343**, 118.
5. P. R. Mussellwhite, *J. Colloid Interface Sci.*, 1966, **21**, 99.
6. J. Castle, E. Dickinson, A. Murray, B. S. Murray, and G. Stainsby, in 'Gums and Stabilisers for the Food Industry', ed. G. P. Phillips, D. J. Wedlock, and P. A. Williams, Elsevier Applied Science, London, 1986, vol. 3, p. 409.
7. E. Dickinson, D. J. Pogson, E. W. Robson, and G. Stainsby, *Colloids Surf.*, 1985, **14**, 135.
8. B. S. Murray, PhD Thesis, University of Leeds, 1987.

Protein Adsorption and Protein–Monoglyceride Interactions at Fluid–Fluid Interfaces

By J. M. Rodríguez Patino and M. R. Rodríguez Niño

DEPARTAMENTO DE INGENIERÍA QUÍMICA, FACULTAD DE QUÍMICA, UNIVERSIDAD DE SEVILLA, 41012 SEVILLA, SPAIN

1 Introduction

Many food products are foams or emulsions (or both), and emulsifiers often play an important role in stabilizing these systems. Proteins and low-molecular-weight surfactants comprise the two main classes of emulsifiers. Surfactants are present in addition to protein in many food colloids, and they can affect the adsorption behaviour of the protein. The relationship between surface tension lowering and colloid stability is well documented in the literature.[1] The dynamic adsorption behaviour is even more important in high-speed processes such as foam formation.[2] The understanding of adsorption behaviour of proteins and protein–surfactant interactions at fluid–fluid interfaces could help improve process design, select emulsifier components in commercial formulations, and lead to enhanced product performance. In this work we present preliminary data on both the adsorption and the interactions between bovine serum albumin (BSA) and a selected lipid contained in several food formulations (monostearin), at the oil–water and air–water interfaces, from time-dependent surface ($\sigma_{A/W}$) or interfacial ($\sigma_{O/W}$) tension measurements.

2 Experimental

Surface and interfacial tensions as a function of time were measured by the Wilhelmy plate method, using a roughened platinum plate attached to a Krüss Digital Tensiometer K10. In order to obtain reproducible results, it was found to be essential to make the measurements according to a set schedule. BSA (Sigma, >99%) was dissolved in standard phosphate buffer solution (pH 7, ionic strength 0.05). Stock protein solutions (10 or 1 wt%), stored at 5 °C for no more than 1 week, were previously diluted before use. Two millilitres of

103

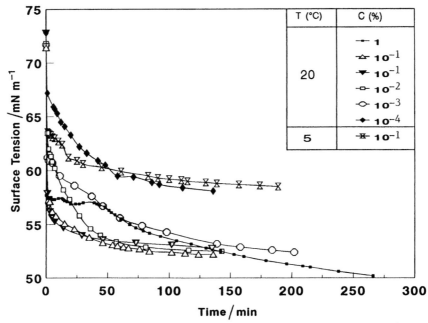

Figure 1   *Variation of surface tension with time at 5 and 20 °C during BSA adsorption at the air–water interface at various protein bulk concentrations*

diluted protein solution were injected into 18 ml of previously cleaned phosphate buffer – by suction through a narrow pipette – to give the final concentration over the range $1\text{--}10^{-4}$ wt%. The oil phase was *n*-tetradecane (Sigma, >99%). A similar ratio (20:18 of buffer solution/oil, by volume) was used in all interfacial tension measurements. Ethanol (Merck, >99.8%), hexane (Merck, 99%), sucrose (Fluka, >99.5%), and monostearin (Sigma, 99%) were used as supplied. The water used was purified by means of a Millipore filtration device. To study protein–lipid interactions at the air–water interface, monostearin (5.2×10^{-4} M dissolved in a mixture of hexane and ethanol, 9:1 by volume) was spread on a film of protein previously adsorbed from the subphase bulk. To study the protein–lipid interactions at the oil–water interface the monostearin was dissolved in *n*-tetradecane (10^{-2} M) and carefully added to the protein previously adsorbed on the aqueous interface. Surface and interfacial tensions were measured at different temperatures (5–40 °C). The aqueous phase composition (ethanol 1 M and sucrose 0.5 M) was also treated as a variable.

3 Results and Discussion

The rate of BSA adsorption at the air–water interface increases when both the BSA concentration in the aqueous phase and the temperature are increased (Figure 1), which agrees with previous data in the literature.[3] At the oil–water

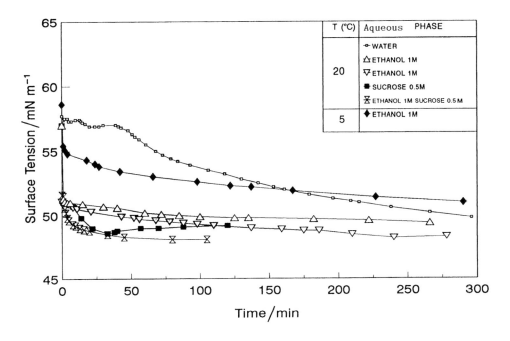

Figure 2 *Variation of surface tension with time at 5 and 20 °C during BSA adsorption at the air–water interface at various subphase compositions*

interface a similar trend can be observed (results not shown). The decreasing surface tensions reflect an increase in the number of adsorbing units and can be considered as being made up from three main processes:[4] (a) diffusion cf protein molecules to the interface, (b) spreading or unfolding of adsorbed protein molecules, and (c) molecular rearrangements of adsorbed molecules. Obviously, in practice, adsorption and rearrangement of proteins occur simultaneously at the interface. The kinetics of protein adsorption is initially determined by diffusion, but at longer times penetration into the film and rearrangement of protein at the interface can control the process. This could be the case with globular proteins – such as BSA studied in this work – at the higher concentrations in the aqueous phase. In Figure 1 there can be observed an induction period in the adsorption of 1 wt% BSA, after the initial step. Moreover, the steady surface tension σ^∞ decreases when both the BSA concentration in the aqueous phase and the temperature are increased.

Solutes in the subphase have an effect on the rate of BSA adsorption (Figure 2). With ethanol the induction period disappears, whereas with sucrose the induction period decreases as does the time to reach $\sigma^\infty_{A/W}$. With a mixture of ethanol and sucrose the behaviour is similar to that of sucrose alone. However, solutes in the subphase have no influence on the $\sigma^\infty_{A/W}$ values. The results could reflect the existence of BSA–solute interactions in the aqueous bulk

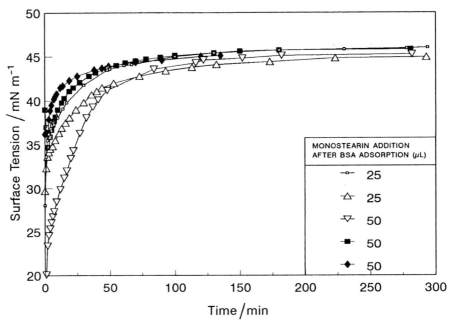

Figure 3   *Variation of surface tension with time at 20 °C after addition of successive amounts of monostearin to a BSA film previously adsorbed at the air–water interface (BSA concentration = 1 wt%)*

phase and at the interface. It is possible that ethanol forms a network of hydrogen or hydrophobic bonds with protein both at the interface and in the bulk phase. These clusters could inhibit protein adsorption. With sucrose in the subphase the unfolding of molecules in bulk solution is possible with a subsequent rapid adsorption on denatured protein.[5]

At the air–water interface the spreading of monostearin on a film of adsorbed protein causes a fast reduction in the surface tension (Figure 3). After the initial period $\sigma_{A/W}$ increases with time θ until $\sigma^{\infty}_{A/W}$ is attained. The $\sigma_{A/W}$–θ curves of monostearin spread either on water or on a BSA film are similar, and the $\sigma^{\infty}_{A/W}$ values attained have the same values. These values are lower than those corresponding to the pure BSA film. So, it could be concluded that the protein is displaced by monostearin from the interface. The BSA–monostearin interactions at the interface and in the aqueous phase increase with increasing amounts of monostearin spread on the interface. It is apparent that the aqueous phase opacity increases with the addition of monostearin. These results agree with those of Dickinson *et al.*[6]. They observed an increase in the monoglyceride content of the aqueous phase by analysis of the subnatant solution after centrifugation of a cream liqueur.

The BSA–monostearin interactions depend on temperature, BSA concentration, and the aqueous phase composition. With ethanol in the subphase we

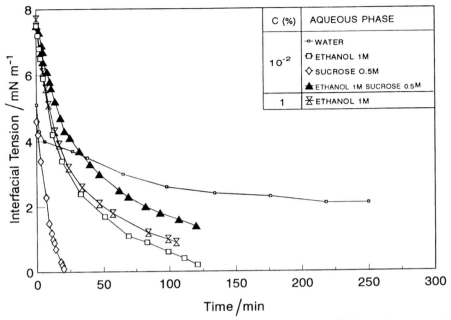

Figure 4   *Variation of interfacial tension with time after addition of monostearin dissolved in* n-*tetradecane on a film of protein previously adsorbed at the air–aqueous interface*

have $\sigma^{\infty}_{A/W}$ (BSA) $\approx \sigma^{\infty}_{A/W}$ (BSA–monostearin) $> \sigma^{e}_{A/W}$ (monostearin), whereas with sucrose in the subphase we have $\sigma^{\infty}_{A/W}$ (BSA) $> \sigma^{\infty}_{A/W}$ (BSA–monostearin) $> \sigma^{e}_{A/W}$ (monostearin), where σ^{e} is the equilibrium surface tension.[7] These results suggest that BSA and monostearin can form a complex with the solute at the interface as well as in the aqueous bulk phase.

At the oil–water interface the monostearin diffusion from the bulk oil phase may displace the BSA molecules that were previously adsorbed at the air–water interface (Figure 4). From measurements of interfacial tension with time, it can be concluded that the rate of BSA displacement by monostearin increases with increasing temperature (data not shown) or decreasing BSA concentration in the subphase. Solutes in the subphase (ethanol and sucrose) also appear to affect the BSA–monostearin interactions at the oil–water interface.

References

1. P. J. Halling, *CRC Crit. Rev. Food Sci. Nutr.*, 1981, **12**, 155.
2. N. Kitabatake and E. Doi, *J. Food Sci.*, 1982, **47**, 1218; 1988, **53**, 1542.
3. G. Serrien, G. Geeraerts, L. Ghosh, and P. Joos, *Colloids Surf.*, 1992, **68**, 219; P. Suttiprasit, V. Krisdhasima, and J. McGuire, *J. Colloid Interface Sci.*, 1992, **154**, 316.

4. E. Tornberg, *J. Sci. Food Agric.*, 1978, **39**, 762; D. E. Graham and M. C. Phillips, *J. Colloid Interface Sci.*, 1979, **70**, 403.

5. J. Lee and S. Timasheft, *J. Biol. Chem.*, 1981, **246**, 7193; T. Arakawa and S. Timasheft, *Biochemistry*, 1982, **21**, 6536; S. Poole, S. West, and C. Walters, *J. Sci. Food Agric.*, 1984, **35**, 701.

6. E. Dickinson, S. K. Narhan, and G. Stainsby, *J. Food Sci.*, 1989, **54**, 77.

7. J. M. Rodríguez Patino and R. Martín Martínez, *J. Colloid Interface Sci.*, 1994, **167**, 150.

Destabilization of Monoglyceride Monolayers at the Air–Water Interface: Structure and Stability Relationships

By J. M. Rodríguez Patino and J. de la Fuente Feria

DEPARTAMENTO DE INGENIERÍA QUÍMICA, FACULTAD DE QUÍMICA, UNI-
VERSIDAD DE SEVILLA, SEVILLA, SPAIN

1 Introduction

The stability of emulsifier films at the air–water interface is a phenomenon related to both cohesive forces in the monolayer and the interactions between molecules in the monolayer and in the sub-surface region. These interactions exert an influence on the mechanisms that control the film instability, such as dissolution and collapse.[1] The film elasticity could be a parameter which quantifies the intermolecular and film–subphase interactions. The elasticity is a measure of the resistance to a change in the film area. High film elasticity is associated with a film which has a strong cohesive structure at the surface.[2]

The aim of this work is to establish a relationship between the destabilization of monomyristin monolayers and the surface pressure, with temperature and subphase composition as variables. This monoglyceride – used as an emulsifier – is slightly soluble in the aqueous subphases studied, as has been observed from the hysteresis loop in the Π–A isotherms. This phenomenon has also been found previously in our laboratory with other monoglyceride monolayers.[3] The monolayer molecular loss was quantified by relaxation measurements at constant surface pressure.[4]

2 Experimental

A commercial, fully automated, Langmuir-type film balance (Lauda) was used[3-5] to obtain Π–A isotherms or to make relaxation measurements. The results of relaxation experiments have been expressed as N/N_0, where N and N_0 are the number of molecules that exist on the surface at time θ and at the initial moment, respectively. The film elasticity is expressed by the modulus $(-\mathrm{d}\Pi/\mathrm{d}A)$, which was obtained from the slope of the Π–A isotherms. The solutes chosen as subphase components—ethanol, glucose, and sucrose—are typical ingredients in food formulations.

3 Results

Monomyristin Monolayer Structure

Solutes in the subphase can modify the monolayer structural parameters. The monolayer structure depends on the surface pressure, the temperature, and the subphase composition. If follows that the elasiticity modulus also depends on these variables. In Table 1, it can be seen that at 35 mN m^{-1} and 20 °C the values of the modulus lie in the order: ethanol 0.5 M > water > glucose 1 M > sucrose 0.5 M. That means that the presence of ethanol in the subphase makes the monolayer less compressible, whereas the sugars exert the opposite effect. A similar behaviour has been found with other monoglycerides[3] and fatty acids.[5]

Table 1   *Structural characteristics of monomyristin monolayers at 20 °C as a function of subphase composition*

Subphase	$\dfrac{\Pi}{\text{mN m}^{-1}}$	$\dfrac{A}{\text{nm}^2}$	*Monolayer structure*	$\dfrac{-d\Pi/dA}{10^2\,\text{mN nm}^{-2}\,\text{m}^{-1}}$
Water	35	0.202	LC	2.9
Ethanol 0.5 M	35	0.160	LC	3.7
Glucose 1 M	35	0.234	LC	1.3
Sucrose 0.5 M	35	0.234	LC	0.9

LC = liquid-condensed.

Monomyristin Monolayer Stability

The effect of aqueous subphase composition on the monomyristin monolayer instability, measured in experiments in which surface pressure and temperature were kept constant at 35 mN m^{-1} and 20 °C, is shown in Figure 1. The monolayer molecular loss was quantified as the variation of N/N_0 *versus* time. Two relaxation equations fit the experimental data:

$$-\log\,(N/N_0) = a\theta^{\frac{1}{2}} \qquad (1)$$

$$-\log\,(N/N_0) = b\theta \qquad (2)$$

The plots of experimental points in Figure 1 verify these equations. Equation (1), the discontinuous line on the plot, fits the rate of film molecular loss during an initial non-steady-state period of desorption (dissolution mechanism).[6] Equation (2), the continuous line on the plots, fits either the rate of molecular loss reached in a steady state (diffusion mechanism) or the transformation of a homogeneous monolayer phase into a heterogeneous monolayer-collapse phase system due to the formation of nuclei.[7] This

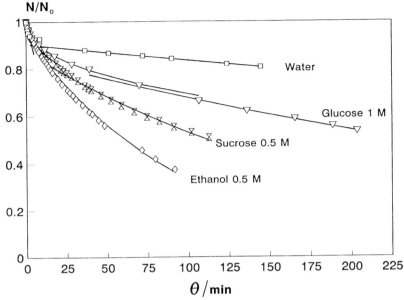

Figure 1 *Relaxation of a monomyristin monolayer spread on aqueous subphases (20 °C, 35 mN m^{-1}). The ratio N/N$_0$ is plotted against time θ*

equation also represents the destabilization due to evaporation.[8] (In this work the monolayer instability due to evaporation can be ignored.)

For short times (θ ≤ 10 min), the film molecular loss is similar no matter what the subphase composition is (Figure 1). However, after this period the number of molecules that remain on the surface for a value of θ follows the order: water > glucose 1 M > sucrose 0.5 M > ethanol 0.5 M. That is, the monolayer stability is highest on water.

The values of the kinetic coefficients (*a* and *b*), and the characteristic time of change (θ*) are shown in Table 2. At low values of the modulus, with a liquid-expanded structure (LE) in the monolayer, either by increasing temperature or by decreasing the surface pressure, the rate of film molecular loss can be quantified using eqn (2). Only at the highest values of ($-dΠ/dA$),

Table 2 *Characteristic parameters for destabilization of monomyristin mono-layers with LC structure at 20 °C and 35 mN m^{-1}*

Subphase	Sucrose 0.5 M	Glucose 1 M	Water	Ethanol 0.5 M
θ (min)	112.6	203.5	145	48.4
N/N$_0$	0.510	0.535	0.808	0.561
$a \times 10^3$	20	16	17	--
LR	0.990	0.995	0.990	--
$b \times 10^5$	210	97	34	500
LR	0.995	0.998	0.999	0.994
θ* (min)	18.45	38.95	6.4	--

with liquid-condensed (LC) and LE structures, is 'collapse' the mechanism that controls the film molecular instability.

4 Discussion

The data in Tables 1 and 2 show that the mechanism that controls the film molecular loss and the kinetic coefficients depends on the elasticity values.

With sugars in the subphase the modulus value decreases and the monolayer instability decreases. As with water in the subphase, the rate of monolayer molecular loss follows a mechanism with two steps related to dissolution and diffusion in the subphase, with an LC structure. The presence of sugars in the subphase does not affect the molecular dissolution from the interface (the value of the *a* coefficient is independent of the sugar content in the subphase) but increases the diffusion towards the subphase (the value of the *b* coefficient decreases when the modulus increases).

The hydrophobic interactions between ethanol and monomyristin, both at the interface and in the subphase, could be related to the diffusion of molecules. The diffusion process [represented by eqn (2)], always occurs with ethanol in the subphase regardless of the modulus value. When the elasticity modulus is very high, due to the rigidity of the monolayer, the lipid is diffusing towards the subphase and at the same time forming nuclei. So, two steps appear in the global process, both of them following eqn (2) with different values of the coefficients.

When the monolayer adopts an expanded structure, the destabilization occurs through diffusion of the molecules (data not shown), due to a decrease in the attraction between molecules in the film and between film molecules and subphase molecules.

Table 3   *Sub-surface zone depth ε and sub-surface concentration C_a in desorption of monomyristin films by a diffusion-controlled mechanism*

Subphase	Sucrose 0.5 M	Glucose 1 M	Water
T (°C)	20	20	20
Π (mN m^{-1})	35	35	35
ε (mm)	0.823	1.217	5.498
C_a (10^{-14} mm^{-3})	1.36	1.14	1.27

For those cases in which the kinetics of desorption follows a diffusion-controlled mechanism with a non-steady-state period, there is a relationship between the values of the kinetic coefficients and the parameters that characterize these kinetics,[6,9] *i.e.* the depth of sub-surface zone (ε) and the sub-surface concentration (C_a). Their values for aqueous sugar solutions and water as subphases are shown in Table 3. It can be seen that, because the sugar molecules in the subphase help the monolayer molecules to diffuse, the depth of the sub-surface zone is lower but the sub-surface concentration is higher. This last effect can be attributed to the possible interactions that can appear between sugar and monoglyceride molecules. The hydrogen bonds between the polar head of the monoglyceride and the –OH groups in the glucose or sucrose molecules can be the principal cause of these interactions which can stabilize the monomyristin monolayer near the surface zone. As a consequence of both effects the characteristic time to change the kinetic order is higher with sugars in the subphase than in the absence of solutes in the subphase.

References

1. P. Baglioni, G. Gabrielli, and G. Guarini, *J. Colloid Interface Sci.*, 1980, **78**, 347.
2. S. H. Kim and J. E. Kinsella, *J. Food Sci.*, 1985, **50**, 1526.
3. J. M. Rodriguez Patino, M. Ruíz Domínguez, and J. de la Fuente Feria, *J. Colloid Interface Sci.*, 1992, **154**, 146.
4. J. de la Fuente Feria and J.M. Rodríguez Patino, *Langmuir*, 1994, **10**, 2317.
5. J. M. Rodríguez Patino, J. de la Fuente Feria, and C. Gómez Herrera, *J. Colloid Interface Sci.*, 1992, **148**, 223.
6. L. Ter Minassian-Saraga, *J. Colloid Interface Sci.*, 1956, **11**, 398.
7. G. L. Gaines, 'Insoluble Monolayers at Liquid-Gas Interfaces', Interscience, New York, 1966.
8. R. Motomura, A. Shibata, M. Nakamura, and R. Matuura, *J. Colloid Interface Sci.*, 1969, **29**, 623.
9. D. J. Chaiko and K. Osseo-Asare, *J. Colloid Interface Sci.*, 1988, **121**, 13.

Competitive Adsorption of Spherical Particles of Different Sizes by Molecular Dynamics

By Eddie G. Pelan and Eric Dickinson[1]

UNILEVER RESEARCH, COLWORTH HOUSE, SHARNBROOK, BEDFORD MK44 1LQ, UK
[1]PROCTER DEPARTMENT OF FOOD SCIENCE, UNIVERSITY OF LEEDS, LEEDS LS2 9JT, UK

1 Introduction

Competitive adsorption between species of different sizes is a common phenomenon in complex colloidal systems of biological and industrial importance. In the field of food colloids, for instance, the competitive adsorption of large protein molecules and small surfactant molecules at the oil–water interface is a key factor controlling the droplet-size distribution and the stability of oil-in-water emulsions.[1,2] The structure and composition of the interfacial layer is such systems is dependent not only on the relative surface activities of the competing species but also on the nature and strength of the protein–surfactant interactions.[3]

This short paper presents the results of a molecular dynamics simulation[4] of competitive adsorption at a fluid interface from a binary mixture of spherical particles of different sizes but with the same particle–surface adsorption strength. The approach is a straightforward extension of the simulation models presented previously for the formation of adsorbed layers from a single-component Lennard–Jones fluid[5] and a binary mixture of equal-sized Lennard–Jones spheres.[6] Though the simulation gives both dynamic and equilibrium information, we direct attention here to just some dynamic aspects of the adsorption process.

2 Molecular Dynamics Model and Methodology

A binary mixture of Lennard–Jones particles was confined to a cubic simulation cell having conventional periodic boundary conditions in the x, y, and z directions.[7] Pairs of like particles with centre-to-centre separation r interact with energy

$$\phi(r) = 4\epsilon\left[(\sigma_i/r)^{12} - (\sigma_i/r)^6\right] \qquad (1)$$

where ϵ is the well depth, taken to be the same for both components, and σ_i is the collision diameter of species i. The collision diameter for unlike particles is taken to be an arithmetic mean, *i.e.*

$$\sigma_{12} = (\sigma_1 + \sigma_2)/2 \qquad (2)$$

where the subscripts 1 and 2 refer to components 1 and 2, respectively. Pairs of unlike particles are characterized by the interaction potential

$$\phi(r) = 4\,\epsilon\,\xi\left[(\sigma_{12}/r)^{12} - (\sigma_{12}/r)^6\right] \qquad (3)$$

where the parameter ξ can be adjusted to allow for unlike interactions that are weaker ($\xi < 1$) or stronger ($\xi > 1$) than the norm. A value of ξ which is substantially greater than unity implies complex formation between the two molecular species, whereas a value of ξ which is substantially less than unity implies a tendency towards liquid–liquid phase separation.[8] It is convenient to express simulated quantities in terms of Lennard–Jones reduced units (denoted *): σ_1 for length (where $\sigma_1 < \sigma_2$), ϵ for energy, ϵ/k for temperature, and $N_1\sigma_1^3/V$ for density, where k is the Boltzmann constant and N_i ($i = 1,2$) is the number of particles of component i in the basic simulation cell of volume V.

An absorbed layer was created in the middle of the simulation box, located at $z = 0$, by subjecting each particle in the system to a symmetrical external potential field, perpendicular to the z direction, of the Gaussian form,

$$u(r_z) = -d_i \exp\left(-r_z^2/2w_i^2\right) \qquad (i = 1, 2) \qquad (4)$$

where r_z is the z coordinate of each particle, d_i is the maximum binding affinity (adsorption strength) of component i, and w_i is the Gaussian half-width of the external potential acting on a component i particle. This attractive potential $u(r_z)$ in eqn (4) represents the effect of a liquid–liquid interface (*e.g.* an oil–water interface) by trapping particles in a narrow band in the middle of the simulation cell parallel to the xy-plane. The physical thickness of the adsorbed layer for each component i is defined by the magnitude of parameters d_i and w_i which can be selectively controlled. Particle trajectories were calculated using a modified Verlet leapfrog algorithm.[7]

3 Numerical Results

The results presented here focus on the dynamic aspects of the creation of the adsorbed layer as a function of reduced temperature and density, binding affinity, and bulk concentration of adsorbate for a single-component Lennard–Jones fluid. Absorption of a binary Lennard-Jones mixture is also investigated as a function of binding affinity and composition.

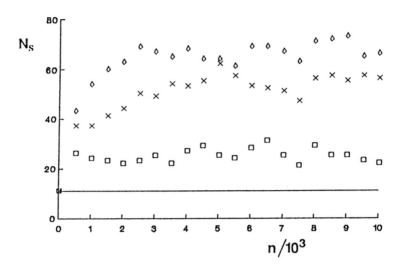

Figure 1   *Initial adsorption of a single-component Lennard–Jones fluid. Number N_s of particles in the monolayer ($|z| \leq \sigma_1/4$) against the number n of time-steps for $T^* = 1.42$, $\rho^* = 0.4$, $\xi = 1.0$: $\square$, d = 2 kT; $\times$, d = 6 kT; $\diamond$, d = 10 kT; the baseline is the number of particles present at d = 0 kT*

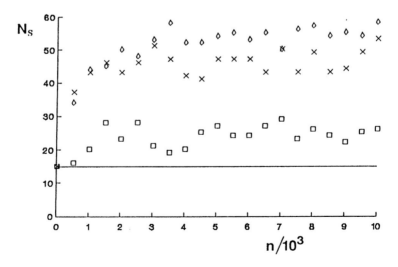

Figure 2   *Initial adsorption of a single-component Lennard–Jones fluid. Number N_s of particles in the monolayer ($|z| \leq \sigma_1/4$) against the number n of time-steps for $T^* = 1.42$, $\rho^* = 0.8$, $\xi = 1.0$: $\square$, d = 2 kT; $\times$, d = 6 kT; $\diamond$, d = 10 kT; the baseline is the number of particles present at d = 0 kT*

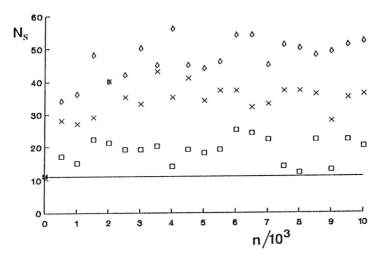

Figure 3   *Initial adsorption of a single-component Lennard–Jones fluid. Number N_s of particles in the monolayer ($|z| \leq \sigma_1/4$) against the number n of time-steps for $T^* = 2.65$, $\rho^* = 0.4$, $\xi = 1.0$: $\square$, d = 2 kT; ×, d = 6 kT; $\diamondsuit$, d = 10 kT; the baseline is the number of particles present at d = 0 kT*

Figure 1 shows the initial adsorption process for the *single-component* fluid at a reduced temperature of $T^* = 1.42$ and density $\rho^* = 0.4$ for three values of the adsorption strength, namely $d = 2\,kT$, $6\,kT$ and $10\,kT$. The plot shows the number of particles in the adsorbed layer ($|z| \leq \sigma/4$) against the number of simulation time-steps. Figure 2 shows the same system as that of Figure 1 but at a higher reduced density of $\rho^* = 0.8$, and Figure 3 shows the same system as Figure 1 but at a reduced temperature T^* of 2.65. All plots show the baseline value (defined as the number of particles in the interface at $t = 0$ and $d = 0$ kT).

The effect of particle number density on the kinetics of adsorption can be seen by comparing Figure 1 and 2. The baseline at $T^* = 1.42$ for $\rho^* = 0.4$ is $N_s \approx 11$, whereas for $\rho^* = 0.8$ it is $N_s \approx 15$, due to the different values of σ used to scale the (unit) cubic simulation volume at different densities. The effect of switching on the binding affinity (well depth) is clearly visible in the Figures: as d increases the number of particles in the monolayer increases accordingly. In the case of $d = 2\,kT$, the binding affinity is only just sufficient to increase the steady-state (plateau coverage) values of approximately 21 and 25 for $\rho^* = 0.4$ and 0.8, respectively, above their base levels. For the adsorption strengths of $d = 6$ and $10\,kT$ there is a clear indication of a rapidly rising adsorption 'knee' before a steady plateau value of interfacial coverage is obtained, the final value of which is dependent of the binding strength. Thus all systems adsorb rapidly, but the final level of coverage depends on the binding affinity used. There is also evidence to suggest that the higher values of adsorption strength require the longest times to achieve equilibrium coverage. This initial increase in the

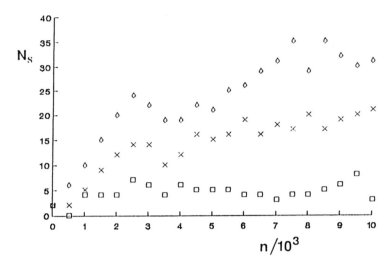

Figure 4   *Initial adsorption of a binary Lennard–Jones mixture. Number N_s of small particles in the monolayer ($|z| \leq \sigma_1/4$) against the number n of time-steps for $T^* = 2.45$, $\rho^* = 0.4$, $\xi = 1.0$, $N_1 = 100$, $N_2 = 40$: □, $d_1 = d_2 = 2$ kT; ×, $d_1 = d_2 = 10$ kT; ◊, $d_1 = d_2 = 20$ kT*

number of particles adsorbed into the layer is fastest for the high-density system (Figure 3).

The effects of overall composition ($N_1 \neq N_2$) and size ratio ($\sigma_1 \neq \sigma_2$) on the kinetics of the adsorption process for a *binary* Lennard–Jones mixture are shown in Figures 4 and 5. Figure 4 shows the number of small particles (for size ratio $S = 3$) present in the monolayer *versus* the time step for a system containing 100 small particles (*i.e.* $N_1 = 100$) and 40 large particles ($N_2 = 40$), with some ($N_{s2} \leq 7$) of the large particles adsorbed at the interface. The reduced temperature and density are $T^* = 2.45$ and $\rho^* = 0.4$ for adsorption strengths of $d_1 = d_2 = 2$, 10 and 20 kT. Figure 5 has conditions identical with Figure 4 with the exception of $N_1 = 450$. Again we see that, as d increases, the steady-state plateau value increases. Increasing the number of small particles in the bulk drives the adsorption process at a faster rate, as shown by the time taken to reach the plateau: in Figure 4 ($N_1 = 100$), it is still arguable whether the plateau has indeed been reached after 10 000 time steps, whereas for Figure 5 the plateau seems to be attained after 5000 steps. It should be noted that, during the adsorption processes presented here, the system heats up: the simulations were therefore rescaled again after equilibrium coverage was obtained to bring them back to the reduced temperature of interest. The simulations were then restarted for 10 000–40 000 steps to generate equilibrium structural and thermodynamic properties of interest which have been reported elsewhere.[4-6]

In conclusion, it can be stated that the initial adsorption process within the current model is diffusion-limited, *i.e.* the rate of adsorption only depends on

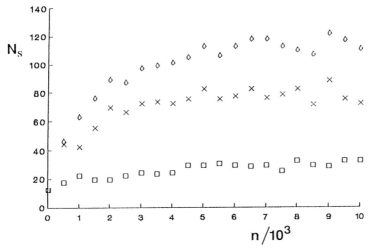

Figure 5 *Initial adsorption of a binary Lennard–Jones mixture. Number N_s of small particles in the monolayer ($|z| \leq \sigma_1/4$) against the number n of time-steps for $T^* = 2.45$, $\rho^* = 0.4$, $\xi = 1.0$, $N_1 = 450$, $N_2 = 40$: $\square$, $d_1 = d_2 = 2$ kT; $\times$, $d_1 = d_2 = 10$ kT; $\diamond$, $d_1 = d_2 = 20$ kT*

the number of particles present in the bulk (the source term in the diffusion process) and is relatively independent of the binding affinity of the adsorbate-species for the interface. However, the equilibrium level of coverage attained is clearly strongly dependent on the value of the binding affinity of the adsorbate for the interface.

References

1. B. A. Bergenståhl and P. Claesson, in 'Food Emulsions', ed. K. Larsson and S. E. Friberg, Marcel Dekker, New York, 2nd edn, 1990, p. 41.
2. E. Dickinson, *J. Chem. Soc., Faraday Trans.*, 1992, **88**, 2973.
3. E. Dickinson, S. R. Euston, and C. M. Woskett, *Prog. Colloid Polym. Sci.*, 1990, **82**, 65.
4. E. Dickinson and E .G. Pelan, *J. Chem. Soc., Faraday Trans.*, 1993, **89**, 3435.
5. E. Dickinson, and E. G. Pelan, *Mol. Phys.*, 1991, **74**, 1115.
6. E. G. Pelan and E. Dickinson, in 'Food Colloids and Polymers: Stability and Mechanical Properties', ed. E. Dickinson and P. Walstra, Special Publication No. 113, The Royal Society of Chemistry, Cambridge, UK, 1993, p. 301.
7. M. P. Allen and D. J. Tildesley, 'Computer Simulation of Liquids', Oxford University Press, 1987.
8. J. S. Rowlinson, 'Liquids and Liquid Mixtures', Butterworths, London, 2nd edn, 1969.

Protein Interactions and Functionality

Protein–Aroma Interactions

By S. Langourieux and J. Crouzet

LABORATOIRE GÉNIE BIOLOGIQUE ET SCIENCES DES ALIMENTS, UNITÉ DE MICROBIOLOGIE ET BIOCHIMIE INDUSTRIELLES ASSOCIÉE À L'INRA, UNIVERSITÉ MONTPELLIER II, 34095 MONTPELLIER CEDEX 05, FRANCE

1 Introduction

The interaction between odour compounds and proteins is a very old story The founding of the perfume industry took place in Grasse at the end of the 17th century in order to mask the disagreeable odour developed by the skin dressings used for glove manufacture.[1] The interactions of odours and perfumes with hairs and furs is well known. More recently, the necessity to develop basic research concerning the interactions between aroma compounds and proteins has been emphasized by several studies related to the difficulties encountered in the removal of off-flavours from products such as soy or fish proteins[2,3] or, on the contrary, related to the flavouring of proteinaceous foods including soy, leaf or single cell proteins.[4] However, such studies are also required for the understanding of the mechanisms implied in flavour retention during processing and storage and in the release of volatile components during food eating.[5]

In 1971 Nawar[6] pointed out the decrease of the headspace concentration of methyl ketones when gelatin was added to dilute water solutions of these compounds and stated that the interaction mechanism was not well understood. This result was supported by several authors[4,7,8] who have given evidence for flavour binding to proteins through, probably, hydrophobic interactions as indicated by the observed decrease of volatility when the chain length of aldehydes or methyl ketones increases or when a denaturation of the protein occurs. More recently, quantitative studies of protein–aroma compound interactions have been performed using different techniques.

2 Equilibrium Methods

Equilibrium methods such as gel filtration, ultrafiltration and more particularly liquid–liquid partition[9–12] or equilibrium dialysis have been used by several authors.[13–17] In this case the interaction between protein and ligand may be treated according to Scatchard:[18]

$$\frac{\bar{\nu}}{[L]} = nK - \bar{\nu}K \qquad (1)$$

where $\bar{\nu}$ is the number of moles of ligand bound per mole of protein, $[L]$ is the molar concentration of free ligand, n is the number of binding sites, and K is the intrinsic binding constant.

A plot of $\bar{\nu}/[L]$ *versus* $\bar{\nu}$ (Figure 1) allows the determination of n and K. The double reciprocal equation (Klotz plot) may be also used:

$$\frac{1}{\bar{\nu}} = \frac{1}{n} + \frac{1}{nK[L]} \qquad (2)$$

The values obtained for 2-nonanone and several proteins are given in Table 1.

The strength of the interactions are dependent on the structure of the aroma compounds. In a homologous series the intrinsic binding constant and the free energy of association increase with the chain length.[10,14,16,19] In the case of the interactions between 2-alkanones and soy protein (Table 2), the binding constant increases three-fold for each methylene group increment in the chain.

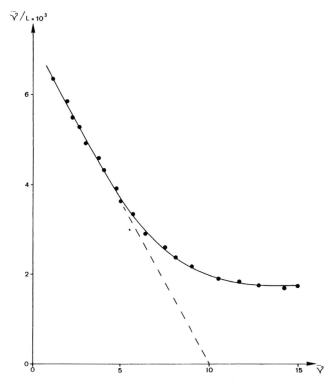

Figure 1   *Scatchard plot of the interaction between limonene and bovine serum albumin determined by liquid–liquid equilibrium (see text for definitions of symbols)*

An increase of this amplitude has been calculated from theoretical considerations.[20-21] These results are in agreement with previous semi-qualitative data, suggesting the hydrophobic nature of the binding.

Table 1 *Binding characteristics of 2-nonanone to proteins at 25 °C*

Protein	n^a	K/M^{-1} ^{b}	Reference
Soy protein	4	930	14
BSA	6	1.8×10^3	10
BSA	6	1.8×10^3	12
Actomyosin	13	386	3
β-Lactoglobulin	1	2440	16

^{a} Number of protein binding sites.
^{b} Intrinsic binding constant.

Table 2 *Thermodynamic constants (K and ΔG) for the binding of methyl ketones to soy protein and β-lactoglobulin at 25 °C[14,16]*

Ligand	Protein	K/M^{-1}	$\Delta G/\text{kcal mol}^{-1}$
2-Heptanone	Soy protein	110	−2.78
2-Octanone	Soy protein	310	−3.40
2-Nonanone	Soy protein	930	−4.05
2-Heptanone	β-Lactoglobulin	150	−2.98
2-Octanone	β-Lactoglobulin	480	−2.66
2-Nonanone	β-Lactoglobulin	2440	−4.62

The binding constant is influenced by the presence of polarizable functional groups such as carbonyls; a linear decrease of the binding constant is noticed when the carbonyl is shifted from the end (nonanal) to the centre of the chain (5-nonanone). A steric hindrance is suggested by the authors,[14] although a decrease of the hydrophobicity of the aliphatic chain cannot be rejected.

The possibility of non-reversible binding of a functionalized aroma compound to a protein is suggested by several authors. Dhont[22] supposes that the aldehyde group of vanillin reacts with amino groups of albumin to produce Schiff's base. According to Dumont[23,24] the interaction between vanillin and bovine serum albumin (BSA) is probably a complex phenomenon involving a slow non-reversible fixation of the aroma compound side by side with a rapid equilibrated reaction. The binding is naturally dependent on the protein conformation and all the factors modifying this conformation are able to affect the fixation of aroma compounds. The partial denaturation of soy protein (1 h at 90 °C) increases the affinity of the protein for 2-nonanone but the site number remains constant. The oligomeric structure may be considered as unmodified, and the observed effect is the consequence of an increase of the hydrophobicity of pre-existent sites.[14] The affinity constants of soy protein and β-lactoglobulin for 2-nonanone decrease when increasing concentrations of urea are used.[15,17] In the case of soy protein the estimation of the number of sites is uncertain whereas in the case of β-lactoglobulin this number remains constant. For this protein the treatment induces an unfolding and a reduction of the hydrophobic region correlated to the decrease of the affinity constant.

Additional binding sites may be revealed by the conformational changes induced by the fixation of aroma compounds on the initial binding sites as indicated by the non-linear Scatchard plot obtained.[10,14] The binding is strongly affected by the presence of lipids: when limonene is used as ligand the site number decreases from 10 to 6 for BSA and from 3 to 2 for caseinate. These results are in good agreement with previously reported data.[4,19] The interactions are also dependent on the characteristics of the medium: temperature,[10,14,15,19] pH and ionic strength,[11,13,15,16,19] as well as chemical modifications of the protein.[14,21,25]

Several limitations are associated with the use of equilibrium methods. The equilibrium between the macromolecules and the aroma compounds is reached after a long time which is generally difficult to determine. During this time degradations of some aroma compounds—oxidation of aldehydes for example—may occur. However, care must be taken concerning the non-specific interactions between aroma compounds and a membrane, the possible coating of the membrane by protein, and the losses of aroma compounds by volatilization.[26] These phenomena have been studied for two flavour compounds: β-ionone and limonene, by dialysis using a Spectrapor membrane without the presence of any receptor. With β-ionone, the equilibrium was reached after only 100 min, but 15–20% of the compound was lost probably by volatilization, since the quantity of β-ionone extracted from the membrane was negligible (Figure 2). In the case of limonene, the losses were large, more than 75%, and were the result of interaction with the membrane and of volatilization. When the liquid–liquid partition method is used, the main

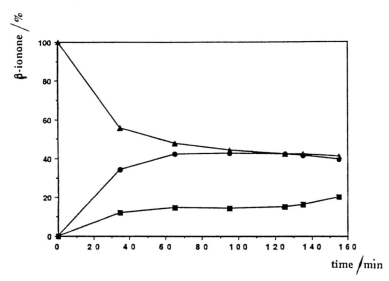

Figure 2 *Kinetics of β-ionone transfer through the dialysis membrane:* ▲, *right half-cell;* ●, *left half-cell;* ■, *losses*

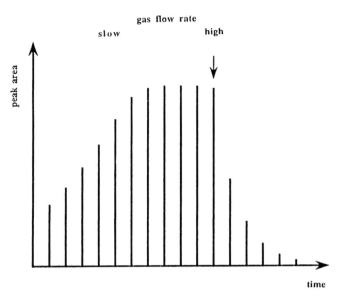

Figure 3   *Variation of the concentration of the solute in the head space as a function of time and gas flow rate*

problems encountered are protein denaturation at the solvent interface and the difficulty of obtaining a quantitative extraction of aroma compounds from the aqueous solution.[2,26]

3 Dynamic Methods

Some of the above disadvantages may be overcome by using dynamic methods such as exponential dilution and dynamic coupled column liquid chromatography. More particularly, the equilibrium may be obtained in a very short time, so that the degradation problems are limited, the transfer and extraction steps are not necessary, and non-specific interactions may be avoided.

Exponential Dilution

When a dilute, aqueous solution of a volatile compound contained in an equilibrium cell is stripped by an inert gas, the concentration of the solute in the gas phase varies as indicated in Figure 3.[25,27–29] Firstly the concentration increases with time, and then a plateau is reached with a constant concentration in the gas phase indicative of the equilibrium. This stage is followed by a decrease of the concentration according to an exponential law. At high gas flow rate, from 30 to 100 ml min[−1], according to the nature of the volatile compound, only the exponential decrease with a short equilibrium phase is observed.

The concentration decrease is related to the infinite dilution activity coefficient γ_i^∞ by the expression

$$\log S = \log S_0 + \frac{D}{RT}\frac{p_i^s}{N}\gamma_i^\infty\, t \qquad (3)$$

where S is the GLC peak area, S_0 is the GLC peak area extrapolated to zero time, D is the carrier gas flow rate (ml min^{-1}), N is the number of moles of solvent, R is the gas constant, T is the temperature, p_i^s is the vapour pressure of the pure solute, and t is the time. The value of γ_i^∞ is calculated from the slope of the straight line obtained by plotting log S *versus* time.

When interactions between volatile compounds and macromolecular systems occur in aqueous solution, the infinite dilution activity coefficient is modified. A decrease is indicative of the fixation of the solute whereas an increase implies a salting-out effect. A reduced activity coefficient,

$$\gamma_{ir}^\infty = \frac{\gamma_{im}^\infty}{\gamma_i^\infty} \qquad (4)$$

where γ_{im}^∞ is the value of γ_i^∞ in the presence of the macromolecular system, may be introduced. This value gives an indication of the nature and strength of the interactions.[12,25,30,31]

The plots giving the variations in γ_{ir}^∞ for several volatile compounds (limonene, linalool, terpenyl acetate and β-ionone) for increasing concentrations

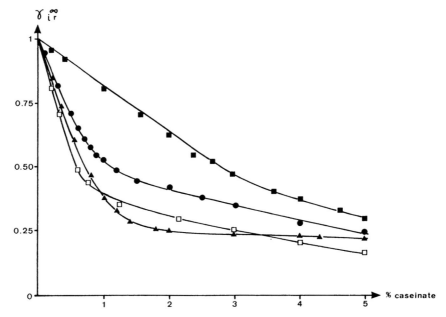

Figure 4   *Variations of the limonene (▲), linalool (●), terpenyl acetate (□) and β-ionone (■) reduced infinite dilution activity coefficient as a function of caseinate concentration*

of caseinate are given in Figure 4. For all the compounds the observed decrease of γ_{ir}^{∞} is indicative of interactions between the ligands and the protein. The shape of the curves are similar, with a rapid decrease to 1.5–2% of caseinate (3% for β-ionone) followed by a slow decrease. The non-linear shape of the curves, similar to that observed when equilibrium techniques are used for the study of interactions between limonene and caseinate or between methyl ketones and BSA,[10] is indicative of co-operative binding behaviour. Information concerning the intensity of the interaction may be given by the decrease of γ_{ir}^{∞} as reported in Table 3. The decrease is more important for the most hydrophobic compounds like limonene and terpenyl acetate than for functionalized compounds like linalool and β-ionone. This point is indicative of the hydrophobic nature of interactions. However, the delipidation of caseinate leads to, as previously stated when a liquid–liquid equilibrium was used, a decrease of the retention of limonene. A faint interaction between limonene and a peptide (trypsin inhibitor) has also been detected.[32]

Table 3   *Values of γ_i^{∞} and change of γ_i^{∞} for volatile components in the presence of sodium caseinate (0.5 wt%) compared with water*

Volatile component	γ_i^{∞}	$\Delta\gamma_i^{\infty}/\%$
Terpenyl acetate	41.16×10^4	39
Limonene	77.7×10^3	33
Linalool	25.95×10^3	26
β-Ionone	1.21×10^6	18

Dynamic Coupled Column Liquid Chromatography (DCCLC)

This method is derived from the Hummel and Dreyer gel filtration technique.[33] After equilibration of a gel filtration column with a ligand solution, an aliquot of a protein–ligand complex produced by dissolution of the protein in a ligand solution is introduced onto the column, which is eluted with the ligand solution, and the concentration of this compound in the eluate is measured. Two peaks are present on the chromatogram: a positive peak corresponding to the protein–ligand complex eluted at the same retention time as the protein, and a negative peak corresponding to the ligand bound to the protein eluted at the retention time of the ligand. This method, which needs a precise standardization, has been transposed in different forms to HPLC.[34–37]

DCCLC was initially introduced by May *et al.*[38,39] for determination of the aqueous solubility of hydrophobic compounds (polycyclic aromatic hydrocarbons, PAH). This method is based on pumping water through a column ('generator column') containing glass beads coated with the aromatic compounds. The concentration of these compounds is determined by HPLC on a C_{18} reverse phase ('analytical') column either directly or after extraction on a short C_{18} ('extractor') column. This method, which avoids several problems such as stability of solutions and losses of compounds to surfaces due to adsorption, allows the determination of the solubility of hydrophobic compounds with a precision and a reproducibility better than ± 3%.

DCCLC has been used for the study of cyclodextrin inclusion complexes and more particularly for the determination of the formation constants.[40] When an aqueous solution of cyclodextrin is used instead of water for the flowing of the generator column, an increase of the solubility of PAH as a result of complex formation is noticed. The formation constant for the complex corresponds to the equilibrium:

$$C_y + P \rightleftharpoons C_yP \tag{5}$$

The formation constant,

$$K_f = \frac{[C_yP]}{[C_y][P]} \tag{6}$$

may be calculated assuming that the cyclodextrin concentration is not depleted to an appreciate extent by complexation (less than 0.1%):

$$K_f = \frac{S_t - S_0}{[C_y]_t S_0} \tag{7}$$

Here S_0 is the solubility in water, S_t is the solubility in the aqueous solution of cyclodextrin, and $[C_y]_t$ is the initial cyclodextrin concentration.

This method was adapted[32] for the determination of the solubility and the study of interactions between some aroma compounds and some biological

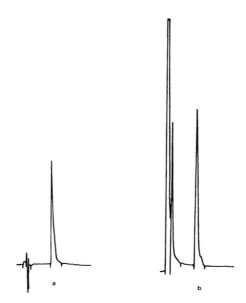

Figure 5   *HPLC determination of limonene using DCCLC: (a) generator column with water; (b) generator column with a 0.5 wt% soybean trypsin inhibitor solution [HPLC elution solvent acetonitrile–water (80:20) at 1 ml min^{-1}]*

molecules: cyclodextrin, dextrin, and soybean trypsin inhibitor (2×10^4 Da). It has been shown that the concentration of limonene and β-ionone eluted from the generator column is constant for sufficiently high water volumes to allow the determination of the solubility by HPLC. When the limonene generator column is eluted with a soybean trypsin inhibitor solution, an increase of the linonene area, as previously indicated, is noticed (Figure 5). The HPLC elution solvent is acetonitrile–water (80:20) used at 1 ml min^{-1}. (It has been demonstrated that in this non-food solvent the limonene has no significantly different behaviour from in food systems.) For a given peptide content the limonene concentration measured by HPLC is found to be constant and independent of the volume of solution flowing through the column. Moreover, when water is substituted by the soybean trypsin inhibitor solution the limonene water solubility is recovered so that the binding of limonene to the peptide may be considered as reversible.

Table 4 *Binding constant of limonene (M^{-1}) as a function of soybean trypsin inhibitor percentage for different elution volumes*

Volume of eluent/ml	% of soybean trypsin inhibitor	
	0.25	0.50
10	2657	2107
50	3117	2698
70	3143	2354
100	2802	–

The values obtained for the binding constant of limonene to trypsin inhibitor for different percentages of the peptide (Table 4) are independent of the peptide content of the solution. This shows that neither limonene nor peptide limiting concentrations are involved. The values obtained for the binding constant for limonene and β-ionone are respectively 2944 ± 784 M^{-1} and 385 ± 105 M^{-1}. These values are indicative of the presence of weak interactions between β-ionone and soybean trypsin inhibitor.

The two dynamic methods tested may be used for the study of the interactions of aroma compounds and globular proteins of well-known conformation. The study of interactions of selected aroma compounds (β-ionone, limonene, linalool, ethyl hexanoate, isoamyl acetate) with β-lactoglobulin is presently in progress. This major whey protein exists as an oligomeric form and as a monomer (18 000 Da) and possesses a well defined conformation[41,42] which is characterized by the presence of anti-parallel β-sheets and of an α-helix chain determining a hydrophobic pocket involved in the binding of several hydrophobic ligands such as retinol,[43] β-ionone,[44] methyl ketones,[16,17] aromatic hydrocarbons,[45] and fatty acids.[46,47] It is necessary to be aware of the limits of the dynamic methods. Only qualitative or semi-quantitative data may be obtained by exponential dilution, but, according to the rapidity of the method, screening of molecules (proteins and aroma compounds) may be performed. Other problems are associated with the solubility of the protein and aroma compounds in the same solvent for HPLC elution, and the UV detection of some aroma compounds.

Acknowledgement

This work was supported by the Ministère de la Recherche et de la Technologie, Grant 90 G 0325, and S.L. received financial support from this source for her thesis.

References

1. Y. Naves and G. Mazoyer, 'Les Parfums Naturels. Essences concrètes. Resinoïdes, Huiles et Pommades', Gauthier Willars, Paris, 1934, p. 22.
2. J. E. Kinsella and S. Damodaran, 'The Analysis and Control of Less Desirable Flavors in Foods and Beverages', Academic Press, New York, 1980, p. 95.
3. S. Damodaran and J. E. Kinsella, *J. Agric. Food Chem.*, 1983, **31**, 856.
4. K. L. F. Franzen and J. E. Kinsella, *J. Agric. Food Chem.*, 1974, **22**, 675.
5. S. M. Van Ruth, J. P. Roozen, and J. L. Cozysen, in 'Trends in Flavour Research', Elsevier, Amsterdam, 1994, p. 59.
6. W. W. Nawar, *J. Agric. Food Chem.*, 1971, **19**, 1057.
7. M. A. Gremli, *J. Am. Oil Chem. Soc.*, 1974, **51**, 95A.
8. S. Arai, M. Noguchi, M. Koji, H. Sako, and M. Fujimaki, *Agric. Biol. Chem.*, 1970, **34**, 1420
9. A. A. Spector, J. Kathryn, and J. E. Fletcher, *J. Lipid Res.*, 1969, **10**, 56.
10. S. Damodaran and J. E. Kinsella, *J. Agric. Food Chem.*, 1980, **28**, 567.
11. S. Damodaran and J. E. Kinsella, *J. Biol. Chem.*, 1980, **255**, 8503.
12. A. Sadafian and J. Crouzet, in 'Progress in Terpene Chemistry', Editions Frontières, Gif-sur-Yvette, France, 1986, p. 165 (in French).
13. M. Beyeler and J. Solms, *Lebensm.-Wiss. Technol.*, 1974, **7**, 217.
14. S. Damodaran and J. E. Kinsella, *J. Agric. Food Chem.*, 1981, **29**, 1249.
15. S. Damodaran and J. E. Kinsella, *J. Agric. Food Chem.*, 1981, **29**, 1253.
16. T. E. O'Neill and J. E. Kinsella, *J. Agric. Food Chem.*, 1987, **35**, 770.
17. T. E. O'Neill and J. E. Kinsella, *J. Food Sci.*, 1988, **53**, 906.
18. G. Scatchard, *Ann. NY Acad. Sci.*, 1949, **51**, 660.
19. O. E. Mills and J. Solms, *Lebensm.-Wiss. Technol.*, 1984, **17**, 331.
20. A. Wishnia, *Proc. Natl. Acad. Sci. USA*, 1962, **48**, 2200.
21. M. H. Abraham, *J. Am. Chem. Soc.*, 1980, **102**, 5910.
22. J. H. Dhont, Proc. Int. Symp. Aroma Research, Pudoc, Wageningen, The Netherlands, 1975, p. 193.
23. J. P. Dumont, 'Flavour Science and Technology', John Wiley, Chichester, 1987, p. 143.
24. J. P. Dumont, unpublished results.
25. F. Sorrentino, A. Voilley, and D. Richon, *AIChE J.*, 1986, **32**, 1988.
26. L. A. Wilson, 'Physical and Chemical Properties of Food', American Society of Agricultural Engineers, St. Joseph, USA, 1986, p. 382
27. J. C. Leroi, J. C. Masson, H. Renon, J. F. Fabries, and H. Sannier, *Ind. Eng. Chem., Process. Des. Dev.*, 1977, **16**, 135.

28. P. Duhem, Contribution à l'étude des mélanges 'eau-hydrocarbures-solvant polaire'. Représentation des équilibres de phases par le modèle NRTL, Thèse, Université de Marseille, 1979.
29. A. Sadafian and J. Crouzet, *Flavour Fragrance J.*, 1987, **2**, 103.
30. A. Lebert and D. Richon, *J. Food Sci.*, 1984, **49**, 1301.
31. A. Lebert and D. Richon, *J. Agric. Food Chem.*, 1984, **32**, 1151.
32. S. Langourieux, 'Interactions Ligand-Récepteur: Cas des Composés d'Arôme en Solutions Aqueuses', Thèse, Université de Montpellier II, 1993.
33. J. P. Hummel and W. J. Dreyer, *Biochim. Biophys. Acta*, 1962, **62**, 530.
34. B. Sebille, N. Thuaud, and J. P. Tillement, *J. Chromatogr.*, 1978, **169**, 159.
35. B. Sebille, N. Thuaud, and J. P. Tillement, *J. Chromatogr.*, 1979, **180**, 103.
36. B. Sebille, N. Thuaud, and J. P. Tillement, *J. Chromatogr.*, 1981, **204**, 285.
37. B. Sebille, N. Thuaud, J. Piquion, and N. Behar, *J. Chromatogr.*, 1987, **409**, 61.
38. W. E. May, S. P. Wasik, and D. H. Freeman, *Anal. Chem.*, 1978, **50**, 175.
39. W. E. May, S. P. Wasik, and D. H. Freeman, *Anal. Chem.*, 1978, **50**, 997.
40. L. A. Blishak, K. Y. Dodson, G. Patonay, I. M. Warner, and W. E. May, *Anal. Chem.*, 1989, **61**, 955.
41. M. Z. Papiz, L. Sawyer, E. E. Eliopoulos, A. C. T. North, J. B. C. Findlay, R. Sivaprasadaroa, T. A. Jones, N. E. Newcomer, and P. J. Kraulis, *Nature (London)*, 1986, **324**, 383.
42. H. L. Monaco, G. Zanotti, P. Spadon, M. Bolognosi, L. Sawyer, and E. E. Eliopoulos, *J. Mol. Biol.*, 1987, **197**, 695.
43. S. Futterman and J. Heller, *J. Biol. Chem.*, 1972, **247**, 5168.
44. E. Dufour and T. Haertle, *J. Agric. Food Chem.*, 1990, **38**, 1961.
45. H. M. Farrell, M. J. Behe, and J. A. Enyart, *J. Dairy Sci.*, 1987, **70**, 252.
46. E. D. Brown, *J. Dairy Sci.*, 1984, **67**, 713.
47. M. C. Diaz de Villegas, R. Oria, F. G. Sala, and M. Calvo, *Milchwissenschaft*, 1987, **42**, 357.

Some Changes to the Properties of Milk Protein Caused by High-Pressure Treatment

By Donald E. Johnston[1] and Robert J. Murphy

FOOD AND AGRICULTURAL CHEMISTRY RESEARCH DIVISION, DEPART-
MENT OF AGRICULTURE FOR NORTHERN IRELAND, NEWFORGE LANE,
BELFAST BT9 5PX, UK
[1]DEPARTMENT OF FOOD SCIENCE, THE QUEEN'S UNIVERSITY, NEWFORGE
LANE, BELFAST BT9 5PX, UK

1 Introduction

The use of high pressure as a means of bringing about change in food may appear at first sight to be a strange concept but it follows simply from the principle of Le Chatelier. When a system is subjected to a constraint then the system will adapt to remove the constraint. When applying high pressure the constraint is volume. Thus reactions with a negative volume change $(-\Delta V)$ will be enhanced and those involving a positive volume change $(+\Delta V)$ will be suppressed. There will also be an effect on the kinetics of reactions, depending upon the size and sign of the activation volume ΔV^*. A reaction with a negative activation volume will be accelerated by pressure application whereas a reaction with a positive activation volume will be retarded.

Conventional food processing by thermal treatment will produce different responses depending upon the free energy change ΔG or activation free energy ΔG^* of the different individual reactions which may take place in the food. Hence some reactions can be selectively carried out while others cannot. As values of ΔV and ΔV^* do not necessarily follow values of ΔG and ΔG^*, high-pressure technology permits an entirely different range of selective changes to be brought about in the food. An interesting example of the difference in selectivity between thermal processing and high pressure is provided by the egg. An egg subjected to a pressure of 800 MPa for 3 min sets solid, just like a hard boiled egg, yet when removed from its shell and eaten tastes just like a raw egg. In addition, differential effects arise in thermal processing of foods due to the temperature difference between the exterior and the interior as heat is transferred by conduction. This can lead to over-processing the exterior in

134

order to give adequate processing in the centre. By contrast, with high-pressure processing, the effect is felt uniformly throughout the food.

The pressure range currently being investigated for use in food processing is roughly 100 MPa to 1 GPa. The areas where high pressure is felt to offer interesting potential are: the reduction of microbial numbers, the control of enzyme reactions, alteration to the conformation of biopolymers, and the control of phase transformations.[1] Reports of findings in all these areas are increasing in the literature.

There has been considerable commercial high-pressure research and development activity in Japan over the last 5 years and as a result a number of high-pressure processed products are already available on retail sale, including a range of low-sugar jams, fruit sauces, and desserts, grapefruit juice and mandarin juice. A number of other products are at various stages of development. All of these products retain a remarkable degree of fresh flavour.

The potential of high-pressure processing in the dairy industry is still relatively unexplored. The aim of our work is to identify changes produced by high-pressure treatment which offer development potential.

2 Materials and Methods

Fresh skim milk samples were obtained daily from a local dairy. Milk samples were placed in polyester screw-capped soft drinks bottles (volume 250 ml) leaving no headspace and sealed. Pressure treatment was carried out in a high-pressure apparatus designed and constructed in the Mechanical and Manufacturing Engineering Department of the University.

The lightness value L^* of milk samples was measured in a 15 mm cell with a Pye Unicam SP 8-200 spectrophotometer with a diffuse reflectance accessory. Readings at 10 nm intervals between 400 and 700 nm were used to calculate L^* by the weighted ordinate method for illuminant D_{65}.[2]

The exposure of surface hydrophobic sites on milk proteins was assessed using the method of Bonomi *et al.*[3] for the fluorescent probe 8-anilino-1-naphthalenesulfonic acid (ANS) with samples at 1:20 dilution.

Gels were prepared by adding sufficient glucono-δ-lactone to the milk to give a final pH of 4.1 and allowing them to set undisturbed at 20 °C. To provide a stirred-yoghurt-type product, gel samples were passed through a stainless-steel sieve (mesh 1 mm).

Apparent viscosities of the stirred-type yoghurt were measured at 5 °C using a Brookfield LVF viscometer fitted with a No. 3 spindle operating at 30 r.p.m.

Sulfhydryl groups and disulfide linkages were measured using Ellman's reagent and the method of Beveridge *et al.*[4] The free sulfhydryl groups were measured using the method of Donovan and Mulvihill.[5]

3 Changes to Skim Milk

After pressure treatment skim milk becomes noticeably less able to scatter light (Table 1).[6] The effect seems to reach a plateau and changes little at pressures

Table 1 *Effects of high-pressure treatment for 1 h on the appearance and protein fractions of milk[a]*

	Control	200 MPa	300 MPa	400 MPa	500 MPa	600 MPa
Lightness (L^*)	85.6	79.5	67.3	66.6	65.4	66.4
Serum N (%) after centrifugation	0.127	0.118	0.106	0.099	0.089	0.086
Non-casein N (%)	0.113	0.101	0.080	0.075	0.067	0.063

[a] Data from reference 6 with permission.

beyond 300 MPa. Casein, the main protein present in milk, occurs in large supramolecular structures called casein micelles, consisting of many molecules of the different individual caseins together with calcium phosphate, which plays an important part in binding the structure together. High hydrostatic pressure is known to have a disruptive effect on hydrogen bonds, ionic interactions and hydrophobic forces due to the volume changes involved in their formation. The change in these non-covalent forces under high pressure allows casein micelle fragments, individual caseins and calcium phosphate to move independently of each other, away from their original casein micelles. When the pressure is released the casein micelles are unable to regain their original structure due to the altered distribution of their components. The decreases in lightness value L^* which we have found, are consistent with micelle fragmentation. The effect on L^* is not as great as can be achieved by adding Ca^{2+} chelating agents to milk, when almost all the casein is rendered soluble.[7] Schmidt and Buchheim have used electron microscopy to examine pressure-treated milk and have concluded that fragmentation, rather than disintegration to individual caseins, is produced by high-pressure treatment.[8] More recently, work in Japan using electron microscopy in conjunction with particle size distribution analysis has also supported this interpretation.[9] The visual effect is lost when the skim milk samples are allowed to form an acid-set gel. Uniformly mixed whole milk samples subjected to high pressure are indistinguishable from unpressurized controls, due to the scattering effect of the fat globules, but when allowed to stand to form a cream layer the effect is seen in the lower layer.

The changes to the milk protein, which is non-sedimentable by centrifugation at 70 000 g, and to the protein soluble at pH 4.6 (Table 1) are suggestive

Table 2 *Effects of duration of high-pressure treatment (400 MPa) on sulfur-containing groups in milk*

	Control	5 min	15 min	30 min	60 min
–SH groups (μM –SH/g dry weight)	1.69	0.60	0.25	0.10	0.06
Total –SH and –S–S–groups expressed as –SH (μM –SH/g dry weight)	21.0	20.6	20.6	20.8	20.8

of changes to the whey protein or interaction between the whey protein and the casein. The –SH groups in milk are present mainly in the whey protein β-lactoglobulin. If the protein is undenatured they are concealed but they can become free and reactive when milk is heated, especially at temperatures above 70 °C.[10] The –SH groups in milk are found to be readily depleted by high-pressure treatment (Table 2). The extent of this decrease is much larger than the decrease caused by heating skim milk.[11] However, by contrast to heated milk, none of the pressure-treated samples display any measurable increase in free –SH groups. The total –SH plus –S–S– groups in pressure-treated milk remains relatively unchanged. It would appear on this basis that the –SH groups are reacting to form –S–S– links. Electrophoresis of the milk protein precipitable at pH 4.6 shows the presence of immobile high-molecular-weight material, which is absent when the electrophoresis is carried out in the presence of mercaptoethanol, giving simultaneously an increase in the intensity of the β-lactoglobulin band. This behaviour is similar to the reported effects of heat treatment.[11]

Fluorescent probes provide a useful means of investigating the unfolding of proteins. The exposure of surface hydrophobic groups on the protein in pressure-treated milk has been measured by binding of the fluorescent probe ANS (Figure 1). Hydrophobic group exposure increased steadily with processing time at 200 MPa, but at higher pressures exposure increases rapidly for up to 30 min of processing time, and extending the treatment beyond this

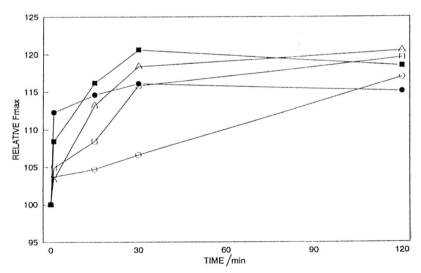

Figure 1   *The effect of duration of pressure treatment on exposure of hydrophobic regions of milk proteins, measured at 1:20 dilution by binding of the fluorescent probe ANS. Relative fluorescence maxima (F_{max}) (control = 100) were determined by the method of Bonomi et al.:[3] ○, 200 MPa; □, 300 MPa; △, 400 MPa; ●, 500 MPa; ■, 600 MPa* (Redrawn from reference 6 with permission)

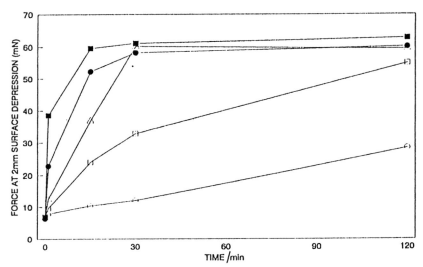

Figure 2   *Gel rigidity measured as force on a 9.9 mm diameter cylindrical probe at 2 mm gel surface depression for acid-set milk gels (pH 4.1) prepared from pressure-treated milks:* ○, *200 MPa;* □, *300 MPa;* △, *400 MPa;* ●, *500 MPa;* ■, *600 MPa*
(Redrawn from reference 13 with permission)

time gives little further increase.[6] The increased exposure could originate from the fragmentation of the micelles and from the unfolding of the protein chains. The relative viscosity of skim milk, measured at 5 °C, increases from a control value of 1.80 as a result of pressure treatment. This would be expected from the micelle fragmentation, but the relative viscosity reaches a plateau value of approx. 2.10 at 30 min treatment time at 200 MPa, and at even shorter times at higher pressures. However, the extent of surface hydrophobic region exposure as shown in Figure 1 continues to increase beyond the point where the viscosity increase has ceased. This would seem to suggest that the unfolding and protein conformation change which is taking place under these treatment conditions is not altering the hydrodynamic properties of the micelle fragments.

Overall there is ample evidence of a variety of changes to milk proteins being caused by high-pressure treatment. Some of the effects are similar to heat-induced alterations and some are not.

4 Acid-Set Gels Prepared from Pressurized Skim Milk

Given the importance of hydrophobic bonding in acid-set gel formation in dairy products,[12] it would be expected that the increased exposure of hydrophobic regions in pressurized milk would provide increased opportunities for favourable protein–protein interactions and hence would give rise to

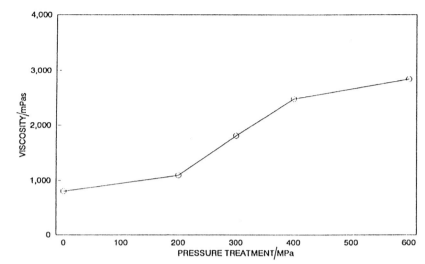

Figure 3 *Viscosity of stirred-type yoghurt (pH 4.1) prepared from skim milk which has been pressure-treated for 15 min. Viscosities were measured at 5 °C using a Brookfield LVF viscometer fitted with a No. 3 spindle operating at 30 r.p.m. after 3 min*

yoghurts with improved properties. Dramatic increases of eight- to nine-fold in gel rigidity can be achieved as shown by the force required to depress the surface by 2 mm (Figure 2), with the effects following a qualitatively similar pattern to the hydrophobic region exposure. It will be recalled that hydrophobic region exposure continues to increase beyond the point where the skim milk relative viscosity has been maximized; so it would appear that protein unfolding is necessary to obtain the maximum improvement in gel properties.

For gels converted to stirred-type yoghurt, viscosity increases of 3.5-fold could be achieved (Figure 3). The state of hydration of the protein in the acid-set gel, measured either as water-holding capacity or protein hydration index, is increased by the pressure treatment[13] but not to the same extent as the increase in gel rigidity or in viscosity of the stirred-type product.

Although protein unfolding, and the changes which accompany it, contribute to the improvement of gel properties, other aspects of gel formation need to be considered also. Measurements taken on electron micrographs of the gels show an increased number of strands in the network.[13] This would imply changes to the process of gel formation after the gel formation point. The mechanism of acid-set gel formation involves a gradual disintegration of the casein micelle structure as the pH drops, releasing calcium phosphate and individual caseins, followed by reaggregation and gel network formation.[14] The reason why the changes to the milk brought about by high-pressure treatment should interact with the acid-set gel-forming mechanism to give such an improvement in gel properties is not yet clear. Controlled denaturation of whey proteins by heat treatment[11] and disintegration of the casein micelle before acidification[7] are both known to give gel improvement, and it is likely that in the case of pressure-treated milks these effects will contribute to the overall improvement. Establishing their relative contributions and the role of increased exposure of surface hydrophobic regions on the protein will require further experimentation.

It is unlikely that high-pressure treatment of milk could compete with conventional pasteurization or u.h.t. as a preservation treatment for large volumes of liquid milk in the short-term future. Apart from the engineering problems which would need to be solved, the changes to legislation would require considerable time. A more likely prospect is the exploitation of a property uniquely provided by high-pressure treatment either to improve an existing product or to create an entirely new one. The changes which we have found to date illustrate the highly interesting effects which can be achieved by high-pressure treatment and the potential for future development.

References

1. D. Farr, *Trends Food Sci. Technol.*, 1990, **1**, 14.
2. D. B. Judd and G. Wyszecki, 'Color in Business, Science and Industry', Wiley, New York, 1975, p. 91.
3. F. Bonomi, S. Iametti, E. Pagliarini, and C. Peri, *Milchwissenschaft*, 1988, **43**, 281.
4. T. Beveridge, S. J. Toma, and S. Nakai, *J. Food Sci.*, 1974, **39**, 49.
5. M. Donovan and D. M. Mulvihill, *Ir. J. Food Sci. Technol.*, 1987, **11**, 87.
6. D. E. Johnston, B. A. Austin, and R. J. Murphy, *Milchwissenschaft*, 1992, **47**, 760.
7. D. E. Johnston and R. J. Murphy, *J. Dairy Res.*, 1992, **59**, 197.
8. D. G. Schmidt and W. Buchheim, *Milchwissenschaft*, 1970, **25**, 596.
9. Y. Shibauchi, H. Yamamoto, and Y. Sagara, in 'High Pressure and Biotechnology', ed. C. Balny, R. Hayashi, K. Heremans, and P. Masson, Colloque INSERM/John Libbey Eurotext, Montrouge,1992, p. 239.
10. R. J. L. Lyster, *J. Dairy Res.*, 1964, **31**, 41.
11. E. Parnell-Clunies, Y. Kakuda, D. Irvine, and K. Mullen, *J. Dairy Sci.*, 1988, **71**, 1472.
12. N. A. Bringe and J. E. Kinsella, in 'Developments in Food Proteins', ed. B. J. F. Hudson, Applied Science, London, 1987, vol. 5, p. 159.
13. D. E. Johnston, B. A. Austin, and R. J. Murphy, *Milchwissenschaft*, 1993, **48**, 206.
14. I. Heertje, J. Visser, and P. Smits, *Food Microstruct.*, 1985, **4**, 267.

Surface Energy at the Ice–Solution Interface for Systems Containing Antifreeze Biopolymers

By David S. Reid, William L. Kerr, June Zhao, and Yuki Wada

DEPARTMENT OF FOOD SCIENCE AND TECHNOLOGY, UNIVERSITY OF CALIFORNIA, DAVIS, CA 95616, USA

1 Introduction

The blood of a variety of fishes from polar oceans has been shown to contain antifreeze glycoproteins (AFGP) or antifreeze proteins (AFP).[1-5] These produce a non-colligative lowering of the freezing temperature of the body fluids. A range of compounds with similar effect can be found in organisms besides fish,[5] and also more recently attempts have been made to synthesize materials with the same properties.[7] There have been many studies on the properties of these biopolymers. The mode of action of the antifreeze biopolymers has been hypothesized to be through a surface adsorption, which interferes with the growth of the ice crystals.[8-12] That being the case, it would be expected that this interaction would manifest itself in other changes in properties. Surface adsorption would also be expected to influence the surface energy at the interface between ice and a solution, and therefore it should be a useful monitor of potential antifreeze activity, and also a useful measure of the extent and type of interaction.

The surface energy at a solid–liquid interface can readily be estimated using a temperature gradient cell by analysing the grain boundary curvature between two crystals in a known temperature gradient. A technique for measuring grain boundary curvatures in a thermal gradient has been developed and described by Jones and Chadwick,[13] and the appropriate method for mathematical analysis has been suggested by Bolling and Tiller.[14] In this work we have constructed a cell based on the design suggested by Jones and Chadwick. By using video techniques, the image of the interface shape can be captured and analysed according to the Bolling and Tiller procedure to yield a value for the surface energy. A variety of interesting materials has been studied. We have previously determined the interfacial energy at ice interfaces in contact with solutions of antifreeze glycoproteins in both active and inactive

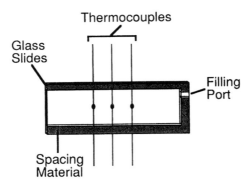

Figure 1   *Schematic diagram of sample cell*

modifications. Now we also study solutions of a series of chimeric antifreeze protein samples manufactured by bacteria containing synthetic genes.[7] These antifreeze proteins have previously been characterized by their effects upon the rates of ice recrystallization in frozen samples. In addition, we quantify here the effect of other hydrocolloid materials at the interface.

2 Materials and Methods

Following the description by Jones and Chadwick,[13] an apparatus for forming and imaging grain boundary grooves has been constructed. Sample cells consist of two microscope slides spaced at approx. 250 μm apart. Three thermocouples, constructed from fine copper and constantan wire are embedded in the cell, as indicated in Figure 1. The junctions are close to the centre line. The cell is held between two temperature-controlled clamps, which can impose a stable temperature gradient upon the cell. The temperature gradient is quantified through measurement of the temperatures at the three thermocouple junctions, whose positional co-ordinates in the cell geometry can be determined. Initially water is introduced into the cell through the filling ports, and then one end of the cell is cooled to around -35 °C while the other is maintained at around 5 °C. In time, a stable freezing interface is established in the central region of the cell, in the region equipped with thermocouples. This interface exhibits grain boundary grooves at the junctions between crystals. The data analysis requires that the shapes of these grooves be analysed. This requires some method of recording the shape of the groove for further analysis.

Transferring and Analysing Images

Grain boundary images were viewed through a Leitz microscope equipped with a 32 $\times$ objective (long working distance, phase contrast). Images were simultaneously recorded with an RCA monochrome video camera (Model TC

1025). The video output was directed to a Macintosh Quadra 660 AV computer, which was equipped with a video board. Video Monitor frame grabber software was used to capture images; these were saved as PICT files for subsequent analysis.

The images were next imported to the NIH 1.54 image analysis program, where the THRESHOLD command was used to help distinguish liquid and solid phases. In some cases the TRACE EDGES routine was used to trace the interface. Here, the drawing tool was used to outline the interface to one side of the grain boundary cusp. The line drawing of the interface was then imported to the analysis program described below.

Analysis was done by a computer program written in Microsoft QuickBasic. The basis of the program is to compare the experimental interface profile with a series of theoretical profiles generated according to the theories of Bolling and Tiller.[14] The program first scans the experimental profile and converts it to numerical (x, y) coordinates. The curve magnification (about 830) and the temperature gradient are input as the critical parameters of the theoretical curves. The operator inputs the range of surface energies, Y_A to Y_B, and the size of the steps between these values to be tested. For example, a first guess might be that the true interfacial energy Y_{SL} lies between 2.0×10^{-2} and 4.0×10^{-2} J m^{-2}; if the size of the iteration step is 0.2×10^{-2} J m^{-2} the program will generate theoretical profiles corresponding to interfacial energies of $2.0 \times 10^{-2}, 2.2 \times 10^{-2}, 2.4 \times 10^{-2}, \ldots 3.8 \times 10^{-2}, 4.0 \times 10^{-2}$ J m^{-2}. At each x coordinate value, the difference between the y coordinates for the theoretical and experimental curves are tabulated and kept as a running sum. Thus, a best fitting theoretical profile can be assessed by comparing the sum-of-squares difference for each of the profiles as compared with the theoretical data.

Experimental Procedure

The cell is filled with water, and placed within the temperature-controlled clamps. After a suitable period, a stable interface is produced. The exact position of this interface within the cell is recorded, using the microscope stage coordinate system. Since the position, and temperatures, of the thermocouple junctions are also recorded, a check of consistency is that, when the temperature gradient on the stage is calculated, the interface position should correspond to the appropriate freezing temperature. Several grain boundary grooves are imaged as described above. While still maintaining the freezing interface, the liquid water is removed from the cell, and replaced with the solution under study. Care has to be taken in this step to avoid the introduction of air bubbles. The cell is then moved a very small distance in the clamps, so that the interface moves. Once the new interface has become stationary, the grain boundary grooves are imaged. The positions of the interface and the thermocouples are once more determined, since these will have moved relative to the clamps. The clamps have a fixed position on the moveable platform of the microscope stage. It is important that the stage temperature gradient be noted for each set of images.

Table 1   *Interfacial energies at the ice–solution interface*

Sample	Concentration	Surface energy	Standard deviation
	mg ml^{-1}	10^2 mN m^{-2}	10^2 mN m^{-2}
Water		3.08	0.18
afa3	10	2.23	0.50
afa3	28	0.85	0.23
afa3$_{SB}$	10	2.76	0.30
afa3$_{SB}$	28	2.68	0.25
afa3$_R$	10	2.04	0.22
afa3$_R$	28	1.28	0.18
afa5	10	2.64	0.21
afa5	28	1.84	0.34
AFGP[a]	2	2.56	0.8
AFGP[a]	4	2.08	0.9
AFGP[a]	11	0.86	0.4
Gelatin	1	4.0	0.5
Guar	1	3.2	0.2
CMC	1	3.5	0.5
BSA	10	3.0	0.2

Polymers Studied

A variety of polymer solutions was prepared for study. Since only 30 mg samples of each antifreeze protein were available, the number of experiments which could be performed was limited. Solutions in the concentration range 8–30 mg AFP per ml of solution were prepared. It took approximately 0.6 ml of solution to load the measuring cell properly. Some 0.1% (*i.e.* 1 mg ml^{-1}) aqueous solutions of carboxymethylcellulose (CMC), gelatin, and guar gum were also studied.

The antifreeze proteins were a gift from DNAP Inc., Oakland, CA. Their designations are spa-afa3, spa-afa3$_{SB}$, spa-afa3$_R$, and spa-afa5. The production of these materials has been described by Mueller *et al.*[7] CMC was Finnfix 700E lot 21089, from Metsaliiton Teollisuus Oy, Finland. Guar gum was lot 101F-0279 from Sigma, and gelatin was Swine skin type 1, lot 72F 0657, also from Sigma. Data for AFGP were taken from reference 11.

3 Results and Discussion

According to the analysis of Bolling and Tiller,[14] if Y_{SL}, the interfacial energy, is assumed to be isotropic, then a grain boundary superimposed in the *x*–*y* plane should be given by

$$x = f(y) = K \ln\{[2K + (4K^2 - y^2)^{1/2}]/y\} - (4K^2 - y^2)^{1/2} + K[\sqrt{2} - \ln(\sqrt{2} + 1)]$$

where $K^2 = Y_{SL}/G\,\Delta S$, ΔS is the entropy of fusion per unit volume of ice and G is the thermal gradient. The value of ΔS is taken to be 1.088 J K^{-1} m^{-3}, and the thermal gradient is measured. A typical match between the experimentally observed groove and the calculated groove shape allows for a discrimination, for any given groove, of approx. 0.2×10^{-2} J m^{-2}. Typically at least five grooves were evaluated for each solution.

The results obtained are summarized in Table 1. The value of Y_{SL} for pure ice–water, at $3.08 \pm 0.18 \times 10^{-2}$ J m^{-2}, compares well with literature values which range from 2.9×10^{-2} to 3.3×10^{-2} J m^{-2}. The presence of AFP and AFGP in the liquid phase causes a significant reduction in Y_{SL}. For all active antifreeze polymers, Y_{SL} decreases as the concentration increases. Hydrocolloids such as gelatin, guar gum and CMC either have only a small effect on Y_{SL} or cause it to increase. The decrease in Y_{SL} in the presence of AFP or AFGP is an indication that these materials are indeed adsorbed at the ice–solution interface. The data collected in reference 11 allow for the estimation of surface excess concentrations. Preliminary calculations, utilizing the limited data we have collected up to this point, suggest that the different AFP materials described by DNAP Inc. have different extents of adsorption, and also that the extent of adsorption is less than that exhibited by the most active AFGP. There may be some relationship between the surface energy, the extent of adsorption, and the differing efficiencies of these AFP materials in inhibiting ice recrystallization. We are currently collecting data to better quantify the recrystallization rates in order to investigate this relationship. We are also collecting data on the influence of some of these AFP materials on the rate of ice crystal nucleation in a well characterized, heterogeneous nucleation model. The mode of action of these materials on ice is clearly distinct from that of the hydrocolloids studied. There is no evidence of a strong interaction between the hydrocolloids and the ice surface. Any effective lowering of recrystallization rates of ice through the addition of hydrocolloids would therefore be expected to result from quite a different mechanism than that which produces the recrystallization inhibition on addition of the AFP materials.

References

1. P. F. Scholander, W. Flagg, V. Walters, and L. Irving, *Physiol. Zool.*, 1953, **26**, 67.
2. M. S. Gordon, B. H. Amdur, and P. F. Scholander, *Biol. Bull.*, 1962, **122**, 52.
3. R. E. Feeney, *Am. Sci.*, 1974, **12**, 712.
4. A. L. DeVries, *Annu. Rev. Physiol.*, 1983, **45**, 245.
5. R. E. Feeney and Y. Yeh, *Food Technol.*, 1993, **47**, 82.
6. J. G. Duman, *J. Exp. Zool.*, 1977, **201**, 85.
7. G. M. Mueller, R. L. McKown, L. V. Corotto, C. Hague, and G. J. Warren, *J. Biol. Chem.*, 1991, **266**, 7339.
8. A. L. DeVries and Y. Lin, *Biochim. Biophys. Acta*, 1977, **495**, 385.
9. J. A. Raymond and A. L. DeVries, *Proc. Natl. Acad. Sci. USA*, 1977, **74**, 2589.
10. R. A. Brown, Y. Yeh, T. S. Burcham, and R. E. Feeney, *Biopolymers*, 1985, **24**, 1265.
11. W. L. Kerr, R. E. Feeney, D. T. Osuga, and D. S. Reid, *Cryoletters*, 1985, **6**, 371.
12. J. A. Raymond, P. Wilson, and A. L. DeVries, *Proc. Natl. Acad. Sci. USA*, 1989, **86**, 881.
13. D. R. H. Jones and G. A. Chadwick, *Philos. Mag.*, 1970, **22**, 291.
14. G. F. Bolling and W. A. Tiller, *J. Appl. Phys.*, 1960, **31**, 1345.

Studies of Interactions between Casein and Phospholipid Vesicles

By Yuan Fang and Douglas G. Dalgleish

DEPARTMENT OF FOOD SCIENCE, UNIVERSITY OF GUELPH, GUELPH, ONTARIO N1G 2W1, CANADA

1 Introduction

The interactions between proteins and phospholipid bilayers have been extensively studied because of their biological importance. Our interest in phospholipids was initially provoked by the widespread use of lecithin in food emulsions.[1] In our previous study on oil-in-water emulsions stabilized by casein,[2] egg-phosphatidylcholine (egg-PC) was found not only to affect the stability of the emulsion, but also to change the structure of the adsorbed casein layer. These findings suggested the importance of studying the interaction of egg-PC and casein alone. In its native form, the casein complex exists in solution and is not associated with biological membranes, although individual caseins have a high affinity for hydrophobic surfaces.[3] For example, casein adsorbs to the surface of polystyrene latices and increases the hydrodynamic radius of the latex particles by about 10 nm,[4] similar to the layers of casein adsorbed to emulsion droplets, which are about 10 nm thick at saturation.[5] There are a few examples of interaction between casein and phospholipid monolayers; β-casein can penetrate a phospholipid choline monolayer and causes a lateral compression of the lipid monolayer,[6] and κ-casein can penetrate a monolayer of dimyristoylphosphatidylcholine.[7] The spontaneous association of protein often causes aggregation and fusion of vesicles.[8–12] The aggregation can be induced by protein bridging and is reversed by adding more protein to the system. Sonication of protein and phospholipid together also leads to lipid–protein complexes.[13,14]

We have studied the interaction between casein and the surface of egg-PC vesicles as well as the effect of homogenizing casein and egg-PC together during vesicle formation. These interactions were studied by photon correlation spectroscopy (PCS) and cryofracture-electron microscopy (EM). These two techniques are complementary because PCS provides information on the average property of the system on a continuous timescale, and EM enables us to gain information on the morphology of the vesicles at a specific moment.

2 Materials and Methods

Egg-PC, *N*-tosyl-*L*-phenylalanine chloromethyl ketone-trypsin and imidazole were purchased from Sigma, St. Louis, MO, and they were used without further purification. All experiments were performed using imidazole buffer at a concentration of 20 mM at pH 7. Whole casein was prepared by precipitating skim milk at pH 4.6 and redissolving the washed precipitate in NaOH at pH 7, after which the solution was freeze-dried.

Vesicle Formation

Vesicles were made by homogenizing dispersions of egg-PC in imidazole buffer with a Model 110S Microfluidizer (Microfluidics Corp., Newton, MA) using an input pressure of 0.3 MPa which corresponds to a pressure drop of 42 MPa in the homogenizing chamber. Egg-PC (20–50 mg) was dispersed in 10 ml of imidazole buffer (20 mM, pH 7), and casein (5–30 mg /ml) was incorporated in the dispersion either before or after homogenization.

Light Scattering

PCS experiments used a Malvern 4700 optical system attached to a 7032 Multi-8 correlator (Malvern Instruments, Southboro, MA) to study the hydrodynamic dimensions of vesicles. Diffusion coefficients of the particles were calculated by the method of cumulants, and the average hydrodynamic dimensions were then calculated, assuming that the vesicles obey the Stokes–Einstein law. Vesicle dispersions were diluted in imidazole buffer (200 μl in 3 ml), which had previously been filtered through a 0.2 μm filter. To investigate the casein–PC interaction, a method based on proteolysis was used, which had been initially developed to study the protein layer adsorbed on the lipid droplets of oil-in-water emulsions.[15,16] Trypsin solution was dissolved to a concentration of 1 mg ml^{-1} in imidazole buffer. After the initial size of vesicles was measured, trypsin was added to the suspension in the light-scattering cell and the proteolysis process was followed by measuring the diameters of the particles every minute for up to 4 h.

Cryofracture and Electron Microscopy (EM)

To follow the morphology change of vesicles during proteolysis, we used a fast spray freezing technique (Balzers Union Spray freezing unit SFU020) combined with cryofracture and EM (Phillips 300). The vesicle dispersion was sprayed under N_2 in very small droplets into liquid propane which was submerged in liquid nitrogen to achieve fast freezing.[17] The frozen sample was then mixed with isobutylbenzene and mounted onto a small gold cup for fracturing. All operations were carried out at low temperatures. These samples were then kept in liquid nitrogen and fractured with a razor blade (Balzers BAF 400T freeze Etch Machine), and the freshly fractured surfaces were then

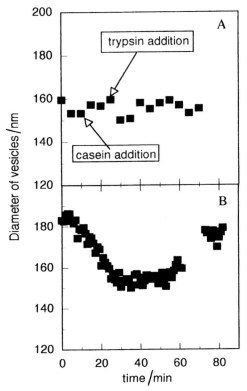

Figure 1   *The changes in hydrodynamic diameter of egg-PC vesicles measured by PCS as a function of time. (A) Diameter of vesicles, and the change in diameter when casein and trypsin are added to the vesicles after formation. (B) Change of diameter of the vesicles formed when egg-PC and casein are homogenized together and trypsin is added*

coated with carbon from an angle of 90° and with platinum from 45°. The replicas were recovered by washing off the sample residue in bleach and were mounted on to EM sample grids which had been previously coated with foamvar.

The vesicles were made at a concentration of 18 mg ml^{-1} of egg-PC and 2 wt% casein. The concentration of trypsin was set at 3 mg ml^{-1}. An aliquot the vesicle dispersion was frozen before adding any trypsin solution, and this sample was used as a control. A 2 ml sample of the vesicular dispersion was then mixed with 100 μl of trypsin. One fraction of the mixture was frozen immediately and more samples were taken at a number of time intervals after the initial mixing of vesicles and trypsin.

3 Results and Discussion

Vesicles made in the absence of casein were shown to be stable with time. The addition of casein solution to the vesicles did not change their sizes; neither did

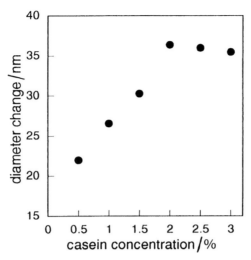

Figure 2   *The change in hydrodynamic diameter of egg-PC vesicles associated with casein, taken as the difference between the size measured before the addition of trypsin and the minimum size obtained during the trypsin treatment. The constant concentration of vesicles was maintained at $6.8 \times 10^{-3} M$, and the casein concentration was varied. Each point is the average of at least two separate experiments*

the addition of trypsin (Figure 1A). When vesicles were made in the presence of casein, the particles once again were stable by themselves, but the vesicles were susceptible to trypsin attack. The average vesicle size decreased at first during trypsin treatment, as has been shown previously for oil-in-water emulsions and model latex–protein complexes.[18] Emulsion droplets maintained a smaller but stable size after treatment with trypsin, but the size of PC–casein vesicles increased after a minimum size was reached (Figure 1B). Figure 2 shows that the average decrease in diameter during trypsinolysis increased with increasing casein concentration up to about 1.5% casein and then levelled off at a value of about 37 nm at higher concentrations of casein. This layer was nearly twice as thick as a saturated casein layer adsorbed to emulsion droplets. We have several possible suggestions to explain this large value. Possibly, only small parts of the casein molecules are involved in the interaction with the bilayer, so that a large portion of the rather flexible polypeptide chains protrude into the solution to give a larger hydrodynamic dimension. Alternatively, the casein molecules may form multilayers which will of course be thicker, although we have little evidence for their formation in oil-in-water emulsions. Third, some casein molecules may interact with more than one vesicle, so that clusters are formed via casein bridging; when the bridging protein is broken down by proteolysis, there is a change in quaternary structure as well as in the diameters of the particles. At present, we cannot rule out the first two possibilities, although the last may be unlikely since bridging

a

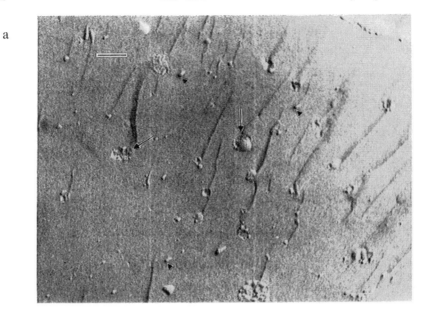

b

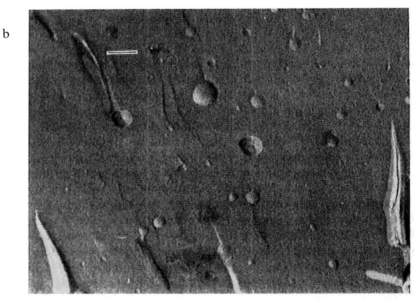

Figure 3   *EM pictures of egg-PC–casein vesicles from freeze-fractured samples. All four*
images have the same magnification, and the bar represents 100 nm. (a)
Vesicles frozen before mixing with trypsin solution; two types of particles can
be seen: vesicles with bumpy surfaces (arrow), and small particles with a long
shadow (arrow head). (b) Vesicles frozen after 30 min incubation with
trypsin; vesicle surfaces are smooth and the smaller particles have disappeared

c

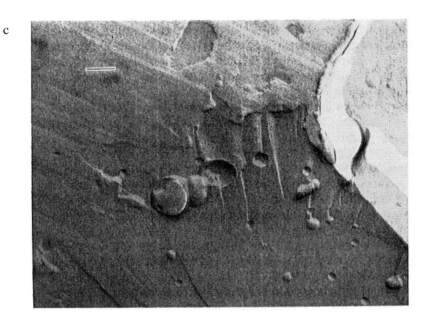

d

Figure 3   *(contd.) (c) Vesicles frozen after 1 h incubation with trypsin. (d) After 2 h incubation with trypsin; a large complex of multilamellar particles co-exists with smaller unilamellar vesicles*

is generally associated with low, rather than high,[8] concentrations of protein, and the results from EM have shown no evidence of vesicle clusters.

The above results indicate that casein does not adsorb to PC vesicle surfaces but that a complex is formed between the two when they are homogenized together. The response of this complex to trypsin treatment suggests that casein is behaving as a transmembrane protein.

Figure 3 shows the EM picture of the egg-PC and casein vesicles before and after the treatment with trypsin. Two distinctive groups of particles can be seen from the picture of the sample without trypsin (Figure 3a). One group represents vesicles, and these appear to have rough edges or bumpy surfaces. The second group includes a large number of much smaller particles with a long shadow. After taking into account the thickness (3 nm) of platinum shadowing,[19] the average diameters of these particles are of the order of 5 nm, and their long shadow (10 nm) indicates that they are not spherical. After trypsin was added to the solution of vesicles, both the vesicles and the small particles underwent significant changes. First of all, the small particles disappeared from the samples which had been incubated with trypsin for half an hour or longer (Figures 3b–3d). Since we know casein is the only component in the vesicle dispersion that is susceptible to trypsin attack, it is certain that these small particles are made largely of casein molecules. From the size of their EM images, the aggregates contain only small number of casein molecules. Surprisingly, these casein aggregates are not spherical but elongated.

As the incubation time with trypsin is increased, the surfaces of the vesicles become smoother, and aggregation and fusion of the vesicles is observed (Figure 3c). Some vesicles become multilamellar liposomes at a longer incubation time (Figure 3d). This is in good agreement with the results shown in Figure 1B, so that the average size increase observed from the PCS measurement is a result of the vesicle fusion mediated by the breakdown of casein associated with vesicles. Because vesicle fusion has to overcome a barrier from hydrophilic to hydrophobic media, the peptide remaining after trypsic proteolysis of the caseins must provide a hydrophobic bridge between the vesicles. This result reinforces the conclusion that the interaction of casein and PC vesicles when homogenized together is transmembrane in nature instead of simple surface association.

4 Conclusions

From the results of dynamic light scattering and EM of cryofracture samples, we can draw the following conclusions. Casein and egg-PC form a complex when they are homogenized together, and the complex exists in the form of vesicles. But casein molecules do not adsorb on already formed vesicles which possess hydrophilic surfaces. The casein layer has a hydrodynamic maximum thickness of 18.5 nm which is nearly double the 10 nm thickness of a saturated casein layer adsorbed on oil droplets of an emulsion and may result from incorporation of multilayers of casein. When the casein–egg-PC vesicles are treated with trypsin, the average size of the vesicles is found to go through a minimum value and then increase again, in contrast to emulsion systems which remain quite stable after casein breakdown. The EM images reveal that the breakdown of casein by trypsin is followed by fusion of the vesicles, the small unilamellar vesicles becoming large multilayer liposomes. The peptides of hydrolysed casein in the lipid bilayer are found to facilitate the vesicle fusion process. When homogenized together with egg-PC, casein behaves as a transmembrane protein.

Acknowledgements

The authors would like to thank Dr P. R. Rand and Mrs Nola Fuller for their generous help with the electron microscopy experiments. This research was funded jointly by the Ontario Dairy Council and the Natural Sciences and Engineering Research Council of Canada.

References

1. T. Graf and L. Meyer, *Int. Flavours Food Additives*, 1976, **7**, 218.
2. Y. Fang and D. G. Dalgleish, *Colloids Surf. B.*, 1993, **1**, 357.
3. D. E. Graham and M. C. Phillips, in 'The Theory and Practice of Emulsion Technology', ed. A. L. Smith, Academic Press, London, 1976, p. 75.
4. D. G. Dalgleish, *Colloids Surf.*, 1990, **46**, 141.
5. Y. Fang and D. G. Dalgleish, *J. Colloid Interface Sci.*, 1993, **156**, 329.
6. M. C. Phillips, M. T. A. Evans, and H. Hauser, *ACS Adv. Chem. Ser.*, 1975, **144**, 217.
7. M. C. A. Griffin, R. B. Infante, and R. A. Klein, *Chem. Phys. Lipids*, 1987, **36**, 91.
8. R. Smith, *Biochim. Biophys. Acta*, 1977, **470**, 170.
9. S. Schenkman, P. S. de Araujo, A. Sesso, F. H. Quina, and H. Chaimovich, *Chem. Phys. Lipids*, 1981, **28**, 165.
10. S. Schenkman, P. S. de Araujo, R. Dijkman, F. H. Quina, and H. Chaimovich, *Biochim. Biophys. Acta*, 1981, **649**, 633.
11. H. P. Haagsman, R. H. Elfring, B. L. M. van Buel, and W. F. Voorhout, *Biochem. J.*, 1991, **275**, 273.
12. R. N. Farias, A. L. Vinals, and R. D. Morero, *Biochim. Biophys., Res. Commun.*, 1985, **128**, 68.
13. L. Huang and S. J. Kennel, *Biochemistry*, 1979, **18**, 1702.
14. E. M. Brown, R. J. Carroll, Ph. E. Pfeffer and J. Sampugna, *Lipids*, 1983, **18**, 111.
15. D. G. Dalgleish and J. Leaver, *J. Colloid Interface Sci.*, 1991, **141**, 288.
16. J. Leaver and D. G. Dalgleish, *J. Colloid Interface Sci.*, 1992, **149**, 49.
17. R. P. Rand, B. Kachar, and T. Reese, *Biophys. J.*, 1985, **47**, 483.
18. D. G. Dalgleish, *Colloids Surf. B.*, 1993, **1**, 1.
19. M. Tihova, B. Tattrie, and P. Nicholls, *Biochem. J.*, 1993, **292**, 933.

Effect of Protein on the Retention and Transfer of Aroma Compounds at the Lipid–Water Interface

By B. A. Harvey, C. Druaux, and A. Voilley

LABORATOIRE DE GENIE DES PROCEDES ALIMENTAIRES ET BIOTECH-
NOLOGIQUES, ENSBANA, UNIVERSITÉ DE BOURGOGNE, 1 ESPLANADE
ERASME, 21000 DIJON, FRANCE

1 Introduction

The physico-chemical behaviour of aroma compounds plays an important role in their release from foods, and also in the perceived flavour, as well as in processes of concentration and extraction from biological products. Generally, food media are multiphase systems, and so the partitioning of small aroma molecules between phases is an important phenomenon; the complexity of such partitioning has been described.[1] In oil–water systems like emulsions, lipid–protein interactions have to be considered, with particular regard to interfacial properties; this is the subject of a recent review.[2]

Much research has been carried out in the last 20 years on interactions directly involving aroma compounds in homogeneous systems. The early work has been summarized in reviews.[3,4] In a number of recent studies, the need has been stated[5,6] for a more detailed and quantitative knowledge of the interaction of aroma compounds with food constituents in order to control flavour release. The present lack of data is particularly pertinent to the binding of aroma compounds to protein, which is rather specific.

Thermodynamic aspects of aroma compound–protein binding—such as the number of binding sites, the binding constants and strengths of interactions—have been investigated by a range of techniques.[7] However, the precise nature of the interactions, the types of forces and the specific protein-binding sites involved have proved more difficult to identify. No overall insight yet exists into the mechanisms.[6] The principal factors influencing the binding of aroma compounds are the natures of the substrate and the volatile compound.[8] Protein conformation at the oil–water interface is sensitive to changes in experimental conditions such as pH, temperature, and salt concentration, and the protein–aroma compound interaction itself depends on conformational

154

change.[9] Such changes are highly influential with respect to subsequent binding, but are difficult to characterize because of their complexity. From the large number of published equilibrium studies on the role of proteins at the lipid–water interface, few conclusions can be drawn about flavour release or the binding processes occurring there.

Our objective is to gain a better overall understanding of the behaviour of aroma molecules in lipid–water systems containing proteins (*i.e.* model food emulsions). Retention is studied by using complementary thermodynamic methods. The transfer is investigated by considering not only molecular diffusion, but also the resistance at the interface in lipid–water systems, using a rotating diffusion cell. This type of cell has been widely used to investigate the kinetic barrier at liquid–liquid interfaces.[10,11] The application of the cell in the study of aroma compound–protein interactions is new, and some preliminary results are presented.

2 Experimental

Lipids used were tributyrin ($C_{15}H_{26}O_6$, Aldrich, 98%) and triolein ($C_{57}H_{104}O_5$, Fluka, 65%, purified in this laboratory to 100%). Two batches of sodium caseinate were used: one of commercial grade, the other purified and freeze-dried (Biochemistry Department, ENSBANA). Aroma compounds were ethyl acetate (Prolabo), ethyl butanoate and 2,5-dimethylpyrazine (Aldrich); all were of 99% purity. Membrane filters were of polytetrafluoroethylene (Sartorius 118 07 47N) of average pore size 0.2 μm.

A dynamic headspace method was used to investigate the vapour–liquid equilibria. Aroma compound vapour above a homogeneous liquid (water or sodium caseinate solution) was sampled periodically (1 ml) in a nitrogen carrier gas stream by an automatic valve into a gas chromatograph (GC) (model Chrompack CP 9000 with a stainless steel column, 3 m × 2.2 mm i.d., containing 100–120 mesh coated with 10% Carbowax 20M; flame ionization detector and injector temperatures were 200 °C). The column temperature, the sampling times, and the intervals between sampling were adjusted according to the aroma compound being investigated.

The rotating diffusion cell is designed hydrodynamically so that stationary diffusion layers of known thickness are created on each side of the oil layer.[12] The cell is shown in Figure 1; the interior is constructed only of glass and Teflon to minimize unwanted aroma compound retention. The oil layer is supported on a porous membrane filter (A) which divides the apparatus into two parts, separating the inner and outer aqueous compartments. The central assembly (B) is rotated by a motor at constant known speed up to 6 Hz. The fixed slotted baffle (C), positioned approximately 2 mm above the filter, ensures a stationary diffusion layer. The Teflon filter is located by a push-fit Teflon ring leaving an exposed circular area of 1.4 cm in diameter. A Teflon lid is screwed down to prevent loss of aroma compound through vaporization. Care must be taken to ensure that all air bubbles are removed from beneath the filter.

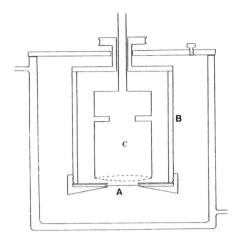

Figure 1   *Schematic diagram of rotating diffusion cell showing (A) filter, (B) rotating assembly, and (C) slotted baffle*

Lipids were pre-saturated with water and the aroma compound was initially contained in the inner compartment only at a concentration of 0.02 M. In experiments with caseinate, equal concentrations were used in the two compartments and the cell was left to stand for 2 h to allow adsorption of caseinate at the lipid–water interface. The flux of aroma compound across the oil layer was measured by periodically sampling with a microsyringe from the solution in the outer compartment through a small hole in the lid fitted with a threaded Teflon cap; the analysis was performed by GC.

3 Theory of Mass Transport in Rotating Diffusion Cell

The hydrodynamics properties of a rotating disc are created inside the cell.[12,13] The thickness of the diffusion layer on each side of the membrane filter is given by the Levich equation

$$Z = 0.643\nu^{\frac{1}{6}}D_{aq}^{\frac{1}{3}}\omega^{-\frac{1}{2}} \tag{1}$$

where D_{aq} is the diffusion coefficient in aqueous solution, ω is the rotation speed, and ν is the kinematic viscosity. The rate of transfer of the aroma compound from the inner to the outer compartment is given by the flux

$$J = kAc \tag{2}$$

where A is the area of the filter, c is the concentration of solute in the inner compartment, and the rate constant k is given by

$$\frac{1}{k} = \frac{2Z}{D_{aq}} = + \frac{2}{\alpha k_i} + \frac{l}{\alpha D_o P} \tag{3}$$

The significance of the three terms on the right-hand side of eqn (3) is as follows:
(1) $2Z/D_{aq}$ describes diffusion through the two stagnant aqueous diffusion layers of thickness Z; (2) $2/\alpha k_i$ is the contribution from interfacial transfer at the two lipid–water boundaries; α is the cross-sectional area of the pores divided by A, and k_i is the rate constant for aroma compound M defined by

$$M(aq) \underset{k_{-i}}{\overset{k_i}{\rightleftharpoons}} M(oil) \qquad \text{where } P = \frac{k_i}{k_{-i}}$$

(3) $l/\alpha D_o P$ is the contribution from diffusion through the lipid in the filter; P is the partition coefficient, l is the filter thickness, and D_o the diffusion coefficient in the lipid.

Substitution of eqn (1) into eqn (3) allows a plot of $1/k$ against $\omega^{-1/2}$ to be made. The intercept is the sum of the second and third terms on the right hand side of eqn (3). The results obtained here are fitted to a Levich plot with a theoretical slope derived from eqn (1). For caseinate solutions, the required viscosity was calculated from an empirical equation.[14] To obtain the rate constant for transfer at the lipid–water interface, k_i, the membrane thickness l and partition coefficient P were measured and the diffusion coefficient D_o in the lipid phase estimated from the Wilke–Chang equation.[15]

4 Results and Discussion

Physico-chemical Interactions between Aroma Compounds and Sodium Caseinate in Water

Interaction of aroma compounds with sodium caseinate was investigated in relation to their physico-chemical properties using two techniques of measurement at equilibrium. Aroma compounds with diverse and known physico-chemical properties were chosen for the study (Table 1).

Table 1 *Physico-chemical characteristics of three aroma compounds*

Aroma compounds	Formula	Molar mass (Da)	Relative density at 20 °C	Solubility in water at 25 °C g l^{-1}	log P_o[a]
Ethyl acetate	$C_4H_8O_2$	88	0.902	86	0.6
Ethyl butanoate	$C_6H_{12}O_2$	116	0.878	5.75	1.7
2,5-Dimethylpyrazine	$C_6H_8N_2$	108	0.990	∞	0.2

[a] Hydrophobicity constant.

Results of vapour–liquid partition coefficients at atmospheric pressure, K^∞, and activity coefficients at infinite dilution, γ^∞, are shown together with values of diffusion coefficients in water, measured using a Stokes cell,[16,17] in Table 2. The retention value of the aroma compound by caseinate, r, is defined by

$$r = (1 - \frac{K_{cas}}{K_{water}}) \times 100\%$$

Table 2   *Vapour–liquid partition, activity and diffusion coefficients of aroma compounds in aqueous solutions at 25 °C*

Aroma compound	$K^{\infty a}$ (in water)	$\gamma^{\infty b}$	$K^{\infty a}$ (in 5% sodium caseinate solution)	r^c	D in water[d] 10^{-10} m^2 s^{-1}
Ethyl acetate	11(9)^e	92	10.6(6)	8	11.6(10)
Ethyl butanoate	27(3)	1170	19(3)	30	9.4(10)
2,5-Dimethylpyrazine	0.10(3)	23	–	–	9.5

[a] Vapour–liquid partition coefficient at 760 mmHg.
[b] Activity coefficient at infinite dilution.
[c] Percentage retention of the aroma compound on the sodium caseinate (number of moles of bound aroma/total number of moles of aroma).
[d] Measured values from Lamer[17].
[e] Values in parentheses are coefficient of variation (%).

Retention was found to increase with increasing aroma compound hydrophobicity, which is consistent with previous work.[18,19] Ethyl butanoate, the most hydrophobic aroma compound for which experiments with caseinate were conducted, showed 30% retention. Values of liquid–liquid partition coefficients expressed as ratios of molar fractions, P_x, and of molar concentrations, P, are shown in Table 3. Lower caseinate concentrations of 0.1 and 0.5 wt% were used in these experiments. Values of P_x are higher between triolein and water than between tributyrin and water, while the reverse is true for values of P; the difference is largely due to the higher molar mass of triolein (885 Da) than of tributyrin (302 Da). Ethyl butanoate shows a decrease in P and P_x of more than 50% between tributyrin and water on increasing the caseinate concentration from 0 to 0.5%, probably largely due to binding of ethyl butanoate to caseinate in the water phase. The corresponding decrease in P and P_x between triolein and water is 17%.

In addition to such thermodynamic information, an explanation of the behaviour of aroma compounds in lipid–water systems requires diffusional data. While diffusion coefficients can be readily measured in a homogeneous phase (Table 2), another approach is needed for considering the effect of an adsorbed layer of caseinate at the lipid–water interface during mass transfer. A kinetic study allowing evaluation of the kinetic resistance of the lipid–water interface to the mass transfer of aroma compounds was therefore made with the rotating diffusion cell.

Table 3　*Liquid–liquid partition coefficients of aroma compounds at infinite dilution and 25 °C*

Aroma compound	Percentage sodium caseinate	$P_x{}^a$	P^b	Var^c	$P_x{}^a$	P^b	Var^c
		Tributyrin–water			Triolein–water		
Ethyl acetate	0	99	6	2	–	–	–
Ethyl butanoate	0	970	60	3	1690	31	0.5
	0.1	650	41	5	–	–	–
	0.5	460	28	14	1400	25.8	1
2,5-Dimethylpyrazine	0	29.9	1.9	0.1	51	1	7
	0.5	30.2	1.9	0.5	53	1	4

a Liquid–liquid partition coefficient expressed as ratio of mole fractions.
b Liquid–liquid partition coefficient expressed as ratio of molar concentrations.
c Coefficient of variation (%).

Mass Transfer at the Lipid–Water Interface

As a test of the performance of the rotating diffusion cell, the transfer of acetone from the inner to the outer compartment was measured through a hydrophilic polyamide membrane filter containing water. As there are no phase boundaries in this system, the Levich intercept represents only the resistance to diffusion within the membrane filter, $1/\alpha D_{aq}$. Reproducibility was good with a standard deviation of total resistance to transfer, $1/k$, of less than 3% at each of four rotation speeds. The mean experimental intercept agreed to within 6% of the value calculated using the literature value of the diffusion coefficient of acetone in water.[16]

Levich plots for the transfer of ethyl butanoate in the absence of caseinate through a layer of tributyrin or triolein are shown in Figures 2 and 3, respectively. The theoretical gradients shown on the plots are a good fit to the

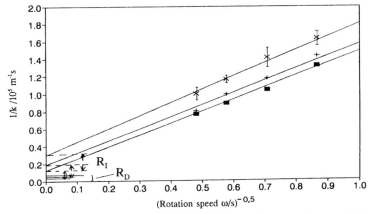

Figure 2　*Levich plot for transfer of ethyl butanoate through a layer of tributyrin at 25 °C:* ■*, no protein;* +*, 0.1 wt% caseinate;* ×*, 0.5 wt% caseinate*

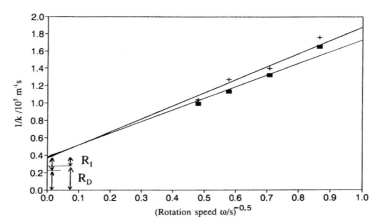

Figure 3   *Levich plot for transfer of ethyl butanoate through a layer of triolein at 25 °C:* ■*, no protein;* +*, 0.5 wt% caseinate*

experimental points, showing that the predicted hydrodynamic conditions were met. The intercepts are close to zero, showing that the rate-limiting step through both tributyrin and triolein is mass transfer through the aqueous diffusion layers. The resistance to transfer through tributyrin is $R_D = 3400$ m^{-1} s, indicating an interfacial resistance R_I of the same order. The corresponding resistance through triolein is 23 000 m^{-1} s; the higher figure is mostly due to the smaller measured value of the partition coefficient P between triolein and water, and to a lesser extent to the smaller calculated value of the diffusion coefficient in triolein than in tributyrin (eqn. 3). Levich plots for ethyl acetate (not shown) were similar to those for ethyl butanoate, showing rate-limiting transfer through the aqueous diffusion layers and small intercepts, despite the lower partition coefficient (Table 2).

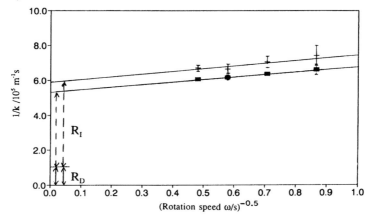

Figure 4   *Levich plot for transfer of 2,5-dimethylpyrazine through a layer of tributyrin at 25 °C:* ■*, no protein;* +*, 0.5 wt% caseinate*

In contrast to the kinetic behaviour of the esters, 2,5-dimethylpyrazine passed through a layer of tributyrin in the absence of caseinate about 5 times more slowly—with a much larger Levich intercept and relatively little resistance in the aqueous diffusion layers (Figure 4). R_D is a minor part of the intercept $(1.0 \times 10^{-5}$ m^{-2} s), while the interfacial resistance is $R_I = 5.1 \times 10^{-5}$ m^{-1} s and therefore the rate-limiting step in the overall transfer. A possible explanation is that both steric hindrance from the fairly rigid cyclic structure of the molecule and electronic interaction from the high electron density on both sides of the molecule contribute to the relatively high value of R_I.

Sodium caseinate of commercial grade was used in the rotating diffusion cell as it showed the same behaviour at a concentration of 0.1 wt% towards the transfer of ethyl acetate through tributyrin as a purified and freeze-dried sample. The solution pH in both compartments was 7.2 in all experiments. Figure 2 shows significantly slower transfer of ethyl butanoate between solutions of sodium caseinate than between water phases, due mainly to a greater resistance in the aqueous diffusion layers $(2Z/D_{aq})$ from the higher viscosity and the effect of viscosity on the diffusion coefficient. Values of R_D are larger with caseinate (5000 and 7200 m^{-1} s at concentrations of 0.1 and 0.5 wt%, respectively) than without, owing to the lower partition coefficient P (Table 3). It is assumed that the caseinate does not significantly penetrate the lipid phase and so affect the value of D_o. There does appear to be a small increase in the interfacial resistance R_I with increasing caseinate concentration, though this result has to be confirmed.

The effect of the caseinate concentration in bulk solution on the thickness of the adsorbed layer can be inferred from investigations with β-casein which adsorbs over a wide range of bulk concentrations at a variety of hydrophobic surfaces, planar and in emulsions,[20] to give a monolayer surface coverage of 2–3 mg m^{-2}. At bulk concentrations above about 0.1 wt%, β-casein is thought to adsorb in a primary layer to a coverage of *ca.* 3 mg m^{-2}, with secondary layers easily washed off.[21] The thickness of adsorbed β-casein on monodisperse polystyrene latex was also found[22] to reach a constant value at a concentration in bulk solution above approximately 0.01 wt%.

The time required for caseinate adsorption at the lipid–water interface must be considered here in the absence of the high shear applied during emulsification, although it is reasonable to think that the caseinate would require a relatively short time to reach a constant surface concentration. It was found in practice that the time the rotating diffusion cell was left to stand after adding caseinate solutions did not measurably affect aroma compound transfer rates for periods between 30 min and 2 h.

The decrease in overall transfer rate through a layer of triolein (Figure 3) when a 0.5 wt% solution of caseinate was used was smaller than that through tributyrin (the intercepts were nearly coincident). Larger values of R_D for triolein (28 000 m^{-1} s with 0.5 wt% caseinate) indicated that the interfacial resistance was small. Figure 4 also shows slower transfer of 2,5-dimethylpyrazine in the presence of 0.5 wt% caseinate, but as the partition coefficient P is unaffected by 0.5 wt% caseinate (Table 3), the diffusional resistance in the lipid phase, R_D, is the same with or without caseinate. The increase in the interfacial resistance (to 6.0×10^{-5} m^{-1} s) is apparently caused by the presence of the adsorbed layer of caseinate at the interface.

5 Conclusions

Application of a new technique for investigating interfacial mass transfer of aroma compounds has been demonstrated using the rotating diffusion cell; this has allowed direct comparison of interfacial kinetic resistances. Different kinetic behaviour of transfer between aqueous phases across a lipid layer was found for aroma compounds with different physico-chemical properties; transfer through the aqueous diffusion layers was the rate-limiting step for esters, while for 2,5-dimethylpyrazine, interfacial transfer was rate-limiting. An apparent effect on the interfacial resistance due to an adsorbed layer of sodium caseinate was observed.

To improve further our understanding of the influence of the lipid–water interface on aroma compound transfer, the nature of the interface could be altered by using different types of surfactant. It would also be of interest to investigate the effect on the kinetic interfacial resistance of competitive adsorption of, for instance, α_{s1}- and β-casein.

References

1. B. L. Wedzicha, in 'Advances in Food Emulsions and Foams', ed. E. Dickinson and G. Stainsby, Elsevier Applied Science, London, 1988, p. 329.
2. M. Le Meste and S. Davidou, in 'Ingredient Interactions: Effects on Food Quality', ed. A. G. Gaonkar, Marcel Dekker, New York, 1994, in the press.
3. J. Solms, 'Interactions of Food Components', ed. G. G. Birch and M. G. Lindley, Elsevier Applied Science, London, UK, 1986, p. 189.
4. P. B. McNulty, in 'Food Structure and Behaviour', ed. J. M. V. Blanshard and P. Lillford, Academic Press, 1987, p. 245.
5. P. Overbosch, W. B. Agterof, and P. G. Haring, *Food Rev. Int.*, 1991, **7**, 137.
6. H. Plug and P. Haring, *Trends Food Sci. Technol.*, 1993, **4**, 150.
7. K. Farès, Thèse 3e cycle, ENSBANA, Université de Bourgogne, 1987.
8. M. Le Thanh, P. Thibeaudeau, M. A. Thibaut, and A. Voilley, *Food Chem.*, 1992, **43**, 129.
9. J. Solms and B. Guggenbuehl, 'Physical Aspects of Flavor', John Wiley, New York, 1990, p. 319.
10. R. H. Guy, T. R. Aquino, and D. H. Honda, *J. Phys. Chem.*, 1982, **86**, 2861.
11. B. A. Harvey, T. M. Herrington, and R. D. Bee, *ACS Symp. Ser.*, 1993, **520**, 220.
12. W. J. Albery, J. F. Burke, E. B. Leffler, and J. Hadgraft, *J. Chem. Soc., Faraday Trans.*, 1976, **72**, 1618.
13. V. G. Levich, 'Physicochemical Hydrodynamics', Prentice-Hall, Englewood Cliffs, NJ, 1962.
14. J. Korolczuk, *N. Z. J. Dairy Sci. Technol.*, 1982, **17**, 135.
15. P. Chang and C. R. Wilke, *J. Phys. Chem.*, 1955, **59**, 592.

16. A. Voilley, Thèse Docteur ès Sciences, ENSBANA, Université de Bourgogne, 1986.
17. T. Lamer, Thése 3ᵉ cycle, ENSBANA, Université de Bourgogne, 1993.
18. S. Damodaran and J. E. Kinsella, *J. Agric. Food Chem.*, 1980, **28**, 567.
19. T. E. O'Neill and J. E. Kinsella, *J. Agric. Food Chem.*, 1987, **35**, 770.
20. E. Dickinson, *J. Chem. Soc., Faraday Trans.*, 1992, **88**, 2973.
21. E. Dickinson, 'An Introduction to Food Colloids', Oxford University Press, 1992, p. 150.
22. D. G. Dalgleish and J. Leaver, in 'Food Polymers, Gels and Colloids', ed. E. Dickinson, Special Publication No. 82, The Royal Society of Chemistry, Cambridge, UK, 1991, p. 113.

Emulsifying and Oil-binding Properties of the Enzymic Hydrolysate of Bovine Serum Albumin

By Masayoshi Saito

NATIONAL FOOD RESEARCH INSTITUTE, MINISTRY OF AGRICULTURE, FORESTRY AND FISHERIES, TSUKUBA, IBARAKI 305, JAPAN

1 Introduction

Emulsifying ability is one of the most important functional properties of food proteins. Many studies have been carried out to determine what characteristics of proteins play the important role in relation to their emulsifying properties. It is now generally accepted that hydrophobicity and flexibility of proteins are especially important factors. However, the details of primary or secondary structure of the protein sites at which protein molecules are adsorbed to the oil globule surface are not yet clear.

In this study, bovine serum albumin (BSA) was hydrolysed by proteases, and the structures of the adsorbing peptide fragments were examined. BSA is the main component of plasma proteins, and extensive information including the primary structure and fatty acid-binding characteristics is available. The hydrolysate of BSA would seem to provide a good model for analysing the emulsifying behaviour of proteins.

2 Hydrolysates of Bovine Serum Albumin

Preparation of hydrolysates was carried out as described previously.[1] BSA was hydrolysed by trypsin or pepsin under various conditions, and the emulsifying activity index (EAI) of the hydrolysates was determined. The EAI of the tryptic hydrolysate (2.5% BSA, 0.04 M Tris/HCl buffer, pH 8.15, trypsin:BSA = 1:100, 37 °C, 60 min) was about 40% higher than that of BSA, suggesting that peptides responsible for the increase in EAI were produced by the hydrolysis.[1]

During emulsification, peptides are adsorbed onto the oil globule surface, which results in a decrease of interfacial tension and a stabilization of the emulsion. Peptides were extracted from the surface and their individual

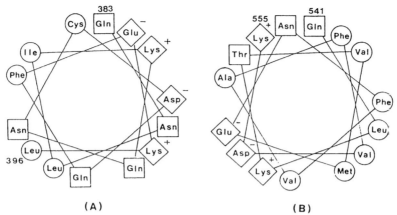

Figure 1 *Helical wheel of (A) 383–396 and (B) 541–555 residues of BSA showing amino acids with (○) hydrophobic R groups, (□) polar but uncharged R groups, and (◇) charged R groups*

components were analysed by sodium dodecyl sulfate/polyacrylamide gel electrophoresis or capillary electrophoresis. Two types of peptide in the hydrolysate were found to contribute to the emulsification: one was a peptide with a molecular mass of about 24 kDa (the 24 kDa peptide), and the other was a mixture of smaller peptides with a molecular mass range of 2–8 kDa.

3 Properties of the Hydrolysate

Amino acid composition and sequence of the 24 kDa peptide were determined, and the 24 kDa peptide was proved to be the same as 377–582 residues of BSA. The BSA molecule is composed of three domains and the 24 kDa peptide is roughly equal to its third domain. This region includes the principal fatty acid binding region.

The hydrophobicity of the tryptic hydrolysate was determined by hydrophobic chromatography. The retention time of the 24 kDa peptide was found to be larger than that of BSA, and the retention times of other peptides were shorter than that of BSA. The preferential adsorption characteristics of the 24 kDa peptide are ascribed to the presence of hydrophobic regions.

The three-dimensional structure of serum albumin (human) has been determined crystallographically by He and Cater.[2] According to them, a large helix exists at the principal fatty acid binding regions in the third domain. When we draw a helical wheel of this region, we find that the helix has an amphiphilic structure (Figure 1).

4 Amphiphilic Structure of Peptides

To clarify the effect of helix amphiphilic structure on emulsification behaviour, three types of peptides (peptides H, S and R) were synthesized and their properties were examined. Each of the three peptides consisted of eight Leu and eight Glu residues, but their sequences were different.

As shown in Table 1, both peptide H and peptide S have an α-helix structure at pH 5.5, but they exhibit a random structure at pH 7.0; in contrast, peptide R has a random structure at pH 5.5 and at pH 7.0. The helix characters of peptide H and peptide S are different; the α-helix of peptide H is an amphiphilic structure, but that of peptide S does not show such a structure. It has been shown that when a peptide takes the α-helical and amphiphilic structure, its EAI and its adsorbed amount at the oil globule surface increases markedly.

Table 1   *Properties of synthesized peptides*

	pH	α-Helix[d]	β-Sheet[d]	EAI $m^2 g^{-1}$	Amount adsorbed $mg\ g^{-1}$ of oil
Peptide H[a]	5.5	+[e]	−	43.1	0.43
	7.0	−	±	26.6	0.00
Peptide S[b]	5.5	+[f]	−	23.4	0.18
	7.0	−	+	61.1	0.48
Peptide R[c]	5.5	−	±	4.7	0.05
	7.0	−	±	12.2	0.00

[a] Leu-Glu-Glu-Leu-Leu-Glu-Glu-Leu-Leu-Glu-Glu-Leu-Leu-Glu-Glu-Leu.
[b] Glu-Leu-Glu-Leu-Glu-Leu-Glu-Leu-Glu-Leu-Glu-Leu-Glu-Leu-Glu-Leu.
[c] Leu-Glu-Leu-Leu-Glu-Glu-Glu-Leu-Leu-Glu-Glu-Glu-Leu-Leu-Glu-Leu.
[d] Evaluated from circular dichroism spectra.
[e] Amphiphilic structure.
[f] No amphiphilic structure.

Although further investigation is needed on this point, we conclude that it is advantageous for peptides to take an amphiphilic α-helical structure in order to possess good emulsifying and oil-binding properties.

5 Acknowledgement

The author is grateful to Dr M. Shimizu (University of Tokyo) for his advice.

References

1. M. Saito, K. Chikuni, M. Monma, and M. Shimizu, *Biosci. Biotech. Biochem.*, 1993, **57**, 952.
2. X. M. He and D. C. Cater, *Nature (London)*, 1992, **358**, 209.

Conformational Stability of Globular Proteins: A Differential Scanning Calorimetry Study of Whey Proteins

By Perla Relkin, Arabelle Muller, and Bernard Launay

SCIENCE DE L'ALIMENT, LABORATOIRE DE BIOPHYSIQUE, ECOLE NATIO-
NALE SUPERIEURE DES INDUSTRIES ALIMENTAIRES, 1 AVENUE DES OLYM-
PIADES, 91305 MASSY, FRANCE

1 Introduction

In food technology, heat treatment is the most widely used process that induces conformational changes on the functional properties of proteins, with desirable and undesirable effects. The heat-induced rupture of non-covalent linkages, which stabilize globular proteins in the compact, folded form, gives rise to a modified tertiary structure—less ordered, with altered surface properties—due to the exposure of previously buried groups.[1] Differential scanning calorimetry (DSC) is a powerful technique for inducing and characterizing thermal conformational perturbations, through the change of a single thermodynamic parameter, the temperature.[2] The shape of the heating curve depends on several factors, such as primary structure of the protein, concentration and chemical environment (pH, ionic strength, presence of lactose, fat or alcohol).

The aim of the present work is to discuss some results for the thermodynamic parameters of heat denaturation, determined by DSC, of three purified whey proteins. The denaturation onset temperature T_s, corrected for the thermal lag of the calorimeter,[3] is determined from the intersection of the ascending part of the peak with the straight baseline. The apparent enthalpy change ΔH_{app} is calculated using a sigmoidal baseline, taking into account not only the specific heat capacity change between folded and denatured states, but also the degree of process co-operativity.[4] The experimental results are discussed in terms of the validity of a two-state model applied to an apparently irreversible denaturation process and also in terms of the tertiary structure of human serum albumin recently determined by high resolution X-ray crystallography.[5]

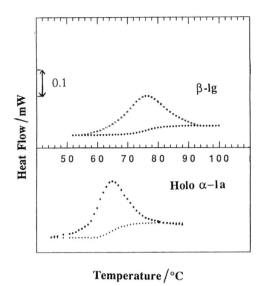

Temperature/°C

Figure 1   *Examples of thermograms (10 °C min⁻¹) observed for β-lactoglobulin
solution (4.95 wt%, pH 7, μ=0.01 M) and holo α-la (4.3 wt%, pH 7,
μ≈0). The sigmoidal baseline is drawn as described in reference 4*

2 Materials and Methods

Calcium-bound α-lactalbumin (holo α-la), calcium-depleted commercial α-lactalbumin (apo α-la) and bovine serum albumin (BSA) were supplied by Sigma (L-6010, L-5835, L-92H0154, respectively). The α-la sample contained less than 0.3 mol of Ca^{2+} mol^{-1}. The β-lactoglobulin (β-lg) sample was obtained by microfiltration.[6] The Na^+ ionic strength μ was adjusted by adding NaCl and the pH was set at 7.0±0.1 with NaOH (2 M). A Perkin Elmer DSC7 microcalorimeter was used for the determination of the thermodynamic parameters of heat denaturation (T_p, T_s, ΔH_{app}, ΔH_{vh}, $\Delta T_{1/2}$) following the methodology described elsewhere.[3]

3 Results and Discussion

An example of the initial heating curve obtained with β-lg or with holo α-la is shown in Figure 1. A further heating run of holo α-la or commercial α-la has shown that the denaturation is about 100% reversible. For β-lg, no peak is observed on the second heating curve. However, for BSA, a weak second peak (80–95 °C) is observed on both the first and second heating curves. This could be attributed to some molecular heterogeneity resulting from an incomplete purification of the sample used.[7] BSA is known to bind a wide variety of molecules and, in particular, it has a strong affinity for fatty acids.[7] The area

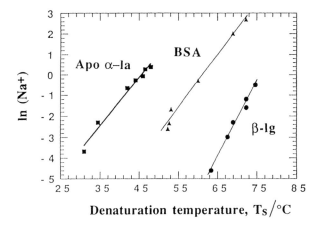

Figure 2   *Relationship on a semi-log scale between* T_s*, the onset unfolding temperature, and the Na$^+$ concentration (pH 7, 4.3 wt% protein)*

under the high-temperature peak is about 9% of that for the low-temperature one. Therefore, we assume that approximately the same amounts of proteins are complexed by fatty acids. The shape of the first and second heating curves of holo α-la is in accordance with a reversible two-state model for the heat unfolding process ($\Delta H_{app}/\Delta H_{vh} \approx 1$). In contrast, commercial α-la presents two reversible peaks. Without addition of salts, the first peak ($T_s = 30$ °C, $\Delta H_{app} = 201$ kJ mol^{-1}) and the second peak ($T_p = 64.9$ °C, $\Delta H_{app} = 30$ kJ mol^{-1}) are attributed to the unfolding of the apo-form (85% of total α-la) and of residual holo-form (15% of total α-la), respectively.[3] The thermodynamic parameters of denaturation (T_s, ΔH_{app}, ΔH_{vh} and $\Delta T_{1/2}$) do not vary with protein concentration (from 2.7 wt% to 9 wt%) for BSA and α-la, but for β-lg they decrease with protein concentration.[8] These results may be explained in the following way: the monomers of BSA and α-la do not interact, while the dimers of β-lg may undergo intermolecular association of increasing degree with increased protein and salt concentrations.

For apo α-la and BSA, T_s increases (Figure 2) while $\Delta T_{1/2}$ decreases with ionic strength μ. We have also observed a linear variation of ΔH_{app} and ΔH_{vh} versus T_s, and ΔC_p^d values determined from the slope of these lines are about 5 kJ mol^{-1} K^{-1} and 10 kJ mol^{-1} K^{-1}, for commercial α-la and BSA, respectively. For β-lg, ΔH_{app} (250 ± 30 kJ mol^{-1}) does not vary, but ΔH_{vh} increases with μ. The increase of T_s (Figure 2) and decrease of $\Delta T_{1/2}$, can be interpreted in terms of the salt ions' effectiveness at stabilizing the tertiary structure and/or in terms of ligand binding. For $\mu < 0.5$ M, the variation of T_s versus μ has a downward curvature (electrostatic and lyotropic effects) and it is linear (lyotropic effect) for $\mu \geq 0.5$ M.[9] In the absence of salt, the hypothetical values of T_s in water are, as expected, in the order α-la < BSA < β-lg, but the chaotropic constants for these three proteins are approximately equal. For β-lg, $\Delta H_{app}/\Delta H_{vh}$ is nearly equal to unity for $\mu < 0.05$ M. On the contrary, for

$\mu \geqslant 0.05$ M, where protein–protein associations are more favoured, its value is about 0.5. $\Delta H_{app}/\Delta H_{vh}$ is between 2 and 3 for BSA, depending on the concentration of added Na$^+$. The existence of a single heat denaturation peak in the temperature range 5–80 °C could also be interpreted as being due to identical unfolding properties of the different domains of BSA.[5,10] In the presence of added salt at low ionic strength ($\mu = 0.1$ M), the three principal domains are denatured independently ($\Delta H_{app}/\Delta H_{vh} \approx 3$) and at high ionic strength ($\mu = 2$ M) only two domains are unfolded ($\Delta H_{app}/\Delta H_{vh} \approx 2$). Although any unfolding peak of the fatty acid-free BSA has been observed in the second heating curves, the two-state model may be applied as for the Na$^+$ binding properties of the Ca^{2+}-free form of α-la.[3] The number n of Na$^+$ ions bound specifically per domain of folded BSA proteins is calculated from the linear variation of ln (Na$^+$) versus T_s (Figure 2) by the following equation:[11]

$$-\Delta H_{vh}/RT_s^2 + n\,d[\ln(Na^+)]/dT = 0 \qquad (1)$$

By using the values of T_s (53.7 °C and 64 °C) and ΔH_{vh} (336 kJ mol^{-1} and 517 kJ mol^{-1}) at $\mu = 0.1$ M and 2 M, respectively, eqn (1) gives $n = 3$ and $n = 2$. Aoki and coworkers[10] also observed at pH $\geqslant 7$ and $\mu \geqslant 0.05$ M a single peak in the DSC heating curves (1 °C min^{-1}) with defatted and SH-blocked BSA (2%). They considered that a crevice in the BSA tertiary structure, responsible for two peaks at $\mu < 0.05$ M, was suppressed by electrostatic screening. In this work, we have not observed three unfolding peaks, as expected at low μ, on account of high scan rate used (10 °C min^{-1}). The same methodology was applied to determine the β-lg unfolding parameters. We obtained $T_s = 69$ °C and 74.7 °C and $\Delta H_{vh} = 366$ kJ mol^{-1} and 405 kJ mol^{-1}, at $\mu = 0.1$ M and 0.6 M, respectively, indicating that, under our experimental conditions, Na$^+$ ions stabilize the protein conformation through specific binding ($n \approx 1$).

References

1. J. E. Kinsella, *CRC Crit. Rev. Food Sci. Nutr.*, 1976, **4**, 219.
2. P. L. Privalov and N. N. Khechinashvili, *J. Mol. Biol.*, 1974, **88**, 665.
3. P. Relkin, B. Launay, and L. Eynard, *J. Diary Sci.*, 1973, **76**, 36.
4. P. Relkin, *Thermochim. Acta.*, 1994, **246**, 371.
5. X. M. He and D. Carter, *Nature (London)*, 1992, **358**, 209.
6. J. L. Maubois, A. Pierre, J. Fauquant, and M. Piot, *Int. Dairy Fed.*, 1987, **212**, 145.
7. T. Peters Jr, *Adv. Protein Chem.*, 1985, **161**, 37.
8. P. Relkin and B. Launay, *Food Hydrocolloids*, 1990, **4**, 19.
9. P. M. Von Hippel and T. Shleish, 'Structure and Stability of Biological Macromolecules', Marcel Dekker, New York, 1969, p. 417.
10. M. Yamasaki, H. Yano, and K. Aoki, *Int. J. Biol. Macromol.*, 1990, **12**, 263.
11. V. Edge, N. M. Allewell, and J. Sturtevant, *Biochemistry*, 1985, **24**, 5899.

Thermal Denaturation and Aggregation of β-Lactoglobulin Studied by Differential Scanning Calorimetry

By M. A. M. Hoffmann, P. J. J. M. van Mil, and C. G. de Kruif

NETHERLANDS INSTITUTE FOR DAIRY RESEARCH (NIZO), PO BOX 20, 6710 BA EDE, THE NETHERLANDS

1 Introduction

Although many Differential Scanning Calorimetry (DSC) studies have been performed on the thermal denaturation of β-lactoglobulin (β-lg), the major whey protein in milk, the detailed mechanism of denaturation is still unclear.[1-4] The denaturation and aggregation of β-lg is followed by an irreversible aggregation. The process of irreversible denaturation and aggregation can be presented schematically as

$$N \rightleftharpoons D \rightarrow A$$

where N is the initial native state, D the denatured state and A the aggregated state. The occurrence of aggregation affects the DSC thermogram and prevents simple analysis of data in kinetic studies.

The objective of this study was to separate the denaturation and aggregation processes by varying the experimental conditions (β-lg concentration, heating rate and ionic strength). As aggregation is an intermolecular process, less aggregation will occur at low concentrations of β-lg. By varying the heating rate we tried to separate the denaturation and aggregation processes in time.

2 Experimental

The thermal behaviour of β-lg solutions in distilled water was monitored by DSC with a Perkin Elmer DSC7 within the temperature interval of 35 to 120 °C at several heating rates. From a DSC thermogram the fractional completion α of the overall reaction can be obtained, i.e.

$$\alpha = \frac{\Delta H_{part}(T)}{\Delta H_{tot}} \tag{1}$$

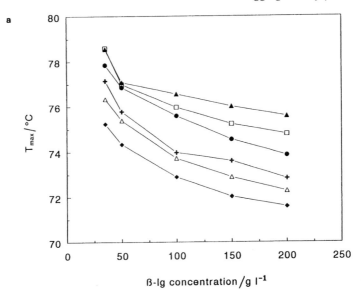

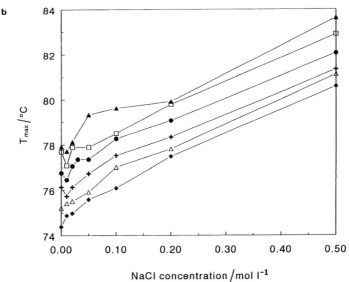

Figure 1   *(a) Peak temperature T_{max} as a function of β-lg concentration. Heating rate ($°C\ min^{-1}$): 1 (◆); 2.5 (△); 5 (+); 10 (●); 15 (□); 20 (▲). (b) Peak temperature T_{max} for 50 g l^{-1} β-lg with various concentrations of NaCl. Heating rate ($°C\ min^{-1}$): 1 (◆); 2.5 (△); 5 (+); 10 (●); 15 (□); 20 (▲)*

where $\Delta H_{\text{part}}(T)$ is the partial enthalpy at temperature T and ΔH_{tot} the total enthalpy. Assuming an Arrhenius temperature dependence, the reaction rate can be written as

$$\frac{d\alpha}{dt} = kc_0^{n-1}(1-\alpha)^n = A\exp\left(-\frac{E_a}{RT}\right)c_0^{n-1}(1-\alpha)^n \qquad (2)$$

where k is the rate constant, E_a the activation energy, A the pre-exponential factor, n the order of the reaction and c_0 the initial reactant concentration. The quantity $\ln k$ for various values of n was plotted against $1/T$ and the order of the reaction was determined by the value of n that gave the best straight line.

In a separate experiment the effect of NaCl on the aggregation was determined by heating a 10 g l^{-1} β-lg solution containing varying amounts of NaCl in test tubes at 75 °C. At regular time intervals a tube was taken, and, after cooling and adjusting the pH to 4.7, the aggregated proteins were separated by centrifugation for 30 min at 20 000 *g*. The native β-lg concentration present in the supernatant was determined by high performance gel permeation chromatography (HP-GPC).[5]

3 Results and Discussion

The DSC peak temperature T_{max} decreases with increasing β-lg concentration (Figure 1a); this decrease may be ascribed to the increasing contribution of aggregation. The aggregation reaction results in a shift of the denaturation equilibrium and increases the conversion of native β-lg molecules on heating, which results in a lower peak temperature.[6]

Adding increasing concentrations of NaCl to 50 g l^{-1} β-lg solutions gives a small decrease in peak temperature at low concentration (0.01 M) and increases the peak temperature at higher concentrations of NaCl (Figure 1b). The increase in thermal stability with increasing NaCl concentration indicates that NaCl stabilizes the native molecule.

A reaction order of 1.5 has led to the best Arrhenius plots. If a single reaction following Arrhenius kinetics is taking place the kinetic parameters are independent of protein concentration and heating rate.[7] The obtained values of the activation energy E_a increase with protein concentration and this increase is more pronounced at low heating rates (Figure 2a). Under these conditions aggregation plays an important role and the assumption of a single reaction is no longer valid. Under conditions where a minimal effect of aggregation is expected, *i.e.* a high heating rate and a low β-lg concentration, the obtained values become more or less constant. Assuming that under these conditions aggregation only has a minor effect, a value of 400 ± 25 kJ mol^{-1} is found for the activation energy of the denaturation reaction. In Figure 2(b) we see that E_a increases with NaCl concentration. Again this effect is more pronounced at lower heating rates, demonstrating the increasing effect of aggregation.

The rate constants k obtained at 75 °C show the same trend as a function of the β-lg concentration as the activation energy; at low concentrations or high heating rate $k(75\ °C)$ is a constant value. At low heating rates $k(75\ °C)$ increases with protein concentration due to the increasing contribution of

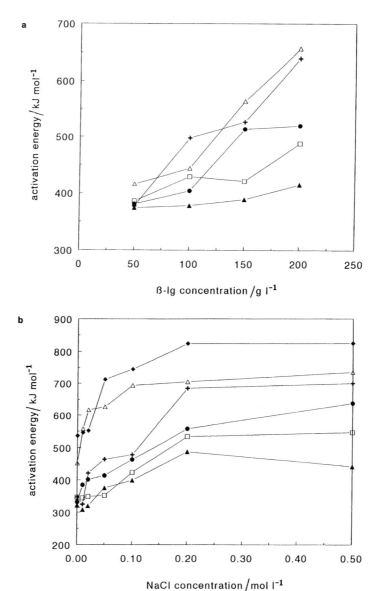

Figure 2   *(a) Activation energy* E_a *as a function of β-lg concentration. Heating rate* $(°C\ min^{-1})$: *2.5 (△); 5 (+); 10 (●); 15 (□); 20 (▲). (b) Activation energy* E_a *for 50 g* l^{-1} *β-lg with various concentrations of NaCl. Heating rate* $(°C\ min^{-1})$: *1 (◆); 2.5 (△); 5 (+); 10 (●); 15 (□); 20 (▲)*

aggregation (Figure 3a). With various NaCl concentrations $k(75\ °C)$ increases from 0 to 0.01 M NaCl and decreases at higher concentrations; this is consistent with the decrease and increase of the peak temperature in the corresponding concentration ranges. The $k(75\ °C)$ values in Figure 3(b) are less dependent on the heating rate than those in Figure 3(a). For various β-lg concentrations 75 °C is very close to the obtained peak temperatures, and both denaturation and aggregation can occur. For several NaCl concentrations, the

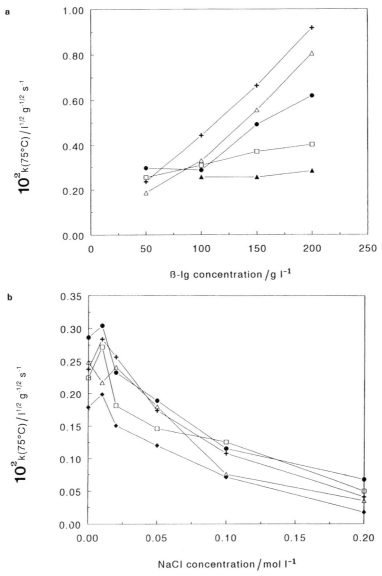

Figure 3   *(a) Reaction rate constants* k*(75 °C) as a function of* β*-lg concentration. Heating rate (°C min⁻¹): 2.5 (△); 5 (+); 10 (●); 15 (□); 20 (▲). (b) Reaction rate constants* k*(75 °C) for 50 g l⁻¹* β*-lg with various concentrations of NaCl. Heating rate (°C min⁻¹): 1 (◆); 2.5 (△); 5 (+); 10 (●); 15 (□)*

peak temperatures are above 75 °C and only the primary stage of the reaction is considered. Aggregation is of less importance and so less effect of varying the heating rate is seen.[6]

More direct information of the effect of NaCl on the aggregation reaction was obtained by studying the influence of several salt concentrations on the conversion of native β-lg in test tubes at 75 °C as measured by HP-GPC. By screening electrical charges, NaCl promotes aggregation: up to 0.05 M, the concentration decrease of native β-lg is accelerated due to facilitated aggregation (Figure 4).[8] With 0.1 and 0.2 M NaCl there is a slow initial concentration decrease of native β-lg; after about 30 min the reaction is accelerated. The slow initial decrease may be due to the slow denaturation reaction under these conditions. Extrapolating the peak temperatures in Figure 1(a) to a β-lg concentration of 10 g l^{-1} and zero heating rate (the conditions of the aggregation experiment) a peak temperature of 75–76 °C is obtained. From Figure 1(b) we see that by addition of NaCl the peak temperature increases strongly. The slow initial concentration decrease observed with 0.1 and 0.2 M NaCl may be due to the fact that the peak temperature is a few degrees above 75 °C whereas for the lower NaCl concentrations it is close to 75 °C. This high peak temperature means longer heating at 75 °C before denaturation can occur, but once the molecules are denatured the accelerated aggregation results in a much faster conversion of native β-lg. At 0.5 M NaCl the peak temperature is so high that even longer heating at 75 °C is required for denaturation and subsequent aggregation.

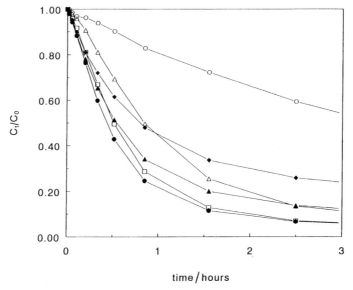

Figure 4   *Effect of NaCl on the fractional decrease of β-lg at 75 °C; initial β-lg concentration $c_0 = 10$ g l^{-1}. NaCl concentration (mol l^{-1}): 0 (◆); 0.01 (▲); 0.05 (●); 0.1 (□); 0.2 (△); 0.5 (○)*

4 Conclusions

The thermal behaviour of β-lg is complex, involving both molecular unfolding and subsequent aggregation. The experimental results confirm that aggregation plays a larger role with increasing β-lg concentration and with decreasing heating rate. At low concentrations and/or high heating rates the effect of aggregation becomes less important. This demonstrates that by varying the experimental conditions, denaturation and aggregation processes can be, at least partially, separated.

The combined effect of denaturation and aggregation as seen with DSC has an effect on the conversion of native β-lg at 75 °C as measured by HP-GPC. Then the rate of denaturation affects the overall aggregation reaction.

Acknowledgements

This research was financially supported by The Ministry of Economic Affairs through the programme IOP-Industrial Proteins, by Friesland Frico Domo/ Dairy Foods, Beilen, and by Coberco Research, Deventer.

References

1. J. N. de Wit and G. A. M. Swinkels, *Biochim. Biophys. Acta*, 1980, **624**, 40.
2. P. Relkin and B. Launay, *Food Hydrocolloids*, 1990, **4**, 19.
3. S. M. Gotham, P. J. Fryer, and A. M. Pritchard, *Int. J. Food Sci. Technol.*, 1992, **27**, 313.
4. M. Paulsson, P.-O. Hegg, and H. B. Castberg, *Thermochim. Acta*, 1985, **95**, 435.
5. J. N. de Wit, *J. Dairy Sci.*, 1990, **73**, 3602.
6. I. V. Sochava, T. V. Belopolskaya, and O. I. Smirnova, *Biophys. Chem.*, 1985, **22**, 323.
7. S. M. Taylor and P. J. Fryer, *Food Hydrocolloids*, 1993, **6**, 543.
8. V. R. Harwalkar and M. Kalab, *Milchwissenschaft*, 1985, **2**, 40.

Changes in Molecular Structure and Functionality during Purification and Denaturation of Faba Bean Proteins

By Harshadrai M. Rawel and Gerald Muschiolik[1]

GERMAN INSTITUTE OF HUMAN NUTRITION, UNIVERSITY OF POTSDAM,
A.-SCHEUNERT-ALLEE 114–116, D-14558 BERGHOLZ-REHBRÜCKE, GERMANY
[1]CENTRE OF ENVIRONMENTAL SCIENCES, UNIVERSITY OF POTSDAM, A.-
SCHEUNERT-ALLEE 114–116, D-14558 BERGHOLZ-REHBRÜCKE, GERMANY

1 Introduction

The extraction, modification and the subsequent use of biopolymers with special surface properties is of importance for the future use of plant proteins, especially those from legumes, in emulsion and foam systems. A physical modification procedure has already been demonstrated to improve the functional properties of soy proteins.[1] The main aim here is to show how to influence the foaming properties of faba bean proteins by reducing the content of residual lipids and by denaturing the proteins using mechanolysis (mechanical working).

2 Experimental

Faba bean proteins FBPI 1 and FBPI 2 were prepared in a similar way to that described in a recent publication,[2] where for FBPI 2 the starting material was partially defatted with isopropanol. Physical modification using mechanolysis was carried out for differing periods of time using a vibrating mill.[1] Mechanolysed and control preparations were characterized for their N-content and solubility at pH 7 using the Kjeldahl method.

The amount of lipids in the investigated samples was determined using the sulfo-phospho-vanillin reaction.[3] The amount of free amino groups in the samples was determined with the TNBS method.[4] A modified procedure of Ellman (1959) was used to measure the content of free available and total sulfhydryl groups in native and denatured protein samples.[5] Reversed-phase high-performance liquid chromatography (RP-

HPLC)[6] was performed to investigate changes in the hydrophobe–hydrophile character of the preparations.

Gel permeation chromatography (GPC) of sodium dodecyl sulfate (SDS) soluble proteins was carried out on Synchropak GPC column (250×4.6 mm i.d., pores = 300 Å, 6.5 mm) using a Shimadzu LC-10AS HPLC system (Japan). The separation was with 0.05 M Tris/HCl, pH 6.8 (0.1% SDS), at 280 nm. SDS–polyacrylamide gel electrophoresis (PAGE) was performed under reducing conditions in the presence of β-mercaptoethanol.[7] Determination of foaming properties was carried out as already described.[1]

Table 1   *Changes induced by defatting and mechanolysis*

Property	FBPI 1			FBPI 2		
	Control	me[a](5 h)	me (30 h)	Control	me (5 h)	me (30 h)
Solubility (NSI-%)(pH 7)	100.0	91.0	61.0	100.0	40.8	71.8
Free amino groups[b]	0.61	0.50	0.88	0.86	1.15	3.37
Sulfhydryl groups[c]						
Free available	2.83	2.10	0.19	2.05	0.82	0.48
Total	3.74	2.33	0.23	3.03	0.98	0.47
RP-HPLC						
Retention time (min)	17.12	17.28	17.67	15.83	15.83	15.82
Concentration of the main peak group	90.3	91.2	91.6	93.8	93.1	93.2
Width (min)	4.6	6.8	8.9	7.7	8.1	10.1
Foaming properties (pH 7)						
Torque (N cm)[d]	0.10	0.25	0.66	0.14	0.68	0.66
Foam volume (ml)	20	40	66	25	50	65
Drained volume (ml)[e]	7.0	4.4	0.5	6.0	0.0	0.0

[a] Duration of modification in hours.
[b] Expressed as μmol free amino groups/mg of protein.
[c] Expressed as nmol of sulfhydryl groups/mg of protein.
[d] Maximum torque (resistance) measured after 10 min whipping.
[e] Drained liquid volume 5 min after ending the foaming process of the protein solution.

3 Results and Discussion

Both samples FBPI 1 and FBPI 2 showed a protein content of *ca.* 85%. Defatting the meal with isopropanol resulted in a decrease of the fat content from 4.2% in FBPI 1 (non-defatted sample) to 2.4% in FBPI 2 (defatted sample). This was accompanied with an increase in solubility and in the amount of free amino groups as shown in Table 1. A parallel decrease in the amount of free, available and total sulfhydryl groups was also observed for the defatted protein (Table 1). SDS/GPC in the absence of reducing agent showed no particular change of the chromatogram pattern for both samples, whereas in SDS/PAGE patterns, a comparatively higher concentration (*ca.* 53.1%) of legumin-specific (α- and β-chains) bands was detected in FBPI 2 compared with 47.6% in FBPI 1. RP-HPLC delivered a higher retention time for FBPI 1

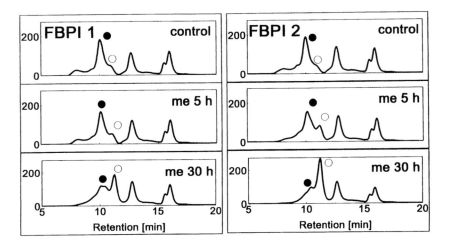

Figure 1   *SDS/GPC of the faba bean protein samples FBPI 1 and FBPI 2.*
Peak 1 = legumin-specific sub-fraction; peak 2 = dissociation product after
mechanolysis

suggesting the presence of a higher hydrophobicity in the sample (Table 1). The foaming properties improved only slightly as a result of the defatting process, although results obtained with a sample defatted with supercritical CO_2 did show a more significant improvement of the surface-active properties.[8]

Further modification of these two samples using vibration milling was found to cause loss of solubility, and a parallel decrease in the amount of sulfhydryl groups was observed (Table 1). Depending on the duration of the modification, an increase in the rate of migration of proteins to the air–water interface as reflected by the increase in foam volume was also noted (Table 1). As a result of mechanical denaturation, a decrease in drainage with parallel changes in molecular mass, hydrophile–hydrophobe character and the proportion of charged groups present were observed. This was generally accompanied by a dissociation of the 11 S protein fraction (70–78 kDa, major component of the isolates) as shown by the GPC analysis (Figure 1). A decreasing concentration of the 11 S protein subfraction (as a result of denaturation; dissociation to 33–34 kDa) correlating with the increasing foam volume was identified (Figure 1). SDS/PAGE results also confirmed that dissociation processes were induced during mechanolysis with a gradual decrease in the concentration of legumin-specific bands largely affecting the α-chains. RP-HPLC chromatograms showed, however, a gradual broadening of the main protein peak, suggesting changes in hydrophile–hydrophobe balance.

The S–S bonds formed are most likely to be of intramolecular nature, since most of the protein is soluble in the presence of SDS (leading to dissociation) suggesting that insolubility results from intermolecular, secondary, reversible interactions (ionic, hydrophobic, hydrogen bonds, *etc.*). All these results reflect the reversible denaturation caused by mechanolysis, this being of a stronger nature for the defatted sample FBPI 2. Further investigations showing the influence of the lipid components on the foaming properties will be published soon.

Acknowledgements

This research project was supported by DAAD and by the FEI (Forschungskreis der Ernährungsindustrie e.V., Bonn), the AiF, and the Ministry of Economics (Project No. 303D).

References

1. H. M. Rawel and G. Muschiolik, *Food Hydrocolloids*, 1994, **8**, 287.
2. G. Muschiolik, H. M. Rawel, Th. zu Höne, and K. Heinzelmann, *Nahrung*, 1994, **38**, 462.
3. N. Zöllner and K. Kisch, *Z. Ges. Exp. Med.*, 1962, **135**, 545.
4. A. K. Hazra, S. P. Chock, and R. W. Albers, *Anal. Biochem.*, 1984, **137**, 437.
5. W. J. Wolf, *J. Agric. Food Chem.*, 1993, **41**, 168.
6. A. V. Flannery, R. J. Beynon, and J. S. Bons, in 'Proteolytic Enzymes', ed. R. J. Beynon and J. S. Bond, IRL Press, Oxford, 1989, p. 159.
7. V. K. Laemmli, *Nature (London)*, 1970, **227**, 680.
8. F. Husband, P. J. Wilde, D. C. Clark, H. M. Rawel, and G. Muschiolik, *Food Hydrocolloids*, 1994, **8**, 455.

Microstructural, Physico-chemical, and Functional Properties of Commercial Caseinates

By Pascale Bastier, Eliane Dumay, and Jean-Claude Cheftel

UNITÉ DE BIOCHIMIE ET TECHNOLOGIE ALIMENTAIRES, GÉNIE BIOLO-
GIQUE ET SCIENCES DES ALIMENTS, UNIVERSITÉ DE MONTPELLIER II,
34095 MONTPELLIER, FRANCE

1 Introduction

The aim of this work was to evaluate the variability in the microstructure, the composition, and the physico-chemical and functional properties of caseinates available on the French market. It was attempted to correlate these properties with one another and to classify the caseinates.

2 Materials and Methods

The traditionally processed caseinates studied were: Ca (1–2–3–4), instant Ca and Na caseinates from Ingredia (62 Arras, France); Ca and instant Ca caseinates from MD Foods (Denmark); Ca, Na (partially hydrolysed), K and Ca/Na caseinates from Armor Proteines (35 Saint Brice en Cogles, France); Ca and low-viscosity Na caseinates from Prolait (79, Niort, France). The extrusion-processed caseinates studied were: Ca, Ca/Na, and low- and high-viscosity Na caseinates from Besnier Bridel Alimentaires (35 Bourgbarre, France); Ca, Na, K and Ca/K caseinates from Unilait (75 Paris, France); and Na and K caseinates from Eurial (44 Nantes, France).

The microstructure of the caseinates was observed on metallized powder (150 Å gold) using a scanning electron microscope JEOL JSM 6300 F. Dry matter was measured on 1 g of powder. Protein content was analysed by the Kjeldahl method ($N \times 6.38$). The pH was measured on a 5% aqueous dispersion of caseinate. Calcium content was evaluated with the calcein method according to Ntailianas and Whitney.[1] Density was measured on 60 g of powder before and after compression. Particle-size distribution was determined on a 0.1% ethanolic dispersion of caseinate using a Malvern Mastersizer E laser diffractometer. The degree of hydrolysis was estimated by formol titration according to Fleming *et al.*[2] Solubility index at pH 7 was determined by drying, until constant weight, the supernatant of a 2% caseinate aqueous dispersion after centrifugation for 15 min at 18 850 *g*. Water absorption capacity was measured on 150 mg of caseinate at 20 °C using a Baumann Apparatus.[3] Apparent viscosity at 150 s^{-1} was measured on a 15% caseinate aqueous dispersion at 25 °C using a Rotovisco Haake double cylinder viscometer.

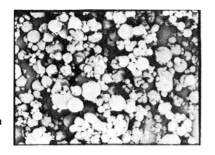

Figure 1 *Microstructure of Ca caseinate 3 from Ingredia (scanning electron microscopy): (a) × 135, (b) × 850*

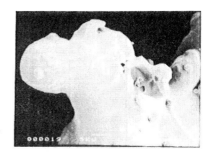

Figure 2 *Microstructure of Instant Ca caseinate from MD Foods (scanning electron microscopy): (a) × 135, (b) × 850*

Figure 3 *Microstructure of Ca/Na caseinate from Besnier Bridel Alimentaire (scanning electron microscopy): (a) × 135, (b) × 850.*

Table 1 Composition and physico-chemical properties of some commercial caseinates

Caseinate	Dry matter (g/g)	Protein content (g/g) N × 6.38	pH[a]	Ca content[b] (g/kg)	Density[c] (g/ml) Before compression	Density[c] (g/ml) After compression	Mean particle diameter[d] (μm) $D_{v0.5}$	Mean particle diameter[d] (μm) $D_{v0.95}$
Ca 1 ING	$0.93 \pm 1 \times 10^{-3}$	$0.86 \pm 0.7 \times 10^{-2}$	6.67 ± 0.00	10.9 ± 0.2	0.26	0.36	44	95
Ca 2 ING	$0.94 \pm 1 \times 10^{-3}$	$0.87 \pm 1.8 \times 10^{-2}$	6.40 ± 0.00	9.5 ± 0.2	0.28	0.40	43	94
Ca 3 ING	0.94 ± 0	$0.88 \pm 0.9 \times 10^{-2}$	6.73 ± 0.00	10.3 ± 0.2	0.28	0.38	42	97
Ca 4 ING	$0.94 \pm 2 \times 10^{-3}$	$0.87 \pm 3.5 \times 10^{-2}$	7.05 ± 0.00	12.7 ± 0.2	0.25	0.33	43	90
Ca inst ING	$0.93 \pm 1 \times 10^{-3}$	$0.84 \pm 1.5 \times 10^{-2}$	6.45 ± 0.01	10.0 ± 0.2	0.21	0.23	NM	NM
Ca MDF	$0.94 \pm 1 \times 10^{-3}$	$0.86 \pm 1.3 \times 10^{-2}$	6.94 ± 0.00	12.7 ± 0.2	0.27	0.41	41	86
Ca inst MDF	$0.94 \pm 2 \times 10^{-3}$	$0.87 \pm 2.7 \times 10^{-2}$	6.95 ± 0.00	13.7 ± 0.2	0.31	0.37	NM	NM
Ca AP	$0.94 \pm 1 \times 10^{-3}$	$0.87 \pm 0.5 \times 10^{-2}$	6.91 ± 0.00	13.0 ± 0.7	0.34	0.46	58	181
Ca PRO	$0.95 \pm 1 \times 10^{-3}$	$0.83 \pm 0.3 \times 10^{-2}$	6.93 ± 0.00	10.9 ± 0.2	0.34	0.48	77	180
Ca BBA	$0.94 \pm 1 \times 10^{-3}$	$0.89 \pm 1.9 \times 10^{-2}$	7.03 ± 0.01	9.6 ± 0.1	0.42	0.53	ND	ND
Ca UNI	$0.94 \pm 1 \times 10^{-3}$	$0.87 \pm 0.3 \times 10^{-2}$	6.89 ± 0.00	10.9 ± 0.2	0.60	0.72	61	152
Na ING	0.93 ± 0	$0.85 \pm 0.1 \times 10^{-2}$	6.72 ± 0.01	NM	0.30	0.42	91	248
Na Hyd AP	$0.92 \pm 1 \times 10^{-3}$	$0.84 \pm 0.2 \times 10^{-2}$	7.35 ± 0.01	NM	0.16	0.21	92	229
Na LV PRO	$0.92 \pm 2 \times 10^{-3}$	$0.83 \pm 2.0 \times 10^{-2}$	6.64 ± 0.01	1.8 ± 0.0	0.26	0.35	ND	ND
Na LV BBA	$0.95 \pm 1 \times 10^{-3}$	$0.89 \pm 0.1 \times 10^{-2}$	6.91 ± 0.01	NM	0.43	0.61	ND	ND
Na HV BBA	$0.94 \pm 1 \times 10^{-3}$	$0.82 \pm 3.8 \times 10^{-2}$	6.97 ± 0.01	5.0 ± 0.0	0.43	0.60	ND	ND
Na UNI	$0.95 \pm 1 \times 10^{-3}$	$0.86 \pm 0.1 \times 10^{-2}$	6.81 ± 0.00	NM	0.48	0.62	61	186
Na EUR	0.93 ± 0	$0.85 \pm 2.7 \times 10^{-2}$	6.70 ± 0.00	NM	ND	ND	72	203

Table 1 *Continued*

Caseinate	Dry matter (g/g) N × 6.38	Protein content (g/g)	pH^a	Ca content^b (g/kg)	Density^c (g/ml) Before compression	Density^c (g/ml) After compression	Mean particle diameter^d (µm) $D_{v0.5}$	Mean particle diameter^d (µm) $D_{v0.95}$
K AP	$0.94 \pm 2 \times 10^{-3}$	$0.83 \pm 0.9 \times 10^{-2}$	6.71 ± 0.00	NM	0.38	0.50	66	186
K UNI	$0.95 \pm 1 \times 10^{-3}$	$0.86 \pm 1.8 \times 10^{-2}$	6.93 ± 0.01	NM	0.52	0.68	80	204
K EUR	$0.93 \pm 1 \times 10^{-3}$	$0.82 \pm 0.7 \times 10^{-2}$	6.65 ± 0.00	NM	ND	ND	64	185
Ca/Na AP	0.92 ± 0	$0.80 \pm 1.0 \times 10^{-2}$	6.85 ± 0.01	7.5 ± 0.1	0.29	0.42	55	196
Ca/Na BBA	$0.94 \pm 1 \times 10^{-3}$	$0.86 \pm 0.1 \times 10^{-2}$	7.10 ± 0.00	10.5 ± 0.2	0.49	0.66	70	240
Ca/K UNI	$0.93 \pm 1 \times 10^{-3}$	$0.84 \pm 0.7 \times 10^{-2}$	7.01 ± 0.01	6.5 ± 0.1	0.54	0.73	49	138

[a] 50 g/kg caseinate solution.
[b] Determined according to reference 1.
[c] Density before and after compression (a 2.5 cm fall for 100 times); all values ±0.01.
[d] $D_{v0.5}$: 50% (v) of the particles with a smaller diameter; $D_{v0.95}$: 95% (v); 1 g/l caseinate dispersions in 0.95 l/l ethanol.
NM, not measurable; ND, not determined.
ING = Ingredia, inst = instant, MDF = MD Foods, AP = Armor Proteines, PRO = Prolait, BBA = Besnier Bridel Alimentaires, UNI = Unilait, EUR = Eurial, Hyd = partially hydrolysed, LV = low viscosity, HV = high viscosity.

Table 2   *Functional properties of commercial caseinates*

Caseinate	Degree of hydrolysis[a] (mg N g^{-1} of protein)	Solubility[b] pH 7 (g/g)	Water absorption[c] capacity (ml/g)	Initial rate of water absorption[d] (ml/g min)	Apparent viscosity[e] (mPa s)
Ca 1 ING	7.1 ± 0.3	$0.93 \pm 2 \times 10^{-3}$	1.27 ± 0.06	$4.3 \times 10^{-2} \pm 0.3 \times 10^{-2}$	320 ± 99
Ca 2 ING	7.9 ± 0.3	$0.97 \pm 5 \times 10^{-3}$	1.40 ± 0.08	$5.7 \times 10^{-2} \pm 0.7 \times 10^{-2}$	630 ± 35
Ca 3 ING	7.4 ± 0.1	$0.91 \pm 3 \times 10^{-3}$	1.28 ± 0.09	$5.1 \times 10^{-2} \pm 0.3 \times 10^{-2}$	310 ± 36
Ca 4 ING	7.3 ± 0.1	$0.81 \pm 2 \times 10^{-3}$	1.70 ± 0.12	$2.8 \times 10^{-2} \pm 0.3 \times 10^{-2}$	40 ± 2
Ca inst ING	7.2 ± 0.3	$0.92 \pm 4 \times 10^{-3}$	1.41 ± 0.04	NM	340 ± 118
Ca MDF	6.8 ± 0.1	$0.93 \pm 6 \times 10^{-3}$	1.68 ± 0.14	$3.5 \times 10^{-2} \pm 0.8 \times 10^{-2}$	1160 ± 186
Ca inst MDF	6.7 ± 0.2	$0.94 \pm 2 \times 10^{-3}$	1.55 ± 0.03	$2.2 \times 10^{-2} \pm 0.4 \times 10^{-2}$	1020 ± 117
Ca AP	6.6 ± 0.1	$0.96 \pm 1 \times 10^{-3}$	1.63 ± 0.05	$2.1 \times 10^{-2} \pm 0.1 \times 10^{-2}$	110 ± 16
Ca PRO	3.7 ± 0.4	$0.51 \pm 4 \times 10^{-3}$	3.49 ± 0.03	NM	70 ± 10
Ca BBA	6.0 ± 0.1	$0.90 \pm 1 \times 10^{-2}$	1.91 ± 0.12	$5.0 \times 10^{-2} \pm 0.5 \times 10^{-2}$	1180 ± 123
Ca UNI	7.9 ± 0.3	$0.90 \pm 1 \times 10^{-3}$	1.90 ± 0.10	$4.9 \times 10^{-2} \pm 0.5 \times 10^{-2}$	1040 ± 169
Na ING	6.3 ± 0.1	$0.97 \pm 2 \times 10^{-3}$	2.25 ± 0.10	$11 \times 10^{-2} \pm 0.5 \times 10^{-2}$	850 ± 71
Na Hyd AP	7.9 ± 0.3	$0.97 \pm 3 \times 10^{-3}$	1.61 ± 0.09	$7.4 \times 10^{-2} \pm 0.3 \times 10^{-2}$	230 ± 80
Na LV PRO	7.1 ± 0.2	$0.96 \pm 1 \times 10^{-2}$	1.51 ± 0.11	$6.0 \times 10^{-2} \pm 0.0 \times 10^{-2}$	20 ± 4
Na LV BBA	7.0 ± 0.2	$0.94 \pm 2 \times 10^{-3}$	2.38 ± 0.02	$12 \times 10^{-2} \pm 0.3 \times 10^{-2}$	1290 ± 384
Na HV BBA	6.6 ± 0.3	$0.87 \pm 2 \times 10^{-3}$	2.99 ± 0.16	$11 \times 10^{-2} \pm 0.4 \times 10^{-2}$	NM
Na UNI	6.2 ± 0.2	$0.89 \pm 2 \times 10^{-3}$	2.73 ± 0.16	$10 \times 10^{-2} \pm 0.5 \times 10^{-2}$	2660 ± 75
Na EUR	6.4 ± 0.4	$0.95 \pm 4 \times 10^{-3}$	3.06 ± 0.15	$13 \times 10^{-2} \pm 0.5 \times 10^{-2}$	2510 ± 139

Table 2 *Continued*

Caseinate	Degree of hydrolysis[a] (mg N g^{-1} of protein)	Solubility[b] (g/g) pH 7	Water absorption[c] capacity (ml/g)	Initial rate of water absorption[d] (ml/g min)	Apparent viscosity[e] (mPa s)
K AP	7.2 ± 0.1	$0.99 \pm 4 \times 10^{-3}$	2.13 ± 0.05	$10 \times 10^{-2} \pm 0.1 \times 10^{-2}$	500 ± 26
K UNI	6.2 ± 0.2	$0.94 \pm 4 \times 10^{-3}$	3.24 ± 0.08	$13 \times 10^{-2} \pm 0.5 \times 10^{-2}$	3230 ± 617
K EUR	8.1 ± 0.6	$0.97 \pm 1 \times 10^{-3}$	3.24 ± 0.15	$15 \times 10^{-2} \pm 0.2 \times 10^{-2}$	2780 ± 238
Ca/Na AP	7.2 ± 0.1	$0.92 \pm 3 \times 10^{-3}$	1.87 ± 0.05	$6.5 \times 10^{-2} \pm 0.2 \times 10^{-2}$	450 ± 80
Ca/Na BBA	5.9 ± 0.1	$0.98 \pm 4 \times 10^{-3}$	1.58 ± 0.02	$2.1 \times 10^{-2} \pm 0.2 \times 10^{-2}$	3800 ± 483
Ca/K UNI	5.5 ± 0.1	$0.90 \pm 3 \times 10^{-3}$	3.00 ± 0.07	$11 \times 10^{-2} \pm 0.4 \times 10^{-2}$	3740 ± 268

[a] Degree of hydrolysis determined by formol titration on 150 mg of caseinate.
[b] Solubility measured by mass, at pH 7.
[c] Baumann apparatus with 150 mg of caseinate powder, at 30 min.
[d] Initial rate, measured after 10 min, Baumann apparatus.
[e] At 150 s^{-1}, on 150 g/kg caseinate dispersions, Rotovisco Haake RV12.
NM, not measurable. For other abbreviations see Table 1.

a   b

Figure 4   *Microstructure of Na caseinate from Unilait (scanning electron microscopy):*
(a) × 135, (b) × 850

3 Results and Discussion

One part of this work has been studied more completely by Bastier *et al.*[4]
Caseinates could be classified into two groups according to the manufacturing
process: traditional caseinates were composed of spherical particles which were
more or less agglomerated (Figures 1 and 2); extruded caseinates were
composed of large, angular and dense aggregates with unfastened fragments
(Figures 3 and 4). Traditional caseinates, without aggregates, had a smaller
mean particle diameter than extruded caseinates, and, in the presence of
aggregates, in the same range as extruded caseinates. Caseinates exhibited a
great variability in physico-chemical and functional properties (Tables 1 and
2). The density of traditional caseinates was lower than that of extruded
caseinates. Caseinates could also be classified according to the nature of the
salt. Calcium caseinates were less soluble, had a lower water absorption
capacity, and were less viscous than sodium caseinates at the same pH.
Potassium caseinates exhibited the highest solubility, highest water absorption
capacity and highest viscosity. Mixed caseinates showed variable behaviour.
For each class of salt, traditional caseinates had a lower water absorption
capacity and were less viscous than extruded ones. However, it was impossible
to correlate the functional properties for all caseinates. Additional biochemical
experiments would be useful to explain the relationships between the protein
and modifications produced by these different treatments and the resulting
physico-chemical properties.

It appears that the choice of a specific caseinate should be done carefully
since a large range of commercial caseinates of different properties are
available.

References

1.   H. A. Ntailianas and R. Mc L. Whitney, *J. Dairy Sci.*, 1964, **47**, 19.
2.   L. K. Fleming, R. Jenness, and H. A. Morris, *Food Microstruct.*, 1985, **4**, 313.
3.   H. Baumann, *Glas Instrumenten-Technik*, 1967, **11**, 540.
4.   P. Bastier, E. Dumay, and J.-C. Cheftel, *Lebensm. Wiss. Technol.*, 1993, **26**, 529.

Biochemical and Physico-chemical Characteristics of the Protein Constituents of Crab Analogues Prepared by Thermal Gelation or Extrusion Cooking

By Maryse Thiebaud, Eliane Dumay, and Jean-Claude Cheftel

UNITÉ DE BIOCHIMIE ET TECHNOLOGIE ALIMENTAIRES, CENTRE DE GÉNIE ET TECHNOLOGIE ALIMENTAIRES, UNIVERSITÉ DE MONTPELLIER II, 34095 MONTPELLIER, FRANCE

1 Introduction

Protein texturization by extrusion cooking at high moisture (60–70% H_2O) is a relatively new process. Recent studies have shown that it is possible to texturize fish surimi (myofibrillar protein concentrate) in the presence of other proteins such as soy protein concentrate, gluten or egg white powders. Cheftel *et al.*[1] have suggested that three stages are essential: (1) fusion–plasticization of protein constituents under high temperature and high shear in the barrel of the extruder, (2) continuous and regular pumping of the plasticized raw material along the screws and in a long cooling die where texturization takes place, and (3) creation of laminar flow in the die in order to form a finely fibrous structure similar to that of crab meat. Such products are already commercialized as new seafood analogues in Japan. Despite the technological advances, the biochemical and physico-chemical mechanisms of protein texturization by high moisture extrusion cooking are not well understood. This work attempts to characterize protein–protein interactions occurring in products texturized by extrusion cooking at high temperature ($> 160\,°C$) and high moisture (63% H_2O). The protein solubility in four dissociating buffers, at two pH values, is determined for a commercial extruded crab analogue and an experimental extruded crab analogue. The soluble protein constituents are analysed by sodium dodecyl sulfate/polyacrylamide gel electrophoresis (SDS/PAGE) in the presence of urea. Results are compared with those for a commercial heat-set crab analogue (prepared by thermal gelation of a thin film, followed by scarification and folding of the sheet).

189

2 Materials and Methods

The products examined are (1) a commercial heat-set crab analogue (11 wt% protein, surimi/egg white protein ratio of 7.2 by weight); (2) a commercial extruded crab analogue produced by Nippon Suisan (14 wt% protein), containing Alaska Pollack surimi, egg white powder and starch; and (3) an experimental extruded crab analogue prepared by extrusion cooking of Alaska Pollack surimi (83.0%), egg white powder (12.4%), potato starch (1.6%), rapeseed oil (1.5%) and NaCl (1.5%). The total protein content is 23% and the surimi/egg white protein ratio is 1.4. Extrusion cooking was carried out with a Clextral BC 45 twin screw extruder (1800 mm long), equipped with a long, rectangular die (400 × 40 × 3 mm), cooled by circulation of water at 10 °C. The temperature of the product before the die was 160 °C. The feeding rate was 30 kg h^{-1} and the screw speed was 100 r.p.m.

Protein solubilization of the samples is performed in four different dissociating solutions (0.5 g of protein/100 g of solution), at two pH values: *Solution Ia*, 30 mM Tris/HCl, pH 8.0; *solution IIa*, solution Ia + 0.5% (w/v) SDS, pH 8.0; *solution IIIa*, solution IIa + 8 M urea, pH 8.0; *solution IVa*, solution IIIa + 10 mM dithiothreitol (DTT), pH 8.0; *solution Ib*, 0.0318 M Na$_2$CO$_3$, 0.0182 M NaHCO$_3$, pH 10.6; *solution IIb*, solution Ib + 0.5% (w/v) SDS, pH 10.6; *solution IIIb*, solution IIb + 8 M urea, pH 10.6; *solution IVb*, solution IIIb + 10 mM DTT, pH 10.6. The protein dispersion is kept for 1 h or 24 h at room temperature, under magnetic stirring, and is then centrifuged (Sorvall RC 2B, Du Pont de Nemours, Paris, France) at 10 000 *g* at 20 °C for 30 min. The protein concentration of the initial dispersion and of the resulting supernatant is determined according to the method of Lowry, modified by Bensadoun and Weinstein.[2] A calibration curve is obtained using a standard solution of bovine serum albumin. Results are given in g of soluble protein per 100 g of total protein.

Electrophoresis gels and protein samples are prepared according to Roussel[3] and Leinot,[4] respectively. Proteins are identified using molecular mass markers (Sigma, St Louis, MO, USA), according to the method proposed by Goll *et al.*[5] Gels are stained overnight in an aqueous solution of Coomassie brilliant blue R at 0.1% (w/v) in methanol (45% v/v) and acetic acid (9% v/v). They are then destained in an aqueous solution of methanol (15% v/v) and acetic acid (7.5% v/v) before being scanned with a Hoeffer GS 300 densitometer (Scientific Instruments, San Francisco, CA, USA).

3 Results and Discussion

Figure 1 shows the solubility of the commercial heat-set analogue and of the commercial and the experimental extruded analogues, at pH 8.0 and 10.6, after 1 or 24 h of solubilization. At pH 8.0, the solubility of the protein constituents of the commercial heat-set analogue increases as the solutions become more dissociating (1 h or 24 h extraction). At pH 10.6, protein constituents are almost totally soluble in the presence of SDS or SDS/urea, after 1 or 24 h of

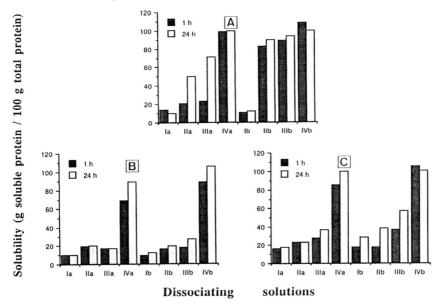

Figure 1 *Solubility in dissociating solutions at pH 8.0 and 10.6, of the protein constituents of the commercial heat-set crab analogue (A), the commercial extruded crab analogue (B), and the experimental extruded analogue (C). Solubility given in g of soluble protein per 100 g of protein in the initial suspension. Protein determination according to Bensadoun and Weinstein.[2] Means of two or three independent determinations (coefficient of variation = 2%). See text for further details*

solubilization. These results may indicate that in the kamaboko gel, protein constituents seem to interact mainly by weak interactions, *i.e.* ionic, hydrophobic and to a lesser extent hydrogen interactions. Disulfide bonds appear to play a minor role. Similar conclusions have been proposed by Lanier *et al.*[6] and Stone and Stanley.[7] Niwa[8] explains that the formation of the protein network seems to depend on interactions between hydrophobic zones exposed during protein unfolding. Roussel and Cheftel[9] have shown the importance of ionic interactions by studying the effect of the addition of $CaCl_2$ or EDTA on the texture of the gel obtained.

Both extruded analogues are sparingly soluble in solutions containing SDS and SDS–urea, at pH 8.0 or 10.6, after 1 h solubilization. A period of 24 h solubilization at pH 10.6, in the presence of DTT, is necessary to achieve a complete recovery of the protein constituents. It is to be noted that the solubility of the experimental extruded analogue is generally higher than for the commercial one. The treatment of extrusion cooking seems to induce the formation of a great number of S–S bonds. The formation of lysinoalanine cross-links does not appear to contribute significantly to texturization by extrusion cooking (results not shown). Weak interactions are less important. This indicates that the mechanism

of protein texturization by extrusion cooking is very different from that
of the classic heat-set gel.

SDS/PAGE of the protein constituents of the commercial heat-set analogue
shows that actin, tropomyosin, troponin, myosin light chains, conalbumin and
ovalbumin are gradually extracted at pH 8.0 in the four buffers (1 h and 24 h
solubilization) whereas at pH 10.6 these proteins are almost completely
extracted after 1 h solubilization in the SDS buffer. Myosin heavy chains (HC)
are very difficult to solubilize at pH 8.0 since the presence of SDS, urea and
DTT together (1 h solubilization) is necessary. At pH 10.6, only 1 h
solubilization is required to extract most of the myosin HC in the presence of
SDS. These results indicate that myosin HC may play an important role in the
creation of the gel network by forming numerous ionic and hydrophobic
interactions. The other protein constituents seem to act as fillers in the
network. It is now commonly admitted that myosin and actomyosin (the
complex of myosin and actin) are largely involved in the thermal gelation of
myofibrillar proteins.[6–8] The present results are in good agreement with these
references.

Figure 2 shows the SDS/PAGE results of protein constituents of both
extruded analogues after 24 h solubilization at pH 10.6. These conditions give
the best separation on the gel. Surprisingly, in all cases, a myosin HC band is
absent from the gel and strong baseline noise is observed. High molecular mass

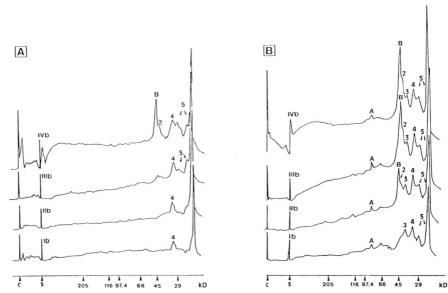

Figure 2   *SDS/PAGE in the presence of urea of the protein constituents of the
commercial (A) and the experimental (B) extruded crab analogues.
Solubilization for 24 h in the pH 10.6 dissociating solutions. Sample 50 µl; c,
stacking gel; s, separating gel. Composition of solutions as in Figure 1. Peak
identification: (1) myosin heavy chains, (2) actin, (3) tropomyosin, (4)
troponin T, (5) troponins I, C and myosin light chains, (A) conalbumin, (B)
ovalbumin*

aggregates can enter neither the stacking gel nor the separating gel. The soluble protein constituents of the experimental extruded analogue can be more easily separated and identified, especially actin and tropomyosin. These analyses attest to an important molecular rearrangement during extrusion cooking. Both random hydrolysis and polymerization of polypeptide chains could explain the baseline noise and the absence of myosin HC. The commercial extruded analogue, which shows a more drastic aggregation than the experimental one, could have been treated at a higher temperature and/or may contain more egg white protein (rich in S–S bonds and SH residues) or other ingredients. The frozen storage of this product could also have enhanced myosin HC aggregation. Kitagawa *et al.*[10] and Miyano *et al.*[11] have already described the same type of aggregation with myofibrillar proteins; and it has been mentioned even for soya proteins,[12] but the phenomenon was not explained in detail. It has been suggested that isopeptide cross-links (glutamyl lysine or aspartyl-lysine types) may be formed.[1] The influence of the composition of the extruded raw material and of the extrusion parameters (temperature, residence time, shearing) on the degree of protein aggregation are not well understood and require further study.

References

1. J.-C. Cheftel, M. Kitagawa, and C. Quéguiner, *Food Rev. Int.*, 1992, **8**, 235.
2. A. Bensadoun and D. Weinstein, *Anal. Biochem.*, 1976, **70**, 241.
3. H. Roussel, Thèse de Docteur ès Sciences, Université des Sciences et Techniques du Languedoc, Montpellier, France, 1988.
4. A. Leinot, Thèse de Docteur ès Sciences, Université des Sciences et Techniques du Languedoc, Montpellier, France, 1990.
5. D. E. Goll, R. M. Robson, and M. H. Stromer, 'Food Proteins', Avi Publishing, Westport, CT, 1977, p. 121.
6. T. C. Lanier, T. S. Lin, V. M. Liu, and D. D. Hamann, *J. Food Sci.*, 1982, **47**, 1951.
7. A. P. Stone and D. W. Stanley, *Food Res. Int.*, 1992, **25**, 381.
8. E. Niwa, 'Surimi Technology', Marcel Dekker, New York, 1992, p. 389.
9. H. Roussel and J.-C. Cheftel, *Int. J. Food Sci. Technol.*, 1990, **25**, 260.
10. M. Kitagawa, T. Iida, and S. Nobuta, *Sci. Rep. of the Hokkaido Fisheries Experimental Station*, 1991, **36**, 81.
11. S. Miyano, K. Satoh, K. Kitazume, N. Nakagawa, and N. Kato, *Nippon Suisan Gakkaishi*, 1992, **58**, 693.
12. A. Noguchi, 'Extrusion Cooking', American Association of Cereal Chemists, St. Paul, MN, 1989, p. 343.

Interactions between Fat Crystals and Proteins at the Oil–Water Interface

By Leanne G. Ogden and Andrew J. Rosenthal

SCHOOL OF BIOLOGICAL AND MOLECULAR SCIENCES, OXFORD BROOKES
UNIVERSITY, GIPSY LANE CAMPUS, HEADINGTON, OXFORD OX3 0BP, UK

1 Introduction

The intention of this study was to investigate the effects of fat crystals on the
interfacial shear viscosity of model systems containing sodium caseinate at the
oil–water interface. Hydrocarbon and triglyceride-based oils were used, the
former because they have been more widely studied in the past, and the latter
because of their relevance to real food systems.

2 Experimental

A Couette-type torsion-wire surface shear viscometer was used.[1] This consisted
of a biconical stainless steel disc suspended from a wire with a torsion constant
of 1.86×10^{-7} N m rad^{-1}, which hung centrally into a water jacketed dish
$(20 \pm 0.1\,°C)$, rotating at a rate of 1.25×10^{-3} rad s^{-1}.

Tristearin (Sigma product number T 4633 recrystallized from diethyl ether)
and n-tetradecane (Sigma product number T 1521) were used. The sodium
caseinate (DeMelkindustrie Veghel) and sunflower oil were food grade. The
sunflower oil was pre-treated with silica to remove surface-active components.
The aqueous phase consisted of a pH 7 phosphate buffer with an ionic
strength of 0.05 M. Sodium caseinate was made up to a concentration of
10^{-3} wt%.

In order to take account of the presence of crystals on the deflection of
the viscometer, control experiments were performed in which tristearin was
crystallized in either n-tetradecane or sunflower oil, which were then
layered over a protein-free buffer solution. Data points on the graphs
presented here are averages of at least two runs, and the data for the
systems containing crystals have already had the controls (2–3 mN m^{-1}s)
subtracted.

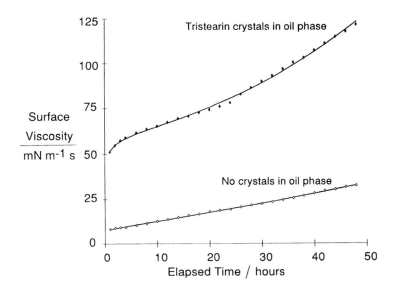

Figure 1 *Apparent interfacial shear viscosity of sodium caseinate (pH 7 phosphate buffer, ionic strength 0.05 M) at the* n-*tetradecane–water interface without (◇) and with (◆) tristearin crystals. Plotted results are the averages of duplicate runs (± 15 mN m^{-1} s)*

3 Results and Discussion

Figure 1 shows results for sodium caseinate at the *n*-tetradecane–water interface. It has been suggested previously that the protein adsorbs at the interface, with loops extending into the oil phase,[2] and with adjacent molecules of protein interacting with each other weakly to produce a low surface viscosity.[3] When the *n*-tetradecane contains tristearin crystals, a higher surface viscosity develops than would be expected from the sum of the individual contributions of caseinate (in a crystal-free control) and crystals (in a protein-free control). The protein loops in the oil phase can perhaps adsorb onto crystals forming aggregates which straddle the interface, thereby resulting in a higher viscosity than in the crystal-free system. Furthermore, the adsorption of protein trains at the interface will reduce the interfacial tension thereby lowering the contact angle. If the contact angle is sufficient to wet the crystals, then the fat crystals might be expected to adsorb at the oil–water interface.

Figure 2 shows equivalent data for sodium caseinate at the sunflower oil–water interface. Even though the oil had been previously treated with silica to remove surface-active components, it is clear that the measured surface viscosity is close to the limit of detection of the viscometer. Other workers[4] have noted the lower surface shear viscosity when proteins are allowed to adsorb at a triglyceride oil–water interface compared with a hydrocarbon oil–water interface. The difference in surface properties of caseinate at the

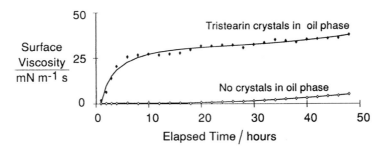

Figure 2   *Apparent interfacial shear viscosity of sodium caseinate (pH 7 phosphate buffer, ionic strength 0.05 M) at the sunflower oil–water interface without (◇) and with (◆) tristearin crystals. Plotted results are the averages of duplicate runs ($\pm 5\,mN\,m^{-1}\,s$)*

sunflower oil–water and *n*-tetradecane–water interfaces may be due to more than one mechanism. (i) The polar regions of the triglycerides are in a plane at the interface, and their hydrocarbon chains project into the oil phase.[5] This orientation of triglyceride molecules results in a liquid crystalline region at the interface, with polar groups forming a layer which could act as a barrier through which the non-polar regions of the caseinate are less able to penetrate. (ii) There may be remnants of surface-active monoglycerides, diglycerides and free fatty acids still present in the sunflower oil after the silica treatment. In the presence of low molecular mass surfactants the structure of the caseinate film is broken up, reducing the strength of the protein–protein interactions (whereas displacement occurs at high concentrations, at low surfactant concentrations the protein remains at the surface).[8]

Whichever mechanism is responsible, the result is a negligible surface viscosity at the food oil–water interface. Yet when tristearin crystals are present in the oil (Figure 2), a substantial surface viscosity is observed. Compared with the *n*-tetradecane system, there is thought to be reduced penetration of the protein into a triglyceride oil phase,[7] and hence we can expect only a limited interaction between fat crystals in the bulk oil phase and hydrophobic groups of the protein. However, the presence of protein at the sunflower oil–water interface will reduce the interfacial tension, resulting in a lower contact angle which would promote the adsorption of the fat crystals at the interface.

In both triglyceride and hydrocarbon oil systems, it may be that the crystals are adsorbed at the oil–water interface protruding into either or both bulk phases.[8] If this were indeed the case, then substantial drag due to the adsorbed crystals might be expected to be conveyed to the interface when sheared. It seems plausible that protein in the aqueous bulk phase is able to adsorb onto such crystals, resulting in agglomerates which span the oil–water interface, thereby elevating the measured surface shear viscosity.

Acknowledgement

We are grateful to Mr Iain Campbell (Unilever Research, Colworth Laboratory) for invaluable discussion and the provision of silica-treated oil and sodium caseinate. This work was made possible by AFRC grant FG101/510 (NI).

References

1. E. Dickinson, B.S. Murray, and G. Stainsby, *J. Colloid Interface Sci.*, 1985, **106**, 259.
2. D. E. Graham and M.C. Phillips, in 'Theory and Practice of Emulsion Technology', ed. A. L. Smith, Academic Press, New York, 1976, p. 75.
3. E. Dickinson, in 'Microemulsions and Emulsions in Foods' ed. M. El-Nokaly and D. Cornell, American Chemical Society, Washington, DC, 1991, p. 114.
4. E. Dickinson and G. Iveson, *Food Hydrocolloids*, 1993, **6**, 533.
5. T. Bursh, K. Larsson, and M. Lundquist, *Chem. Phys. Lipids*, 1968, **2**, 102.
6. E. Dickinson, A. Mauffret, S. E. Rolfe, and C. M. Woskett, *J. Soc. Dairy Technol.*, 1989, **42**, 18.
7. E. Dickinson, 'An Introduction to Food Colloids', Oxford University Press, 1992, p. 155.
8. K. Boode and P. Walstra, *Colloids Surf. A*, 1993, **81**, 121.

Emulsions

Surface Structures and Surface-active Components in Food Emulsions

By Björn Bergenståhl, Pia Fäldt, and Martin Malmsten

INSTITUTE FOR SURFACE CHEMISTRY, PO BOX 5607, S-114 86 STOCKHOLM, SWEDEN

1 The Emulsion Droplet Surface

The stability and flow properties of emulsions are largely determined by interactions between emulsion droplets. The type and magnitude of the interaction depends on the composition of the surface. Food emulsions are complex mixtures, and they usually contain both low-molecular-mass surface-active lipids and a versatile range of more or less surface-active proteins and polysaccharides. The actual chemical composition of the emulsion droplet surface is the key factor determining most of the surface interactions.

In systems containing several surface-active components, three types of adsorbed layers can be identified based on how the layers are formed.[1] In reality, the differences between the three adsorption structures discussed below are not sharp, but this simplified description can provide a useful basis for describing the properties of complex systems.

(i)   *Competitive adsorption.* A monolayer containing one predominant type of molecule at the interface builds up through competition with other less surface-active components that may be replaced at the interface.

(ii) *Associative adsorption.* An adsorbed layer containing a mixture of several different surface-active components is formed.

(iii) *Layer adsorption.* One component adsorbs on top of another.

In this paper, the properties of different adsorption structures are discussed. Examples, primarily from our own research, demonstrating these ideas in different systems are presented.

2 Competitive Adsorption

In a system containing several surface-active components, a homogeneous monolayer is predominantly formed from the most surface-active component. The main driving force for surfactant adsorption onto a hydrophobic surface is the hydrophobic interaction. This is also valid for adsorption of emulsifiers onto oil droplets. In a mixture of two emulsifiers, the most hydrophobic emulsifier has the strongest affinity for the interface. A consequence of this is that, under conditions of competitive adsorption, the component with the lowest water solubility* (*i.e.* the lowest critical micelle concentration[2]) dominates the interface. The character of the adsorbed layer, for instance its ability to generate repulsive interactions, is determined by the dominating compound. The structure of the layer depends on the geometrical shape of the molecules and on lateral interactions between the molecules in the layer. Non-ionic surfactants may form very dense layers due to head-group attraction. Ionic surfactants are able to form extremely loose layers due to inter-head-group repulsion.

Competitive adsorption with macromolecular emulsifiers is somewhat different. The polymer is anchored to the surface at several points. Polymer adsorption is more or less irreversible, partly for energetic reasons and partly due to the more or less independent conformational changes on different segments, which for statistical reasons make desorption an unlikely event for a polymer. Once adsorbed, a polymer may undergo slow conformational changes causing changes in adsorbed amount and surface properties. This is observed as an ageing effect. Ageing causes interfacial tension to decrease slowly over a period of several hours and the surface rheological properties also to change slowly.[3] If a second component is present, changes may allow this component to adsorb and 'segment by segment' to replace the first component at the interface. In the presence of a competing, more surface-active polymer, even polymer adsorption may be reversible.[4]

A special case of competitive adsorption is the observed adsorption sequence of serum proteins to glass surfaces in contact with blood plasma. When a glass surface is placed in contact with blood, small proteins adsorb first as they have a higher molar concentration and exhibit more rapid diffusion. As adsorption proceeds, the composition at the surface changes and smaller molecules are replaced by larger molecules until eventually the largest, most adaptable molecules with the highest affinity dominate the surface (the Vroman series,[5] see Table 1).

Competitive Adsorption at the Air–Water Interface during Spray Drying

During a spray drying operation, liquid is pumped through a nozzle (atomizer) where it is disrupted into small droplets. The liquid droplets meet hot air and they dry rapidly to form powder particles. It has been shown[6] that the adsorbed layer at the air–water interface of the droplets determines the composition of the surface of the powder particles after drying. We have

* This is obviously an oversimplification if we have 'insoluble' components present. 'Insoluble' material is never present in solution and is therefore non-adsorbing. Low molecular mass emulsifiers are insoluble below the chain transition temperature (Kraft temperature).

Table 1   *The order of adsorption to a glass surface in contact with blood plasma (after Vroman and Adams[5])*

	Proteins
1[a]	Albumin
2	IgG
3	Fibronectin
4	Fibrinogen
5[b]	High molecular mass kininogen

[a] First adsorbed. [b] Predominates at very long times.

studied effects on the powder surface due to competitive adsorption from a solution caseinate (or glycine) + lactose. The surface composition of spray dried mixtures of lactose + protein and lactose + glycine were estimated using ESCA (electron spectroscopy for chemical analysis).[7] The surface coverage of caseinate or glycine displayed in Figure 1 shows a pronounced difference between the glycine + lactose system and the caseinate + lactose system. Caseinate already appears at the powder surface at a very low overall concentration (0.01 wt%) and it dominates the surface at a protein/lactose ratio of 1:99. The surface composition of the powders made from the glycine + lactose solutions is the same as the overall composition of the solution.

These results can be understood if the adsorption at the air–water interface is recognized as being competitive. Sodium caseinate displays a pronounced

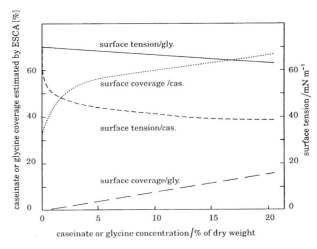

Figure 1   *The surface coverage of caseinate and glycine in powders manufactured by spray drying of caseinate + lactose or glycine + lactose solutions. The surface coverage is given as a function of the dry weight concentration of the solution. The results are compared with the surface tension of the solutions before drying[6]*

ability to lower the interfacial tension, and is therefore able to compete successfully with lactose which has very low surface activity. Glycine affects the surface tension only to a minor extent. These results show that the composition of the air–water interface of the drying droplets is reflected in the surface composition of the dried powder. In addition, scanning electron microscope pictures show that changes in powder structure when protein is added to the solution are associated with the presence of protein in the surface. When the surface coverage of protein increases, dents start to appear on the particles.

3 Associative Adsorption

In associative adsorption, a mixed surface layer is formed. The overall properties displayed by the surface layer are some sort of average of those of the pure components.

A typical associative system may be a long-chain alcohol (for instance *n*-decanol) + a charged surfactant (*e.g.* a soap). The alcohol acts as a spacer between the charged groups, which decreases head-group repulsion within the layer and reduces the surface energy. This increases adsorption and enhances surface activity. Similarly, a lamellar phase is formed in the corresponding three-component phase diagram of water + sodium caprylate + decanol.[8] Mixed layers are commonly formed due to associative adsorption with natural and technical emulsifier blends. This is also a necessary requirement of the common assumption that an average hydrophile–lipophile balance (HLB) value should describe the properties of an emulsifier blend.[9] A common system that is assumed to act in this way is a mixture of sorbitan esters and ethoxylated sorbitan esters where the smaller sorbitan ester molecules can fill the space between the bulky ethoxylated esters.[10] Similarly, a strong associative adsorption at the interface is observed in combinations of anionic and cationic surfactants.[11] Mixed layers are also suggested for several natural protein mixtures such as sodium caseinate[12] or lipid + protein mixtures such as egg yolk.[13]

In the case of associative adsorption, both components are present at the surface. If this situation is to be stable, adsorption of the second component should either be enhanced by the presence of the first component, or not influenced by it. The total amount of adsorbed material should be greater than or equal to the sum of the adsorbed amounts for the two components present alone at the same individual bulk concentrations.

4 Layered Adsorption

Adsorption in layers is possible when different classes of surface-active components are present in a mixture. The two components must be of quite different character to give a structure with a discrete layered character rather than a mixed surface. The second component effectively adsorbs to a particle displaying the characteristic properties of the primary adsorbing emulsifier.

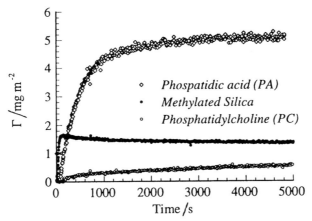

Figure 2   *The amount of apoprotein B adsorbed* Γ *on various lipid surfaces measured using ellipsometry. The lipid model surfaces are applied on silica wafers using a spin coating technique*

This usually means a more hydrophilic surface, which can be expected to reduce the adsorbed amount at the outer layer. However, in some cases, the presence of certain groups increases adsorption of specific substances.

Adsorption of Apoprotein to Phospholipid Surfaces

An example of protein–phospholipid interactions is adsorption of apoprotein B to phospolipid-covered emulsion droplets during formation of synthetic low density lipoprotein (LDL) particles. The background to this interest is that previous studies have shown that leukaemia tumour cells have an increased uptake of LDL, which normally carries cholesterol in the bloodstream. Consequently, there is an interest in using artificial LDL particles for targeting of lipophilic cytostatica to these cells. Our own research in this field has focused on understanding the interfacial behaviour of apoprotein B at phospholipid surfaces and on the preparation of oil-in-water emulsions coated with firmly anchored biologically active apoprotein molecules. The problem involved is likely to be one of layered adsorption, since, for several reasons, emulsification must first be carried out with the phospholipid emulsifiers, whereafter the apoprotein is added. In Figure 2 we show the adsorption of apoprotein B at two phospholipid surfaces, *i.e.* phosphatidylcholine (PC) and phosphatidic acid 25 (as measured using ellipsometry). These phospholipids differ mainly in their head-group composition. For comparison, we show also the adsorption of apoprotein B at a hydrophobic surface (methylated silica to mimic the oil–water interface). As can be seen, adsorption at the phospholipid surface may be either higher or lower than that at the hydrophobic surface, depending on the head-group composition. We have found similar differences for other proteins, *e.g.*, fibrinogen, fibronectin and immunoglobulin (IgG).

These effects are expected to be of great importance, *e.g.* for the optimization of intravenously administered colloidal drug vehicles.

Adsorption of Chitosan to Bile Salt + Phospholipid Surfaces

The second example of adsorption to phospholipid surfaces is the adsorption of chitosan to phospholipid-covered emulsion droplets in the presence of a bile salt. Chitosan is a positively charged polysaccharide conventionally prepared by alkaline deacetylation of chitin. Chitosan has potential use as a hypercholesterolemic agent[14] and one possible mechanism is an interaction between chitosan and the lipid droplets in the intestine. Here, we present results investigating the interactions caused by chitosan adsorbed to a phospholipid + bile salt surface.[15] The affinity of chitosan for a phospholipid surface was first shown using electrophoretic mobility measurements. Mobility changed from negative to positive at very low concentrations of chitosan (Figure 3). The effects were particularly pronounced at pH values between 4.5 and 5.5. The effects of chitosan were much less pronounced in the absence of bile salt. Hence, it was clear that the bile salt gave the chitosan its ability to adsorb to the emulsion droplets. The properties of adsorbed chitosan layers have been investigated by Claesson and Ninham[16] using the surface force technique. Results showed that the interactions at low pH were dominated by an electrostatic repulsion, which at pH 6.7 was replaced by an attraction. At higher pH interactions were dominated by a long-range repulsion of a steric nature (Table 2).

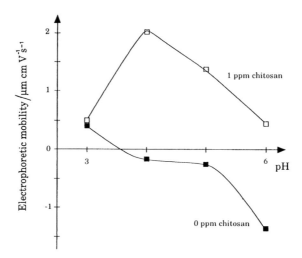

Figure 3   *The pH dependence of the adsorption of chitosan to a phospholipid + bile salt emulsion droplet surface as measured from the change in electrophoretic mobility*[23]

Table 2   *Surface interactions generated by adsorbed layers of chitosan[16]*

pH	Force	Interaction
pH 5	Long-range repulsion	Electrostatic
pH 6.7	Short-range wall-like repulsion; attractive minima	Steric; flat polymer conformation
pH 9	Long-range repulsion	Steric

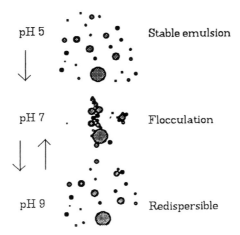

Figure 4   *The flocculation properties of the chitosan + phospholipid + bile salt emulsion[23]*

If chitosan forms the outer layer surrounding the emulsion droplets, it is reasonable to assume that droplet interactions should (to some extent) follow the same pattern as for the surface interactions in the surface force experiment. To investigate this, an emulsion was manufactured with 6 wt% soybean oil, 0.14 wt% soybean lecithin (pure PC) and 0.08 wt% sodium glycocholate. The emulsion was homogenized using a Microfluidizer and 500 ppm chitosan was added to the emulsion (in 0.1 wt% acetic acid solution). The pH was adjusted using NaOH. The stability at various values of pH is shown in Figure 4. It is interesting to note that the results follow the overall behaviour indicated in Table 2. At low pH the emulsion is stable, whereas at pH 6.7 the emulsion flocculates irreversibly. The emulsion is redispersible when the pH is increased further. If the pH of a redispersed emulsion is reduced again, it flocculates at pH 6.7. Hence, the interactions between the emulsion droplets with adsorbed chitosan are analogous to those between two mica surfaces covered by adsorbed chitosan.

Adsorption of Hydrocolloids to Emulsifier Surfaces

In this third set of experiments, we wanted to investigate the emulsion stabilizing properties of hydrocolloids. It is known that most hydrocolloids are not useful for generating or supporting emulsification alone, and they must therefore be used in combination with other emulsifiers. The crucial point for their performance in most food emulsions is whether or not they adsorb onto the emulsion droplet surfaces covered by low-molecular-mass emulsifiers and/or proteins. The aim of some recent experiments has been to study whether hydrocolloids adsorb to emulsifier surfaces. These experiments are based on making simple turbidimetric flocculation rate measurements. A reduced flocculation rate is evidence for formation of an adsorbed layer.[17,18]

Emulsions with reproducible average particle size well below 1 μm were obtained by high-pressure homogenization using a special high-pressure technique called 'microfluidization'.[19] In our experiments, we homogenized emulsions with 3 wt% soybean oil and 3 wt% emulsifier (expressed in terms of the oil phase). The result was usually an emulsion with mean particle sizes of 500 and 600 mn. The concentration used in the experiments was chosen in order to give a final turbidity of about 0.3 (optical density), which had been found to give a satisfactory reproducibility and accuracy (good resolution of the spectrophotometer, linear response, rapid process, small contribution from multiple scattering). This usually meant a dilution of 50 times. The flocculation was induced by the addition of 0.03 M $MgCl_2$. The salt eliminated the electrostatic repulsion, which for most of the emulsions investigated was sufficient to induce the flocculation process.

Figure 5 shows a typical plot of the flocculation rate *versus* concentration of added hydrocolloid. From this graph it is possible to determine a critical concentration at which the flocculation rate is reduced by a factor of 10. In

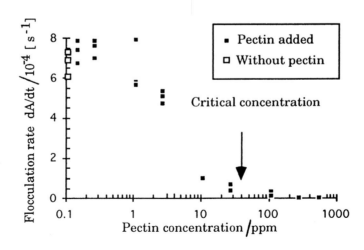

Figure 5   *Example of the concentration dependence of the flocculation rate when a hydrocolloid is added to an emulsion. The plot shows the effect of the addition of pectin (xss 100) (from Grindsted A/S, Denmark) to a soybean oil emulsion emulsified with sodium caprylate*

Table 3 The amount of stabilizing polymer required to stabilize various combinations of hydrocolloid + emulsifier. The critical concentration (the amount necessary to reduce the flocculation rate to 10%) of polymer is given in ppm

Hydrocolloid	Emulsifier				
	Sodium caprylate	Soybean lecithin	Monoolein	Sorbitan oleate	Sucrose ester
Methylcellulose	1	1	300	3	5
Ethyl hydroxyethylcellulose					
E 150	3	5	20	3	5
E 320 G	2	0.5	8	3	5
Pectin					
xss 100	30	30	150	7	500
rs 400	10	50	150	15	300
Gum tragacanth	30	1000	150	3	1
Xanthan	100	300	300	150	>300
Rice starch	1000	>1000	>1000	500	>1000
Dextran	1000	>1000	>1000	>1000	>1000

Table 3 the critical concentrations with different hydrocolloids and different emulsifiers are presented. A few conclusions are evident.

(i)   Several of the hydrocolloids are surface-active, especially the modified celluloses, and they adsorb to the surface of the emulsion droplets.

(ii) Some of the hydrocolloids, especially starch and dextran, display a very low surface activity and may be considered as non-adsorbing in these systems.

(iii) The emulsifiers present in the system have a very large influence on adsorption. With the sucrose group, the adsorption of gum tragacanth is strongly enhanced, but adsorption of modified cellulose is reduced. Monoolein reduces adsorption of hydrocolloids in general. Sodium caprylate and lecithin promote strong adsorption of modified celluloses.

Considering the very different molecular shapes and characters of the emulsifiers and hydrocolloids, as well as the slight ability of hydrocolloids to decrease the oil–water interfacial tension,[18] it seems very unlikely that the hydrocolloids penetrate the emulsifier layer to any significant degree. Therefore, the only reasonable conclusion is that the observed adsorption of these hydrocolloids is an adsorption onto the emulsifier layers, as suggested in Figure 6. One difficulty with these experiments is that they were performed under very dilute conditions and that some of the emulsifiers, for instance sodium caprylate, may have desorbed to some extent. However, several of the emulsifiers included have an extremely low aqueous solubility (lecithin, monoolein, sorbitan oleate) and can be expected to have remained at the interface.

It is reasonable that the hydrophilic/hydrophobic character, the acid/base character and the presence of specific groups all affect the adsorption of the hydrocolloids. A consequence is that a particularly high affinity requires a special combination of emulsifiers and hydrocolloids. For instance, it has been shown above that adsorption of chitosan onto a polar lipid surface is strongly enhanced by the presence of bile salts at the interfaces, and that the presence of phosphatidic acid in a phospholipid layer increases protein adsorption.

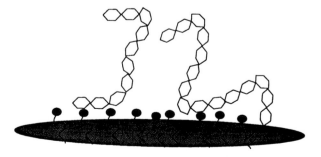

Figure 6   *Suggested model for the adsorbed layer of hydrocolloid on top of an emulsifier layer*

Emulsions containing Emulsifiers, Proteins, and Hydrocolloids

In the experiments described above it was shown that hydrocolloids may adsorb to typical emulsifier surfaces. These results indicate that adsorption phenomena will influence the functionality of the products regardless of whether the hydrocolloid is added to act as a protective colloid or merely to increase the viscosity. Most food systems, however, are more complex than a simple mixture of oil, emulsifiers and thickeners. Proteins are almost always present. They are always fairly surface-active and can be expected to influence the adsorption of hydrocolloids to emulsion droplets and thereby their functionality. In order to study this, we chose a realistic model system to provide a useful example of how a complex technical mixture may alter the function of hydrocolloids in emulsions. The aim of the investigation was to map the adsorption of hydrocolloids to the interfaces in these systems.

Table 4 *The stabilizing addition of hydrocolloid to a mono/diglyceride surface: a mono/diglyceride–protein surface and a pure protein surface*

Hydrocolloid	Emulsion surface		
	Mono/diglyceride	*SMP** *(60 ppm)*	*Mono/diglyceride (360 ppm SMP)*
Na alginate	500	7	20
Carrageenan	500	8	25
Carboxymethylcellulose	>1000	8	40
Locust bean gum	>1000	900	>1000
Milk protein (SMP)*	500	–	–

* Milk protein (ultrafiltrated spray-dried skim milk).

The model system mimics an ice-cream emulsion. Ice-cream is a frozen foam of a dairy emulsion. Hydrocolloids are added to increase the softness and reduce the sandy appearance induced by large ice crystals. The main mechanism for this is probably a thickening of the aqueous phase. The aim of our investigation is to see whether the hydrocolloid adsorbs to the complex surface that may be formed by milk proteins and emulsifiers after the homogenization but before whipping the emulsion. The first hypothesis is that the emulsion droplets are covered by a layer of the emulsifier and that the hydrocolloids may adsorb onto this surface. To test this, an emulsion containing the emulsifier and the fat was produced. The adsorption of protein and hydrocolloids was studied by the flocculation technique. The results are presented in Table 4 and can be compared with the previously described experiment for a monoolein surface (Table 3). It is evident that adsorption to the pure mono/diglyceride surface is weak or absent for all the hydrocolloids investigated. The results are also comparable to those obtained with the monoolein surface.

An obvious adsorption is found only with the milk protein. Adsorption is significant even though the critical concentration is comparably high, indicating a low affinity. The milk protein is also the predominant ingredient in the emulsion. Thus, we can assume that the milk protein is present at the interface when it is present in the emulsion. However, the adsorbed amount may be fairly low, as indicated by other techniques.[20] To investigate the system further we continued the experiments with an investigation of the adsorption of hydrocolloids to a surface that was a combination of the emulsifier and the protein. To keep the protein/fat ratio similar to the ratio in ice-cream, the experiments were performed with 360 ppm protein.

The experiments described in Table 3 were based on induced flocculation, reduced by the presence of adsorbed layers. The protein layer was resistant to the addition of Mg^{2+}. Thus a new flocculation system had to be used. An interesting possibility was found to be the addition of acid. A buffer containing 0.01 M acetic acid and 0.01 M lactic acid was used. The experimental procedure was as follows: 3 wt% coconut butter was emulsified at 50 °C and 800 bar. If the emulsion contained only milk protein, it was added before the homogenization. If the emulsion contained the mono/diglyceride, the protein was added after the emulsion was diluted. A suitable dilution was 360 ppm fat in water. The pH of the buffer was adjusted by addition of NaOH. The pH giving the optimal rate of flocculation was determined.

Table 5   *The influence of hydrocolloids on emulsions*

Effect	Property	Influence on the interdroplet interactions	Influence on emulsions
Influence on the interactions	The polymer is well adsorbed	Steric repulsion	Stabilization of emulsion
	Incomplete or reversible adsorbed layer	Bridging	Flocculation and rapid creaming
	Non-adsorbing polymer	Depletion attraction	Flocculation and rapid creaming
Influence on the solvent	Increased viscosity	No effect	Slower flocculation and creaming

To estimate the consequences of the results with hydrocolloids in emulsions, it is fruitful to discuss how polysaccharides can be expected to influence emulsions. Different possible influences of hydrocolloids on the stability of emulsions are listed in Table 5. Note that non-adsorbing polymers, weak adsorption and incomplete adsorption all destabilize emulsions. This destabilization effect can be expected to be significant in finely dispersed systems ($\lesssim 1$ μm) where flocculation is a significant mechanism in destabilization. Emulsions in this droplet-size range have very high collision frequencies (more than 1 collision per particle per second) and the thickening has to be extremely large in order to prevent or significantly reduce flocculation. However, for high concentrations and larger particle sizes, an increased attraction may even increase the stability due to formation of a network.

The main uses of hydrocolloids are as thickeners and gelling agents. Traditionally, industry also describes its products in this way, providing its customers with the product viscosity or gel strength at a defined concentration. The possible surface activity has usually not been given more than scattered attention. However, for a large number of technical applications, hydrocolloids are used in a more indeterminate way as 'stabilizers', and their function is rather unclear. In the technical textbooks, the effects are usually explained as being due to an increase in viscosity[21-23] with few exceptions, such as gum arabic. The results of this investigation suggest that hydrocolloids should be recognized as potential surface-active molecules whose properties are highly influenced by the presence of emulsifiers and proteins.

Acknowledgement

Norman Burns is acknowledged for a skilful linguistic revision of the manuscript.

References

1. B. Bergenståhl and P. M. Claesson, in 'Food Emulsions', ed. K. Larsson and S. Friberg, Marcel Dekker, New York, 2nd edn, 1990, p. 41.
2. B. Kronberg, *J. Colloid Interface Sci.*, 1983, **96**, 55.
3. E. Dickinson, A. Murray, B. S. Murray, and G. Stainsby, in 'Food Emulsions and Foams', ed. E. Dickinson, Special Publication No. 58, The Royal Society of Chemistry, London, 1987, p. 86.
4. E. Dickinson, *Food Hydrocolloids*, 1986, **1**, 3.
5. L. Vroman and A. L. Adams, *J. Colloid Interface Sci.*, 1986, **111**, 391.
6. P. Fäldt and B. Bergenståhl, *Colloids Surf. A*, submitted.
7. P. Fäldt, B. Bergenståhl, and G. Carlsson, *Food Struct.*, 1993, **12**, 225.
8. K. Fontell, L. Mandell, H. Lehtinen, and P. Ekwall, *Acta Polytech. Scand. Chapter 2, Chemistry Series III*, 1968, **74**, 2.
9. J. T. Davies, Proc. 2nd Int. Congr. Surface Activity, London, 1957, vol. 1, p. 426.
10. J. V. Boyd, N. Krog, and P. Sherman, in 'Theory and Practice of Emulsion Technology', ed. A. L. Smith, Academic Press, London, 1976, p. 123.
11. P. Jokela, B. Jönsson, and H. Wennerström, *Prog. Colloid Polym. Sci.*, 1985, **70**, 17.
12. E. Dickinson, E. W. Robson, and G. Stainsby, *J. Chem. Soc., Faraday Trans 1*, 1983, **79**, 2937.
13. M. C. Phillips, M. T. A. Evans, and H. Mauser, *ACS Adv. Chem. Ser.*, 1978, **144**, 217.
14. I. Furda, US Patent 4 223 023, 1980.
15. P. Fäldt, B. Bergenståhl, and P. M. Claesson, *Colloid Surf. A*, 1993, **71**, 187.
16. P. M. Claesson and B. Ninham, *Langmuir*, 1992, **8**, 406.
17. B. Bergenståhl, S. Fogler, and P. Stenius, in 'Gums and Stabilisers for the Food Industry', ed. G. O. Phillips, D. J. Wedlock, and P. A. Williams, Elsevier Applied Science, London, 1986, vol. 3, p. 286.

18. B. Bergenståhl, in 'Gums and Stabilisers for the Food Industry', ed. G. O. Phillips, D. J. Wedlock, and P. A. Williams, IRL Press, Oxford, 1988, vol. 4, p. 363.
19. E. Cook and A. P. Lagace, US Patent 4 533 254, 1985.
20. N. Krog. N. M. Barfod, and R. M. Sanchez, *J. Dispersion Sci. Technol.*, 1988, **10**, 483.
21. G. R. Sanderson, *Food Technol.*, 1981, **35(7)**, 50.
22. M. Glicksman, 'Gum Technology in the Food Industry', Academic Press, New York, 1969.
23. P. A. Sandford and J. Baird, in 'The Polysaccharides', ed. G. O. Aspinall, Academic Press, New York, vol. 2, 1983.

On the Stability of Milk Protein-Stabilized Concentrated Oil-in-Water Food Emulsions

By Barbara van Dam, Karen Watts, Iain Campbell, and Alex Lips

UNILEVER RESEARCH, COLWORTH LABORATORY, SHARNBROOK,
BEDFORD MK44 1LQ, UK

1 Introduction

In many oil-in-water (O/W) food emulsions, milk proteins are used to form protective adsorbed layers around the oil droplets stabilizing them with respect to flocculation and coalescence.[1] Several studies have investigated (competitive) adsorption of pure proteins components,[2,3] and the effect of casein mixtures, such as sodium caseinate, on interfacial composition.[4-6] There is less information available on the adsorption behaviour of skimmed milk powder (SMP),[7] which contains whey proteins in addition to a mixture of caseins. An example of a system in which the interface provides controlled instability is whipping cream, where the droplets partially coalesce (clump) as a result of mechanical shear and air incorporation.[8,9] Only a few studies have tried to relate the adsorption characteristics of the protein on the interface to the stability against coalescence of the oil droplets.[10] There are still gaps in the understanding of how interfacial and stability properties are related in real O/W food emulsions, where complex protein mixtures are being used.

This work shows some interesting features concerning the adsorption of SMP and caseinate in model O/W emulsions, and their orthokinetic and quiescent stability. In particular, the study covers a wide range of protein levels, from the minimum required to produce stable droplets to levels well above that required for full surface coverage, and the role of fat phase volume is examined. Complications from the presence of low molecular mass amphiphilic emulsifiers have been deliberately avoided.

2 Experimental

Emulsions were prepared in 4 kg batches by circulating through a Crepaco homogenizer for 5 min at an operating pressure of 100 bar and a temperature of 60 °C. After homogenizing, the emulsions were cooled and stored at 5 °C for 72 h prior to experimentation. The model oil phase was an equal mixture (by mass) of coconut oil and fully hardened palm kernel oil, at a phase volume of either 20 wt% or 30 wt%. The resulting fat solids level was 70% at 5 °C (after 3 days cooling), as measured with nuclear magnetic resonance (NMR) using a Bruker Microspec. The SMP (Dairy Crest) contained 36% protein (composed of 85% caseins and 15% whey proteins) with the remaining 64% consisting mainly of carbohydrates. Sodium caseinate (DMV Spray Bland) consisted of *ca.* 90% caseins. Throughout this paper the levels of caseinate and SMP have been expressed in terms of their intrinsic protein content. Total added protein levels studied ranged from 0.2 to 10 wt%.

Model emulsions were centrifuged (Beckman Centrifuge J2–21) at 48 500 *g* for 2 h at 5 °C. The protein content of the continuous phase was determined by nitrogen analysis (using a Foss–Heraeus Macro Nitrogen Analyzer) yielding an estimate of the amount of protein adsorbed per unit weight of fat.[7] The surface area of the fat in the emulsion was measured by small-angle laser light scattering using a Malvern Mastersizer. Complementary observations of aggregation were made by light and electron microscopy.

When an emulsion containing (solid) fat droplets is subjected to a mechanical shear field, such as during the processes of whipping or churning, the fat droplets will partially coalesce into larger aggregates. This ultimately results in a churned system containing a fat (butter) phase separated completely from the aqueous phase. The time taken for the churning to occur is related to the stability of the emulsion under shear. The orthokinetic studies carried out here involve the shearing of model emulsions (200 ml) at a constant speed (900 r.p.m.) and at fixed temperature (15 °C) in the presence of air, using the apparatus shown in Figure 1. Preliminary experiments have shown that the destabilization process could be accelerated by shearing the emulsions in the presence of air, by increasing the temperature from 5 to 15 °C, or by increasing the stirrer speed (tested in the range 200–900 r.p.m.). Most of the emulsions investigated were found to be very stable and thus they could not be churned within 1 h of shearing (as indicated by the constant torque value). Consequently, these emulsions were sheared for 50 min and *relative* stabilities

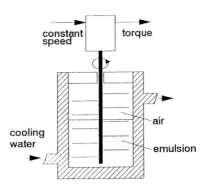

Figure 1   *Schematic picture of the apparatus used for measuring the orthokinetic stability of emulsions (temperature 15 °C, speed 900 r.p.m., sample volume 200 ml)*

were assessed by a fat extraction technique[11] involving agitation of 50 ml of petroleum spirit (40:60) with 10 ml of sheared emulsion in a glass cylinder for 3 min. The solvent (containing fat) was decanted and, after evaporation, the weight of fat left gave the percentage of fat extracted from the initial emulsion. The technique was validated experimentally, where it was found that little fat could be extracted with solvent from an emulsion prior to shear, with the amount of extractable fat increasing for the sheared systems as a function cf shearing time. In the case of a churned system all the fat could be extracted. The method was applied to compare the stability of emulsions containing different types and levels of protein at the same fat levels and shear history.

3 Results

State of Dispersion of Emulsions

Under the chosen homogenization conditions, the minimum attainable break-up diameter $D_{3,2}$ was found to be 0.45 ± 0.05 μm for caseinate emulsions and 0.55 ± 0.05 μm for SMP emulsions. Emulsions coated with SMP showed pronounced homogenization-induced clustering at low protein levels attributed to bridging of casein micelles.[8] At the lowest added protein level of 0.2 wt%, the clusters present were aggregates of large droplets (*ca.* 5 μm) indicating substantial coalescence of droplets within clusters. The individual fat droplet size within clusters decreased with increasing SMP level to the minimum break-up size. The extent of clustering was more dependent on fat phase volume than on the fat-to-protein ratio. It required *ca.* 1.5% protein at 20 vol% fat and *ca.* 3.5 wt% protein at 30 vol% fat to eliminate clustering. Clustering was not obviously observable in the caseinate emulsions, and, independent of fat phase volume, added protein levels greater than 0.5 wt% sufficed to form stable globules at their minimum break-up size.

Surface Coverage

Figure 2 compares levels of protein adsorbed per unit weight of fat for caseinate and SMP emulsions as a function of the equilibrium protein solution concentration. Although both protein mixtures mainly consisted of caseins (in SMP emulsions 93% of the protein at the interface is casein[8]), the level of adsorbed protein in the SMP emulsions was found to be much higher than in the caseinate emulsions. Whereas levels of adsorbed protein appeared independent of the fat phase volume in caseinate emulsions, a weak dependence was seen for SMP emulsions.

The 'isotherms' in Figure 3 illustrate corresponding estimates of surface coverage, surface areas having been determined by small-angle laser light scattering. Since the surface area for clustered samples could not be measured reliably, the data in Figure 3 relate only to systems with a narrow droplet size distribution and surface areas close to those expected for the minimum break-up size (*i.e.* 0.55 ± 0.05 μm). For caseinate emulsions a maximum surface

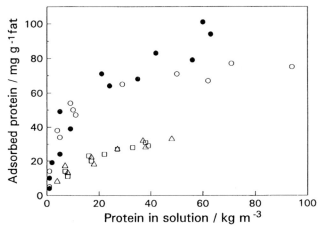

Figure 2   *Amount of adsorbed protein* versus *the protein solution concentration for caseinate emulsions (△, 20% fat; □, 30% fat) and SMP emulsions (○, 20% fat; ●, 30% fat)*

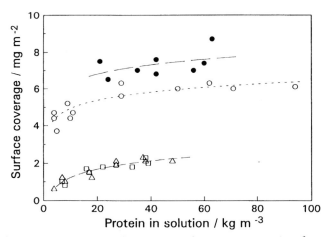

Figure 3   *Surface coverage* versus *the protein solution concentration for caseinate emulsions (△, 20% fat; □, 30% fat) and SMP emulsions (○, 20% fat; ●, 30% fat)*

coverage of 2–2.5 mg m^{-2} was measured for protein concentrations above 3 wt%. In contrast, the plateau in the isotherm for SMP emulsions was at much higher coverages, *i.e.* at 6 ± 1 mg m^{-2} for 20 wt% fat and 8 ± 1 mg m^{-2} for 30 wt% fat.

Emulsion Stability

The results in Figures 4 and 5 confirm the expected increase in orthokinetic stability of the emulsions with increasing surface coverage. The filled points

region. The response to post-addition (Figure 6), following a general adsorption isotherm, is consistent with thermodynamic (low affinity) control; however, the lack of reversibility on dilution argues against this. The conclusions of reference 4 may be of relevance, demonstrating the possibility of total protein load in caseinate emulsions remaining constant on dilution, whilst the interfacial balance of α_{s1}-casein and β-casein may change. A pseudo-plateau coverage (with no obvious multi-layer behaviour) at a level of 2–2.5 mg m^{-2} was observed above a protein concentration of *ca.* 2 wt%. Previous studies[3] have found similar plateau levels for sodium caseinate systems, and also have suggested that the individual caseins adsorb onto the interface as monomers rather than in an aggregated or micellar structure.

The observation of enhanced creaming was unexpected. The fact that it could be observed in the pseudo-plateau region, and that it then increased with protein level, argued against a 'polymer bridging' mechanism.[12] At present we have no satisfactory explanation. A possible factor could be the self-assembly of sodium caseinate in bulk solution. Micellar structures formed at higher concentrations, if non-adsorbing, could perhaps act as depletants in a manner similar to that reported for sodium dodecyl sulfate micelles in O/W emulsions.[13] Osmometry and other studies are in progress to define more clearly the solution behaviour of sodium caseinate. The measurements of the relative orthokinetic stability of the caseinate emulsions indicate that stability increases with increasing amount of adsorbed protein below monolayer coverage.

Substantially higher levels of adsorbed protein for SMP-stabilized emulsions (in the range of 5–8 mg m^{-2}), suggest a greater involvement of micellar aggregates in the adsorption than is the case with caseinate, for which monomeric adsorption is postulated. The SMP isotherms at low solution concentrations display a higher affinity than their caseinate equivalents. The observable desorption upon dilution indicates a substantial degree of reversibility of protein adsorption. It is possible that some of the protein, presumably micellar species, is only weakly attached to the interface. It has been demonstrated[2] that adsorption at a planar oil–water interface from high bulk concentrations of β-casein results in thick adsorbed layers with a primary layer of irreversibly adsorbed protein (*ca.* 3 mg m^{-2}) and secondary layers of reversibly adsorbed protein, which can easily be washed off the interface. The dilution experiments (Figure 6) lend support to the hypothesis that adsorption isotherms for SMP depend on fat phase volume. It appears that the adsorption of the protein is substantially under thermodynamic control as opposed to just kinetic control. The observation of the adsorbed amount increasing with fat phase volume (at constant equilibrium protein solution concentration) is consistent with an adsorption mechanism of labile micellar species 'bridging' fat droplets. This possibility is being investigated further.

Regarding the stability behaviour of SMP emulsions, creaming was not observed. Electron microscopy has confirmed that the casein micelles formed from SMP adsorb on the fat droplets, and are unlikely therefore to act as depletants. Under shear, the SMP emulsions (at their minimum break-up size) are very stable, more so than the caseinate emulsions. For both systems it appears that levels of binding should be very close to saturation coverage in order to achieve good orthokinetic stability.

References

1. E. Dickinson, 'An Introduction to Food Colloids', Oxford University Press, 1992.
2. D. E. Graham and M. C. Phillips, *J. Colloid Interface Sci.*, 1979, **70**, 415.
3. E. Dickinson, E. W. Robson, and G. Stainsby, *J. Chem. Soc., Faraday Trans. 1*, 1983, **79**, 2937.
4. E. W. Robson and D. G. Dalgleish, *J. Food Sci.*, 1987, **52**, 1694.
5. S. L. Yong and C. F. Shoemaker, *Food Hydrocolloids*, 1990, **4**, 33.
6. Y. Fang and D. G. Dalgleish, *J. Colloid Interface Sci.*, 1993, **156**, 329.
7. H. Oortwijn and P. Walstra, *Neth. Milk Dairy J.*, 1979, **33**, 134.
8. P. Walstra and R. Jenness, 'Dairy Chemistry and Physics', Wiley, New York, 1984.
9. D. F. Darling and R. J. Birkett, in 'Food Emulsions and Foams', ed. E. Dickinson, Special Publication No. 58, The Royal Society of Chemistry, London, 1987, p. 1.
10. K. P. Das and J. E. Kinsella, *J. Colloid Interface Sci.*, 1990, **139**, 551.
11. A. Fink and H. G. Kessler, *Milchwissenschaft*, 1983, **38**, 330.
12. E. Dickinson, *J. Chem. Soc., Faraday Trans.*, 1992, **88**, 2973.
13. J. Bibette, D. Roux, and B. Pouligny, *J. Phys. II (Paris)*, 1992, **2**, 401.

Ultrasonic Studies of the Creaming of Concentrated Emulsions

By Eric Dickinson, Jian Guo Ma, Valerie J. Pinfield, and
Malcolm J. W. Povey

PROCTER DEPARTMENT OF FOOD SCIENCE, UNIVERSITY OF LEEDS, LEEDS
LS2 9JT, UK

1 Introduction

Creaming is a significant destabilizing process in food emulsions, and the
study of the creaming process is therefore relevant to the shelf-life and
stability of foods. The velocity of propagation of ultrasound through such
systems may be used as a probe of the kinetic processes involved,[1-5] and
also as a way of monitoring crystallization kinetics in emulsions.[6-7] The
interpretation of the ultrasound velocity in terms of the physical properties
of the system is vital in order to gain a correct understanding of these
processes. Much work on creaming in emulsions has employed the Urick
equation to relate the ultrasound velocity to the volume fraction of the oil
phase in oil-in-water emulsions.[2-5] However, this relationship is invalid in
many instances.[8-9] This paper seeks to address the issue of the application
of the Urick equation to emulsion systems, both from a theoretical and an
experimental standpoint.

Table 1 *Physical properties of the oil and water phases at 30°C used in the
analysis*

	Water[8]	Sunflower oil[a]
Density/kg m^{-3}	998.2	912.9
Ultrasound velocity/m s^{-1}	1482.3	1437.5
Shear viscosity/Pa s	0.001	0.054
Thermal conductivity/J m^{-1} s^{-1} K^{-1}	0.591	0.17
Specific heat capacity at constant pressure/J kg^{-1} K^{-1}	4182	1980
Thermal expansivity/K^{-1}	2.1×10^{-4}	7.3×10^{-4}

[a] Values for velocity, density and thermal expansivity from reference 20, pp. 216, 77, and 78. Other
parameters are taken to be the same as at 20°C.[8]

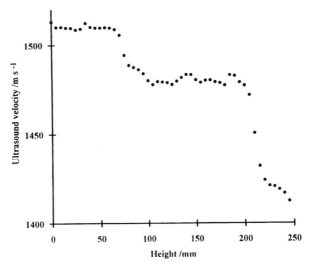

Figure 1   *Measured ultrasound velocity against height for an 18 vol% mineral oil-in-water emulsion containing 2 wt% Tween 20 and 0.035 wt% xanthan after 189 h*

2 Experimental

Recent experiments[1] have investigated the creaming stability of oil-in-water emulsions in the presence of the microbial polysaccharide xanthan.[5,10,11] The emulsions studied were 18 vol% mineral oil, with 2 wt% of the water-soluble non-ionic surfactant Tween 20. The effects of ionic strength on the creaming stability have been determined by the addition of sodium chloride to the aqueous phase. The particle size distribution had a mean diameter $d_{32} = 0.65 \pm 0.02$ μm and the experiments were carried out at 30 °C in samples of height 250 mm. The creaming process was observed by the measurement of the velocity of ultrasound of frequency 1.2 MHz in each sample as a function of height at intervals during the creaming period.

At low xanthan concentrations, with no added salt, the creaming rate increased with increasing xanthan concentration. The instability was manifested in the rapid development of a serum layer at the bottom of the emulsion. At higher xanthan concentrations, a cream layer was formed at the top of the emulsion. The destabilization process is due to depletion flocculation,[10–13] although at higher xanthan concentrations creaming is restricted by the formation of a gel-like network. The addition of salt to the emulsion reduces its stability, possibly due to a modification of the inter-droplet depletion interaction. The results of this work are reported in more detail elsewhere.[1]

Figure 1 shows an example of an ultrasound velocity profile which results after 189 h for a xanthan concentration of 0.035 wt% and no added sodium chloride. This particular sample shows the development of an oil-deficient serum layer at the bottom of the sample, in addition to an oil-rich layer at the top. The velocity of ultrasound is lower in the oil phase than in the aqueous phase, and therefore a lower ultrasound velocity indicates a higher concentration of oil. The velocity in the serum layer is close to the velocity in the aqueous phase since the oil concentration is low there. In order to obtain greater insight into the creaming process, it is necessary to interpret these data in terms of the volume fraction of the oil phase as a function of height and time. Figure 2 shows the volume fraction profile for the same system, calculated from the commonly used Urick equation[14] for the ultrasound velocity v:

$$v = \frac{1}{\sqrt{\rho_0 \kappa_0}} \tag{1}$$

where $\rho_0 = (1-\phi)\rho_1 + \phi\rho_2$, $\kappa_0 = (1-\phi)\kappa_1 + \phi\kappa_2$, ρ is the density, κ the adiabatic compressibility and ϕ the dispersed phase volume fraction. The subscripts denote the continuous phase (1) and the dispersed phase (2), respectively. It can be seen in this plot that the Urick equation predicts a volume fraction of *ca.* 70% in the cream layer, which is significantly higher than the 63% expected for random close packing of monodisperse hard spheres. In addition, the integrated sum of the volume fraction of dispersed phase over the height of the sample is equivalent to a 21% average throughout the sample, which is larger than the initial volume fraction (18%). It may be concluded that the Urick equation overestimates the volume fraction in the sample, particularly in the cream layer ($\phi \geq 0.5$). The plots of the measured velocity and the Urick theoretical velocity as a function of volume fraction are shown in Figure 3 for comparison.

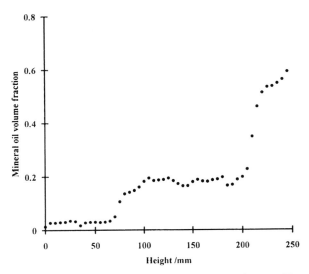

Figure 2   *Volume fraction profile corresponding to the emulsion in Figure 1, calculated using the Urick equation. Note the high volume fraction ($\approx 70\%$) in the cream layer*

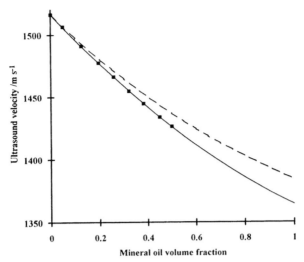

Figure 3   *Measured ultrasound velocity (■) as a function of oil volume fraction for known emulsion samples. The fitted curve (———) acts as a calibration curve for measurements on creaming emulsions. Also shown (– – – –) is the Urick velocity calculated for this system*

3 Effect of Ultrasound Scattering

The results above indicate that the Urick equation is not universally applicable to emulsion systems, and that an alternative interpretation is required. Nevertheless, the Urick method has been applied with considerable success to many emulsions.[2-5] It is therefore necessary to investigate the reasons for the failure of the Urick equation and in what circumstances it can be successfully applied.

The Urick equation [eqn (1)] is equivalent to the Wood equation for the propagation of sound through a pure bulk fluid, using volume-averaged values of density and adiabatic compressibility for the mixture. This is a static description of the emulsion. It ignores the dynamic processes, namely scattering, which occur at the interfaces of the dispersed phase droplets. When a plane wave is incident on a particle, the outgoing waves are not necessarily planar, nor in phase with the incident wave. Thus, in the forward direction, the wave undergoes a phase change (observed as a modification to the velocity) and some energy loss (seen as increased attenuation). In the long-wavelength regime (where the wavelength of the ultrasound is much larger than the particle radius) there are two dominant scattering modes, thermal and visco-inertial.[9] The former is due to the differences in thermal properties between the two media, which cause heat flow across the surface of the particle, out of phase with the incident wave. Visco-inertial scattering is due to the density (inertia) difference between the two media, which therefore move differently in response to the incident pressure wave.

The ultrasound velocity through a dispersion in which scattering occurs may be calculated from scattering theories.[8–9,15–18] The scattering theories assume that the thermal and shear waves produced at each particle are strongly attenuated and have no effect on adjacent particles. The scattering in the dispersion can be considered as the sum of the effects of individual particles (single scattering) or as a more complicated system in which multiple scattering can occur. The theoretical ultrasound velocity in a monodisperse 18% sunflower oil-in-water emulsion at 30 °C is shown in Figure 4. The velocity is only calculated in the long-wavelength region (where these theories are useful), and is a function of the parameter $r\sqrt{f}$ where r is the radius of the particle and f the frequency. The Urick equation does not incorporate any explicit dependence on either of these variables, and therefore it cannot provide an accurate description of ultrasound velocity when scattering is significant.

Range of Application of the Urick Equation

The ultrasound velocity may be written in a modified Urick form [eqn (~)] in the long-wavelength region when multiple scattering is negligible (for example, at small volume fractions). The 'density' and 'compressibility' which must then be used to calculate the velocity are the *dynamic* values rather than the volume-averaged static values.[8] These dynamic properties are based on the single-particle scattering coefficients,[8,15,16] and therefore include the dependence on the frequency and particle radius. However, in practice, the dynamic values are often unknown. Even when the physical properties of the two phases are known, the particle radius may not be; or the average particle size may change

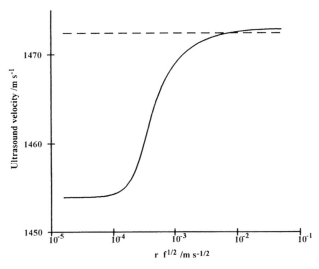

Figure 4 *Ultrasound velocity as a function of the composite parameter $r\sqrt{f}$ for an 18 vol% sunflower oil-in-water emulsion at 30 °C, calculated from multiple scattering theory*

during an experiment. This is especially true in creaming experiments, as will be seen later in this paper, where the particle-size distribution is shown to vary with height and may be quite different in some places from the initial distribution.

It is clear that, in order to apply a modified Urick type of equation, experiments must be conducted in the region of the velocity curve (see Figure 4) where the velocity does not change significantly with radius. There are therefore two ranges of $r\sqrt{f}$ corresponding to the 'flat' regions on the velocity curve (Figure 4) in which simple interpretation of the ultrasound velocity results may be achieved. The velocity in neither of these regions is equal to the Urick velocity, but the difference between the velocity and the Urick velocity is constant, though unknown in an experimental situation. There are therefore two methods which enable the Urick equation to be applied in these regions, calibration and 'renormalization'. Since the velocity is independent of radius, it is a unique function of the volume fraction of the oil phase. Measurements of the ultrasound velocity in emulsions of known concentration can be used to produce a calibration curve which incorporates the scattering correction (the deviation from the Urick velocity) implicitly. Thereafter, the oil volume fraction can be accurately determined from the ultrasound velocity. The only assumption which needs to be made is the one assuming independence of the velocity with radius. If the particle sizes in the system are such that the velocity does vary with radius, then some error in the calculated volume fraction will ensue from using a single calibration curve. Unless the mean particle radius is known at each height and time, systems such as these do not allow easy interpretation of ultrasound velocity data.

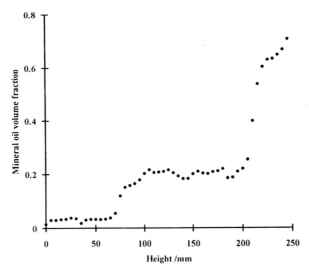

Figure 5   *Volume fraction profile for the emulsion shown in Figures 1 and 2 as calculated from the ultrasound velocity profile using the calibration curve shown in Figure 3*

The second method is called 'renormalization' and is based on the principle of conservation of volume of the oil phase. The integral (or sum) of the calculated volume fraction over the height of the sample is related to the total volume of oil in the system which is known and is constant throughout the experiment. It has already been observed (Figure 2) that the volume fractions calculated from the Urick equation are likely to be overestimates, and that the error is largest at high concentrations. Therefore, if scattering is significant, the sum of the calculated volume fraction in the system will be greater than the total volume of oil, and may increase with time as more concentrated regions develop. Thus, calculation of the integrated volume fraction is a simple check on whether ultrasound scattering is significant in the sample. It may be used in conjunction with the calibration method to check that the concentration profiles are consistent. This will indicate whether the ultrasound velocity is in fact independent of particle radius, as assumed by the calibration technique.

The renormalization process involves the scaling of each volume fraction (calculated from the Urick equation) by a factor which ensures that the integrated volume fraction over the whole system remains equal to the real physical value. The renormalization factor is calculated on the initial scan. On subsequent measurements, the Urick volume fraction is scaled by the renormalization factor. This empirical procedure will produce the correct results for the volume fraction in the sample, as long as the velocity varies linearly with volume fraction. If this correction is insufficient, a second-order term can be used to adjust the volume fraction, assuming a quadratic functionality of velocity on volume fraction.

Corrected Experimental Results

The measurements on the mineral oil-in-water system described earlier have been corrected using the calibration method. The calibration curve for the ultrasound velocity measured in a series of emulsions of known volume fraction is plotted in Figure 3, which also includes the Urick velocity for comparison. Figure 5 shows the corrected creaming profile for the same system as illustrated in Figures 1 and 2. The volume fraction in the cream layer is now 59%, which is not unreasonable. As an additional check, the sum of the volume fraction over the height of the sample was calculated (equivalent to 18.6% throughout the sample) and found to be consistent with the initial volume fraction in the sample (18%). It may therefore be concluded that for the particle radii in this particular sample, the velocity is relatively independent of radius. In addition, the calibration and renormalization methods for the application of the Urick equation appear to produce acceptable results. The concentration profiles may then be used to interpret the creaming and kinetic processes which are occurring in the sample.[1]

4 Creaming Model

In order to better understand the physical processes taking place in the creaming of a concentrated polydisperse emulsion, and the effects of scattering on the corresponding ultrasound velocity profiles, a simple numerical model has been constructed. It is a macroscopic model of the creaming process, which incorporates the hydrodynamic interactions between the particles implicitly through a 'hindrance' factor (which is a function of the local overall volume fraction) in the particle velocity. The effect of particle diffusion is included in the model but the direct effect of particle interactions (and associated phenomena such as depletion flocculation) is neglected. The model analyses the creaming process in a sample by considering the system as a number of discrete layers, with discrete (but narrow) size fractions of particle radius, and using discrete time intervals to model the time evolution of the system. Creaming proceeds by the movement of a calculated volume of particles of each size between adjacent layers in each time interval, according to the creaming velocity and diffusive flux.

The results of the creaming model are presented in the form of concentration profiles or as size distributions. These results can be transformed into ultrasound velocity profiles for direct correlation with experimental measurements. The ultrasound velocity has been calculated in each instance by three methods: the Urick equation, the single scattering theory and the multiple scattering theory. Previous comparison of measured ultrasound velocity with theoretically calculated results in known systems has shown that multiple

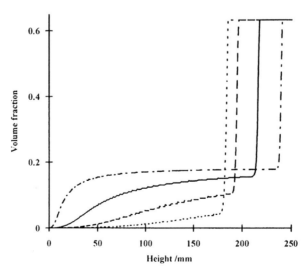

Figure 6   *Total volume fraction as a function of height as predicted by the creaming model for a system corresponding to an 18 vol% sunflower oil-in-water emulsion at 30 °C after various times: - - - - -, 25 days; ————, 100 days; – – – –, 250 days; - - - - - - -, 500 days*

scattering theory provides the best results.[18] In order to illustrate the results derived from the model, a system was analysed corresponding to an 18 vol% sunflower oil-in-water emulsion at 30 °C. A (natural) log–normal particle-size distribution was used, centred on a diameter of 0.72 μm, with a width defined by a standard deviation of 0.59. The ultrasound frequency was 1 MHz. A sample corresponding to the mineral oil-in-water emulsion which was analysed experimentally was not chosen for the model because of the lack of reliable data on the mineral oil which was required for the ultrasound scattering calculations.

The plot of total volume fraction against height (Figure 6) shows features which are qualitatively familiar from the experimental results. The cream layer is denoted by a maximum concentration of 63% and no compression is allowed in the model. The serum layer has a rather diffuse interface due to the polydispersity in the sample. Two features of the results are of particular interest. Particles of diameter smaller than 0.5 μm move downwards from the top of the system, so that the upper part of the cream contains a reduced number of particles of this size. This is due to the smaller particles being entrained in the downward flow of the continuous phase. The second feature of interest is a fractionation effect which occurs in the cream layer, so that the particle-size distribution becomes increasingly skewed to the larger particles towards the top of the cream layer (Figure 7). This is due both to the larger particles moving fastest, so that they occupy a greater proportion of the upper layers, and also to the downward flow of the smaller particles. The computational details and the predictions of the model are discussed in more detail in reference 19.

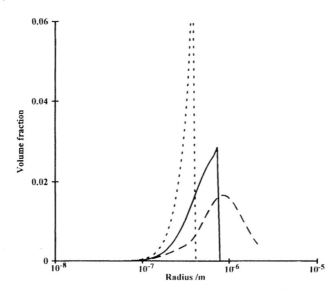

Figure 7   *Particle-size distributions in the cream layer, after 500 days, as predicted by the creaming model for the emulsion in Figure 6, at three heights: - - - - - -, 220 mm; ——————, 225 mm; – – – –, 250 mm*

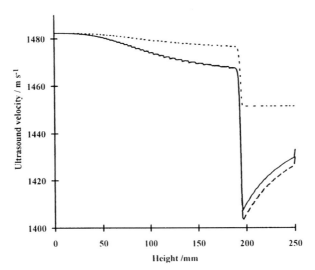

Figure 8   *Prediction from the creaming model of the ultrasound velocity as a function of height for the emulsion of Figures 6 and 7 after 250 days. The velocity is calculated in three ways: - - - - - -, from the Urick equation; – – – –, from single scattering theory; ————, from multiple scattering theory*

Figure 8 shows the ultrasound velocity profile of the sunflower oil-in-water system after 250 days, based on the Urick equation and the single and multiple scattering theories. It is clear that a direct application of the Urick equation to such a system yields incorrect results for the volume fraction, especially in the cream layer where the scattering velocity falls outside the possible range of the Urick velocity for any physical volume fraction.

It is observed in the simulated volume fraction and size distribution profiles that there exists a fractionation process in the cream layer, in which the size distribution becomes strongly skewed towards the larger particles near the top of the cream. This effect is clearly seen in the ultrasound velocity profiles which show a strong variation in velocity in the cream layer. Figure 4 indicates that larger particles have a higher velocity than smaller ones in the mid-region of the curve. Thus the variation of the velocity in the cream reflects the change in size distribution in this region. Observation of this effect in measured ultrasound velocity profiles is a good indication of the presence of scattering in the system, which may not be eliminated by the methods proposed in the previous section. These methods rely on the velocity being independent of radius, and are not applicable where the velocity varies significantly with particle size. Unfortunately, the range of particle sizes in which the velocity is strongly dependent on radius occurs, in this instance, in the colloidal size regime (for ultrasound frequencies of the order of a few MHz). The scattering theories may therefore not be simply used to calculate volume fraction from ultrasound velocity data when scattering is significant because the local size distribution is generally unknown.

5 Conclusions

We have confronted the difficulties and restrictions encountered in applying the Urick equation to the interpretation of ultrasound velocity measurements in emulsions. Experimental results demonstrate that the volume fraction profiles predicted by the Urick equation can be inconsistent with known facts about the system being measured. Ultrasound scattering theory shows that, in the long-wavelength regime, there are two ranges of particle radius (or, equivalently, of frequency) for which the velocity varies little with radius. Two practical methods are proposed for a simple interpretation of ultrasound velocity data in these regions, closely related to the Urick formulation. In the first method, a calibration curve is constructed from measurements of the velocity in samples of known volume fraction. Secondly, a renormalization method is proposed in which the volume fractions calculated from the Urick equation are scaled by a factor which ensures that the total calculated volume of oil in the system remains constant and equal to the known physical value. These techniques have been successfully applied to the experimental samples in the study.

A model is presented which predicts the creaming behaviour in idealized stable emulsion systems. The results for an emulsion of 18 vol% sunflower oil in water at 30 °C show features of particle back flow and fractionation in the cream region. These results are presented in the form of ultrasound velocity profiles to demonstrate the effects of ultrasound scattering on the interpretation of measured profiles. A variation of velocity in the cream layer is observed, corresponding to the varying size distribution there. The particle sizes in the model system fall into the region of the ultrasound velocity curve in which the velocity varies strongly with radius, and a similar velocity variation with height in an experimental system is an indication that this is indeed the case. The modifications to the Urick formulation proposed earlier in this paper cannot be applied successfully to such a system.

Acknowledgement

E. D. and M. J. W. P. acknowledge receipt of an AFRC grant which funded this work.

References

1. E. Dickinson, J. G. Ma, and M. J. W. Povey, *Food Hydrocolloids*, 1994, **8**, 481.
2. A. J. Fillery-Travis, P. A. Gunning, D. J. Hibberd, and M. M. Robins, *J. Colloid Interface Sci.*, 1993, **159**, 189.
3. E. Dickinson, M. I. Goller, and D. J. Wedlock, *Colloids Surf. A*, 1993, **75**, 195.
4. Y. Cao, E. Dickinson, and D. J. Wedlock, *Food Hydrocolloids*, 1991, **5**, 443.
5. P. A. Gunning, D. J. Hibberd, A. M. Howe, and M. M. Robins, *Food Hydrocolloids*, 1988, **2**, 119.
6. E. Dickinson, D. J. McClements, and M. J. W. Povey, *J. Colloid Interface Sci.*, 1991, **142**, 103.

7. D. J. McClements, M. J. W. Povey, and E. Dickinson, *Ultrasonics*, 1993, **31**, 433.
8. D. J. McClements and M. J. W. Povey, *J. Phys. D*, 1989, **22**, 38.
9. D. J. McClements, *Adv. Colloid Interface Sci.*, 1991, **37**, 33.
10. Y. Cao, E. Dickinson, and D. J. Wedlock, *Food Hydrocolloids*, 1990, **4**, 185.
11. H. Luyten, M. Jonkman, W. Kloek, and T. Van Vliet, in 'Food Colloids and Polymers: Stability and Mechanical Properties', ed. E. Dickinson and P. Walstra, Special Publication No. 113, The Royal Society of Chemistry, Cambridge, UK, 1993, p. 224.
12. S. Asakura and F. Oosawa, *J. Chem. Phys.*, 1954, **22**, 1255.
13. A. Vrij, *Pure Appl. Chem.*, 1976, **48**, 471.
14. R. J. Urick, *J. Appl. Phys.*, 1947, **18**, 983.
15. P. S. Epstein and R. R. Carhart, *J. Acoust. Soc. Am.*, 1953, **25**, 553.
16. J. R. Allegra and S. A. Hawley, *J. Acoust. Soc. Am.*, 1972, **51**, 1545.
17. P. Lloyd and M. V. Berry, *Proc. Phys. Soc.*, 1967, **91**, 678.
18. D. J. McClements, *J. Acoust. Soc. Am.*, 1992, **91**, 849.
19. V. J. Pinfield, E. Dickinson, and M. J. W. Povey, *J. Colloid Interface Sci.*, 1994, **166**, 363.
20. D. J. McClements, PhD Thesis, University of Leeds, 1988.

Formulation and Properties of Protein-Stabilized Water-in-Oil-in-Water Multiple Emulsions

By Jane Evison, Eric Dickinson, Richard K. Owusu Apenten, and Andrea Williams

PROCTER DEPARTMENT OF FOOD SCIENCE, UNIVERSITY OF LEEDS, LEEDS LS2 9JT, UK

1 Introduction

A water-in-oil-in-water (W/O/W) multiple emulsion is a colloidal system in which oil droplets containing inner aqueous droplets are dispersed within a second aqueous phase. It is usually prepared by adopting a two-stage emulsification procedure whereby a primary water-in-oil (W/O) emulsion is prepared initially with a lipophilic emulsifier which is then re-emulsified in a solution containing a hydrophilic emulsifier.[1-3]

Potential applications of multiple emulsions within pharmaceutical and cosmetic products have been well documented.[4,5] In the pharmaceutical area, the encapsulating properties of multiple emulsions can be exploited for controlling the release of drugs or even enzymes to a predetermined target. In the food industry, multiple emulsions have potential applications to encapsulate flavours, enzymes, nutrients, *etc.*[3,6-9] In addition, there tend to be lower levels of oil in W/O/W multiple emulsions than in equivalent ordinary oil-in-water (O/W) emulsions, due to the presence of the inner aqueous phase; this feature could be exploited in the development of low calorie products.[3]

Multiple emulsions are, however, extremely susceptible to destabilization. The breakdown mechanisms of multiple emulsions can be categorized into two major groups: there are processes which can lead to rupture of the oil layer causing the loss of inner aqueous droplets, and then there are the more common processes of creaming, flocculation and coalescence which affect the outer (oil) droplets.[6,10] Instability problems with W/O/W systems are aggravated by the presence of both hydrophobic and hydrophilic emulsifiers which have the potential for migrating between the two oil–water interfaces. In order to ensure good stability, it is usual practice to use large (excess) amounts of lipophilic emulsifiers when formulating multiple emulsions.[1,2] However, large amounts of surfactants in foods are undesirable because of possible toxicological implications, or because they tend to have a detrimental effect on the palatability of the food. A possible alternative method of achieving a high degree of stability in multiple emulsions is to encapsulate a macromolecular stabilizing agent such as a protein or hydrocolloid within the inner droplets. Both gelatin[4] and bovine serum albumin (BSA)[11] have been shown to improve markedly the yield and stability of multiple emulsions. The increased stability is thought to be partly due to an interfacial interaction between Span 80 and the macromolecules.[7,11] A similar improvement in yield and stability has also been observed with an interfacial barrier formed from microcrystalline cellulose.[12]

In previous research,[6,13] we have demonstrated the feasibility of preparing fine multiple emulsions with a high yield (95%) using *n*-tetradecane, kerosene or soybean oil as the oil phase, Span 80 (sorbitan monolaurate) or Admul Wol as the primary emulsifier, and sodium caseinate as the secondary emulsifier. These emulsions were found to possess good stability with respect to loss of inner aqueous phase, but the formulations had the disadvantage of containing rather large amounts of lipophilic emulsifier. In this paper, we attempt to investigate the possible stabilizing effect of encapsulated BSA, gelatin or xanthan gum on multiple emulsions formulated using lower concentrations of lipophilic emulsifier. The effects on the yield of the inner aqueous droplets on the shearing stability of the multiple emulsions and on the quiescent long-term W/O/W emulsion stability have been examined. Electrophoretic mobility measurements have also been carried out to observe the effect of encapsulating BSA within the inner droplets on the effective surface charge density of the outer multiple emulsion droplets.

2 Materials and Methods

Span 80 and kerosene oil were obtained from Fluka. The soybean oil, Tween 20, Poly R-478 (violet dye, 5×10^4 Da), lysozyme (derived from chicken egg-white) and BSA (99% purity, crystallized and lyophilized) were bought from Sigma. Admul Wol 1403, which was obtained from Quest International (Food Ingredients), consists of a mixture of partial esters of polyglycerol (mainly di-, tri- and tetraglycerol) with linearly interesterified castor oil fatty acids (predominantly ricinoleic). The sodium caseinate (Dutch) was a gift from the Hannah Research Institute. The gelatin [pI = 4.8, Bloom strength (6.67 wt%) = 221 g] originated from P. Leiner and Sons Ltd. and the xanthan gum was a gift from Kelco Inc. The buffer salts and sodium azide were AnalaR-grade from BDH Chemicals.

Multiple emulsions were prepared using a two-stage emulsification procedure. Initial primary W/O emulsions (20 vol% aqueous phase) were prepared by emulsifying 1–8 wt% Span 80 in kerosene or 1–8 wt% Admul Wol in soybean oil together with phosphate buffer (0.05 M, pH 7) using a jet homogenizer (jet hole size 0.33 mm, pressure 400 bar) operating at room temperature.[14] In the second stage, the W/O emulsion (20 vol% aqueous phase) was re-emulsified at room temperature in the jet homogenizer with

0.125 wt% sodium caseinate or 0.35 wt% Tween 20 dissolved in the pH 7 phosphate buffer using a larger jet hole of 0.81 mm and a lower pressure of 100 bar. A small amount of sodium azide (0.1 wt%) was incorporated into the buffer solutions to prevent microbial growth. In some of the formulations, BSA, gelatin or xanthan gum was encapsulated in the inner aqueous droplets. The BSA or gelatin was dissolved in the primary aqueous phase, whilst the xanthan gum was dispersed in the *oil* prior to the first stage of homogenization. In the preparations containing gelatin, the primary W/O emulsions were chilled in ice for 30 min prior to the second emulsification step in an attempt to promote gelatin gelation.

A Malvern Mastersizer (static multi-angle light-scattering apparatus) was used to measure the time-dependent changes of the droplet size of the W/O/W emulsions. Average droplet sizes (d_{32}, d_{43}) were determined when the emulsions were freshly prepared and after storage for 1 week at 25 °C. The yield of the multiple emulsions was assessed by measuring the amount of dye (Poly R-478) released from the inner aqueous droplets following the second emulsification step.[15] The experimental procedure adopted for the determination of the yield has been described elsewhere.[6] Using yield analysis, the effect on the stability of encapsulating BSA (0–2 wt%), gelatin (0–2 wt%) or xanthan gum (0, 0.2 or 1 wt%) in the inner droplets was assessed over a period of 7 days. The orthokinetic stability was assessed by subjecting the multiple emulsions (75 ml samples) to a turbulent flow field at 2000 r.p.m. in a Couette device with inner and outer cylinder diameters of 30 mm and 40 mm, respectively.[16] During the shearing process, small aliquots were removed at short time intervals for the determination of the average droplet size. Electrophoretic measurements were conducted using laser Doppler electrophoresis (Zetasizer 4, Malvern Instruments). W/O/W samples stabilized with 0.35 wt% Tween 20 were formulated using 5 mM imidazole buffer (pH 7) with 0 or 1 wt% BSA encapsulated in the inner droplets. These emulsion samples were diluted (1:10000) with imidazole buffer, pH 7, and then immediately injected into the electrophoretic cell. Electrophoresis measurements were carried out using a cell voltage of 135 V (sensed) and a modulator frequency of 250 Hz.

3 Results and Discussion

Figure 1 shows the effect of reducing the level of lipophilic emulsifier (Admul Wol or Span 80) on the multiple emulsion yield. In both the soybean oil and kerosene formulations, the multiple emulsion yield falls very rapidly when the concentration of lipophilic emulsifier is reduced below 3 wt%. Greater loss of yield over time is observed in formulations which contain lower amounts of lipophilic emulsifier (Figure 2). The loss of inner aqueous droplets also accounts for observed reductions in the average droplet size of the multiple emulsions. Typically, a kerosene multiple emulsion formulated with 8 wt% Span 80 has an initial mean droplet size d_{43} of 1.40 μm which decreases over the 7 day period to 1.20 μm. In contrast, a multiple emulsion formulated with 2 wt% Span 80 has an initial d_{43} of 5.43 μm which falls to 3.19 μm after 7 days.

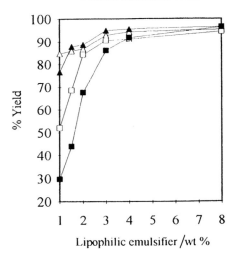

Figure 1   *The effect of lipophilic emulsifier concentration on the multiple emulsion yield:* ▲,△, *fresh W/O/W emulsion;* ■,□, *after 7 days at 25 °C. Solid and open symbols refer to multiple emulsions made with soybean oil (Admul Wol, 0.125 wt% sodium caseinate) and kerosene (Span 80, 0.125 wt% sodium caseinate), respectively. Lines have been drawn through the symbols simply to guide the eye*

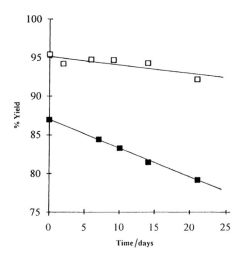

Figure 2   *The effect of Span 80 on the time-dependent yield of kerosene multiple emulsions stabilized with 0.125 wt% sodium caseinate:* □, *8 wt% Span 80;* ■, *2 wt% Span 80*

The shrinkage rate of the 2 wt% Span 80 system is so severe that it is even detectable in terms of the less sensitive d_{32} parameter (2.14 to 1.57 μm). The loss of large amounts of inner aqueous phase can be mainly attributed to the instability of the primary emulsion. At very low levels of Span 80, the kerosene W/O emulsions tend to contain large droplets which can be either too large to

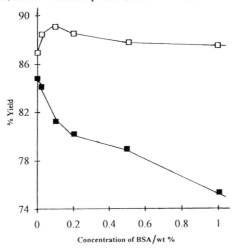

Figure 3   *The effect on the yield of encapsulating BSA in the inner aqueous droplets of a kerosene W/O/W multiple emulsion stabilized with 0.125 wt% sodium caseinate: □, freshly prepared; ■, after 7 days at 25 °C*

withstand the shearing forces in the second stage of homogenization, or may coalesce with the outer oil–water interface shortly after being encapsulated. At high Admul Wol concentrations, the soybean formulations tend to have slightly higher initial yields than the kerosene preparations. However, when the concentration of Admul Wol is lowered, the resultant formulations tend to have lower yields, and are less stable over the 7 day period than similarly formulated kerosene emulsions (Figure 1). The higher initial yield in the soybean emulsions may be due to the higher viscosity of the soybean oil which could reduce the effects of the disruptive shearing forces (in the second homogenization stage) that are mainly responsible for breaking the primary emulsion.

Figure 3 shows the effect of encapsulating BSA within the inner aqueous droplets of a kerosene multiple emulsion formulated with 2 wt% Span 80. This particular multiple emulsion formulation has been used here because it exhibits a relatively low level of stability without compromising the reproducibility of the yield assay or droplet-size distribution determination. (The experimental error of the yield assay for this particular preparation is less than 1%). It is interesting to note that there is a slight increase in the multiple emulsion yield up to 0.1 wt% BSA. Similar trends with multiple emulsions containing BSA have been reported elsewhere;[7,11] this increase in yield is thought to be due to the presence of a complex at the inner oil–water interface formed by interaction of Span 80 and BSA molecules. However, we see from Figure 3 that the presence of BSA has a detrimental effect on the long-term stability of these multiple emulsion formulations. A similar trend in the measured yield with added macromolecule in the inner droplets is found for multiple emulsions containing gelatin (Figure 4). A low level incorporation of gelatin into the inner aqueous phase appears to enhance the yield, whilst the

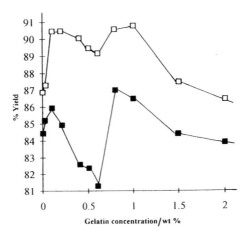

Figure 4   *The effect on the yield of encapsulating gelatin in the inner aqueous droplets of a kerosene W/O/W multiple emulsion stabilized with 0.125 wt% sodium caseinate:* □, *freshly prepared;* ■, *after 7 days at 25 °C*

Table 1   *The effect on the yield of encapsulating xanthan gum in a W/O/W multiple emulsion (0.125 wt% sodium caseinate, 8 wt% Span 80 in kerosene)*

Concentration of xanthan gum	% Yield	
wt%	*Freshly prepared*	*After 7 days*
0	86.9 ± 0.6	83.9 ± 0.5
0.2	89.5 ± 0.5	84.7 ± 2.6
1.0	90.8 ± 0.6	86.1 ± 0.4

addition of larger amounts – but small enough not to cause gelation (*i.e.* <0.5 wt%) – seems to have a detrimental effect on the long-term yield stability. However, further additional amounts of gelatin to levels such that gelation occurs in the inner aqueous droplets seems to improve again slightly the initial yield and the retention of the inner droplets over time. Observations with xanthan gum are qualitatively similar (Table 1), although these yield results are subject to greater experimental errors which possibly arise from the fact that not all of the gum migrates from the oil into the aqueous droplets after the first emulsification step. Improvements in yield and stability in W/O/W emulsions with xanthan present are presumably due to the large low-stress viscosity increase caused within the inner aqueous droplets.

Shear-induced coalescence of W/O/W multiple emulsions, unlike that observed for simple O/W systems,[16] takes the form of a gradual initial steady increase in the mean droplet size, d_{32}, followed by a rapid rise in mean droplet size.[6] Table 2 shows the results obtained when W/O/W multiple emulsions encapsulating BSA or gelatin are sheared continuously in turbulent flow. The

Table 2 *The effect of encapsulated protein in a W/O/W multiple emulsion (0.125 wt% caseinate, kerosene, 8 wt% Span 80) on the orthokinetic stability: the critical coalescence time t_c refers to the time at which the maximum rate of coalescence is obtained[16] when the multiple emulsion is subjected to a turbulent flow field at 2000 r.p.m.*

Type and concentration of protein	t_c/min	% Yield[a]
No protein	215	86.9 ± 0.6
0.2 wt% BSA	170	88.2 ± 0.2
1.0 wt% BSA	175	87.3 ± 0.2
0.2 wt% gelatin	150	89.8 ± 1.1
1.0 wt% gelatin	250	90.2 ± 1.7

[a] Obtained for multiple emulsions formulated using 2 wt% Span 80.

Table 3 *The effect of encapsulating protein in multiple emulsion droplets on the electrophoretic mobility. Emulsions were dispersed in imidazole buffer (5 mM, pH 7) at 25 °C. The zeta potential was calculated from the mobility measurements using the Smoluchowski equation for electrophoresis of non-conducting spheres*

	Kerosene		Soybean oil	
	Mobility	Zeta potential	Mobility	Zeta potential
Nature of inner aqueous phase	10^{-8} m^2 V^{-1} s^{-1}	mV	10^{-8} m^2 V^{-1} s^{-1}	mV
No protein	−4.28 ± 0.14	−54.5 ± 1.9	−4.81 ± 0.14	−61.3 ± 1.8
1 wt% BSA	−3.97 ± 0.20	−50.5 ± 0.2	−4.61 ± 0.07	−58.7 ± 0.8
1 wt% Lysozyme	−4.23 ± 0.13	−53.7 ± 1.8	−4.34 ± 0.12	−55.1 ± 3.1

effect of the encapsulated protein in the inner aqueous droplets of the multiple emulsions appears to be rather small. However, there is a slight tendency for the critical coalescence time to be lower in the cases where the inner droplets contain protein. Presumably, this could be connected in some way to the slightly higher yields that tend to prevail in multiple emulsions with encapsulated protein. In a previous paper,[6] we have reported that the shearing stability behaviour of a W/O/W multiple emulsion differed only marginally from that of a similarly formulated O/W emulsion. The amount and composition of the emulsifier on the outer oil–water interface was found to play a much more crucial role than the presence of inner aqueous droplets in determining the shear stability. It is possible in the present study that the high-yielding multiple emulsion systems have less surface-active material (Span 80) at the oil–water interface and are consequently less stable. What is interesting to note from the food industry point of view is that the multiple emulsion containing gelled inner aqueous droplets (1 wt% gelatin) is more resistant to shear flow.

The effect on the electrophoretic mobility of encapsulating protein (BSA or lysozyme) within the inner droplets of multiple emulsions is shown in Table 3. The multiple emulsions formulated here were prepared using Tween 20 as the hydrophilic emulsifier. Tween 20 was used in preference to sodium caseinate as the secondary emulsifier in these experiments so that any surface charge effects caused by the encapsulated protein could be more readily detected. The Tween 20 molecules tend to desorb from the outer oil–water interface when the emulsion is diluted prior to injection into the electrophoresis cell. This becomes apparent when the electrophoretic mobility for the kerosene emulsion after dilution in a 0.35 wt% solution of Tween 20 in buffer is compared with that for the emulsion diluted in buffer only (Table 4). As we had rather expected, there appears to be no significant change in the electrophoretic mobility when protein is encapsulated in the inner droplets. In the soybean oil multiple emulsions, where the electrophoretic mobility is larger, the presence of encapsulated protein seems slightly to reduce the mobility. There is a tendency for the mobility of multiple droplets containing lysozyme to be lower than that for multiple droplets encapsulating BSA. It would seem that this is inextricably linked to the pI values of the proteins (BSA, pI = 4.5; lysozyme, pI = 10.8).

In previous research conducted by Ueda and Matsumoto,[17] the electrophoretic mobility of multiple emulsion droplets containing protein has been found to be related to the pI value of the encapsulated protein. The protein molecules in the inner droplets are considered by Ueda and Matsumoto[17] to have protruded across the oil layer into the continuous aqueous phase to produce an effect on the mobility. Another possible explanation, however, for their observed reduction in electrophoretic mobility is that it could be the result of changes in the surfactant composition at the outer oil–water interface. The yield of the multiple emulsion and the pI value of the encapsulated protein are both likely to influence the distribution of surfactant molecules between the inner and outer oil–water interfaces.

In conclusion, it appears that proteins or other macromolecular stabilizers are unlikely to be able completely to replace lipophilic emulsifiers in multiple emulsions. However, proteins in combination with stabilizers do have the capacity to confer some enhanced degree of stability on a multiple emulsion system when the lipophilic emulsifier concentration is substantially reduced.

Table 4   *The effect on the electrophoretic mobility of dispersing a kerosene W/O/W emulsion or O/W emulsion in imidazole buffer (5 mM, pH 7) containing Tween 20*

System	Dispersion medium	Mobility $10^{-8}\,\mathrm{m^2\,V^{-1}\,s^{-1}}$	Zeta Potential mV
Kerosene W/O/W	Buffer only	-4.28 ± 0.14	-54.5 ± 1.9
Kerosene W/O/W	0.35 wt% Tween 20	-1.80 ± 0.10	-22.4 ± 0.5
Kerosene W/O/W	0.10 wt% Tween 20	-2.23 ± 0.05	-28.5 ± 0.7
Kerosene O/W	0.35 wt% Tween 20	-1.72 ± 0.04	-22.1 ± 0.3

Acknowledgements

Financial support to E. D. and R. K. O. A. for this research from the Agricultural and Food Research Council is gratefully acknowledged. J. E. thanks AFRC for the award of a Research Studentship.

References

1. S. Matsumoto, in 'Non-Ionic Surfactants: Physical Chemistry', ed. M. J. Schick, Marcel Dekker, New York, 1987, p. 549.
2. S. Matsumoto, Y. Kita, and D. Yonezawa, *J. Colloid Interface Sci.*, 1976, **57**, 353.
3. B. de Cindio, G. Grasso, and D. Cacace, *Food Hydrocolloids*, 1991, **4**, 339.
4. J. A. Omotosho, *Int. J. Pharm.*, 1990, **62**, 81.
5. Th. F. Tadros, *Int. J. Cosmet. Sci.*, 1992, **14**, 93.
6. E. Dickinson, J. Evison, R. K. Owusu, and A. Williams, in 'Gums and Stabilisers for the Food Industry', ed. G. O. Phillips, D. J. Wedlock, and P. A. Williams, Oxford University Press, vol. 7, 1994, p. 91.
7. N. Garti, *Food Struct.*, 1994, in the press.
8. R. K. Owusu, Q. Zhu, and E. Dickinson, *Food Hydrocolloids*, 1992, **6**, 443.
9. E. Dickinson, J. Evison, J. W. Gramshaw, and D. Schwope, *Food Hydrocolloids*, 1994, **8**, 63.
10. A. T. Florence and D. Whitehill, *J. Colloid Interface Sci.*, 1981, **79**, 23.
11. J. A. Omotosho, T. K. Law, T. L. Whateley, and A. T. Florence, *Colloids Surf.*, 1986, **20**, 133.
12. K. P. Oza and S. G. Frank, *J. Dispersion Sci. Technol.*, 1989, **10**, 163.
13. E. Dickinson, J. Evison, and R. K. Owusu, *Food Hydrocolloids*, 1991, **5**, 481.
14. I. Burgaud, E. Dickinson, and P. V. Nelson, *Int. J. Food Sci. Technol.*, 1990, **25**, 39.
15. J. L. Zatz and G. C. Cueman, *J. Soc. Cosmet. Chem.*, 1988, **39**, 211.
16. E. Dickinson and A. Williams, *Colloids Surf. A*, 1994, **88**, 317.
17. K. Ueda and S. Matsumoto, *Bull. Chem. Soc. Jpn.*, 1991, **64**, 3163.

Effect of Non-Starch Polysaccharide on the Stability of Model Physiological Emulsions

By Annette Fillery-Travis, Lucy Foster, Sarah Moulson, Martin Garrood, Sybil Clark, and Margaret Robins

INSTITUTE OF FOOD RESEARCH, NORWICH LABORATORY, NORWICH RE-SEARCH PARK, COLNEY, NORWICH NR4 7UA, UK

1 Introduction

This work is part of a study to investigate the mechanism of interaction of non-starch polysaccharide (NSP) with the lipid phases formed within the stomach and small intestine. Prior to absorption by the gut, ingested lipids must undergo a series of biotransformations to increase their solubility in the aqueous environment of the body.[1] The first stage is the formation of an emulsion during passage through the stomach and small intestine. The emulsion is stabilized by surfactants which are either produced by the gut or intrinsic to the diet.[2] Initially we have characterized, *in vitro*, the behaviour of two representative surfactants present during digestion, L-α-phosphatidylcholine (the principal surfactant within the stomach) and sodium taurocholate (a bile salt present in the small intestine). They demonstrate synergistic behaviour, under conditions of physiological relevance to the duodenum. This is in agreement with our observations *in vivo* that the emulsions in the duodenum have a higher surface area and are more stable than those isolated from the stomach. The effect of NSP preparations (cabbage pectin and guar gum) on the stability of the emulsions has also been investigated in order to identify the mechanisms of interaction.

2 Materials and Methods

The L-α-phosphatidylcholine, (PC, chromatographically pure, grade 1 derived from egg yolk, Lipid Products, Surrey) was used as a dispersion of liposomes at pH 6.6. Sodium taurocholate (NaT; Sigma) and olive oil (low acidity, Sigma) were used without further purification.

For each emulsifier a concentrated premix emulsion (8 wt% olive oil) was prepared (using a Waring blender on a fixed shear cycle) with a weight mean droplet diameter of approximately 1 μm. The premix was diluted to give an emulsion of 2 wt% oil, 0.15 M NaCl, 0–0.1 wt% NSP and surfactant concentrations of 0–13 mM in a citric acid buffer of pH 6.6 with 0.3 wt% sodium azide as preservative. For all mixed surfactant emulsions the pre-mix was prepared with PC only and the NaT was added with the continuous phase diluent. The emulsions were kept constantly agitated at 37 °C and their stability to coalescence monitored by regular determination of mean droplet size and oil content. Droplet-size distributions of emulsions were measured using a Malvern Mastersizer. The oil content was determined from density measurements using an Anton Paar DMA60 vibrating tube density meter with a DMAA602 measuring cell. A Malvern Instruments Zetasizer 3 with a AZ4 capillary electrophoresis cell was used to determine droplet mobility using a diluent prepared to exactly the same aqueous phase composition as the emulsion.

3 Results

For all emulsions prepared with NaT, either as the sole surfactant or in combination with PC, no persistent flocculation was observed by optical microscopy. Thus, it was possible to estimate the number of droplets present in the emulsion by combining the droplet-size distribution measured at a given time with the corresponding oil volume fraction determination. The rate of decrease in total droplet number was then used to monitor emulsion instability to coalescence.

Severe coalescence was observed for both single surfactant emulsions within 150 h. Those prepared with PC did tend to form flocs readily, and the presence of a liquid crystalline interface controlled the rate of coalescence. The limited

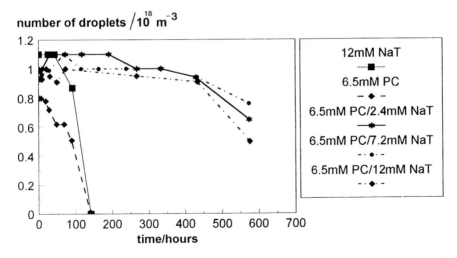

Figure 1   *Coalescence rates of emulsions stabilized by single surfactants or PC-stabilized emulsions exposed to varying concentrations of NaT*

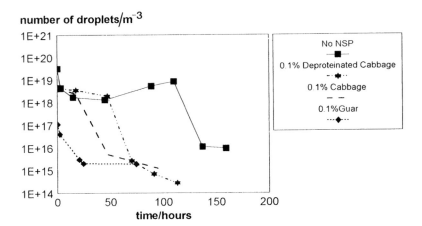

number of droplets/m⁻³

Figure 2   *Coalescence rates of PC-stabilized emulsions in the presence of cabbage fibre preparations and guar gum. NSP contains some protein impurities unless otherwise stated*

stability of NaT-stabilized emulsions occurred by virtue of the high surface charge imparted by the surfactant which prevents rapid flocculation. However, upon exposing the phospholipid-stabilized emulsions to NaT at a range of concentrations, a significant increase in stability was observed (Figure 1). The droplet microelectrophoretic mobility of the mixed surfactant emulsion was found to be significantly greater than that of the PC-stabilized system,[3] but only 50% of that observed for droplets stabilized only by NaT. These results have suggested that the enhanced stability to coalescence of the mixed systems is due to an increase in the electrostatic repulsive interaction between droplets resulting from the co-operative adsorption of charged NaT without significant change in the liquid crystalline interface.

For PC-stabilized emulsions (stomach model), the presence of NSP in the form of either cabbage pectin[4] or guar gum, has resulted in significant destabilization of the emulsion (Figure 2). Possible mechanisms for this destabilization are: (1) competitive adsorption of the NSP preparations at the interface by surface-active protein impurities (de-proteinated cabbage pectin was nevertheless still an effective destabilizer), and (2) loss of surfactant due to direct binding with NSP. For the mixed surfactant system the presence of NSP had no effect on emulsion stability.

4 Conclusions

The presence of NaT has the effect of enhancing significantly the stability to coalescence of phospholipid-stabilized emulsions. The migration of NaT to the liquid crystalline PC interface imparts an increased surface charge to the droplets whilst maintaining, at least in part, the stabilizing character of the interface. NSP preparations of cabbage pectin and guar gum are found to destabilize emulsions containing PC only, but have no effect in the presence of NaT. These results are in good agreement with results of parallel *in vivo* studies on lipid emulsification and stability in the stomach and duodenum of rats fed a cabbage diet.

Acknowledgements

We wish to acknowledge many helpful discussions with Dr Ian Johnson of our Nutrition, Diet and Health Department and Dr Keith Waldron of our Food Molecular Biochemistry Department (who supplied the NSP preparations). This work was funded by the Ministry of Agriculture, Fisheries and Food.

References

1. M. C. Carey, *Annu. Rev. Physiol.*, 1983, **45**, 651.
2. M. Armand, P. Borel, P. Ythier, G. Dutot, C. Melin, M. Senft, H. Lafont, and D. Lairon, *J. Nutr. Biochem.*, 1992, **3**, 333.
3. A. Fillery-Travis, L. Foster, S. Moulson, M. Garrood, S. Clark, and M. Robins, to be published.
4. G. Moates, R. Selvendran, and K. Waldron, to be published.

Investigation of the Function of Whey Protein Preparations in Oil-in-Water Emulsions

By Gerald Muschiolik, Silke Dräger,[1] Harshadrai M. Rawel,[1] Paul Gunning,[2] and David C. Clark[2]

CENTRE OF ENVIRONMENTAL SCIENCES, UNIVERSITY OF POTSDAM, 14558 BERGHOLZ-REHBRÜCKE, GERMANY
[1] GERMAN INSTITUTE OF HUMAN NUTRITION, 14558 BERGHOLZ-REHBRÜCKE, GERMANY
[2] FOOD BIOPHYSICS DEPARTMENT, INSTITUTE OF FOOD RESEARCH, NORWICH LABORATORY, NORWICH NR4 7UA, UK

1 Introduction

It is necessary to determine the effect of processing and environmental factors in order to compare the effect of different protein preparations on emulsion behaviour. The aims of the present study are to prepare protein-stabilized oil-in-water (O/W) emulsions and to study the effects of acidification and addition of NaCl on average particle size, rheological behaviour and water-holding properties for two whey protein concentrates (WPC) containing protein at different levels of denaturation.

2 Materials and Methods

The properties of the whey protein concentrates were as follows.

WPC 1 with 75% protein. Solubility (15 000 g, 10 min): at pH 4 = 56%, at pH 7 = 89%, in 0.44 M NaCl (pH 7) = 92%; secondary structure: 34% α-helix, 16% β-structure, 50% random coil; unfolding: relative fluorescence at 338 nm = 60; free SH groups = 0, total SH groups = 21 nmol mg^{-1} of protein; molecular mass of native (HPLC): 34% high molecular mass (more than 79 kDa); hydrophobicity [reversed-phase (RP)-HPLC]: 3% of total protein after 6.1 min retention time.

WPC 2 with 80% protein. Solubility: at pH 4 = 18%, at pH 7 = 72%, in 0.44 M NaCl (pH 7) = 72%; secondary structure: 18% α-helix, 32% β-structure, 50% random coil; unfolding: relative fluorescence at 338 nm = 46; free SH groups $= 6$ nmol mg^{-1} of protein, total SH groups $= 10$ nmol mg^{-1} of protein; molecular mass native (HPLC): 83.7% high molecular mass (more than 79 kDa); hydrophobicity (RP-HPLC): 22% of total protein after 6.1 min retention time.

O/W emulsions (dispersed phase 30 vol%) stabilized with 2 wt% WPC were prepared by high-pressure homogenization (HH 20) with or without 2 wt% NaCl (added either to the protein solution or to the emulsion). Emulsions were acidified with 0.25 wt% citric acid to pH 3.9. The separable water phase of the emulsions (WH, %) was estimated by gentle centrifugation (1000 g, 10 min). Rheological behaviour (Ostwald factor k, flow index n) was characterized with a Rheolab RV III (Brookfield, concentric cylinder). Mean particle size d_{43} of single and flocculated oil droplets was determined using a Coulter LS 100 (Micro Volume Module). The molecular mass distribution of protein in the separated water phase of the emulsions (15000 g, 10 min) was estimated by gel permeation HPLC.

3 Results and Discussion

Protein sample WPC 1, containing the lower level of denatured protein, shows similar emulsifying properties to WPC 2 which contains highly denatured protein. WPC 1 shows in neutral O/W emulsions a lower sensitivity to NaCl (a lower degree of particle flocculation on adding NaCl; see Table 1). WPC 1 in acidified emulsions without NaCl causes a different type of particle flocculation resulting in a higher viscosity than WPC 2. Neutral emulsions with WPC 2 become more viscous than those with WPC 1 following addition of NaCl to the protein solution.

When salt is added after homogenization to neutral emulsions containing WPC 1 or WPC 2, the increase in viscosity is much lower and the water phase volume obtained following centrifugation is reduced in neutral and acidified emulsions (Table 1). In comparison with neutral emulsions without NaCl, the

Table 1 *Particle-size distribution of single and flocculated oil droplets, the Ostwald-factor* k, *the flow rate* n *and the separable water phase (WH) of O/W emulsions made with 2 wt% WPC*

Protein sample	[NaCl] %	Neutral emulsion (pH 6.8–7.7)				Acidified emulsions (pH 3.9)			
		d_{43} μm	k Pa s^n	n	[WH] %	d_{43} μm	k Pa s^n	n	[WH] %
WPC1	0	1.47	0.002	1.00	0	23.86	3.3	0.24	3.0
WPC2	0	1.81	0.003	1.10	0	21.45	0.1	0.90	2.0
WPC1	2[a]	8.79	10.0	0.19	13.0	8.72	4.7	0.30	17.0
WPC2	2[a]	153.05	16.0	0.15	23.0	41.04	1.8	0.50	21.0
WPC1	2[b]	3.76	0.025	0.80	0	8.52	10.0	0.18	7.0
WPC2	2[b]	5.92	0.041	0.90	0	31.55	3.5	0.46	6.5

[a] NaCl added to the WPC solution.
[b] NaCl added to the emulsion.

water phase volume following centrifugation contains less high molecular protein fraction when NaCl is present and no high molecular mass fraction after acidification. When NaCl is added before emulsification, the high molecular mass fraction in the separated phase is reduced (Table 2).

The primary effect of NaCl will be to reduce charge repulsion at acidic and neutral pH. This will enhance protein packing at the interface, and possibly decrease the thickness of the adsorbed layer by reducing chain extension. This will decrease the effectiveness of steric stabilization (giving droplet flocculation) and reduce electrostatic stabilization (again giving flocculation). The thickness of the adsorbed layer and the size of droplet flocs could be enhanced by increasing adsorption of the high-molecular mass fraction. When salt is added after homogenization, it seems to be that there is no increased packing of the adsorbed protein (but a higher content of high molecular mass protein fraction is observed in the separated water phase, Table 2).

The stability of the flocculated droplets (*i.e.* the stability to the shear of the Coulter MVM stirrer) seems to correlate with the changed rheological behaviour, *i.e.* the increased Ostwald factor k (with increasing d_{43} factor k increases).

Table 2   *Content of high-molecular mass fraction (HMF) of WPC in the separated water phase (15 000 g, 10 min) of the O/W emulsions with 2 wt% WPC 1*

		Emulsion		
Neutral	Acidified	2 wt% NaCl to WPC solution	2wt% NaCl to the emulsion	[HMF] %
$+$ [a]	$-$	$-$	$-$	100
$-$	$+$	$+$	$-$	0
$+$	$-$	$+$	$-$	ca. 30
$+$	$-$	$-$	$+$	ca. 60
$-$	$+$	$-$	$+$	0

[a] The symbol '$+$' means the emulsion formulation used for the test (neutral or acidified, with or without NaCl).

4 Conclusions

Addition of salt to WPC emulsions apparently reduces electrostatic repulsion leading to oil droplet flocculation and associated changes in the rheological behaviour. Highly denatured WPC containing more high molecular mass species is more susceptible to changes in rheological behaviour in neutral and acidified emulsions associated with NaCl-induced droplet flocculation.

The functionality of whey protein concentrate and its influence on the rheological behaviour of O/W emulsions can therefore be influenced by changing the degree of protein denaturation and by adjusting the method of preparing the emulsion systems.[1]

Acknowledgements

The authors wish to thank Danmark Protein A/S for providing the whey proteins. The research was partly funded by the MWMT Land Brandenburg and by the DAAD (Germany).

Reference

1. G. Muschiolik, S. Dräger, and A. Sutton, *Dt. Milchwirtschaft*, 1993, **44**, 1054.

Shear-Induced Instability of Oil-in-Water Emulsions

By Andrea Williams and Eric Dickinson

PROCTER DEPARTMENT OF FOOD SCIENCE, UNIVERSITY OF LEEDS, LEEDS LS2 9JT, UK

1 Introduction

When an emulsion is subjected to a flow field both droplet disruption and droplet aggregation may occur. Which of these processes happens predominantly is dependent upon: the size distribution of the droplets, the viscosities of the two phases, the nature of the interface, and the shear-rate. In this study we investigate (i) the effect of protein concentration and (ii) the effect of initial droplet-size distribution on the stability of a model system in a controlled turbulent flow field. We compare experimental findings with results obtained from a numerical model.

2 Methods

Experiments

The amount of protein adsorbed at the interface in fine β-lactoglobulin (β-lg)-stabilized emulsions (20 vol% n-tetradecane, pH 7) was measured as a function of the concentration of protein used to make the emulsion. This was calculated using combined results from protein analysis experiments on the serum phase and measurements of the specific surface area of the emulsion.[1] The emulsion was subjected to a turbulent flow field in a concentric cylinder shearing device. The droplet-size distribution was measured as a function of time using a Malvern Mastersizer. From a plot of average volume–surface droplet diameter d_{32} versus time, a critical destabilization time t_d corresponding to the time of the maximum rate of increase of d_{32} was extracted.[2]

Numerical Modelling

We have developed a phenomenological model in order to predict the change in droplet-size distribution of an emulsion with time in a shearing experiment. It is based upon the assumption that droplet disruption kinetics are first-order and droplet aggregation kinetics are second-order.[3] The initial droplet-size distribution was taken to be of a standard log–normal form. Certain parameters had to be set in the program: these were the simulation time interval Δt, the fractal dimensionality D, the critical size below which droplets could not be disrupted, d_c, and the droplet disruption and coalescence rate constants, k'_d and k'_c. When full immediate coalescence of the colliding droplets occurred, then D had a value of 3. A value of D of less than 3 implied that the clusters initially formed had ramified (fractal-type) structures. Further details concerning the assumptions of the model may be found elsewhere.[4]

Table 1 *Influence of protein concentration on the initial average droplet diameter* d_{32}^0, *the protein surface coverage* Γ, *and the critical destabilization time* t_d. *Oil-in-water emulsions (20 vol% n-tetradecane, protein dissolved in 0.05 M NaCl, pH 7) were made with a jet homogenizer (0.81 mm jet hole, 100 bar)*

β-lg conc./wt%	$d_{32}/\mu m$	$\Gamma/mg\ m^{-2}$	t_d/min
0.125	1.97	1.19	110
0.25	1.55	1.33	380
0.375	1.21	1.39	520
0.5	0.96	1.44	700

3 Results and Discussion

Table 1 shows that, as expected, when the concentration of β-lg used to make the emulsion is increased, the initial average droplet diameter d_{32}^0 decreases; and the concentration of protein adsorbed at the interface and the orthokinetic stability both increase. According to Smoluchowski[5] the rate of orthokinetic coalescence is proportional to the cube of the sum of the diameters of the colliding droplets. Therefore a smaller initial droplet diameter will cause the emulsion to be more stable. Also a higher concentration of protein at the droplet interface will probably better protect against coalescence.

Figure 1 shows the predicted change in d_{32} with time for two different values of d_{32}^0. It shows that, if all other conditions are equal, then the steady-state droplet diameter achieved at the end of the simulation is the same irrespective of whether the initial droplet diameter is high or low. If the plots in Figures 1 and 2 for a low value of d_{32} are compared, it can be seen that the shape of the graph predicted by the computer model is qualitatively the same as that given by the experiment, with an extended lag stage at the beginning of the time period where d_{32} does not change much, followed by a period when d_{32} increases rapidly, until it levels off again to a steady-state value.

Figure 2 shows the results from experiments designed to measure the change in d_{32} with time for emulsions starting with either a high or a low d_{32} value. It can be seen that the final d_{32} value is rather higher after long shearing times for the coarse initial emulsion than for the fine initial emulsion. For coarse and fine emulsions made with a small-molecule surfactant (Tween 20, 0.5 wt%),

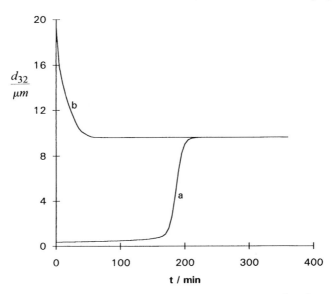

Figure 1   *Influence of initial droplet-size distribution on computed evolution of average droplet diameter d_{32} as a function of shearing time t. The two assumed initial average droplet diameters are (a) $d_{32}^0 = 0.36\ \mu m$ and (b) $d_{32}^0 = 19.3\ \mu m$. Other parameter values are set as follows: $\Delta t = 2\ s$, $D = 2.0$, $d_c = 10\ \mu m$, $k'_d = 0.1\ s^{-1}$, $k'_c = 300\ m^{-\frac{3}{2}}\ s^{-1}$*

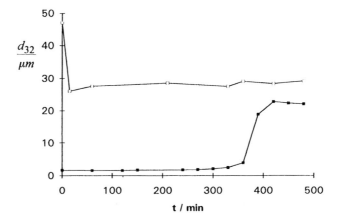

Figure 2   *Influence of the initial average droplet diameter on the experimental change in d_{32} with time t for oil-in-water emulsions (20 vol% n-tetradecane, 0.25 wt% β-lg in 0.05 M NaCl). Open and solid symbols refer respectively to high and low initial average droplet diameters*

the final steady-state average droplet diameter is found to be the same (*ca.* 11 μm). It is quite likely that the structure of the protein layer is altered during the experimental procedure, and that it is rather different for the coarse and fine emulsions. This could possibly be due to the homogenization process itself, and/or the effect of the flow field and the interaction with air of protein adsorbed at the macroscopic emulsion interface.

4 Conclusions

The orthokinetic stability of the β-lg emulsion increases as the concentration of protein used to make it is increased. This is thought to be associated with the smaller average droplet size and with the amount of protein adsorbed at the interface increasing as the amount of protein used to make the emulsions is increased.

When other conditions are kept constant, the final steady-state average droplet size for an emulsion subjected to a flow field is expected to be the same, whatever the initial droplet-size distribution. This is not true for the β-lg emulsion. It would seem, therefore, that the structure of the adsorbed layer of globular protein molecules at the oil–water interface is affected by factors such as the homogenization process and the shearing action.

References

1. J.-L. Courthaudon, E. Dickinson, Y. Matsumura, and D. C. Clark, *Colloids Surf.*, 1991, **56**, 293.
2. E. Dickinson, R. K. Owusu, and A. Williams, *J. Chem. Soc., Faraday Trans.*, 1993, **89**, 865.
3. P. Becher and M. J. McCann, *Langmuir*, 1991, **7**, 1325.
4. E. Dickinson and A. Williams, *Colloids Surf. A*, 1994, **88**, 317.
5. M. Smoluchowski, *Z. Phys. Chem.*, 1917, **92**, 129.

Surfactant–Protein Competitive Adsorption and Electrophoretic Mobility of Oil-in-Water Emulsions

By Jianshe Chen, Jane Evison, and Eric Dickinson

PROCTER DEPARTMENT OF FOOD SCIENCE, UNIVERSITY OF LEEDS, LEEDS LS2 9JT, UK

1 Introduction

Most food emulsions of the oil-in-water type are stabilized primarily by an adsorbed layer of protein forming a protective steric barrier around the dispersed droplets.[1] The composition and structure of stabilizing layers in oil-in-water emulsions are significantly affected by the competitive displacement of protein from the interface by surfactant.[2] Understanding the structure and dynamics of mixed protein–surfactant films is necessary for controlling the formation, stability and rheology of these food colloids.

It has been found that the surface viscoelasticity of an adsorbed protein film develops slowly with time due to the gradual protein unfolding and the development of new protein–protein interactions at the interface.[3] Our recent research has shown that this macromolecular rearrangement and the associated strengthening of interfacial intermolecular interaction has a strong effect on the competitive adsorption behaviour of milk proteins with water-soluble surfactants.[4,5] Therefore, the question arises as to whether this kind of interfacial change has any effect on the electrophoretic mobility of the particles. Our main aims here are to investigate the change of mobility with respect to the change of protein surface concentration and to explore the ageing effect on the competitive adsorption with this new technique.

2 Materials and Methods

The proteins β-lactoglobulin (1.84×10^4 Da) and β-casein (2.4×10^4 Da) were obtained from Sigma, as was the commercial-grade emulsifier Tween 20 [polyoxyethylene(20) sorbitan monolaurate, 1.23×10^3 Da]. The research-grade $C_{12}E_2$ (diethylene glycol n-dodecyl ether, 2.74×10^3 Da) was obtained from Fluka. Hydrocarbon oils and buffer salts were AnalaR-grade reagents.

The protein-stabilized emulsion (0.4 wt% protein, 20 wt% oil, 20 mM buffer, pH 7.0) was made by using a mini-homogenizer at 300 bar. The particle size distribution was measured with a Malvern Mastersizer S2.01. The protein concentration in the serum phase was determined by fast protein liquid chromatography. The protein surface concentration was calculated from the known surface area of the emulsion and the difference in bulk protein concentration between that determined in the serum and that used to make the emulsion.[6]

The electrophoretic mobility and the calculated zeta potential were determined with the Malvern Zetasizer 4. The instrument was aligned daily with standard latex solution. The measurement was carried out by diluting 10 μl of fresh or aged emulsion in 30 or 50 ml Tween 20 buffer solution keeping the count rate at about 1500. The quoted mobility was the average of four measurements. The ZET5104 sample cell was used for all the experiments.

3 Results and Discussion

Figure 1 illustrates the competitive adsorption of protein and Tween 20 in β-lactoglobulin- and β-casein-stabilized emulsions at pH 7.0. The electrophoretic mobility decreases gradually with the increase of Tween 20 concentration and finally reaches the value of the Tween 20-stabilized emulsion at 0.1 wt% Tween 20 concentration. The complete displacement is believed to occur here. The competitive adsorption of β-lactoglobulin- and β-casein-stabilized

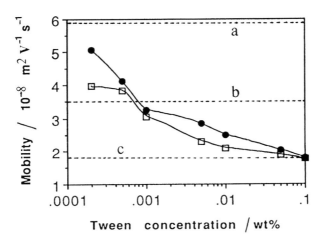

Figure 1   *The competitive adsorption in a protein-stabilized emulsion (0.4 wt% protein, 20 wt% n-tetradecane, 5 mM imidazole buffer, pH 7.0) with Tween 20 added to the fresh emulsion. The electrophoretic mobility is plotted against the Tween 20 concentration: □, β-casein-stabilized emulsion mixed with Tween 20; ●, β-lactoglobulin-stabilized emulsion mixed with Tween 20. The broken lines (a), (b), and (c) represent the electrophoretic mobilities of the fresh emulsions stabilized by β-lactoglobulin, β-casein, and Tween 20, respectively*

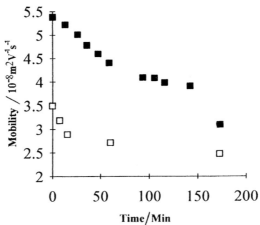

Figure 2   *The effect of ageing an oil-in-water emulsion (0.4 wt% β-lactoglobulin, 20 wt% n-tetradecane, 5 mM imidazole buffer, pH 7.0) on the ability of Tween 20 to displace β-lactoglobulin. The experiment was carried out at 10^{-3} wt% Tween 20 concentration. The electrophoretic mobility is plotted against the time after mixing with Tween 20 solution: □, fresh emulsion; ■, 24 h old emulsion*

emulsions have been studied previously in terms of the surface concentration.[4,5] Both the surface concentrations of β-lactoglobulin and β-casein were decreased to zero at 0.4 wt% Tween 20 concentration. The earlier results are quite consistent with the electrophoretic mobility data reported here. The electrophoretic mobility of the β-lactoglobulin-stabilized emulsion is higher than that of the β-casein-stabilized emulsion even though the β-lactoglobulin is present at a lower surface concentration than is the β-casein. This is probably because the thicker β-casein layer pushes the shear plane out further from the interface.

Figure 2 shows the time-dependent mobility of a fresh emulsion and a 24 h old emulsion dispersed in 10^{-3} wt% Tween 20 solution. The much higher electrophoretic mobility for the aged emulsion implies a poorer ability of Tween 20 to displace β-lactoglobulin molecules from the aged interface. We also see an ageing effect on the electrophoretic mobility of an oil-in-water emulsion (0.4 wt% β-lactoglobulin, 20 wt% n-hexadecane, 20 mM bistris propane buffer, pH 7.0) without Tween 20. The calculated zeta potential is decreased from about -22 mV for a fresh emulsion to about -18 mV for a 24 h old emulsion. The gradual change in calculated zeta potential suggests a slow build up of β-lactoglobulin molecules at the interface.

An ageing effect was also clearly observed in the competitive adsorption of an emulsion containing 0.1 wt% oil-soluble surfactant $C_{12}E_2$. The change of the (negative) zeta potential agrees quite well with the increase of the surface concentration of β-lactoglobulin as can be seen in Figure 3. Figure 4 shows the calculated zeta potential change due to the protein competitive displacement in the fresh and 24 h old emulsions. It was found that the complete displacement

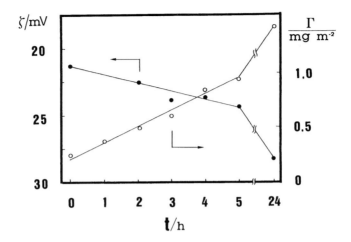

Figure 3   *The effect of ageing an oil-in-water emulsion (0.4 wt% β-lactoglobulin, 20 wt% n-hexadecane, 0.1 wt% $C_{12}E_2$, 20 mM bistris propane buffer, pH 7.0) on the competitive displacement of β-lactoglobulin by Tween 20. The electrophoretic mobility was measured in 2×10^{-4} wt% Tween 20 solution. The surface coverage experiment was carried out at 0.4 wt% Tween 20 concentration. The calculated (negative) zeta potential ζ (●) and the surface coverage Γ (○) are plotted against the ageing time of the emulsion before mixing with Tween 20 solution*

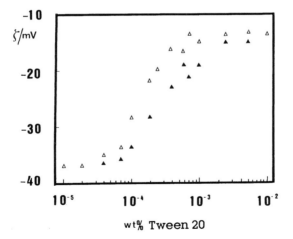

Figure 4   *The ageing effect of an oil-in-water emulsion (0.4 wt% β-lactoglobulin, 20 wt% n-hexadecane, 0.1 wt% $C_{12}E_2$, 20 mM bistris propane buffer, pH 7.0). The zeta potential ζ is plotted against the Tween 20 concentration: △, fresh emulsion; ▲, 1 day old emulsion*

of β-lactoglobulin for the aged emulsion containing 0.1 wt% $C_{12}E_2$ was more difficult than for the fresh emulsion. The results are also in agreement with earlier surface concentration results.[4] This suggests that the structural development of β-lactoglobulin at the oil–water interface does not only affect the competitive adsorption with low molecular mass surfactant, but may also affect the emulsion stability due to the change in droplet zeta potential.

References

1. E. Dickinson and G. Stainsby, 'Colloids in Food', Applied Science, London, 1982.
2. J. A. de Feijter, J. Benjamins, and M. Tamboer, *Colloids Surf.*, 1987, **27**, 243.
3. E. Dickinson, S. E. Rolfe, and D. G. Dalgleish, *Int. J. Biol. Macromol.*, 1990, **12**, 189.
4. J. Chen and E. Dickinson, *J. Sci. Food Agric.*, 1993, **62**, 283.
5. J. Chen, E. Dickinson, and G. Iveson, *Food Struct.*, 1993, **12**, 135.
6. J.-L. Courthaudon, E. Dickinson, and D. G. Dalgleish, *J. Colloid Interface Sci.*, 1991, **145**, 390.

Osmotic Pressure of Emulsions Containing Polysaccharide + Non-ionic or Anionic Surfactants

By Eric Dickinson, Michael I. Goller, and David J. Wedlock[1]

PROCTER DEPARTMENT OF FOOD SCIENCE, UNIVERSITY OF LEEDS, LEEDS LS2 9JT, UK
[1]SHELL RESEARCH, THORNTON RESEARCH CENTRE, THORNTON, CHESTER CH1 3SH, UK

1 Introduction

In many colloidal systems of commercial and industrial importance, the stability of the suspension or emulsion is controlled by a combination of a low molecular mass surfactant species (non-ionic, anionic or cationic) and a high molecular mass polymeric species, often a polysaccharide.[1] In technological terms, the role of the surfactant is to act as a dispersant or emulsifying agent, while the role of added biopolymer is to inhibit sedimentation or creaming by modifying the rheological behaviour of the aqueous dispersion medium.[2,3]

Our current research aims to investigate the influence of surfactant–surfactant and surfactant–biopolymer interactions on the structure and stability of various oil-in-water emulsions. This paper focuses on results from a new apparatus for measuring the osmotic pressure of emulsions emulsified by either the non-ionic surfactant polyoxyethylene sorbitan monolaurate (Tween 20) or the anionic surfactant sodium dodecyl sulfate (SDS). Rheological and creaming measurements have been carried out on the same systems.

2 The Osmometer

An instrument has been developed for measuring the osmotic pressure of solutions and emulsions. The apparatus involves the direct measurement of osmotic pressure using an electronic pressure-sensing technique. A sensitive transducer measures the pressure difference existing between a solution and a solvent, separated by a differentially permeable membrane. The output from the transducer is interfaced to a computer.

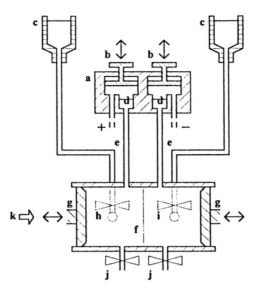

Figure 1   *Schematic diagram of apparatus used to measure osmotic pressure: (a) perspex block; (b) oil level adjustment piston; (c) reservoir; (d) oil–solution/solvent interface; (e) flexible poly(vinyl chloride) tubing; (f) differentially permeable membrane; (g) chamber volume adjustment piston; (h) solution inlet valve; (i) solvent inlet valve; (j) chamber outlet valve; (k) motor*

Figure 1 shows a schematic diagram of the osmometer. In this design the solution (or colloidal dispersion) and solvent are held in two stainless steel chambers, separated vertically by a membrane. The left-hand chamber contains the solution (or emulsion) and the right-hand chamber the solvent. Both sides are connected to the transducer; the positive side of the transducer is connected to the chamber containing the solution. The membrane is held between two rubber 'O' rings to prevent chamber–chamber leakage and is sandwiched between two stainless steel sieves to confer mechanical support. The two chambers are tightly screwed together; this clamps the membrane and its supports firmly in place. The osmometer is kept in a temperature-controlled cabinet at 25 °C.

MF-Millipore filters [Millipore (UK), Watford, Hertfordshire] with nominal pore sizes 0.025 or 0.1 μm are used as the membranes. They are made from biologically inert mixtures of cellulose acetate and cellulose nitrate. The membrane differential permeability is based on a surface screen sieve mechanism.

The transducer is a differential pressure transmitter ARA200 (Hartmann and Braun, Germany). It has a 4–20 mA output, equivalent to 0–10 mbar pressure. The current is converted by a 195.7 Ω resistor to a potential difference range 0.78–3.91 V ($\equiv$0–10 mbar). A high-speed analogue-to-digital converter PC27 (Amplicon Liveline, Brighton, Sussex) is used to interface the

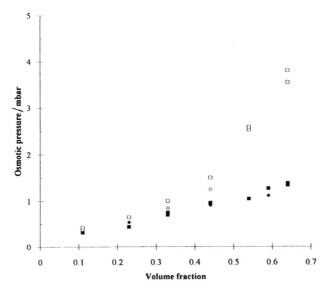

Figure 2 *Graph of equilibrium osmotic pressure versus volume fraction for a number of oil-in-water emulsions (oil:surfactant = 30:1, pH 7, ionic strength 0.05 M, 25 °C) emulsified by:* ■, *Tween 20 (0.025 μm pore size);* ◆, *Tween 20 (0.1 μm pore size);* □, *SDS (0.025 μm pore size);* ◇, *SDS (0.1 μm pore size)*

transducer output to a computer, where the data are transferred to a PC-based spreadsheet. The transducer is filled with mineral oil, and the chambers containing the solution and the solvent are connected to the transducer by a direct liquid link, *via* a connection contained within a transparent perspex block (see Figure 1). This block, like the transducer, is filled with oil The hydrostatic connection between the block and the chambers involves a direct oil–solution/solvent junction being set up. The interfaces are adjusted into a wide region of the perspex block and their heights equalized using the oil level adjustment piston. To measure the osmotic pressure of the solution (or emulsion), all the connections are made, and then the solution inlet valve (h) is closed, whilst the solvent inlet valve (i) is left open. This has the effect of applying atmospheric pressure equally to both sides.

3 Emulsions Without Added Polysaccharide

A number of different oil-in-water emulsions were prepared. Two stock 60 wt% oil emulsions (emulsified by either Tween 20 or SDS) were diluted down with phosphate buffer (pH 7, ionic strength 0.05 M) to produce emulsions with different oil concentrations, while the oil:surfactant mass ratio was kept constant (30:1). The Tween 20 emulsions had average size $d_{43} = 0.68 \pm 0.03$ μm whereas the SDS-emulsified emulsions had $d_{43} = 0.56 \pm 0.02$ μm. The osmotic pressure of the emulsions was measured over a 10 h period using a membrane with pore size 0.025 μm.

Figure 2 is a plot of equilibrium osmotic pressure *versus* volume fraction for oil-in-water emulsions emulsified by either Tween 20 or SDS. It shows that at low volume fraction $\phi = 0.11$ the Tween 20 and SDS emulsions have approximately the same osmotic pressures, but as the volume fraction is increased the osmotic pressures of the two types of emulsion start to diverge. At $\phi = 0.64$ the SDS emulsion has an osmotic pressure three times greater than that of the Tween 20 emulsion. The osmotic pressure of the SDS emulsion increases sharply above $\phi = 0.44$ due presumably to an increase in the droplet–droplet repulsive forces as the negatively charged droplets are forced into close proximity at the high volume fraction. The presence of the non-ionic surfactant Tween 20 at the droplet surface produces a more steady rise in osmotic pressure over the volume fraction range 0.13–0.64. The Tween 20-coated droplets do not seem to exhibit such strong inter-droplet repulsive forces as the SDS-coated droplets. This result is corroborated in Figure 3 by a graph of relative viscosity η_{rel} *versus* volume fraction for a number of oil-in-water emulsions emulsified by either Tween 20 or SDS. Figure 3 indicates that the Tween 20 emulsion and the SDS emulsion have virtually identical η_{rel} values below $\phi \approx 0.45$. Above $\phi \approx 0.45$, both sets of emulsion samples show a rapid increase in η_{rel}, especially the SDS emulsion which has significantly higher values of η_{rel} than the Tween 20 emulsion. This result demonstrates a difference in droplet–droplet interaction between the two types of emulsion, with the SDS droplets forming a less readily deformable structure than the Tween 20 droplets.

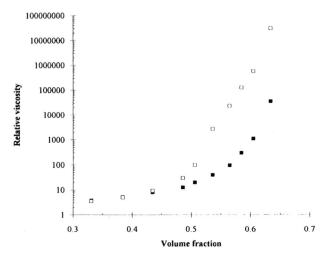

Figure 3   *Graph of relative viscosity* versus *volume fraction of oil-in-water emulsion (oil:surfactant = 30:1, pH 7, ionic strength 0.05 M, 25 °C) emulsified by:* ■, *Tween 20;* □, *SDS. The relative viscosity is calculated from the zero shear-rate viscosities of emulsion and solvent*

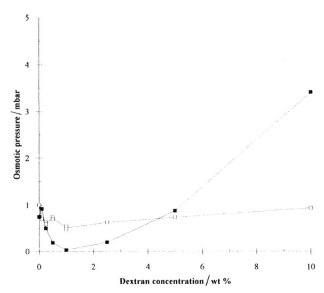

Figure 4 *Graph of equilibrium osmotic pressure* versus *dextran concentration for oil-in-water emulsions (30 wt% oil, pH 7, ionic strength 0.05 M, 25 °C) emulsified by:* ■, *Tween 20;* □, *SDS. The solvent for each experiment has a dextran concentration equal to the dextran concentration of the emulsion continuous phase*

The osmometer appears therefore to offer a highly sensitive technique for studying droplet–droplet interactions in emulsions. Osmotic pressures as low as 0.2 mbar can be differentiated and at relatively high ionic strengths (0.05 M).[4,5]

4 Emulsions Containing Dextran

Osmotic pressure measurements have been made on oil-in-water emulsions to observe the effect of replacing the phosphate buffer continuous phase with dextran solutions of various concentrations. The droplets remain as the 'solute', but the 'solvent' now becomes the aqueous dextran solution. The membrane has to take on a dual role: the pore size must be small enough to confine the oil droplets to the left-hand side of the chamber, but large enough to allow the dextran molecules to pass through. Both the Tween 20 and the SDS emulsions have a minimum droplet size > 0.1 μm, and the average diameter of the dextran molecules is 0.04 μm. A membrane with a pore size of 0.1 μm was therefore chosen as this is predominantly impermeable to the droplets and at the same time is freely permeable to the dextran molecules.

A number of different emulsions was made by blending together the stock 60 wt% emulsions with the dextran solutions in a 1:1 mass ratio using a magnetic stirrer. The resulting emulsions contained 30 wt% oil, 1 wt% emulsifier (either Tween 20 or SDS) and various concentrations of dextran. Each experiment

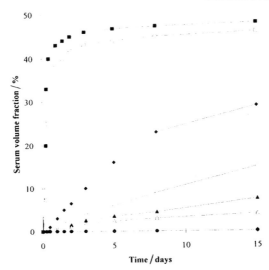

Figure 5   *Visual creaming measurement. The serum volume fraction (expressed as a percentage of total volume) is plotted against quiescent storage time for an oil-in-water emulsion (30 wt% oil, 1 wt% Tween 20, pH 7, ionic strength 0.05 M, 25 °C) containing various concentrations of dextran:* ●, ≤0.1 wt%; ■, 0.3 wt%; □, 0.5 wt%; ◆, 1.0 wt%; ◇, 2.0 wt%; ▲, 5.0 wt%; △, 10 wt%

had its own individual solvent, a solution with a dextran concentration equal to the dextran concentration of the emulsion continuous phase. This allowed a known, constant dextran concentration to be present at the start of each experiment. The osmotic pressures of the Tween 20 and SDS emulsions were measured over a 10 h time period.

Figure 4 is a graph of osmotic pressure *versus* dextran concentration for oil-in-water emulsions (30 wt% oil) emulsified by either Tween 20 or SDS. Both plots show minimum osmotic pressures at 1 wt% dextran, with the minima formed by the Tween 20 emulsion being particularly prominent. The minima appear to coincide with dextran concentrations that produce the greatest enhancement of creaming rates (see Figures 5 and 6). The enhanced creaming rates are caused by depletion flocculation,[6] which induces a tendency towards segregation of the droplets and the biopolymer solution. This phase separation results in poor miscibility, a negative second osmotic virial coefficient,[7] and, consequently, a lower osmotic pressure. The system of non-ionic surfactant (Tween 20) + non-ionic dextran is known to show segregative-type phase separation,[8] whereas the system anionic surfactant (SDS) + dextran is less incompatible.

5 Conclusions

Osmotic pressure measurements have been shown usefully to distinguish between systems emulsified by a non-ionic surfactant (Tween 20) and an

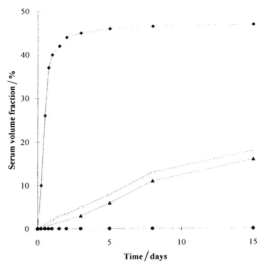

Figure 6   *Visual creaming measurement. The serum volume fraction (expressed as a percentage of total volume) is plotted against quiescent storage time for an oil-in-water emulsion (30 wt% oil, 1 wt% SDS, pH 7, ionic strength 0.05 M, 25 °C) containing various concentrations of dextran:* ●, ≤0.5 wt% and 10 wt%; ◆, 1.0 wt%; ◇, 2.0 wt% and ▲, 5.0 wt%

anionic surfactant (SDS). There is a good correlation between the difference in the limiting low-stress apparent shear viscosity of Tween 20 and SDS emulsions without added polysaccharide and the difference in the osmotic pressure of the same emulsions. The osmotic pressures of concentrated systems of oil droplets in the presence of dissolved dextran show minima that coincide with the dextran concentration that produces the strongest depletion attraction, as indicated by creaming measurements.

Acknowledgements

We thank Mr P. V. Nelson for expert technical support in the design and construction of the osmometer. M. I. G. acknowledges receipt of an SERC Cooperative Studentship in conjunction with Shell Research (Sittingbourne).

References

1.   E. Dickinson, M. I. Goller, and D. J. Wedlock, *Colloids Surf. A*, 1993, **75**, 105.
2.   Y. Cao, E. Dickinson, and D. J. Wedlock, *Food Hydrocolloids*, 1990, **4**, 185.
3.   Y. Cao, E. Dickinson, and D. J. Wedlock, *Food Hydrocolloids*, 1991, **5**, 443.
4.   R. H. Ottewill, *Faraday Discuss. Chem. Soc.*, 1990, **90**, 1.
5.   J. W. Goodwin, R. H. Ottewill, and A. Parentich, *Colloid Polym. Sci.*, 1990, **268**, 1131.
6.   G. J. Fleer, J. H. M. H. Scheutjens, and B. Vincent, *ACS Symp. Ser.*, 1984, **240**, 245.

7. P. J. Flory, 'Principles of Polymer Chemistry', Cornell University Press, Ithaca, NY, 1953.

8. B. Lindman, A. Carlson, S. Gerdes, G. Karlström, L. Piculell, K. Thalberg, and K. Zhang, in 'Food Colloids and Polymers: Stability and Mechanical Properties', ed. E. Dickinson and P. Walstra, Special Publication No. 113, The Royal Society of Chemistry, Cambridge, UK, 1993, p. 113.

Interfacial and Stability Properties of Emulsions: Influence of Protein Heat Treatment and Emulsifiers

By Eric Dickinson and Soon-Taek Hong

PROCTER DEPARTMENT OF FOOD SCIENCE, UNIVERSITY OF LEEDS, LEEDS LS2 9JT, UK

1 Introduction

In food emulsions there are usually both proteins and low molecular mass surfactants present.[1] These components take part in forming the surface film around the oil droplets, and the structure and composition of the film are largely dependent on competitive adsorption between these species,[2] with consequences for the surface rheology and emulsion stability.[3] β-Lactoglobulin, the major whey protein, is of considerable interest in both fundamental and applied research. In experiments involving surface shear viscosity measurements,[4] competitive adsorption,[3] and orthokinetic stability[3] we have found useful information on the interfacial and stability properties of emulsions containing β-lactoglobulin in the presence and absence of small-molecule emulsifiers. From the practical point of view, most food emulsions are manufactured under severe conditions such as high temperatures, high intensity shear fields, and so on. In the present study an attempt has been made to investigate the effect of heat on competitive adsorption, surface shear viscosity and orthokinetic stability of an emulsion containing β-lactoglobulin.

2 Experimental

Bovine β-lactoglobulin, Tween 20 and n-tetradecane (purity > 99%) were obtained from Sigma. Commercial-grade DATEM was donated by Grindsted Products (Denmark). Buffer salts were AnalaR-grade reagent.

The native protein solution (0.5 wt% β-lactoglobulin in 20 mM bis-tris buffer, pH 7) prepared at room temperature was placed in a 100 ml flask in a water bath at 70 °C for 30 min, and then cooled immediately to room temperature to produce the heat-treated sample. The experimental procedures for making the emulsions (20 wt% oil, 0.4 wt% protein, pH 7) and

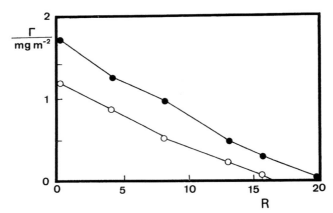

Figure 1   *Competitive displacement of β-lactoglobulin by Tween 20 from the oil–water interface in emulsions (0.4 wt% protein, 20 wt% oil, 20 mM bis-tris buffer, pH 7) stored for 1 h at two different temperatures. Protein surface concentration Γ is plotted against surfactant:protein molar ratio R:○, room temperature; ●, 70 °C*

determining the protein surface coverage have been described in detail elsewhere.[3] Competitive adsorption of Tween 20 and β-lactoglobulin was studied at room temperature and 70 °C. The surface shear viscosity at the interface between *n*-tetradecane and the dilute aqueous solution (2×10^{-3} wt% protein in 2 mM bis-tris buffer, pH 7) was determined using a Couette-type surface rheometer.[4] Orthokinetic stability of emulsions in turbulent shear flow was investigated as described previously.[5] The native protein solution (0.5 wt% β-lactoglobulin in 20 mM bis-tris buffer, pH 7) prepared at room temperature (approx. 20 °C) was placed in a 100 ml flask in a water bath at 40, 60, 70 or 80 °C for 30 min, and then cooled immediately to room temperature to produce the heat-treated sample. The experimental procedure for shearing the emulsions (20 wt% oil, 0.4 wt% heat-treated protein, pH 7) and determining the mean droplet-size are described in detail elsewhere.[5]

3 Results and Discussion

Figure 1 shows the effect of heat treatment of β-lactoglobulin on its competitive adsorption with Tween 20 in *n*-tetradecane-in-water emulsions. We see that, at each value of *R*, the protein surface concentration is higher in the system containing the heat-treated β-lactoglobulin. Das and Kinsella[6] attributed the increased amount of partially denatured heat-treated β-lactoglobulin adsorbed on the droplets to an increase in protein surface hydrophobicity.

   Figure 2 shows the time-dependent surface shear viscosity of β-lactoglobulin at the planar *n*-tetradecane–water interface. It is observed that the surface shear viscosity of the β-lactoglobulin film at 70 °C is higher and more

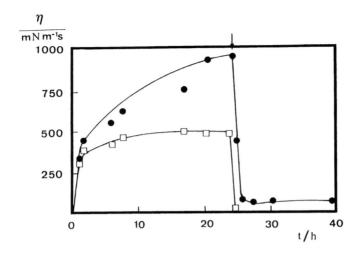

Figure 2 *Time-dependent surface shear viscosity of β-lactoglobulin at the n-tetra-decane–water interface (2 × 10⁻³ wt% protein, 2 mM bis-tris, pH 7) treated at two different temperatures. The apparent surface shear viscosity η is plotted against time t: □, 25 °C; ●, 70 °C. Tween 20 is added to the 24 h old interfacial film at surfactant:protein molar ratio R = 1. The arrow denotes the point at which Tween 20 (R = 1) is added to the aqueous subphase*

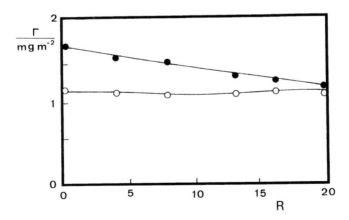

Figure 3 *Competitive displacement of β-lactoglobulin by DATEM from the oil–water interface in emulsions (0.4 wt% protein, 20 wt% oil, 20 mM bis-tris buffer, pH 7) stored for 1 h at two different temperatures. Protein surface concentration Γ is plotted against surfactant:protein molar ratio R: ○, room temperature; ●, 70 °C*

time-dependent than that at 25 °C, and that the addition of surfactant at 70 °C ($R=1$) leads to a sudden drop in surface shear viscosity which then levels off after a few hours to a finite substantial value of $\eta \approx 70$ mN m^{-1} s. From Figures 1 and 2, it is concluded that heat-treated β-lactoglobulin is more difficult than the native protein to displace from the oil–water interface by Tween 20.

Figure 3 shows the experimental results for competitive protein displacement as a function of amount of added emulsifier DATEM (diacetyl tartaric acid ester of monoglyceride). In contrast to Tween 20, DATEM produces rather little displacement of β-lactoglobulin at room temperature or 70 °C. Figure 4 shows the time-dependent surface shear viscosity of β-lactoglobulin at the planar n-tetradecane–water interface with DATEM. It is observed that there is a rapid fall in surface shear viscosity after addition of DATEM but to a lesser extent than that found with Tween 20 (Figure 2); this is then followed by a slow recovery over the next 10–15 h. From Figures 3 and 4, we infer that complexation of DATEM with β-lactoglobulin could be the explanation of the poor competitive displacing power of this emulsifier.

Figure 5 shows the influence of heat treatment of β-lactoglobulin on shear-induced coalescence stability of emulsion droplets at room temperature. We see that the emulsion containing β-lactoglobulin which had been heated at 70 °C is destabilized more rapidly than the emulsions containing β-lactoglo-bulin heated at 20, 40, and 60 °C. Chen *et al.*[3] have shown that orthokinetic stability is probably triggered not simply by extensive protein displacement, but by some change in adsorbed layer properties associated with the surfactant

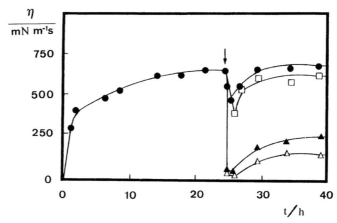

Figure 4   *Time-dependent surface shear viscosity of β-lactoglobulin at the n-tetra-decane–water interface (2×10^{-3} wt% protein, 2 mM bis-tris, pH 7, 50 °C). DATEM is added to the 24 h old interfacial film at surfactant:pro-tein molar ratio R in the range 1–16:* ●, R = 1; □, R = 4; ▲, R = 8; △, R = 16. *The apparent viscosity η is plotted against time t. The arrow denotes the point at which DATEM (R = 1) is added to the aqueous subphase*

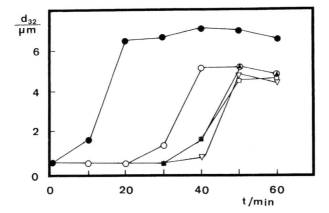

Figure 5 *Influence of heat treatment of β-lactoglobulin on shear-induced coalescence of emulsion droplets (0.4 wt% protein, 20 wt% oil, 20 mM bis-tris, pH 7). Protein solutions (0.5 wt%, 20 mM bis-tris, pH 7) were heated at different temperatures: ▲, room temperature; ▽, 40 °C; □, 60 °C; ○, 70 °C; ●, 80 °C. The emulsions were sheared with a Silverson blender at room temperature*

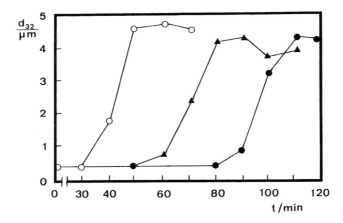

Figure 6 *Influence of surfactant DATEM on shear-induced coalescence of emulsion droplets (0.4 wt% protein, 20 wt% oil, 20 mM bis-tris, pH 7). DATEM dissolved at 45 °C is added to the fresh emulsion at room temperature at surfactant:protein molar ratio R: ○, R = 0; ▲, R = 1; ●, R = 2. The emulsions were sheared with a Silverson blender at room temperature*

added. Decreased orthokinetic stability above 70 °C could therefore be attributable to some changes in inter-droplet interactions or surface rheology associated with heat denaturation of the β-lactoglobulin.

Figure 6 shows the influence of the ionic surfactant DATEM added after homogenization on the shear-induced coalescence stability of emulsion droplets. We see that the emulsions containing DATEM show much better orthokinetic stability than the pure protein-stabilized emulsion. The likely explanation for this finding involves some kind of interfacial complexation of DATEM with the β-lactoglobulin.

References

1. E. Dickinson and G. Stainsby, 'Colloids in Food', Applied Science, London, 1982, ch. 1.
2. E. Dickinson, A. Mauffret, S. E. Rolfe, and C. M. Woskett, *J. Soc. Dairy Technol.*, 1989, **42**, 18.
3. J. Chen, E. Dickinson, and G. Iveson, *Food Struct.*, 1993, **12**, 135.
4. E. Dickinson, S. E. Rolfe, and D. G. Dalgleish, *Int. J. Biol. Macromol.*, 1990, **12**, 189.
5. E. Dickinson, R. K. Owusu, and A. Williams, *J. Chem. Soc., Faraday Trans.*, 1993, **89**, 865.
6. K. P. Das and J. E. Kinsella, *J. Colloid Interface Sci.*, 1990, **191**, 551.

Foams

Surface and Bulk Properties in Relation to Bubble Stability in Bread Dough

By J. J. Kokelaar, T. van Vliet, and A. Prins

DEPARTMENT OF FOOD SCIENCE, WAGENINGEN AGRICULTURAL UNIVERSITY, PO BOX 8129, 6700 EV WAGENINGEN, THE NETHERLANDS

1 Introduction

The surface and bulk rheological properties of bread dough are relevant to the behaviour of gas bubbles in dough throughout the breadmaking process. The entrapped number of bubbles and their physical stability during the breadmaking process is crucial for the final appearance of the baked loaf, which should have a high bread volume and a fine and regular crumb structure. After mixing and first proof, a certain number of gas bubbles, with a relatively narrow bubble size distribution, should be present in the dough. They should be stable towards coalescence until the dough is transformed into a solid-like structure (bread) due to gelatinization of the starch.

A discussion will be given in this paper as to which stage of the breadmaking process (mixing, fermentation or baking) surface properties primarily dominate bubble behaviour, and during which stage bulk properties are more important. Moreover, the effect of different dough components on the surface properties will be described.

2 Materials and Methods

For the determination of the surface properties, flour, gluten and lipids of a mixture of American wheat cultivars with relatively good baking properties (Spring) were used. A dough dispersion was made by freeze-drying a dough, grinding it, and dispersing it in water at a concentration of 0.5 wt%. Gluten was obtained by washing out a flour dough according to the procedure of Weegels et al.;[1] it was then freeze-dried and ground. For each experiment, 10 mg was sprinkled on a water surface (area ca. 80 cm^2). Total wheat lipids were extracted from flour with water-saturated 1-butanol at 90 °C. Surface properties were determined for two different amounts of lipid spread at an air–water surface (7.6 mg m^{-2} and 15.2 mg m^{-2}, 'low' and 'high' concentrations, respectively). Gliadins were isolated by shaking gluten powder with 70%

ethanol and freeze-drying afterwards. Surface rheological measurements were performed on a 10 mg l^{-1} gliadin suspension in water. Food-grade sodium stearoyl-2-lactylate (SSL) suspensions were made at a concentration of 0.1 wt% ('high' concentration) and of approximately 0.004 wt% ('low' concentration). These SSL suspensions were pre-treated by heating or mechanical treatment in order to induce an ordered structure at the air–water surface resulting in a high surface dilational modulus (>200 mN m^{-1}).[2]

The surface dilational modulus ($E = d\gamma/d\ln A$) was determined as a function of frequency using a so-called ring trough.[3] Biaxial extensional bulk properties of flour and gluten doughs were determined by lubricated uniaxial compression.[4]

3 Results and Discussion

Mixing

During mixing of the dough ingredients, the gluten proteins form a cohesive network in which air bubbles can be entrapped. These may be subdivided into smaller ones if the stress σ exerted on them is larger than the Laplace pressure inside the bubble. The ratio of the exerted stress over the Laplace pressure is called the Weber number We.[5] This is expressed by the equation

$$We = \sigma \frac{r}{\gamma} = \eta_c G \frac{r}{\gamma} \tag{1}$$

where η_c is the (apparent) viscosity of the continuous phase, G is the strain rate, γ the surface tension, and r the radius of the bubble. If We exceeds a critical value We_{cr} the bubble will break up. We_{cr} depends on the type of flow and on the ratio of the viscosity of the gas in the bubble (air) to that of the continuous phase (dough).[5] The apparent viscosity of dough varies from approximately 470 to 300 Pa s at shear rates from 10 to 50 s^{-1} (which are representative of mixing conditions).[6] The ratio of the viscosity of air (10^{-5} Pa s) to that of dough is very small, indicating a very high Weber number for shear flow.[5] Therefore bubbles will be disrupted by a combination of shear, and for the smaller ones mainly elongational flow. An order of magnitude calculation using $\gamma \approx 40$ mN m^{-1} for a dough dispersion[2,7,8] results in final bubble radii of approximately 5–50 μm. These sorts of values are also reported by Bloksma.[6]

The surface tension of a water surface containing either spread wheat lipids or the surfactant SSL is approximately 27 mN m^{-1}.[2] In a dough system the tension can be somewhat higher since other components will be present at the gas–dough interface as well, and the equilibrium situation is probably not obtained. However, in principle, lipids and added surfactants will lower the minimum (and probably also the mean) gas bubble radius of the air bubbles formed during mixing. This may lead to more and smaller bubbles in the dough resulting in a finer crumb structure.

In relation to the gas bubbles, the amount, the mean radius, and also the size distribution, are all important, because in later stages of the process bubbles should grow uniformly in order to obtain a regular crumb. Therefore mixing should result in a fairly narrow bubble-size distribution. During this stage gas cells will be stable towards coalescence as a result of the high hydrodynamic resistance between droplets approaching each other, caused by the high (apparent) viscosity of the continuous phase.

Fermentation

For the purpose of simplification, we assume that fermentation starts as soon as the mixing step ends. In fact, we know that fermentation reactions already start during mixing because then the yeast cells are already active. After some time some of the gas bubbles start to grow due to CO_2 production. During fermentation the relevant instability mechanisms of the gas bubbles are disproportionation and coalescence.[9,10] Disproportionation will tend to be more important directly after mixing while coalescence starts to play a significant role when gas bubbles are approaching each other to such an extent that thin dough films are formed.

Directly after mixing, disproportionation of gas bubbles will occur.[11] Disproportionation is counteracted by the presence of a finite surface dilational modulus E over the timescale that the process takes place. If the surface dilational modulus exceeds half the surface tension ($E > \gamma/2$), and if the

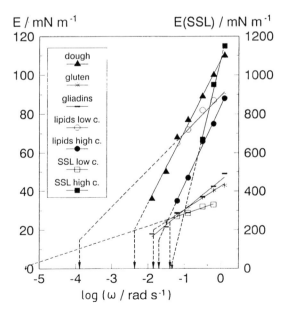

Figure 1   *Surface dilational modulus* E *of different dough components as a function of the logarithm of the frequency* ω. *For each component the critical frequency at which* E = γ/2 *is indicated. The values of* E *for the SSL suspensions are plotted on the right axis. (See text for further details)*

surface rheological behaviour is purely elastic over the relevant time-scale, the process will stop.[12] By plotting the measured E values of different dough components as a function of the logarithm of the radial frequency (Figure 1), these components can be compared regarding their ability to satisfy the surface properties required to give a substantial slowing down of disproportionation. Indicated are the values of the critical frequency ω_{cr} at which $E=\gamma/2$. Using these frequency values, rough estimates can be made of the shrinking rates of bubbles during disproportionation, because the requirement of Lucassen[12] does not hold over longer timescales ($>1/\omega_{cr}$). The values of $E=\gamma/2$ are for gliadins 22.5 mN m^{-1}, for dough and gluten 20 mN m^{-1}, and for lipids and SSL 14 mN m^{-1}.

The biaxial strain rate $\dot{\epsilon}_B$ during shrinking of a bubble can be written according to van Vliet *et al.*[10] as

$$\dot{\epsilon}_B = \frac{\mathrm{d}\ln(r_t/r_0)}{\mathrm{d}t} = \frac{\mathrm{d}\ln r}{\mathrm{d}t} \tag{2}$$

and the local strain ϵ_B as

$$\epsilon_B = \ln\frac{r_t}{r_0} \tag{3}$$

Equations (2) and (3) imply that, as a first-order approximation, ϵ_B is unity after a time of $1/\omega_{cr}$ and thus $r_t/r_0=1/e$ after this time. In Table 1 the results obtained for $1/\omega_{cr}$ are given as well as the consequences for retarding disproportionation of gas bubbles in bread dough. From Table 1 it is clear that wheat lipids and the surfactant SSL at the right concentrations can retard disproportionation to a large extent. The total dough system retards the process to some extent. The exact composition of the surface is not known but it will be a mixture of lipids and proteins. Gluten and gliadins retard disproportionation hardly or not at all under the experimental conditions.

Table 1   *Global estimation of the time in which small gas bubbles shrink from r_0 to r_0/e during disproportionation for different surface compositions*

Component at the surface	$\log \omega \ (E=\gamma/2)$	Time $(r_t=r_0/e)$ $=1/\omega_{cr}$ (s)	Retarding disproportionation?
Total dough	−2.32	209 (3.5 min)	To some extent
Gluten	−1.79	62 (1 min)	Hardly
Gliadins	−1.34	22 (0.3 min)	Hardly
Lipids: low conc.	−3.81	6460 (2 h)	Very much
Lipids: high conc.	−1.65	45 (0.75 min)	Hardly
SSL: low conc.	−5.34	2×10^5 (60 h)	Very much
SSL: high conc.	−1.31	20 (0.3 min)	Hardly

The development of a biaxial stress around growing gas cells (before and during the disproportionation process) resisting their growth has not yet been taken into consideration. However, at a certain strain the biaxial stress around the expanding larger bubbles can become so high that it exceeds the Laplace pressure and thus counteracts disproportionation as well. In an earlier publication[11] it was shown that for Spring dough the biaxial stress around an expanding gas bubble exceeds the Laplace pressure in gas bubbles of roughly the same size already in the early stage of fermentation. So, a rather high resistance to biaxial extension of dough films will be favourable for stopping the disproportionation process. Besides, pronounced strain hardening of the dough during growth of gas bubbles is favourable for uniform growth of the gas cells and for stopping disproportionation. This will be discussed further in another paper.[13] Doughs of wheat cultivars exhibiting most pronounced strain hardening, which are in general those with good baking properties,[2] will tend to have the most regular bubble-size distribution at the end of the fermentation stage.

During fermentation the gas cells may expand to a relative volume of 4 (volume fraction of gas in dough ≈ 0.75) if the wheat flour used has good baking characteristics. (For poorly baking wheat cultivars this can be much lower.) Bloksma[9] has calculated, assuming spherical gas cells arranged in a cubical array, the relation between the volume of gas and the number and distance between gas cells. He has concluded that, at the end of 'tin proof', gas cells will become strongly deformed. In case of a broad bubble-size distribution, the larger cells will certainly approach each other very closely and some will also press against others to form a polyhedral foam structure. For a wheat flour with poor baking characteristics, however, the bubble structure may probably still be spherical with no flat films present between them.

The resistance of a thin film against further extension will arise from surface and/or bulk rheological properties. The contributions of these two components can be estimated by using the equation[2]

$$2[\gamma + E(\Delta A)] = h\sigma_B \qquad (4)$$

where $E(\Delta A)$ indicates that E is dependent on ΔA, and h is the thickness of the film. If, for the unknown parameters, data obtained for Spring dough (dispersion) are taken ($\gamma = 42$ mN m^{-1}, $E = 35$ mN m^{-1}, $\sigma_B = 2 \times 10^3$ N m^{-2}),[2,11] and assuming linear behaviour, a value of h of approximately 75 μm is obtained. This is probably an overestimation because E will be lower for a large change in surface area, as at the end of final proof, where the surface area has been enlarged by a factor of 5.4.[10]

Surfactants like SSL can give a water surface a high dilational modulus.[2] However, it is questionable whether these high surface moduli values will still exist after the large bubble surface expansions at relatively low ϵ_B as take place during fermentation and 'oven rise'. It is known that surfactants like SSL provide a fermenting dough with a certain shock resistance[14] indicating a high modulus over short time-scales which is in agreement with Figure 1. If biaxial stresses are higher or lower than those found for Spring dough, then bulk properties are dominating the behaviour in films that are, respectively, thinner or thicker than 75 μm.

The relevant bulk rheological properties for preventing coalescence are (i) the extent of biaxial strain-hardening and (ii) the extensibility of the dough films.[2,10] It has been shown[2] that, for the different wheat cultivars tested, those which have the most pronounced strain-hardening properties and the highest extensibilities exhibit the best baking performance.

For very thin films (smaller than the diameter of a starch granule), it is probably more relevant to consider the properties of the material between the starch granules which is mainly a gluten network. Differences in biaxial stress levels, as well as in strain-hardening, are even more pronounced for gluten doughs than for flour doughs for wheat cultivars varying in baking performance.[2] This indicates that, for cultivars with good baking properties, bulk properties are important down to very thin films.

Baking

During the final stage in the breadmaking process some major changes occur. Firstly, the gas cells expand fast in a relatively short time, the so-called oven rise. The relative volume of the gas in the dough changes from *ca.* 4 to 6.5 for high quality loaves.[6] For low quality loaves, the end relative volume can be *ca.* 4. Secondly, the fluid dough mass is transformed into a solid bread structure which is caused for the larger part by the gelatinization of the starch. Thirdly, the foam structure of the dough is transformed into a sponge structure with interconnected gas cells. Because changes occur relatively rapidly during baking, this stage is the most critical of the whole process, and so the properties of the dough at elevated temperatures should be fully considered in order to relate rheological properties to baking performance.

Stability of the dough membranes towards rupture before gelatinization of starch is the essential property required to get a high loaf volume and a regular crumb structure in the final baked product. As was discussed above, the calculated film thickness at room temperature of the membranes, where surface and bulk properties contribute equally to film resistance, is *ca.* 75 μm or lower. However, at 45 °C the surface dilational modulus of a dough suspension is about a factor of five lower than at 20 °C.[2] Biaxial stresses of flour doughs are roughly similar at 55 °C compared with those at 20 °C at the same values of ϵ_B and $\dot{\epsilon}_B$, but for (Spring) gluten dough the biaxial stress is 50% lower at 55 °C. Strain-hardening behaviour of both flour and gluten doughs does not change very much due to a temperature increase from 20 to 55 °C.[2] Overall, and again assuming linear behaviour, it follows that, for films down to about 15–30 μm, the bulk properties of (gluten) dough determine for the most part the stability of the film towards extension at more elevated temperatures. As mentioned above, the essential properties to prevent coalescence of gas bubbles are strain-hardening at a sufficient level and a large fracture strain of the film between the bubbles. These properties must also be present at higher temperatures.

In films thinner than those mentioned above, surface properties may become important in stabilizing the films against further extension. However, it is doubtful whether surface properties are responsible for differences in baking quality between various wheat cultivars, because no difference in surface properties between dough suspensions and gluten from different wheat cultivars differing in baking performance can be detected.[2]

4 Conclusions

By considering both the surface and the bulk rheological properties of bread dough and its different components, and by comparing the values of the parameters that play a role in determining the behaviour of the gas bubbles in dough, the following conclusions can be drawn (see Table 2).

Table 2 *Relevance of surface and bulk properties to breadmaking*[a]

Stage in breadmaking process	Surface properties		Strain-hardening	Resistance to extension	Extensibility
	γ	E			
Mixing	+	−	−	−	−
First proof	+ +	+ +	+ +	+ +	−
Final proof	−	−	+ +	+	+
Baking	?	?	+ +	+	+ +

[a] Explanation of symbols: + + very relevant; + relevant; − not relevant

The surface tension of the dough and its apparent viscosity during mixing determine, for a given mixing procedure, the (minimum) gas bubble size after mixing. Adding a surfactant like SSL may cause a decrease in the bubble radius through its effect on γ.

Directly after mixing, disproportionation of gas bubbles will occur. This process can be retarded very much by wheat lipids and added surfactants, when they are present at the right concentration. If the dough has sufficient strain-hardening properties, the gas bubbles that are large enough to expand at the existing carbon dioxide partial pressure will have a narrow bubble-size distribution. This leads to a regular crumb structure in the final product. In more advanced stages of 'tin proof', bulk rheological properties such as strain-hardening and extensibility of the dough determines the stability of the dough membrane towards rupture. These properties differ for wheat cultivars of different breadmaking performance.[2]

Also, during the baking process, bulk properties mainly determine the coalescence stability of the gas bubbles in dough. Only in films thinner than the diameter of a starch granule may surface properties become important for stabilizing the dough film. However, no differences in surface properties between wheat cultivars with pronounced differences in baking performance could be found. This indicates that the main differences in baking performance between wheat cultivars cannot be explained by differences in surface rheological properties.

References

1. P. L. Weegels, J. P. Marseille, and R. J. Hamer, *Starch*, 1988, **40**, 439.
2. J. J. Kokelaar, 'Physics of breadmaking', PhD Thesis, Wageningen Agricultural University, Wageningen, The Netherlands, 1994.
3. J. J. Kokelaar, A. Prins, and M. de Gee, *J. Colloid Interface Sci.*, 1991, **146**, 507.
4. S. H. Chatraei, C. W. Makosko, and H. H. Winter, *J. Rheol.*, 1981, **25**, 433.
5. P. Walstra, in 'Encyclopedia of Emulsion Technology', ed. P. Becher, Marcel Dekker, New York, 1984, vol. 1, p. 58.
6. A. H. Bloksma, *Cereal Foods World*, 1990, **35**, 228, 959.
7. A. D. Evers, H. R. Kerr, and J. Castle, *J. Cereal Sci.*, 1990, **12**, 207.
8. A.-C. Eliasson, J. Silverio, and E. Tjerneld, *J. Cereal Sci.*, 1991, **13**, 27.
9. A. H. Bloksma, *Cereal Foods World*, 1990, **34**, 237, 260.
10. T. van Vliet, A. M. Janssen, A. H. Bloksma, and P. Walstra, *J. Texture Stud.*, 1992, **23**, 439.
11. J. J. Kokelaar, T. van Vliet, and A. Prins, in 'Food Colloids and Polymers: Stability and Mechanical Properties', ed. E. Dickinson and P. Walstra, Special Publication No. 113, The Royal Society of Chemistry, Cambridge, UK, 1993, p. 407.
12. J. Lucassen, in 'Anionic Surfactants', ed. E. H. Lucassen-Reynders, Marcel Dekker, New York, 1981, p. 217.
13. T. van Vliet, and J. J. Kokelaar, in Proc. 4th Eur. Rheol. Congress, Sevilla, 1994, (published by Steinkopff, Darmstadt, Germany, 1994, p. 201).
14. N. J. Krog, in 'Food Emulsions', ed. K. Larsson and S. E. Friberg, Marcel Dekker, New York, 2nd edn, 1990, p. 127.

Comparison of the Foaming and Interfacial Properties of Two Related Lipid-binding Proteins from Wheat in the Presence of a Competing Surfactant

By F. Husband, P. J. Wilde, D. Marion,[1] and D. C. Clark

INSTITUTE OF FOOD RESEARCH, NORWICH LABORATORY, NORWICH RE-
SEARCH PARK, COLNEY, NORWICH NR4 7UA, UK
[1]LABORATOIRE DE BIOCHIMIE ET TECHNOLOGIE DES PROTÉINES, INRA, BP
527, RUE DE LA GÉRAUDIÈRE, 44026 NANTES CEDEX 03, FRANCE

1 Introduction

Puroindolines are members of a new family of lipid-binding proteins which have been recently isolated from wheat seeds. Two isoforms, named puroindoline-*a* and puroindoline-*b* have been characterized (Figure 1). These proteins are basic and their folded conformation is stabilized by five disulfide bridges. Puroindoline-*a* has a unique tryptophan-rich domain which is partly truncated in the *b*-isoform. The proteins are synthesized as higher molecular mass precursors composed of N- and C-terminal extensions which are cleaved during the maturation of the protein in the cell.[1,2] This post-synthetic processing is carried out to variable extents *in vivo* as demonstrated by mass spectrometry analysis of purified puroindoline-*a*. This has revealed that some entities lack the first N-terminal amino acid and the last two C-terminal amino acids.[1] The physiological and functional consequence of this processing are still unknown.

Puroindolines exhibit a high affinity for lipids and expecially for membrane lipids. This affinity has been exploited in the isolation of puroindolines by use of phase partitioning in non-ionic detergents, as for transmembrane proteins.[1,3] It has been shown *in vitro* that puroindolines bind a phospholipid surfactant, lysophosphatidylcholine, strongly; and this binding exerts a quite favourable effect on the foaming properties of both protein and surfactant.[4] It is suggested that such interfacial properties could have technological relevance. For example, these proteins may be implicated in gas retention in bread doughs,[5] and it has been shown that puroindoline can protect against the lipid-induced destabilization of beer foams.[6] In this paper, we compare the

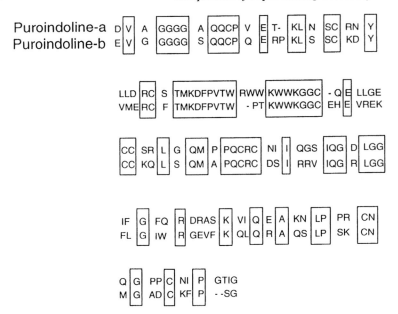

Figure 1   *The primary sequences of puroindoline-*a *and -*b *with sequence alignment for maximum homology*

interfacial behaviour of puroindoline-*a* and -*b* and try to relate the observed differences to the lipid-binding properties and structure of these proteins.

2 Materials

L-α-Lysophosphatidylcholine caproyl (LPCC) (product no. L3010), L-α-Lysophosphatidylcholine lauryl (LPLC) (product no. L5629), L-α-Lysophosphatidylcholine palmitoyl (LPPC) (product no. L5254), L-α-Lysophosphatidylcholine stearoyl (LPSC) (product no. L2131), phosphorylcholine chloride, calcium salt (product no. P0378), egg white (product no. E-0500) and Florisil (product no. F6875) were purchased from Sigma. 5-*N*-(dodecanoyl)-aminofluorescein (DDAF) (product no. D109) was purchased from Molecular Probes. Pyrene (product no. 18 551-5) was purchased from Aldrich. Sodium dodecyl sulfate (product no. 28332) and Tween 20 (product no. 28320) were purchased from Pierce and Warriner. Soya oil was purchased from the local supermarket; it was purified by passage over a Florisil column and stored under nitrogen at 4 °C until required. This treatment resulted in an increase in surface tension and a decrease in the colour of the oil. All other chemicals used were of 'AnalaR' grade from BDH Chemicals and were used without further purification. Surface chemically pure water (surface tension more than 72.8 mN m^{-1} at 20 °C) obtained by steam distillation of deionized water from potassium permanganate was used throughout this study. All experiments were carried out in 10 mM sodium phosphate buffer and at a protein concentration of 0.1 mg ml^{-1} unless stated otherwise.

Purification of puroindoline-*a* was carried out according to the method described previously.[1] Puroindoline-*b* elutes from the ion exchange column as a distinct peak immediately after the puroindoline-*a* peak. In this work we have used a wheat flour obtained from the French variety Etoile de Choisy which is especially rich in both puroindoline-*a* and -*b* isoforms.

The concentrations of puroindoline-*a* and -*b* were measured spectrophotometrically using absorption coefficients of 1.94 mg ml^{-1} cm^{-1} and 1.66 mg ml^{-1} cm^{-1}, respectively.[4]

3 Methods

The relative stability of foams generated from 0.1 mg ml^{-1} solutions of puroindoline-*a* and -*b* was measured by a foam micro-conductivity method which has been fully described previously.[7] In the studies involving the competitive adsorption of the various chain lengths of lysophosphatidyl choline (LPC), the lipid analogue was added to the sample immediately prior to the foaming measurement. In the experiments involving soya oil, 0.2 ml of oil was added to 4.3 ml of protein solution (0.1 mg ml^{-1}) and mixed gently with a Pasteur pipette. The decay in conductivity of the foam was monitored for 5 min. The initial conductivity value (C_0) provides an index of sample foamability, whereas foam stability is related to the conductivity remaining after 5 min drainage (C_{300}). Provided that the C_0 value does not change significantly from sample to sample, it is acceptable to compare C_{300} values directly. However, the large changes in C_0 observed in this study necessitated the normalization of C_{300} values. The percentage foam stability (FS%) was calculated from the expression (C_{300}/C_0) × 100% to give the relative decay in foam conductivity. The foamability and foam stability data presented here are the averages of at least four measurements. Typically the variation between measurements obtained from the same sample was 5%.

The interaction of LPC with puroindoline caused an enhancement of the intrinsic fluorescence of the protein, which was used to study the binding process. Details of the fluorescence titration method used have been reported previously.[4] The measurements were carried out at 20 °C. Scatchard plots[8] of the results demonstrated that the binding process was positively co-operative (*i.e.* the binding of each ligand made the binding of subsequent ligands easier). Therefore, the data were analysed using the Hill equation constrained to five binding sites.[9] (The number of LPC binding sites per mole of protein has been determined previously by equilibrium dialysis measurements.[4])

To measure the critical micelle concentration (CMC), a fluorescence method, involving the strongly hydrophobic fluorophore pyrene, was used.[10] The fluorescence data obtained were expressed as the ratio of the intensities of the vibronic bands III and I in the pyrene emission spectrum. A plot of this ratio against LPLC concentration showed a sharp increase at the onset of micellization and the CMC was determined as the concentration at which a steady-state plateau value was reached.

Air-suspended thin liquid films (approximately 100 μm diameter) were formed in a ground glass annulus (approximately 3 mm diameter) as described previously.[11] The drainage properties of the thin films were observed using a Nikon Diaphot inverted microscope. The surface diffusion in thin films formed from mixtures of puroindoline-a and LPC was investigated by fluorescence recovery after photobleaching (FRAP) as previously described[11,12] employing 5-N-(dodecanoyl)aminofluorescein (DDAF) as the fluorescent probe. This amphiphilic molecule accumulated in the interfacial layer. It was not covalently linked to the absorbed protein, but rather acted as a probe of interactions and molecular packing in the adsorbed layer. DDAF was added to the samples at a final concentration of 0.3 μM. FRAP experiments were performed on air–water thin films that had reached equilibrium thickness. The presence of DDAF did not alter the drainage properties of the thin films.

The surface dilational properties were measured according to the method of Kokelaar *et al.*[13] In the apparatus used, a periodic interfacial expansion and compression was achieved by dipping and raising a ground glass cylinder (diameter = 10 cm) into a vessel using a sinusoidal drive. The percentage area change was 13% which was determined to be within the linear region. Experiments were carried out on separate solutions of puroindoline-a and -b at a concentration of 5 mg l^{-1}.

Far-UV circular dichroism (CD) spectra were measured using a Jasco spectropolarimeter with samples contained in a 0.1 mm demountable cell. Two spectra were obtained for each sample at a scan speed of 20 nm min^{-1} with a time constant of 4 s. All CD measurements were performed at a protein concentration of 1 mg ml^{-1}. The spectra were analysed using the CONTIN program.[14]

4 Results

Comparison of the Foaming Properties of Puroindoline-*a* and -*b*

The foam stability of puroindoline-a and -b was investigated as a function of concentration and the results are shown in Figure 2. Both samples behave in a very similar manner, with a characteristic plateau in foam stability occurring at a protein concentration of approximately 0.1 mg ml^{-1} in both samples. The foaming properties of puroindoline-a and -b in the presence of LPPC (C_{16}) were also investigated and the results are shown in Figure 3. The presence of LPPC caused an increase in the foam stability in both puroindoline samples. However, the enhancement in foam stability was greatest with puroindoline-a. The foaming properties of LPPC alone are also shown in Figure 3. Foams of measurable stability were first obtained at an LPPC concentration of 50 μM. The foam stability of LPPC alone was always less than that observed in the presence of either protein fraction in the concentration range investigated.

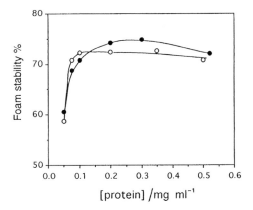

Figure 2   *A comparison of the stability of foams of (○) puroindoline-a and (●) -b as a function of protein concentration*

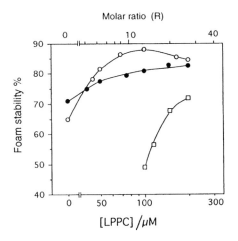

Figure 3   *The stability of foams formed from solutions containing 0.1 mg ml^{-1} (○) puroindoline-a and (●) -b as a function of the concentration of added LPPC. The concentration dependence of the stability of foams formed from LPPC alone (□) is also shown. R refers to the molar ratio of LPPC to protein*

The structural properties of puroindoline-*a* and -*b* were compared to try to find an explanation for the differences in the foaming properties of these samples in the presence of LPPC. Secondary structure analysis of far-UV CD spectra using the CONTIN program[14] showed that both puroindoline-*a* and -*b* have comparable secondary structure comprising approximately 25% α-helix, 50% β-sheet and 25% aperiodic structure. Hence even though the primary structure was only (52%) homologous, the relative amounts of α-helix:β-sheet:aperiodic structure were very similar. Surface dilational measurements revealed that the interfacial properties of both proteins did not differ significantly. The values of the surface elastic modulus of puroindoline-*a* and -*b* after 15 min were 18.3 mN m^{-1} and 18.2 mN m^{-1} respectively.

Binding Isotherms

The binding curves for LPPC with puroindoline-*a* and -*b* are shown in Figure 4. The symbols represent individual data points in the fluorescence titration. The solid lines are the fits to the Hill equation constrained to five binding sites[4] and *v* is the fractional occupation of the binding sites. Both puroindoline-*a* and -*b* have shown positively co-operative binding with LPPC with Hill coefficients (N_h) of 1.23 and 1.42, respectively. The dissociation constant (K_d) for LPPC with puroindoline-*a* was 55 μM, compared with 1.04 mM for puroindoline-*b*. The smaller dissociation constant observed with the former indicated stronger binding.

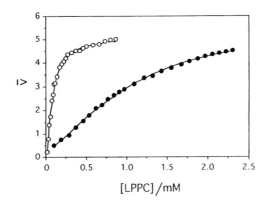

[LPPC]/mM

Figure 4   *Binding of LPPC to ($\bigcirc$) puroindoline-*a* and ($\bullet$) puroindoline-*b*, where $\bar{v}$ is the fractional occupation of the binding sites on the protein. The points represent experimental data in the fluorescence titration; the solid line is the fit to the Hill equation (see text for details)*

Effect of LPC Structure on Foam Stability

Studies were initiated to identify whether the enhancement in foam stability observed in the presence of LPC was related to the structure of this molecule. Foam conductivity decay curves obtained for puroindoline in the presence and absence of phosphorylcholine (1.11 mM) were indistinguishable and confirmed that the presence of the zwitterionic head group of LPC did not influence foaming properties.

The effect of the length of the fatty acyl chain of LPC on foaming of puroindoline-*a* was investigated and the results are summarized in Figure 5. The data are expressed as efficiency of foaming which relates to the maximum change in foam stability properties (*i.e.* FS% C_{300} value) compared with puroindoline-*a* alone, which has an efficiency of 100%. LPCC (C_6) alone did not produce a stable foam up to a concentration of 200 μM. In the presence of puroindoline, LPCC (C_6) caused a decrease in foam stability with increasing concentration. In contrast, LPLC (C_{12}) alone produced a foam of measurable stability at a concentration of 100 μM and stability increased steadily across the concentration range 100–400 μM. Mixtures of LPLC + puroindoline produced a slight enhancement of foam stability compared with puroindoline alone in the LPLC concentration range 100–300 μM. At 400 μM LPLC, the values of foam stability for LPLC and LPLC + puroindoline mixtures were similar. LPPC alone formed a foam with measurable stability at a concentration of 50 μM. An enhancement in stability with puroindoline was seen when LPPC (C_{16}) was added to puroindoline at low concentrations (50–150 μM). LPSC (C_{18}) alone did not produce a stable foam up to a concentration of 1 mM. The addition of 100 μM LPSC to puroindoline-*a* produced a measurable enhancement of foam stability. Further additions of 100–600 μM produced further enhancements to foam stability. Some variability in values was observed due to limited precipitation of the LPSC.

Data showing the concentration of LPC which gave maximum foam enhancement are shown in Figure 5 as a function of LPC acyl chain length. No data are available for LPCC, since this molecule reduced foam stability. However, between C_{12} and C_{18}, the optimum LPC concentration decreased with increasing chain length.

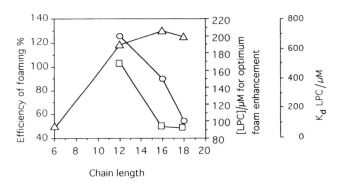

Figure 5   *The influence of LPC chain length on:* △, *the efficiency of foaming;* ○, *the concentration of LPC required for maximum foam enhancement;* □, *the dissociation constant* K_d *of the complex with puroindoline-*a

Binding of LPC to Puroindoline as a Function of Acyl Chain Length

The effect of LPC acyl chain length on K_d for puroindoline-*a* is presented in Figure 5. The binding of LPC becomes tighter with increasing acyl chain length. LPCC does not bind to puroindoline at all. LPLC binds loosely to the

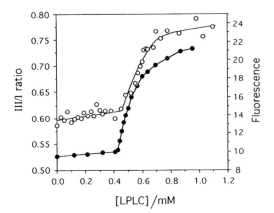

Figure 6   *A comparison of (●) the fluorescence titration curve for the binding of LPLC to puroindoline-a and (○) the band III/I ratio of pyrene fluorescence used to determine the CMC of LPLC*

protein giving a rather unusual titration curve (Figure 6) characterized by a K_d value of 555 μM. The K_d values for the puroindoline-a complexes with LPPC and LPSC are 55 μM and 48 μM, respectively. The correlation between the K_d values and the LPC concentration which gives maximum foam enhancement strongly implicates the role of the LPC–puroindoline complex in foam enhancement.

The raw data from a fluorescence titration of LPLC and puroindoline-a are presented in Figure 6. The initial delay in the enhancement of the intrinsic fluorescence of puroindoline-a indicated by the lag in the intensity of the signal between 0 and 400 μM LPLC is significant. At higher concentrations of LPLC, an increase in fluorescence occurs, which indicates the onset of binding. LPLC binding with puroindoline-b is even weaker.

The critical micelle concentration of LPLC was determined by the pyrene method.[14] The results of the experiment are expressed as the III/I ratio (also shown in Figure 6) as a function of the LPLC concentration. The initial plateau in the fluorescence signal signifies the presence of only monomeric LPLC; it is observed up to an LPLC concentration of approximately 400 μM. An increase in fluorescence occurs at higher LPLC concentrations, indicating the appearance of micelles in the solution. It is noteworthy that the binding data and the data from the CMC determination are superimposable.

Thin Films

The effect of the presence of LPCC, LPLC, and LPPC on the appearance and drainage properties of air-suspended thin films of puroindoline-a is summarized in Table 1.

Table 1   *The drainage behaviour of thin films stabilized by puroindoline*-a *as a function of LPC concentration and acyl chain length*

[LPC]		
μM	R	Comments
Puroindoline-a + *LPCC*		
0	0	highly aggregated protein-type film
200	25	aggregates; protein-type film; slightly faster drainage
400	50	aggregates; protein-type film; faster drainage
783	98	fewer aggregates; faster drainage
1165	146	more rapid protein-type drainage (does not drain completely to black)
Puroindoline + *LPLC*		
0	0	highly aggregated
100	12.5	semi-mobile; very few aggregates; draining to black
260	32.5	semi-mobile; very few aggregates; draining to black
400	50	surfactant-like mobility; draining to black
Puroindoline + *LPPC*		
0	0	highly aggregated
12	1.5	semi-mobile; film drains to black
24	3.0	surfactant-like film drainage

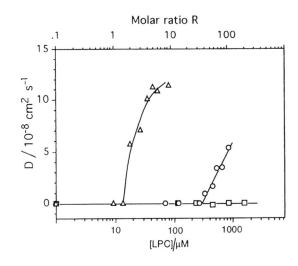

Figure 7   *The surface diffusion coefficient of DDAF in the adsorbed layers of air-suspended liquid thin films stabilized by mixtures of puroindoline-a (0.1 mg ml⁻¹) as a function of LPC concentration:* △, *LPPC;* ○, *LPLC;* □, *LPCC*

The effect of the presence of LPCC, LPLC and LPPC on protein–protein interactions in the adsorbed layer of the thin films, as measured by the FRAP technique, is presented in Figure 7. The presence of LPCC, even at a molar ratio R of 205 (1645 μM LPCC), does not promote the onset of surface diffusion of the fluorescent probe molecule, DDAF, in the adsorbed layer of the thin film. In contrast, the presence of comparatively high concentrations of LPLC (300 μM) induces the onset of surface diffusion at $R = 30$. The onset of surface diffusion induced by adsorption of LPPC is observed at $R = 2$ (17.6 μM LPPC).

Practical Application in a Real Food System

The deleterious effect of lipids on protein-stabilized foams is well recognized. One possible means of controlling foam destabilization by lipids could involve exploiting molecules that selectively bind lipids. The effectiveness of puroindoline-*a* and -*b* in this regard has been tested using a model system comprising soya oil-destabilized egg-white. The results are shown in Figure 8. The stability of egg-white alone (FS% = 66%) was decreased by 45% in the presence of soya oil. Addition of 0.1 mg ml^{-1} puroindoline-*a*, -*b* and further egg-white all produced an increase in foam stability. The presence of puroindoline-*a* in the model system negated the effect of soya oil significantly, restabilizing the mixture to give a foam stability value of 61%. Puroindoline-*b* was found to be less effective at restoring the foam stability, but still gave a very significant improvement to the foam. In contrast, the addition of an equivalent amount of egg-white protein only produced a very small increase in measured foam stability.

4 Discussion

The surface properties, in particular the foaming behaviour, of puroindoline-*a* and -*b* are analogous, with both proteins displaying similar foamability, foam stability and surface elastic modulus. The proteins share a primary structure with 52% homology (Figure 1) and indistinguishable secondary structures as determined by CD. However, the foaming properties of the two proteins are distinguishable in the presence of LPC. Our earlier studies[4] have suggested that the effect LPC on the foaming properties of purindoline-*a* was due to the formation of a lipid–protein complex and this hypothesis is confirmed by the results presented in this paper. The observed differences in the foaming properties of these proteins in the presence of LPC correlates with the strength of interaction between LPC and the protein isoform. The origin of these differences is most probably dependent on the hydrophobicity or the overall shape of the binding site. This is probably altered in puroindoline-*b* since Trp-41 and Trp-42 are replaced by proline and threonine residues (Figure 1). It has been shown that aromatic residues and especially tryptophan are both important for lipid–protein interactions[15] and the surface activities of proteins and peptides.[16]

The zwitterionic phosphorylcholine headgroup of LPC does not appear to

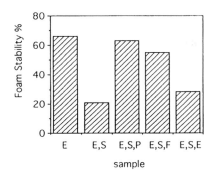

Figure 8   *The foam stability of solutions of: (E) egg-white protein (1 mg ml^{-1}) alone; and in the presence of (E,S), 4.3 vol% soya oil; (E,S,P), 4.3 vol% soya oil and 0.1 mg ml^{-1} puroindoline-a; (E,S,F), 4.3 vol% soya oil and 0.1 mg ml^{-1} puroindoline-b; and (E,S,E), 4.3 vol% soya oil and 0.1 mg ml^{-1} egg-white protein*

contribute in any significant way to the foaming properties of the mixed system. In contrast, the length of the acyl chain has a marked effect. Firstly, clear evidence is presented, in the form of the superimposition of the binding curve and the CMC determination for LPLC, which demonstrates irrefutably that the LPC moiety that binds or initiates binding to the protein is the LPC micelle rather than the monomer. Secondly, the strength of interaction as determined by the K_d value of the LPC–puroindoline complex is strongly correlated with chain length and therefore the hydrophobicity of the LPC.

The origin of the foam enhancement observed with LPC molecules which bind tightly to both puroindoline molecules (*e.g.* LPPC) still eludes us. We are reduced to speculation about the possibility of a conformational change in the protein in the complexed form which significantly enhances the surface properties. However, the improved surface properties will be counterbalanced by the effect of high concentrations of free LPC monomer in the system. The latter will act as a competitive surfactant and will tend to break protein–protein interactions in the adsorbed layer, ultimately leading to displacement of the adsorbed protein. This in turn will depend on the relative surface activity of the monomeric LPC compared with the protein. Thus in the case of LPLC, there is a small enhancement in foaming properties at high LPLC concentrations (200 μM). However, the binding of LPLC is very weak leading to high concentrations of LPLC monomer in solution. However, the onset of surface diffusion which occurs in the LPLC system does not occur until the LPLC concentration reaches 300 μM ($R = 37.5$) LPLC. In contrast, enhancement in foam stability occurs at much lower concentrations with LPPC due to the tighter binding properties of this molecule and its lower CMC of 6 μM. However, the enhancement in foam stability reaches a plateau coincident with the onset of surface diffusion at $R = 1.5$ (10 μM LPPC).

The complex nature of the interaction of puroindoline with lipids may broaden its technological application. We have previously demonstrated the effectiveness of this protein in the protection of beer foam[6] from destabilization by free fatty acids, phospholipid and triglycerides. However, beer is unusual amongst food systems, insofar as the pH is characteristically in the range 4.0–4.5, and the foam is stabilized by a range of polypeptides of ill-determined molecular mass and structure. Here, we have demonstrated the effectiveness of puroindoline-*a* in the control of the lipid-induced destabilization of egg-white foam, which is stabilized by globular proteins at neutral pH. The results have demonstrated the versatility of these lipid-binding proteins, particularly in applications which involve destabilization of food foams by insoluble lipids such as the triglycerides, which form a major component of vegetable oil. The effectiveness of this approach for the control of lipid destabilization of foam merits further study.

References

1. J. E. Blochet, C. Chevalier, E. Forest, E. Pebay-Peyroula, M.-F. Gautier, P. Joudrier, M. Pezolet, and D. Marion, *FEBS Lett.*, 1993, **329**, 336.
2. M.-F. Gautier, *Plant Mol. Biol.*, 1994, **25**, 43.
3. J. E. Blochet, A. Kaboulou, J. P. Compoint, and D. Marion, in 'Gluten Proteins 1990', ed. W. Bushuk and R. Tkachuk, American Association of Cereal Chemists, St. Paul, MN, 1991, p. 314.
4. P. J. Wilde, D. C. Clark, and D. Marion, *J. Agric. Food Chem.*, 1993, **41**, 1570.
5. D. Marion, in 'Cereal Chemistry and Technology: A Long Past and a Bright Future', ed. P. Feillet, IRTAC, Paris, 1992, p. 57.
6. D. C. Clark, P. J. Wilde, and D. Marion, *J. Inst. Brew., London*, 1994, **100**, 23.
7. D. C. Clark, P. J. Wilde, and D. R. Wilson, *Colloids Surf.*, 1991, **59**, 209.
8. G. Scatchard, *Ann. N. Y. Acad. Sci.*, 1949, **51**, 660.
9. A. V. Hill, *J. Physiol.*, 1910, **40**, 40P.
10. K. Kalyanasundaram and J. K. Thomas, *J. Am. Chem. Soc.*, 1977, **99**, 2039.
11. M. Coke, P. J. Wilde, E. J. Russell, and D. C. Clark, *J. Colloid Interface Sci.*, 1990, **138**, 489.
12. D. C. Clark, M. Coke, A. R. Mackie, A. C. Pinder, and D. R. Wilson, *J. Colloid Interface Sci.*, 1990, **138**, 195.
13. J. J. Kokelaar, A. Prins, and M. De Gee, *J. Colloid Interface Sci.*, 1991, **146**, 507.
14. S. W. Provencher and J. Glockner, *Biochemistry*, 1981, **20**, 33.
15. M. Schiffer, C. H. Chang, and F. J. Stevens, *Protein Eng.*, 1992, **5**, 213.
16. M. Enser, G. B. Bloomberg, C. Brock, and D. C. Clark, *Int. J. Biol. Macromol.*, 1990, **12**, 118.

Reflectance Studies on Ice-Cream Models

By Rodney D. Bee and Richard J. Birkett

UNILEVER RESEARCH, COLWORTH HOUSE, SHARNBROOK, BEDFORD
MK44 1LQ, UK

1 Background

Ice-cream has a complex physical and colloid chemistry. It contains four
principal phases: ice crystals, fat droplets and air cells, dispersed at high phase
volume in a matrix of concentrated sugars, polymers and salts. It is normally
kept more than 30 °C below ambient. Its properties depend on the state of this
dispersion (*e.g.* particle size, phase volume, state of aggregation of the phases),
which changes as a function of temperature.

Many valuable measurements can be made on whole ice-cream, but there is
also much to be gained from working with simple composite models. These
avoid the sensitivity to temperature of whole ice-cream and are not
complicated by the wide range of poorly defined species. One such model, used
to understand aeration and the contribution of surface-active species, has been
reported elsewhere.[1] This paper develops the model and illustrates how, using
reflectance measurements, the constituent phases of ice-cream contribute to the
scattering of light from its surface.

The appearance of ice-cream is widely recognized to be important. In
Britain, yellow ice-cream is associated with creamy perception, whereas white
ice-creams are widely favoured elsewhere in Europe. Whilst the perceived
colour results from a chromophore absorbing a component of the incident
radiation, the amount of incident light scattered determines how bright (and
white) the ice-cream appears. It is therefore of significant interest to know (i)
why ice-cream is so white, (*i.e.* scatters light so well), (ii) how this is determined
by the structure, and (iii) what measurements of the light scattering of ice-
cream can tell us about ice-cream structure.

2 Techniques and Approach

Experimental Samples

Scattering coefficients have been measured on models of increasing complexity
from simple oil-in-water emulsions to full ice-cream formulations. The ice

phase in industrially produced ice-cream is not dendritic, but consists of rounded crystals having the appearance of pebbles. We have therefore used small solid glass spheres (ballotini) to represent the ice phase, thus avoiding the extreme temperature sensitivity typically noted in such experiments. Models of intermediate complexity are dispersions of a single phase (ice, fat, air cells, or ballotini) in a concentrated maltodextrin matrix; the more complex models are composites of two disperse phases in this matrix. Finally we shall discuss scattering coefficients measured on whole ice-creams in which the disperse phase volumes have been systematically varied.

The emulsions for the simplest models were prepared by multiple passes (Rannie Laboratory homogenizer) of crude pre-dispersions. They were characterized for particle size by both Coulter Counter and photomicrography. Homogenization pressures were varied to produce samples with a range of particle sizes. Ice dispersions in maltodextrin were prepared using a small-scale continuous freezer, the water content being adjusted to give the required range of phase volumes of ice at the temperature of measurement. Various phase volumes of ballotini, air and fat were made by diluting concentrated preparations with continuous phase which had first been equilibrated with the same disperse phase. The experimental samples were handled and prepared as described previously.[1]

Ballotini/air systems were made by aerating various ballotini dispersions in maltodextrin solutions containing the hydrolysed protein aerating agent (0.1 wt%). Whole ice-cream samples were prepared on a research-type continuous freezer using a 'model' formulation representative of typical commercial ice-cream.

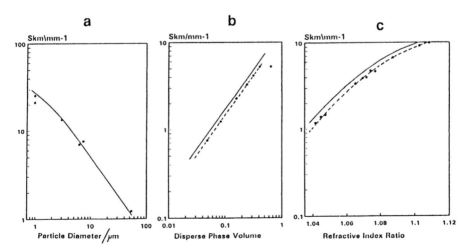

Figure 1   *Experimental Kubelka–Munk scattering coefficients (broken lines) at*
$\lambda = 550$ *nm: (a) paraffin oil-in-water emulsions ($n_1/n_2 = 1.10$; $\phi = 0.5$) as a function of droplet diameter; (b) disperse phase volume of paraffin oil ($n_1/n_2 = 1.10$); (c) refractive index ratio. Solid lines are calculated*

Materials

Liquid paraffin BP (Evans Medical) was used for the simple oil-in-water emulsions. The dispersions were stabilized with Triton X-100 (1 wt%, Rohm and Haas) and polymeric surfactant (1 wt% polyvinyl alcohol, 98% hydrolysed, molecular mass 126 000 Da, Aldrich). Xanthan gum (0.2 wt%) and sodium azide (0.01 wt%) were added, with gentle stirring, to give stability against creaming and microbial spoilage, respectively. The continuous phase of ice-cream was modelled using 63 DE maltodextrin solution (Sweetose, Ragus Sugars). Gas cell dispersions in this matrix were prepared using a hydrolysed whipping agent (Gunther D-100) as described elsewhere.[1] Small solid glass spheres (Ballotini, grade 20, size range 5–80 μm, Jencons) were used to model ice in the reflectance studies of the aeration process and for the single largest particle size on which scattering determinations were made (Figure 1a). Measurements of the (essentially solid) fat in the model ice-cream matrix were made on emulsions of hardened palm kernel oil in sodium caseinate (1 wt%) solutions.

Optical Measurements

Diffuse reflectance spectra were obtained[*] using either a Pretema Spectromat (for the simple dispersions) or an ICS Micromatch 2000 (ICS, Newbury, UK) for thin layers of sample (path length 2 mm) with a glass coverslip. Readings were corrected for errors arising from external diffuse, and total internal, reflectance at the air–glass cover slip interface.[3] Scattering coefficients were calculated at specific wavelengths from the reflectance data according to the Kubelka–Munk model of reflectance and transmission in thin layers.[4] This model required reflectance measurements to be made in pairs, with each sample measured against a matt black backing, and also against a white backing, both of known reflectance. In practice, a single cell with a divided backing was used for the measurements.

Determination of Scattering Coefficients

An explicit hyperbolic solution[5] of the Kubelka–Munk model was used to calculate the scattering coefficient S_{KM} for the experimental samples, *i.e.*

$$S_{KM}d = (1/b)\{1/\text{ctgh}[(1 - aR_B)/(bR_B)]\} \tag{1}$$

where

$$a = \tfrac{1}{2}\{R_W + [(R_B - R_W + R_C)/R_B R_C]\} \tag{2}$$
$$b = (a^2 - 1)^{1/2} \tag{3}$$

[*] The specimen was illuminated by diffuse illumination. The viewing angle was less than 10° from normal to the surface with no ray of the viewing beam greater than 5° from the beam axis. The port area was less than 10% of the internal reflecting surface area.[2]

S_{KM} is the Kubelka–Munk scattering coefficient of unit thickness of sample, d

S_{KM} is the Kubelka–Munk scattering coefficient of unit thickness of sample, d is the sample thickness, R_W is the reflectance over white backing, R_B is the reflectance over black backing, R_C is the reference reflectance of the white backing alone, and cgth x is the hyperbolic cotangent of x. The reflectance of the black backing alone was very close to zero; calculations showed that it could be neglected. The absorption coefficient K is given by

$$K = S_{KM}(a - 1) \tag{4}$$

As expected for these samples, measured absorption coefficients were always small compared with the corresponding scattering coefficient (by at least two orders of magnitude) and for all subsequent discussions they are assumed to be negligible. The wavelength dependence of scattering for these systems is very slight; accordingly only scattering data for 550 nm are given.

Theoretical values of the scattering coefficient, S_{Mie}, have been calculated for the simple dispersion experimental systems on the basis of classical Mie scattering theory.[6] Total scattering functions are calculated from a knowledge of the particle diameter, refractive index ratio, disperse phase volume and wavelength. This approach relates to isotropic scattering and is therefore not comparable with the experimental scattering coefficients derived from the Kubelka–Munk model which assesses back-scattered light. The function is therefore modified by an angular term[7] to yield a scattering coefficient equivalent to the experimentally derived values, *i.e.*

$$S_{KM} = \frac{3}{4}(1 - \cos\theta)S_{Mie} \tag{5}$$

where θ is the scattering angle for constant particle diameter, wavelength and refractive index ratio.

The Mie theory makes the assumptions that the system is monodisperse, that the particles are spherical, and that only a single scattering event occurs. In our calculation, polydispersity is incorporated by integrating the scattering coefficient over the range of droplet diameters in the sample (measured by the Coulter Counter) weighted by the frequency of occurrence. Multiple scattering and the divergence of the disperse phase particles from spherical are not taken into account. Despite this, the correlation between the experimental and theoretical scattering coefficients is surprisingly good up to volumes of $\phi = 0.6$ (Figure 1b).

3 Results and Discussion

Simple Dispersions

A paraffin oil-in-water emulsion ($n_1/n_2 = 1.10$; $\phi = 0.5$) was used to establish the comparability between the theoretically predicted and measured scattering coefficients. Data for a range of mean droplet sizes from 0.4 to 8 μm, measured at 550 nm, are given in Figure 1a. The single data point at 60 μm was obtained from a suspension of ballotini with the refractive index ratio adjusted to 1.10.

The theoretical prediction gives good agreement with practical measurements. Similarly, the predicted scattering coefficient holds reasonably well over a broad phase volume range for a given mean particle size (8.0 μm, $\lambda = 550$ nm) (Figure 1b); the experimental range is extended in this case by both dilution of the initial preparation or centrifugal concentration. Only at extremely high disperse phase volumes ($\phi > 0.6$) does the experimentally determined scattering deviate from a linear log–log relationship with ϕ. Figure 1(c) shows the theoretically predicted and experimentally determined relationship between scattering and refractive index ratio ($\phi = 0.5$, $\lambda = 550$ nm), the refractive index ratio being manipulated experimentally, in this case by addition of urea to the aqueous phase of the emulsion. Again reasonably good agreement is observed.

The above data show that, for simple oil-in-water emulsions, the experimentally determined scattering coefficients agree reasonably well with the predicted Kubelka–Munk scattering behaviour for particle size, phase volume and refractive index ratio as variables.

Ice-cream Models

At a refractive index ratio of 1.10, the scattering data in our model emulsion in Figure 1(c) approximate to that of double dairy cream. However, in relating this model to ice-cream we are concerned with the scattering from not one, but three, disperse phases. Furthermore, the continuous phase is not a (dilute)

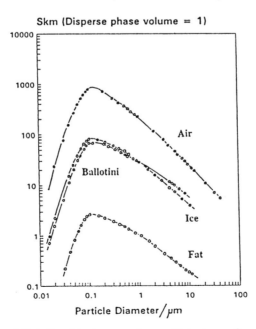

Figure 2   *Predicted Kubelka–Munk scattering coefficient as a function of particle size for air, ice, ballotini, and fat in maltodextrin matrix (70 wt% solids). Data are extrapolated to a phase volume of unity*

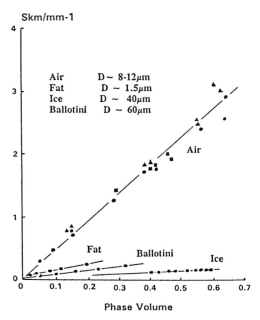

Figure 3   *Experimental Kubelka–Munk scattering coefficients as a function of phase*
volume in maltodextrin solution (70 wt%). Ice data at − 20 °C; *ice content*
adjusted by solids concentration

aqueous solution; at for example − 20 °C it has a refractive index of 1.47
having become concentrated in sugars by the removal of water as ice. To
mimic this using a model maltodextrin solution requires a solution of 70 wt%
solids; this has therefore been used in the following model work comparing the
scattering of air, ice, ballotini and fat disperse phases.

With a continuous phase of refractive index $n_{cont.} = 1.47$, predictions of
Kubelka–Munk scattering coefficients for these phases ($n_{ice} = 1.31$, $n_{air} = 1.00$,
$n_{ballotini} = 1.60$, $n_{fat} = 1.46$) produce the data in Figure 2. Each disperse phase
gives a similar overall relationship. For a given particle size and phase volume,
the scattering power of the different phases falls clearly in the order
air≫ice ≡ ballotini≫fat. It is, however, unrealistic to compare equivalent
particle sizes for each of the disperse phases. Fat droplets are typically 1–2 μm
in ice-cream premix whereas ice crystals, in the final product, are likely to be
approximately 40 μm. The gas cells in ice-cream are less well defined in size. A
wide range, for example 10–500 μm, may be anticipated depending on
formulation, processing variables, and age of the sample. The gas cell
dispersion used here to measure scattering coefficients has been prepared as
described elsewhere.[1] For this particular preparation, the gas cell dispersion is
at the smallest end of the normal range with a $d_{3.0}$ value of 8–12 μm. Figure 3
shows measured scattering coefficients for these phases and ballotini as a
function of phase volume. It can be seen that the predictions in Figure 2 do

hold, at least qualitatively. Fat present as small droplets (1–2 μm) scatters slightly more per unit phase volume than ice or ballotini, the size effect outweighing the small refractive index difference between fat and the matrix phase. Ice and ballotini particles, of approximately equal sizes, scatter almost equally, confirming that ballotini can be used as a substitute for ice in this simple model. As expected, with by far the largest refractive index ratio, the scattering from the air outweighs that from any other component.

In passing, it is interesting to note that for each of these phases there is a linear dependence of scattering coefficient on phase volume.

Measurements during an Aeration Sequence

Work with the present, simple ice-cream model[1] has pointed out the importance of the combination of the gaseous and solid disperse phases to the rheological properties. The presence of a solid disperse phase also has a major influence on the amount and nature of the gas phase which is incorporated. Figure 4 shows the aeration of a simple maltodextrin solution and demonstrates that gas incorporation is significantly diminished by low-phase volumes of solid particles. The sequence of gas phase incorporation is believed to be identical to that described previously.[1] The reduced gas phase volume results from the increased viscosity due both to the presence of a

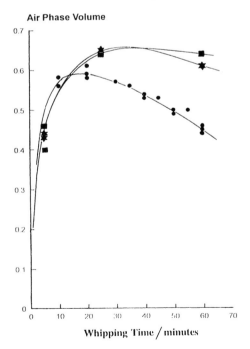

Air Phase Volume

Whipping Time / minutes

Figure 4   *Effect of ballotini on air phase volume as a function of time during an aeration experiment:* ■, *0%;* ★, *5%;* ●, *10%*

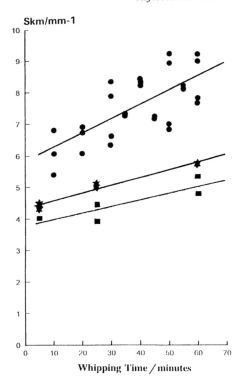

Figure 5   *Measured Kubelka–Munk scattering coefficient during the aeration process described in Figure 4:* ■, *0%;* ★, *5%;* ●, *10%*

solid disperse phase and the more effective comminution of the gas cells. This is supported by the scattering data of Figure 5. Here, scattering increases through the aeration experiment (without or with ballotini), which is consistent with diminishing gas cell size. It even outweighs the effect of significantly diminishing gas-phase volume which occurs when 10 wt% of ballotini is present.

At given aeration times, the scattering, at unit gas phase volume, increases significantly when 5–10 wt% additions of ballotini are made. This cannot be explained in terms of scattering from ballotini *per se*, but it must arise from the much smaller gas cells which are formed when the ballotini are dispersed in the aqueous phase. There is a clear analogy here with the manufacture of ice-cream. In the freezer, a solid phase (ice) is produced under shear, whilst at the same time air is incorporated and becomes comminuted into fine bubbles in the shear field.

If both ballotini and gas cells can be assumed to be dispersed homogeneously through the aqueous phase, and the scattering from each assumed to be additive, Figure 2 may be used to estimate the relative air cell sizes in Figure 5. Taking the scattering coefficient of a unit phase volume at $t = 30$ min we have: no ballotini, $S_{KM} = 4.3$, $d = 50$ μm; 5 wt% ballotini, $S_{KM} = 5.1$, $d = 40$ μm;

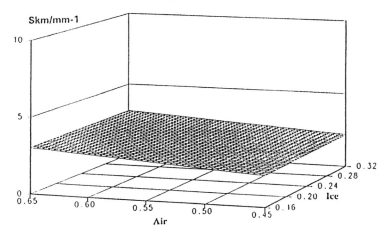

Figure 6 *Predicted Kubelka–Munk scattering coefficients for a linear addition of ice and air in a matrix with fat content fixed at zero. Matrix varies as ice and air volume fractions vary*

10 wt% ballotini, $S_{KM} = 7.2$, $d = 28$ μm. Whilst as yet we have here only a qualitative technique for cell sizing in gas dispersions, in this phase volume range, it is clear that the method offers potential for development, particularly for changes taking place during a given process.

Comparison of the Simple Model with a Whole Ice-cream

As a first step towards understanding how the component phases scatter light when combined, we have taken a simple, linear addition of the effects of the ice and the gas phase (Figure 6). For comparison, we have measured scattering coefficients on 87 full ice-creams comprised of a wide range of phase volumes of ice, air, fat, and matrix. The empirical relationship

$$S_{KM} = -17.5\Phi_{ice} + 3.9\Phi_{air} - 57.2\Phi_{matrix} + 2.4\Phi_{air}\Phi_{ice} + 89\Phi_{air}\Phi_{matrix} + 150.6\Phi_{ice}\Phi_{matrix} \tag{6}$$

describes the scattering from a range of combinations of phase volumes of these phases. Figure 7 shows the surface which describes just the combination of ice and air as disperse phases (*i.e.* fat content has been extrapolated to zero to permit comparison with the simple linear combination of the two phases given in Figure 6). There is a reasonable qualitative agreement with simple linear addition, *i.e.* the combinations have coefficients generally just below 5, but at both high and low combinations of phases the scattering for 'whole' ice-cream (with zero fat) diminishes quickly. Looking at the data another way, at high ice content, adding more air diminishes the scattering. This is the converse of the simple notion reached earlier that increasing the phase volume, particularly of air, would be expected to increase the scattering coefficient. In short, a linear

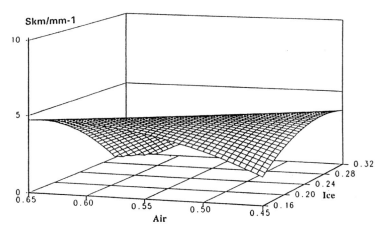

Figure 7   *Scattering coefficient surface plot derived from eqn (6), with fat content fixed at zero. Matrix varies as ice and air volume fractions vary*

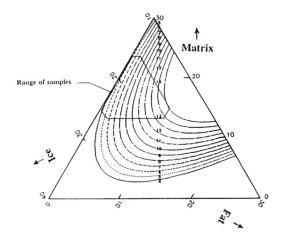

Figure 8   *Predicted scattering coefficient relationship for ice, air, and matrix for a given value of the gas phase volume (0.60)*

combination of scattering coefficients is quite inadequate even for simple combinations.

Given the problem of wanting to optimize the scattering from a practical sample, the effects of three of the four phases are represented advantageously in a triangular form. Figure 8 illustrates this for a combination of phases of ice, matrix and fat totalling 40% of the volume (*i.e.* $\Phi_{air} = 0.6$). The contours join points of constant scattering. Scattering is optimized by selecting the direction, from any point, in which lines of increasing scattering are crossed most quickly.

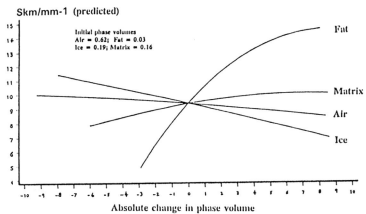

Figure 9   *Scattering coefficients derived from eqn (6) for absolute changes in phase volume of ice, air, matrix, and fat in a model whole ice-cream*

Finally, the scattering, as predicted from the above-mentioned relationship as a function of change in absolute phase volume, can be represented as in Figure 9. Here the lines indicate the effect of removal or addition of each phase whilst keeping constant the relative volumes of the other three. As noted from Figure 8, it can be seen that, once a certain level of the dispersed phases in an ice-cream has been reached, further addition of gas phase or ice phase decreases the scattering. More surprisingly, as we have seen that fat alone scatters only very weakly when dispersed in the matrix, Figure 9 shows that small increases in fat level produce a dramatic increase in scattering. It is well established that, in ice cream, the fat droplets are strongly adsorbed to the air–matrix interface. The increased scattering cannot stem from the mismatch between the gas cells and the emulsion droplets, as it is much the same as that between air and matrix. Rather it is postulated that the increase results from the small effective size of the fat droplets adsorbed at the air–water interface in a full ice-cream. Alternatively, in the presence of higher levels of fat, the gas cells are smaller when produced, or they are stabilized against increase in size with time when higher levels of fat are present.

Acknowledgements

The authors are grateful to Marc Maste, Ellen de Jong, Steve Dyks and Ian Evans for use of their unpublished data and to Allan Clark and Graham Cleaver for lending their computational skills.

References

1.   R. D. Bee, A. Clement, and A. Prins, in 'Food Emulsions and Foams', ed. E. Dickinson, Special Publication No. 58, The Royal Society of Chemistry, London, 1987, p. 128.

2.   D. B. Judd and G. Wyszecki, 'Colour in Business, Science and Industry', 3rd edn, 1975, p. 123.
3.   R. J. Birkett, A. Clarke, and G. H. Meeten, *Colloids Surf.*, 1987, **24**, 259.
4.   P. Kubelka and F. Munk, *Z. Tech. Phys.*, 1931, **12**, 593.
5.   P. Kubelka, *J. Opt. Soc. Am.*, 1984, **38**, 448.
6.   G. Mie, *Ann. Phys.*, 1908, **25**, 377.
7.   P. S. Mudgett and L. W. Richards, *Appl. Opt.*, 1971, **10**, 1985.

Disproportionation in Aerosol Whipped Cream

By M. E. Wijnen and A. Prins

DEPARTMENT OF FOOD SCIENCE, WAGENINGEN AGRICULTURAL UNIVERSITY, BOMENWEG 2, 6703 HD WAGENINGEN, THE NETHERLANDS

1 Introduction

Aerosol whipped cream is an instant substitute for fresh whipped cream. Fresh whipped cream is formed by beating air bubbles, mainly consisting of nitrogen, into the cream. The foaming of aerosol whipping cream is caused by a different mechanism. Due to the high pressure in an aerosol can (9 bar) the propellant nitrous oxide ('laughing gas') is for the most part dissolved in the cream. When the cream leaves the can, the nitrous oxide comes out of solution, causing bubble growth and foam formation.

Fresh whipped cream has an overrun of about 100% and forms a stable and firm aerated produce. Aerosol whipped cream has an overrun of about 600%, but the foam is not stable; it collapses very fast and alters appearance becoming shiny within 30 min after foam formation (Figure 1).

2 Theory

Disproportionation, the growth of larger bubbles at the expense of smaller bubbles, is one of the mechanisms that can affect the stability of foam. It is the result of gas diffusion between bubbles of different size in the foam and between bubbles and the atmosphere, driven by the Laplace pressure difference. This process is strongly promoted by a better solubility of the dispersed gas in the surrounding liquid.

The solubility of nitrous oxide in the cream is approximately 50 times higher than the solubility of air in the cream. It is expected that aerosol whipped cream is very susceptible to disproportionation, particularly because of the high solubility of the nitrous oxide in the cream. The aim of this research is to investigate whether the theory for stopping disproportionation is applicable to aerosol whipped cream.

According to Lucassen[1] disproportionation can be stopped when, over a

Figure 1   *Aerosol whipped cream (1) immediately after foam formation and (2) after 30 min*

relevant time-scale, the bubble surface is *purely* elastic and the surface dilational elasticity E_s is larger than half the surface tension γ, *i.e.*

$$E_s \geq \frac{\gamma}{2} \qquad (1)$$

Pure elastic behaviour means the absence of relaxation processes, *i.e.*

$$\eta_s = 0 \qquad (2)$$

where η_s is the surface dilational viscosity.

3 Experimental

The surface dilational elasticity and viscosity of aerosol whipping cream with a fat content of 32 wt% were measured by means of a sinusoidal oscillation method using a special Langmuir trough with a cylindrical barrier called a 'ring trough'.[2] The dynamic surface properties of the cream were measured at different angular velocities between 0.013 and 1.3 rad s^{-1}, resulting in time-scales of the order of 1 to 100 s. Measurements were carried out at 6 °C; a relative deformation of 1.6% was applied.

4 Results and Discussion

Figure 2 shows the dynamic surface properties of aerosol whipping cream as a function of the angular velocity. For the sake of comparison η_s is multiplied by the angular velocity ω (in rad s^{-1}). The conditions for stability against the process of disproportionation [eqns (1) and (2)] are also represented. The surface tension of the cream at 6 °C is 40 mN m^{-1}. The measurements show that we have $E_s > \gamma/2$. In this context the surface tension at equilibrium is considered. For stabilization against disproportionation, the shrinking of the smaller bubbles in the foam has to be stopped. During the process of shrinking, compression of the bubble surface occurs. This will result in a lower

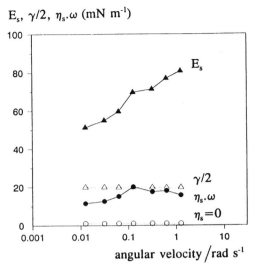

Figure 2 *The dynamic surface properties of aerosol whipping cream as a function of the angular velocity*

surface tension than 40 mN m^{-1} and presumably a higher value for E_s, thereby satisfying the inquality in eqn (1) even better.

The cream surface, is, however, visco-elastic over the time-scale of the experiment ($\eta_s > 0$) indicating that the bubble surface is not *purely* elastic; so eqn (2) is not satisfied. The bubble surface will therefore not be able to stabilize the aerated product against disproportionation over the time-scale of the experiment, although the dynamic surface properties can possibly slow down the process.

References

1. J. Lucassen, in 'Anionic Surfactants', ed. E. H. Lucassen-Reynders, Marcel Dekker, New York, 1981, p. 217.
2. J. J. Kokelaar, A. Prins, and M. de Gee, *J. Colloid Interface Sci.*, 1991, **146**, 507.

Determination of Protein Foam Stability in the Presence of Polysaccharide

By Esra Izgi and Eric Dickinson

PROCTER DEPARTMENT OF FOOD SCIENCE, UNIVERSITY OF LEEDS, LEEDS LS2 9JT, UK

1 Introduction

Proteins and polysaccharides are the main biopolymers involved in food foam stabilization.[1] A reliable new method developed to investigate the physico-chemical factors that affect formation and stability of foams involves monitoring the rate of decay of foam structure by determining the rate of build-up of pressure at constant volume and temperature.

The pressure difference Δp between the inside and outside of a spherical bubble of radius r is given by Laplace equation[2]

$$\Delta p = 2\gamma/r$$

where γ is the surface tension. The pressure difference between bubbles of different sizes gives rise to gas transfer from smaller to larger bubbles as a result of diffusion through the aqueous phase (which causes the smaller bubbles to shrink). The loss of bubbles by disproportionation and coalescence results in an increase in pressure in a closed system containing foam. So, by monitoring the small increase in pressure above a protein foam, the rate of foam collapse can in principle be followed.[3]

The pressure monitoring technique is used here to investigate the effect of anionic polysaccharide dextran sulfate (DS) on the stability of foams containing bovine serum albumin (BSA) or β-lactoglobulin at pH 6 and 7. Previous studies carried out on emulsions at pH 7 have suggested[4] that the formation of an electrostatic complex between BSA and DS occurs so that enhanced emulsion stability is obtained.[4] However, β-lactoglobulin + DS mixtures have not shown much improvement in emulsion stability.[5]

2 Materials and Methods

The operation of the differential pressure transducer apparatus is described elsewhere in detail.[6] Protein + polysaccharide mixtures were prepared by

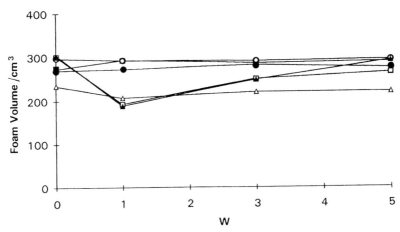

Figure 1   *Plot of initial foam volume, measured visually, against polysaccharide:protein weight ratio* W: □, *0.05 wt% BSA at pH 7;* ●, *0.3 wt% β-lactoglobulin at pH 7;* △, *0.1 wt% BSA at pH 6;* ■, *0.15 wt% β-lactoglobulin at pH 7;* ○, *0.15 wt% β-lactoglobulin at pH 6;* ▲, *0.05 wt% BSA at pH 6*

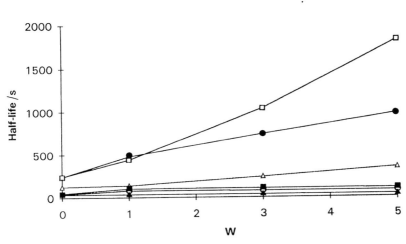

Figure 2   *Plot of half-life, determined by visual observation, against polysaccharide:-protein weight ratio* W: □, *0.05 wt% BSA at pH 7;* ●, *0.1 wt% BSA at pH 6;* ■, *0.15 wt% β-lactoglobulin at pH 6;* △, *0.05 wt% BSA at pH 6;* ○, *0.15 wt% β-lactoglobulin at pH 7;* ▲, *0.3 wt% β-lactoglobulin at pH 7*

dissolving BSA (0.05 wt%, 0.1 wt%) or β-lactoglobulin (0.15 wt%, 0.3 wt%) together with DS (ratio 1:1, 1:3, 1:5 by weight) in phosphate buffer (0.005 mol dm^{-3}). Foam was generated by sparging a 15 ml sample of solution with nitrogen gas for 30 s (or 20 s for 0.1 wt% BSA) at a constant flow rate of

0.8 1 min^{-1} at 30 °C for β-lactoglobulin. The pressure change was recorded, immediately after switching off the nitrogen supply, by means of a voltage output connected to a computer. Visual observations of the initial foam volume, and the time for the protein foam to collapse to 50% of its initial volume, defined as the half-life, were also recorded.

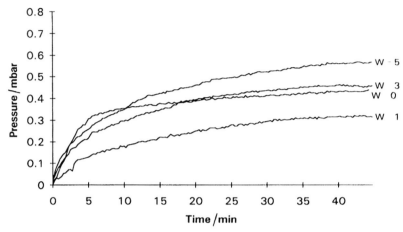

Figure 3   *Time-dependent pressure change for the BSA + DS system at different values of polysaccharide:protein weight ratio W at pH 7*

3 Results and Discussion

Figure 1 shows the initial foam volume for BSA + DS and β-lactoglobulin + DS as a function of polysaccharide:protein weight ratio W. For BSA + DS mixtures the presence of DS leads to a slight reduction in the foaming properties of the protein at low molecular mass ratios ($W \approx 1$) and a slight increase at high ratios, although there is no change obtained in initial foam volume with addition of DS to β-lactoglobulin foams.

Figure 2 shows a plot of foam half-life against W for the same set of systems. The BSA + DS mixture at pH 7 indicates a dramatic increase in foam stability. However, the effect on stability is reduced at pH 6. For the β-lactoglobulin + DS system, the presence or absence of DS does not affect foam stability to any measurable extent.

Figure 3 shows time-dependent pressure changes for BSA + DS foams at pH 7 recorded with the pressure transducer apparatus at different DS concentrations. If the foam decay follows a simple first-order kinetic law, then we can write

$$\Delta p_\infty - \Delta p(t) = (\Delta p_\infty - \Delta p_0) \, \exp(-Bt)$$

where Δp_∞ is Δp at $t = \infty$, Δp_0 is Δp at $t = 0$, and B is a rate constant for the foam decay. A plot of $\log[\Delta p_\infty - \Delta p(t)]$ against t gives B. The larger the value of B, the less stable is the foam. Table 1 shows calculated B values for the β-lactoglobulin + DS mixtures and for the BSA + DS mixtures. An increase in stability, which is consistent with the half-life determined by simple visual observation, is observed for the BSA + DS system. For the β-lactoglobulin + DS system, the data show a slight improvement in stability, which is hardly observable in the half-life measurements estimated visually.

Table 1 *Calculated values of the rate constant* B $(min^{-1})^a$ *for the BSA + DS foam system and for the β-lactoglobulin (β-lact.) + DS foam system for different values of the polysaccharide:protein weight ratio* W

W	0.05 wt% BSA at pH 7	0.05 wt% BSA at pH 6	0.1 wt% BSA at pH 6	0.15 wt% β-lact. at pH 7	0.15 wt% β-lact. at pH 6	0.3 wt% β-lact. at pH 7
0	0.224	0.138	0.309	1.446	1.378	1.318
1	0.096	0.117	0.107	0.852	0.972	1.220
3	0.088	0.106	0.097	0.805	0.686	0.740
5	0.074	0.078	0.071	0.691	0.658	0.722

^{a} Estimated error $\pm$ 0.005 min^{-1}.

4 Conclusions

Mixing the anionic polysaccharide DS with BSA improves foam stability dramatically at pH 7 and slightly at pH 6. Addition of DS to β-lactoglobulin does not show any improvement in foamability of the protein, but a slight increase in stability is observed on addition of DS. The results from the pressure monitoring technique are consistent with foam volume and stability results obtained visually. Apparently, this new instrumental method is capable of detecting slight changes in functional properties of macromolecules affecting foam formation and stability which perhaps cannot be differentiated by simple visual observation. Sensitive pressure monitoring may therefore be a valuable method for studying the influence of various physical and chemical factors on the foaming properties of biopolymers.

References

1. P. J. Halling, *CRC Crit. Rev. Food Sci. Nutr.*, 1981, **15**, 55.
2. E. Dickinson, 'An Introduction to Food Colloids', Oxford University Press, 1992, ch. 5.
3. M.-A. Yu and S. Damodaran, *J. Agric. Food Chem.*, 1991, **39**, 1555.
4. E. Dickinson and V. B. Galazka, in 'Gums and Stabilisers for the Food Industry', ed. G. O. Phillips, P. A. Williams, and D. J. Wedlock, IRL Press, Oxford, 1992, vol. 6, p. 351.
5. E. Dickinson and V. B. Galazka, *Food Hydrocolloids*, 1991, **5**, 281.
6. E. Dickinson, E. Izgi, and P. V. Nelson, *Food Hydrocolloids*, 1993, **7**, 307.

Bubble Growth on an Active Site: Effect of the Cavity Volume

By Antien F. Zuidberg and A. Prins

DEPARTMENT OF FOOD SCIENCE, DAIRYING AND FOOD PHYSICS GROUP, BIOTECHNION, BOMENWEG 2, 6703 HD WAGENINGEN, THE NETHERLANDS

1 Introduction

Heterogeneous bubble growth is a phenomenon which occurs in gas-supersaturated liquids such as champagne, soft drinks and beer. In studying the kinetics of bubble growth the properties of the active site at which the bubbles are formed must be taken into account. For example, both the cavity mouth radius and the wetting behaviour can directly affect the bubble size at detachment. The rate of bubble formation can be influenced not only by the shape of the cavity but also by the cavity volume.

2 Theory

It is hypothesized that a large cavity volume in an active surface can (i) increase the growth rate of a single bubble, and (ii) lower the formation frequency of a bubble train.

Due to the Laplace pressure difference across the gas–liquid interface, the pressure in the gas-filled cavity after bubble detachment (see Figure 1a) is given by

$$P_{in} = P_{out} + \frac{2\gamma}{R} \tag{1}$$

where P_{in} is the pressure in the bubble, P_{out} is the pressure outside the bubble, γ is the surface tension, and R is the radius of the detaching bubble. Depending on the wetting behaviour of the liquid on the surface, R will change, as will the pressure, and therefore so will the volume according to Boyle's law for an ideal gas,

$$nPV = \text{constant} \tag{2}$$

where n is the number of moles of gas, P is the pressure, and V is the volume.

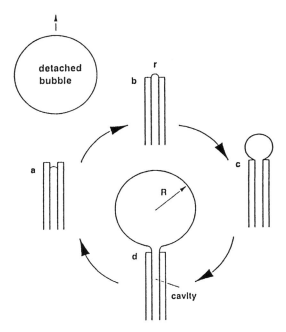

Figure 1 *The cycle of bubble growth on a cavity with a finite volume. (a) After bubble detachment the interface is retreated and the radius of curvature is large. (b) The gas cavity is 'pumped up' by gas diffusion; the radius of curvature is small. (c) The bubble is 'blown up' by the expansion of gas. (d) The bubble grows by gas diffusion until the buoyancy forces the bubble to detach*

Here we assume perfect wettability of the liquid on the cavity surface. Before a new bubble can be formed the pressure has to reach (Figure 1b)

$$P_{in} = P_{out} + \frac{2\gamma}{r} \qquad (3)$$

where r is capillary radius. Since for a capillary tube we have $r < R$, the required *increase* in pressure has to be brought about by gas transport from the bulk liquid to the gas–liquid interface. The amount of gas transport needed to pressurize a greater cavity volume is larger, and, as the area of 'retreated' interface is also relatively small, the whole process requires more time. This will effectively lower the frequency with which bubbles are consecutively formed from one active surface.

After the 'half-bubble' is formed, the bubble may grow relatively fast as the pressure decreases with increase in bubble radius R (Figure 1b→1c). Again using Boyle's law the volume expands with a decrease in pressure. This gas expansion occurs relatively fast, after which the bubble fills up to its final size by diffusion from the supersaturated liquid (Figure 1c→1d). The absolute increase in volume due to gas expansion is greater for a *larger* cavity volume; therefore the growth rate from half bubble is faster.

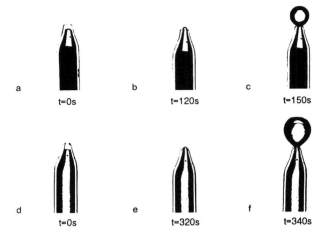

Figure 2   *Photographs of the cycle of a bubble growing on a capillary tube. Two capillary volumes are shown: (a), (b) and (c) 2 μl; (d), (e) and (f) 22 μl.*

3 Experimental Method and Results

A glass capillary (a precision micropipette from Drummond Scientific Company, USA) was used to form the cavity. The cavity volume was varied with a teflon plunger (2 and 22 μl) and bubble growth was photographically recorded at time intervals of 10 s. Figure 2 shows various moments in the cycle of bubble growth and detachment from a capillary of mouth radius $r \approx 150$ μm. From Figure 2 the main difference observed between the two cavity volumes is the time taken between the stages of bubble growth. On a cavity with a larger volume (Figures 2e, 2f, and 2g) the formed bubble detaches in a shorter time, but the time required for the accumulation of enough gas phase to start the growth of a new bubble is much longer.

4 Conclusions

The bubble growth rate is higher for a larger cavity volume. At the same time, however, the frequency of consecutive gas bubble formation is greatly reduced. For a relatively small cavity volume, both effects become less pronounced. Bubble growth will then be primarily governed by gas diffusion.

Mixed Biopolymer Systems

Thermal Behaviour of Kappa-Carrageenan + Galactomannan Mixed Systems

By Paulo B. Fernandes,[*] M. P. Gonçalves,[1] and Jean-Louis Doublier[2]

ESCOLA SUPERIOR DE BIOTECNOLOGIA, UNIVERSIDADE CATÓLICA PORTUGUESA, RUA DR. ANTÓNIO BERNARDINO DE ALMEIDA, 4200 PORTO, PORTUGAL
[1] DEPARTAMENTO DE ENGENHARIA QUÍMICA, FACULDADE DE ENGENHARIA, UNIVERSIDADE DO PORTO, RUA DOS BRAGAS 4099, PORTO CODEX, PORTUGAL
[2] INRA-LPCM, BP 527, 44026 NANTES CÉDEX 03, FRANCE

1 Introduction

Locust bean gum is known to form synergistic gels upon mixing with kappa-carrageenan (κ-car).[1,2] Although less spectacular, κ-car + guar gum mixtures also may develop synergistic properties in certain situations.[2,3] These effects have found many different applications, particularly in the food industry.[4] In previous work, we have reported on the viscoelastic properties of κ-car + galactomannan mixtures as a function of temperature.[2] A thermal hysteresis was observed for the mixed systems prepared with three types of galactomannans [guar gum (GG), tara gum and locust bean gum (LBG)] differing in their mannose/galactose (M/G) ratio. This hysteresis depended upon the type of galactomannan used. Furthermore, for LBG, a shoulder systematically appeared on the melting curve between 20 and 30 °C.

In the present work, in a first set of experiments, we describe the thermal behaviour of κ-car + galactomannan mixtures, the galactomannans being fractions of LBG differing in their M/G ratio (ranging from 3.0 to 4.9). The total polysaccharide concentration is kept constant at 1% and the κ-car/galactomannan ratio is fixed at 4:1. Then, we describe the behaviour of κ-car + LBG and of κ-car + GG mixtures, the κ-car concentration being fixed at 0.75% and the galactomannan content ranging from 0 to 1.2%. The overall aim is to obtain accurate description of the melting curves of such systems.

[*] Present address: Nestec Ltd., Nestlé Research Centre, Vers-chez-les-Blanc, PO Box 44, CH-1000 Lausanne 26, Switzerland

2 Materials and Methods

A Portuguese carob flour was obtained as described by Gonçalves *et al.*[5] and was purified as described by Fernandes *et al.*[6] The purified LBG was fractionated according to its solubility in water at temperatures 10, 35, 60 and 85 °C. This was carried out as described by Gonçalves and Silva.[7] A commercial grade guar sample (purified) was kindly provided by SBI (France). The κ-car sample extracted from *Eucheuma cotonii* was supplied by SBI in the K^+ form and was used without further purification.[8]

The samples, galactomannans and the κ-car, were first dispersed in water under strong stirring. The dispersion was stirred for 1 h at room temperature and then heated at 90 °C for 30 min with continuous stirring. The mixed solutions were prepared by mixing the pure solutions of galactomannan and κ-car at 90 °C. The hot mixture was then poured directly onto the plate of the rheometer. In these experiments, no KCl was added; the only K^+ ions present were those from the salt form of the κ-car.

Dynamic measurements were performed using a Carri-Med CS-50 controlled-stress rheometer fitted with a parallel plate geometry (gap 4 mm; plate diameter 6 cm) with radial grooves in order to avoid gel slippage. The storage modulus G' and the loss modulus G'' were obtained from temperature sweep experiments by cooling the systems from 55 to 5 °C and then reheating to 55 °C. Colling and heating were carried out at the rate of 18 °C h^{-1}. Measurements were performed at a constant frequency of 1.0 Hz. The strain amplitude was fixed at 0.02. The experimental conditions used were the same as those of Rochas,[9] Plashchina *et al.*[10] and Fernandes *et al.*[2] The sol–gel transition was defined by the cross-over of the viscoelastic moduli, G' and G'', as described by Cuvelier *et al.*[11] and Lin *et al.*[12]

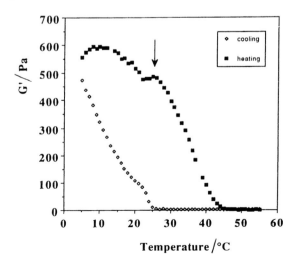

Figure 1   *Variation of* G' *as a function of temperature for a 4:1 κ-car + LBG mixture (total polysaccharide concentration = 1.0%)*[2]

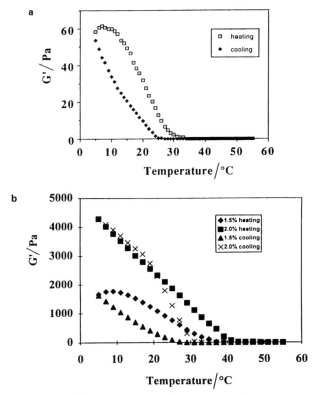

Figure 2   *Variation of G' as a function of temperature for κ-car concentration of (a)*
1.0%;[2] $T_g = 24$ °C; $T_m = 32$ °C; (b) 1.5% and 2.0%

3 Results

The stability of κ-car aggregates is enhanced by the presence of the added
galactomannan.[1,2] Clearly, the *melting* temperature T_m of the mixed gels is
influenced by the galactomannan: in contrast, the *gelation* temperature T_g is
not.[2] In this paper, we have studied the behaviour of the melting curves of the
κ-car + galactomannan mixed systems, since these curves show interesting
special details that may be important in understanding of the role of the
polysaccharides in the gel structure.

Table 1   *Transition temperatures T_g and T_m of κ-car systems*

κ-car concentration/%	T_g/°C	T_m/°C
1.0	24	32
1.5	28	36
2.0	32	41

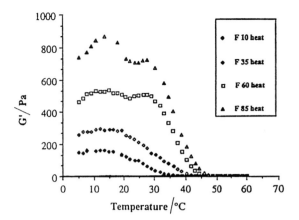

Figure 3   *Variation of* G′ *as a function of temperature (only the melting curve is shown) for 4:1 κ-car +galactomannan mixed systems (total polysaccharide concentration = 1.0%) for LBG fractions with different M/G ratios ranging from 3.0 to 4.9*

The G' variation during the cooling–heating cycle of the 1% 4:1 κ-car + LBG mixture is presented in Figure 1.[2] The thermal hysteresis is clearly shown. The T_g value is seen at 26 °C; T_m at 46 °C. We also see clearly a shoulder on the melting curve at around 20–30 °C. The cooling–heating cycles of 1%, 1.5% and 2% κ-car systems are displayed in Figure 2. The transition temperatures of these gels are presented in Table 1. As expected, T_g and T_m increase as the κ-car concentration increases. These values are in good agreement with Rochas[9] and Fernandes *et al.*[13] In the case of the blends the hysteresis is more pronounced than in case of κ-car. Another difference lies in the presence of a shoulder (shown by an arrow in Figure 1) for mixtures at approx. 20–30 °C in contrast with κ-car alone (the decrease of G' on the melting curve is smooth). In order to study the effect of LBG fractions on the thermal behaviour of κ-car gels, a LBG sample was fractionated and the fractions obtained were mixed with κ-car in the ratio 4:1 (carrageenan: galactomannan) and at 1 wt% total polymer concentration. The thermo-rheological behaviour of these blends is presented in Figure 3. The results displayed in Figure 3 are summarized in Table 2. We see that the melting temperature increases as the M/G ratio increases. Similarly, G' at low temperature increases with the ratio M/G. The shoulder at 20–30 °C is evident for M/G ratios higher than approx. 4. The effect of the addition of different amounts of LBG to a κ-car gel is shown in Figure 4. The melting temperature remains constant (44 °C) beyond a LBG content of 0.15%. The value of G' at low temperature increases strongly with the LBG content. The shoulder is still evident in the range 20–30 °C at a LBG content higher than 0.25%. Similarly, the effect of addition of different amounts of GG to a κ-car gel is displayed in Figure 5. The melting temperature remains constant (30 °C) beyond a GG

Table 2   *Parameters related to the gel→sol transition for mixed systems of κ-car + LBG fractions*

LBG fraction	M/G	T_m/°C	G'(at 5 °C)/Pa
F10	3.0	35	155
F35	3.8	42	264
F60	4.2	46	467
F85	4.9	47	755

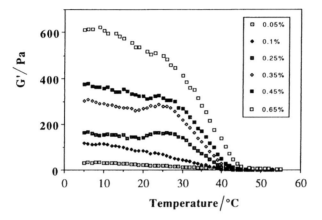

Figure 4   *Effect of the addition of different amounts of LBG (M/G = 4.0) to a 0.75% κ-car gel (only the melting curve is shown)*

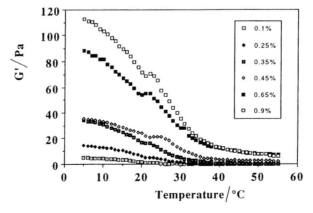

Figure 5   *Effect of the addition of different amounts of GG (M/G = 1.7) to a 0.75% κ-car gel (only the melting curve is shown)*

content of 0.4%. Values of G' at low temperature follow a pattern quite comparable to that of LBG. The shoulder is clearly seen at 20–25 °C for GG contents higher than 0.35%.

4 Conclusion

This set of data confirms our previous observations on the thermal behaviour of κ-car + galactomannan mixtures.[2] The melting temperature of κ-car gels seems mostly ascribed to the concentration of aggregates of κ-car helices.[9] The stability of κ-car aggregates is enhanced by the presence of the added galactomannan. Clearly, the melting temperature of the mixed gels is influenced by the galactomannan, particularly by the M/G ratio. In contrast, the gelation temperature is not. We have discussed this point previously:[2] these phenomena were interpreted on the basis of two interpenetrating networks, the major one arising from κ-car gelation, and the secondary one originating from galactomannan aggregation. The presence of a shoulder at approx. 20–30 °C was systematically observed and should not be considered fortuitous. This temperature range coincides with that of the gelation temperature from the gelling curves. This may provide an indication of the physical origin of such a phenomenon. If we assume that gelation takes place as soon as κ-car adopts a helical conformation, the shoulder may originate from the loss of the helical order of a few κ-car helices. Further studies are needed to elucidate the molecular interpretation of the shoulder in the melting curves of κ-car + galactomannan mixed gels.

References

1. D. A. Rees, *Adv. Carbohydr. Chem. Biochem.*, 1969, **24**, 276.
2. P. B. Fernandes, M. P. Gonçalves, and J.-L. Doublier, *Carbohydr. Polym.*, 1992, **19**, 261.
3. P. B. Fernandes, M. P. Gonçalves, and J.-L. Doublier, *J. Texture Stud.*, 1994, **25**, 267.
4. J. K. Seaman, in 'Handbook of Water-soluble Gums and Resins', ed. R. J. Davidson, McGraw-Hill, New York, 1980, p. 14.
5. M. P. Gonçalves, P. B. Fernandes, and J. Lefebvre, in 'Proc. II Int. Carob Symp.', ed. P. Fito and E. Mulet, Generalitat Valenciana, Conselleria i Pesca, Valencia, Spain, 1988, p. 407.
6. P. B. Fernandes, M. P. Gonçalves, and J.-L. Doublier, *Carbohydr. Polym.*, 1991, **16**, 253.
7. M. P. Gonçalves and J. A. L. Silva, *Food Hydrocolloids*, 1990, **4**, 277.
8. P. B. Fernandes, PhD Thesis, Portuguese Catholic University, Porto, Portugal, 1992.
9. C. Rochas, PhD Thesis, Scientific and Medical University of Grenoble, Grenoble, France, 1982.
10. I. G. Plashchina, I. R. Muratalieva, E. E. Braudo, and V. B. Tolstoguzov, *Carbohydr. Polym.*, 1986, **6**, 15.
11. G. Cuvelier, C. Peigney-Nourry, and B. Launa, in 'Gums and Stabilisers for the

Food Industry', ed. G. O. Phillips, D. J. Wedlock, and P. A. Williams, IRL Press, Oxford, 1990, vol. 5, 549.

12. Y. G. Lin, D. T. Mallin, J. C. W. Chien, and H. H. Winter, *Macromolecules*, 1991, **24**, 850.
13. P. B. Fernandes, M. P. Gonçalves, and J.-L. Doublier, *Food Hydrocolloids*, 1991, **5**, 71.

Whey Protein + Polysaccharide Mixtures: Polymer Incompatibility and Its Application

By A. Syrbe, P. B. Fernandes, F. Dannenberg, W. Bauer, and H. Klostermeyer[1]

NESTLÉ RESEARCH CENTRE, BP 44, VERS-CHEZ-LES-BLANC, 1000 LAUSANNE 26, SWITZERLAND
[1] FORSCHUNGSZENTRUM MILCH UND LEBENSMITTEL WEIHENSTEPHAN, TU MÜNCHEN, INSTITUT FÜR CHEMIE UND PHYSIK, VÖTTINGER STRASSE 45, 85354 FREISING, GERMANY

1 Introduction

The separation of polymer solutions into distinct liquid phases is a well-known phenomenon in synthetic polymer chemistry. A theoretical description in terms of statistical thermodynamics has been discussed in standard works.[1,2] There are two basic types of separation processes: 'complex coacervation' and 'incompatibility'. The former refers to an associative type of polymer interaction, concentrating both polymers in one phase, whereas the latter refers to polymer segregation into separate phases.[3] The balance between polymer–polymer and polymer–solvent interactions, along with the polymer molecular masses, controls the thermodynamic behaviour of ternary mixtures. Unequal polymer–solvent interactions generally increase polymer incompatibility and asymmetry of binodal and tie lines.[4] Higher molecular masses of the polymers also promote incompatibility.[5] The thermodynamic equilibrium of a ternary system is characterized by its phase diagram. Each point of a phase diagram represents one of the possible compositions of the three-component system. A binodal line marks the range of compositions leading to separation into two equilibrium phases. Tie lines connect pairs of equilibrium phases.

One of the first applications of incompatibility in bioscience was the use of aqueous two-phase systems of (semi)-synthetic polymers like dextran and polyethyleneglycol for the mild extraction of subcellular material.[6] 'Natural' ternary systems formed by water and proteins of polysaccharides may also show incompatibility. Phase diagrams of protein + polysaccharide and protein + protein mixtures have been evaluated by several analytical techniques.[6–11] Some empirical rules concerning biopolymer incompatibility are known.[7,8] Globular proteins behave very differently in their native and denatured states. Denaturation changes protein–protein interactions as well as the interactions with solvent or other biopolymers. Thus, incompatibility can arise with biopolymers that are miscible with the native protein.[12]

On mixing, the equilibrium phases of an incompatible polymer solution behave like water and oil, *i.e.* they form an unstable 'emulsion'. The flow pattern in the mixing device, the viscosity ratio and the interfacial tension of the two phases all determine the size and shape of the dispersed droplets.[13,14] When at least one of the phases can be gelled, the liquid emulsion may act as a precursor for an engineered food structure.[15] Recently, the food industry has shown an increased interest in biopolymer-based incompatible mixtures as a tool for food texture improvement. This approach offers an alternative to the mechanical structuration of precipitates as suggested by heated whey protein solutions.[16]

Studies on whey protein + polysaccharide mixtures have so far mainly focused on complex formation with anionic polysaccharides, *i.e.* on complex coacervation.[17] The present work establishes phase diagrams of incompatible aqueous whey protein + polysaccharide systems, and evaluates such systems with a view to 'structure engineering'.

2 Experimental

Whey protein isolate (WPI) was obtained from Le Sueur Isolates (USA). The α-lactalbumin (α-LA) and β-lactoglobulin (β-LG) fractions were purchased from Protose (Canada). All protein isolates were almost free of lactose, non-protein nitrogen, and fat. Insoluble protein was precipitated from the aqueous protein solution at pH 5 and separated by centrifuging (4 h, k-factor $\approx$ 200, $\langle RCF \rangle \approx 10^5\ g$). The top layer was also discarded and the remaining clear solution was freeze-dried and resolubilized as needed. The purified whey protein isolate contained 93 wt% soluble protein and 2 wt% ash. HPLC data indicated the following protein composition: whey protein isolate (80% β-LG; 15% α-LA; 5% bovine serum albumin, BSA), α-fraction (> 90% α-LA), β-fraction (> 95% β-LG). In some of the trials (whey protein isolate at different pH values, α-LA *versus* β-LG) the effect of varying ionic strengths was minimized by adjusting the protein solutions to the desired pH, dialysing them against distilled water, freeze-drying and resolubilizing them in a 1 wt% NaCl solution. The polysaccharides used were maltodextrin Glucidex 6 with a DE $\approx$ 6 from Roquette (France), dextran T2000 from Pharmacia (Sweden) and methylcellulose A4C and A15LV from Dow Chemical (USA). The κ-carrageenan in its K$^+$ form and guar gum were obtained from Sanofi Bio-Industries (France), xanthan gum from Jungbunzlauer (Austria) and arabinogalactan from Sigma (USA). Apple pectin and medium-viscosity sodium carboxymethylcellulose were purchased from Fluka (Switzerland). Methylcellulose (MC) solutions were prepared by dispersing the polysaccharide in hot distilled water ($\approx$ 90 °C) and cooling the suspension to 4 °C under stirring. All other polysaccharides were dispersed in distilled water at room temperature for about 1 h and the slurries heated to $\approx$ 90 °C for at least 30 min under stirring. No further purification was carried out. Whey protein will be further denoted as WP, polysaccharide as PS, dextran as D and maltodextrin as MD.

All solutions are expressed on a wt% basis. Clear stock solutions of WP and PS were mixed in different ratios and thoroughly stirred at constant temperature in isolated flasks. The mixtures became turbid as a result of emulsion droplet formation. Form formation had to be avoided to ensure good contact of the two phases. Samples were taken after *ca.* 5 h and *ca.* 15 h. The two phases were separated in a Beckman centrifuge (1–5 h, k-factor $\approx$ 2000, $\langle RCF \rangle \approx 10^4\ g$). (Too high a centrifugal field may lead to gelation of protein caused by gradient formation or even to the fractionation of the polydisperse system.) The tubes were sliced to avoid remixing of the two phases. Total protein content was determined by Kjeldahl analysis,[18] and total PS content by the phenol–sulfuric acid method.[19] The mass ratio of the equilibrium phases was also recorded for most of the samples to confirm the data of the chemical analysis. In agreement with earlier results,[20] no significant differences were found between samples taken at different intervals, indicating that phase equilibrium was already reached after 5 h. All measurements were carried out at least in duplicate. Relative deviation between duplicates was of the order of 3–5%.

According to the mass balance, points in a phase diagram representing a mixture and its associated equilibrium phases lie on a straight line. To account for slight deviations from this ideal behaviour, a linear fit was performed for each set of points. The binodal lines were fitted manually to the equilibrium phase compositions. The amount of WP fractions in the equilibrium phases and stock solutions was determined by reversed-phase-HPLC.[21]

Viscoelastic measurements were done on a Carri-Med CSL 100 rheometer using a parallel plate device (gap width 2 mm, plate diameter 6 cm). Heating and cooling was performed in the rheometer. Various amounts of PS were added to a 12 wt% solution of WP isolate. The liquid WP + PS mixtures were then poured onto the rheometer plate, covered with silicone oil, heated for 10 min from 25 to 80 °C, held at 80 °C for 20 min, and then rapidly cooled to 25 °C. After 5 h setting time, the G' and G'' moduli were determined at a frequency of 1 Hz and a strain amplitude of 0.01.

Light microscopy was performed using phase contrast optics. For transmission electron microscopy of heated mixtures, solid samples were cut into small cubes and liquid samples were encapsulated in agar gel tubes. Samples were fixed in phosphate-buffered 2.5% glutaraldehyde, post-fixed in 2% osmium tetroxide, dehydrated in a graded ethanol series, and embedded in an epoxy resis (Spurr). Ultra-thin sections were stained with uranyl acetate and lead citrate and observed in a Philips CM12 TEM.

3 Screening Incompatibility

The incompatibility of WP + PS mixtures was investigated at 20 °C for various neutral and anionic polysaccharides, pH values and polymer concentrations. As already known for other proteins,[7,8] a pH near the isoelectric point (pI) enhances protein–protein interaction, renders the protein less hydrophilic, and favours incompatibility. The ionic strength can enhance or reduce incompatibility by influencing the electrostatic interaction between similar (protein + protein, PS + PS) and dissimilar (protein + PS) molecules. Hence incompatibility of the native WP + PS mixture was first checked at pH 5 at an ionic strength given by the WPI ash content. When the system remained homogeneous, the pH test range was extended down to 3 or up to 8, and NaCl added to a maximum level of about 1 wt%. As the demixing of ternary polymer solutions occurs only above a specific polymer concentration threshold, the WP and PS stock solutions were prepared at the highest possible concentrations which avoided mixing problems. The following concentrations were used: native WP 20–25%, MC 5–10%, D 20%, MD 50% (or added directly as powder into the protein solution), and guar, pectin, κ-carrageenan, carboxymethylcellulose and xanthan 2%. The concentration and pH range for the trials with heat-denatured WP were further limited because of gel formation in the protein phase before mixing. We chose a protein stock solution of 8% WP slowly heated to 90 °C within 1 h at pH 7. As a consequence of the limited polymer concentration, the phase separation threshold may not have been reached in some of the mixtures. But mixtures that are even more concentrated and more viscous are of little interest from a practical point of view.

With native WP, incompatibility was observed for arabinogalactan, MD, D and MC at pH values near the pI. Increasing or lowering the pH prevented phase separation. Guar gum gave homogeneous mixtures. Homogeneous mixtures were also formed by pectin, κ-carrageenan, carboxymethylcellulose and xanthan at pH > pI. Precipitation occured with these anionic polysaccharides at pH < pI. Heat-denatured WP was incompatible with MC, carboxymethylcellulose, guar and κ-carrageenan at pH 7. Arabinogalactan, D and MD gave homogeneous mixtures in the accessible concentration range at pH 7.

4 Phase Diagrams of Whey Protein + Polysaccharide Systems

Figures 1(a) and 1(b) illustrate the emulsion-like structure of an incompatible WP + MC mixture. When the pH changes and the composition of the mixture comes close to the critical point, precipitates form as shown in Figures 1(c) and 1(d). This behaviour is accompanied by problems with separating the liquid phases which makes it impossible to obtain reliable analytical data around the critical point. Binodal lines are therefore fitted only to the range of real liquid–liquid phase separation (the mid-part is extrapolated).

The importance of PS excluded volume to WP + PS incompatibility is illustrated in Figure 2(a). The phase diagrams of three polysaccharides consisting of glucose monomers show two essential features: the binodal lines shift to the low concentration range, and the tie lines get flatter in the order MD > D > MC. The molecular dimensions of these biopolymers in aqueous

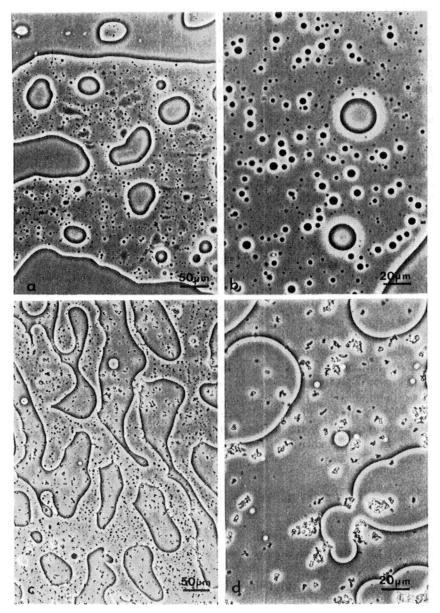

Figure 1   *Emulsion-like structure of an incompatible WP + MC A4C mixture after stirring: (a, b) liquid–liquid system at pH 5.1, (c, d) precipitates occuring at pH 4.45*

solution and their excluded volumes increase in the order MD < D < MC, as indicated by their intrinsic viscosities. A higher excluded volume leads to a greater loss in translational entropy on mixing, and this causes

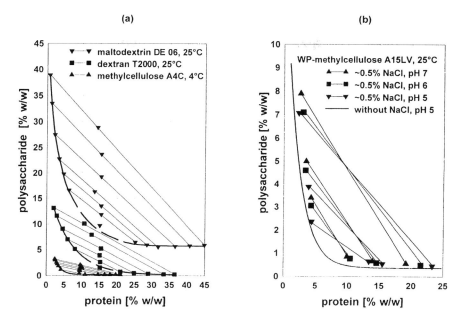

Figure 2   *(a) Phase diagrams of WP + PS systems: excluded volume effect of three different glucose polymers (pH 5). (b) Tie lines of WP + MC A15LV systems: solvent distribution at different pH values (pH 5–7, 25 °C, 0.5% NaCl)*

incompatibility to occur at lower polymer concentrations. At the same time, polymers with greater excluded volume 'imbibe' more solvent and show higher relative hydrophilicity and therefore flatter tie lines. The influence of pH on the phase separation behaviour of the WP is shown in Figure 2(b). As the pH is moved away from the pI, the proteins carry a higher net charge and so become better competitors for water. This is indicated by the increasing slope of the tie lines at higher pH.

Since pH is one of the key parameters for incompatibility of WP + PS mixtures, the WP main fractions, α-LA and β-LG, with their quite different pI values, are expected to exhibit different phase separation behaviour. Figure 3(a) gives an example of partition coefficients of α-LA and β-LG in an incompatible WP + MC mixture at pH 5. As the partition coefficient is a measure of the distribution of components among pairs of equilibrium phases, the lower partition coefficient of α-LA indicates a more symmetrical distribution between the two separated phases. Phase diagrams of mixtures of the individual whey proteins with the same polysaccharide at almost identical ionic strength, temperature and pH in Figures 3(b) and 3(c) actually show that the incompatibility of α-LA is less pronounced. The region of phase separation is smaller and the tie lines are steeper compared with β-LG. This conclusion may not apply to other pH values. Further results (not shown) suggest that the

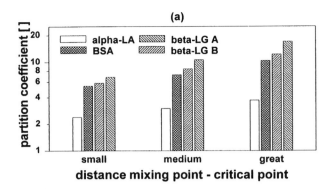

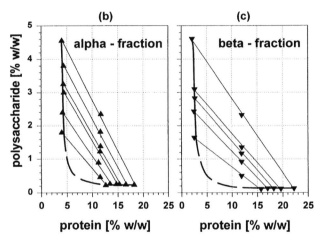

Figure 3   *Phase separation behaviour of WP fractions: (a) Partition coefficients of the main WP fractions in a WP + MC A4C system (pH 5, 4 °C); (b, c) phase diagrams of α-LA and β-LG + MC A4C systems (pH 5, 25 °C, 0.5% NaCl)*

difference between the pH of the mixture and the pI of the protein determines the degree of incompatibility. A pH of 5, which is close to the pI of β-LG, will favour incompatibility of β-LG. At pH 3.8 (closer to the pI of α-LA than to the pI of β-LG), α-LA is incompatible with MC, wheras β-LG gives homogeneous mixtures.

The fast polymerization and gelation of α-LA in concentrated aqueous solutions at pH < 4.5 complicates the observation of liquid–liquid phase separation in α-LA + PS mixtures: the phase separation process accelerates the gelation of α-LA by a concentration increase in the α-LA-rich phrase. If the gelation rate gets too high, solid-like protein precipitates are formed instead of dispersible liquid droplets.

5 Heating Whey Protein + Polysaccharide Mixtures

The different degree of incompatibility of native and denatured WP with polysaccharides can be demonstrated '*in situ*' when single-phase native WP + PS mixtures change into phase-separated systems on heating. Figure 4 shows

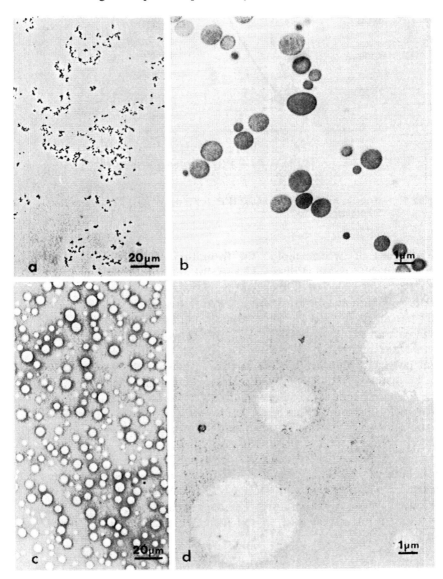

Figure 4   *Microstructure of one-phase WP + PS mixtures after heat treatment: (a, b) equilibrium phase of WP + MC A4C mixture (2.4% WP and 3.2% MC, pH 5, 2 min at 120 °C); (c, d) WP + κ-carrageenan mixture (10.7% WP and 0.1% κ-carrageenan, pH 7, 30 min at 90 °C)*

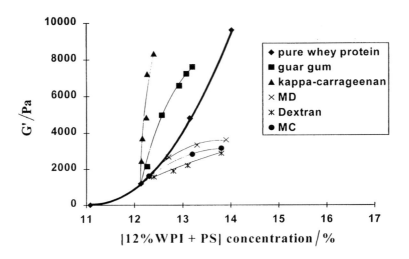

Figure 5   *Storage moduli of one-phase WP + PS mixtures after heat treatment (pH 7, 20 min at 80 °C)*

micrographs of two examples: the formation of gelled WP droplets in a continuous MC phase (Figures 4a and 4b) and the formation of a filled gel with PS beads in a continuous WP gel network, (Figures 4c and 4d). Both mixtures were homogeneous before the heat treatment. Phase separation not only has an impact on the microstructure, but also on the low deformation rheological data of mixed WP + PS gels. This is shown in Figure 5, where the storage moduli G' of WP + PS gels and a pure WP gel at the same total polymer concentration are compared. Two groups of polysaccharides can be distinguished. The heated mixtures containing κ-carrageenan or guar give a sharp increase of G' even at low concentrations. The micrographs show inclusions of PS-rich beads in the protein network. The water bound in these inclusions is withdrawn from the protein-rich phase and increases the WP concentration in the continuous phase. This and the very viscous or even gelled PS inclusions reinforce the overall structure. For MC, D and MD, however, G' remains well below the G' of a pure WP gel. No phase separation was observed with D and MD. In the case of MC, the inclusions are liquid at room temperature. The slight reinforcing effect of these three polysaccharides is only due to the increased protein concentration in the continuous phase.

Mechanical energy applied to a phase-separating system may modify the structural and rheological properties of the mixture. Figure 6 illustrates this effect. Droplets forming on gentle stirring of an incompatible native WP + D mixture start to coalesce and gel when the mixture is heated. Large aggregates give a gritty texture, (Figures 6a and 6b). If stirring is continued during heat treatment the maximum size of these aggregates is reduced and a smooth

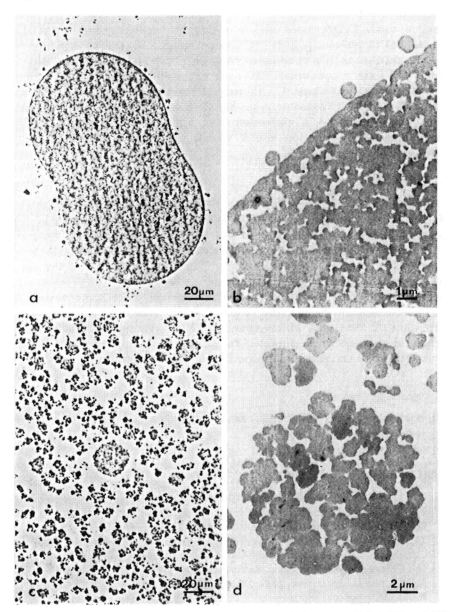

Figure 6   *Microstructure of a two-phase WP + D T2000 mixture (15% WP and 10% D) after combined heat and shear treatment at pH 5: (a, b) quiescent conditions; (c, d) low shear conditions*

texture is obtained (Figures 6c and 6d). The basic structural units, *i.e.* the original emulsion droplets, remain unchanged.

With regard to 'structure engineering', WP + MC mixtures offer several advantages; incompatibility occurs at low concentrations and MC may be used in food. Furthermore MC is a thermoreversible gelling agent. This means that at sufficiently high MC concentration a heated incompatible WP + MC mixture with a continuous MS-rich phase and a dispersed WP-rich phase will form a continuous MC-matrix, embedding the WP-rich droplets and preserving their shape. Continued heating also leads to gelation of WP. On cooling, the MC phase becomes liquid again and releases the irreversibly gelled protein particles. A suspension of smooth protein particles in a viscous shear thinning MC solution results, with a minimum of mechanical energy input needed.

6 Conclusions

The existence of incompatible WP + PS systems has been demonstrated. Native and heat-denatured whey proteins show strong differences in their interactions with polysaccharides. Native WP shows incompatibility only with neutral polysaccharides whereas heat-denatured WP was found to be incompatible also with anionic polysaccharides. The high net charge of the native protein at pH values far from the pI favours compatibility. The proteins α-LA and β-LG show different phase separation behaviour. When homogeneous WP + PS mixtures are heated, the onset of phase separation can lead to the formation of filled gels. Phase-separated WP + PS mixtures show microstructural changes when subjected to shear while being heated, and seem therefore to be a valuable tool for food structure engineering.

Acknowledgements

The authors thank C. Appolonia, M.-L. Dillmann, and Dr R. Jost and his team for their substantial assistance

References

1. P. J. Flory, 'Principles of Polymer Chemistry', Cornell University Press, Ithaca, NY, 1953, chs. 12 and 13.
2. H. Tompa, 'Polymer Solutions', Butterworths, London, 1956, chs. 4 and 7.
3. J. Th. G. Overbeek, 'Colloid and Surface Chemistry', MIT Video Course, MIT Center for Advanced Engineering Studies, Cambridge, MA, USA, 1974–1985.
4. C. C. Hsu and J. M. Prausnitz, *Macromolecules*, 1974, **7**, 320.
5. R. L. Scott, *J. Chem. Phys.*, 1949, **17**, 279.
6. P.-A. Albertsson, 'Partition of Cell Particles and Macromolecules', Wiley-Interscience, New York, 1971.
7. Ju. A. Antonov, V. Ya. Grinberg, and V. B. Tolstoguzov, *Nahrung*, 1979, **23**, 207, 597, 847.
8. V. I. Polyakov, O. K. Kireyeva, V. Ya. Grinberg, and V. B. Tolstoguzov, *Nahrung*, 1985, **29**, 153, 323.
9. V. I. Polyakov, V. Ya. Grinberg, and V. B. Tolstoguzov, *Polym. Bull.*, 1980, **2**, 757.
10. C. M. Durrani, D. A. Prystupa, and A. M. Donald, *Macromolecules*, 1993, **26**, 981.
11. S. Kasapis and E. R. Morris, *Carbohydr. Polym.*, 1993, **21**, 249.

12. V. I. Polyakov, I. A. Popello, V. Ya. Grinberg, and V. B. Tolstoguzov, *Nahrung*, 1986, **30**, 365.
13. J. O. Hinze, *AIChE J.*, 1955, **1**, 289.
14. H. P. Grace, *Chem. Eng. Commun.*, 1982, **14**, 225.
15. V. B. Tolstoguzov, in 'Food Structure—Its Creation and Evaluation', ed. M. V. Blanshard and J. R. Mitchell, Butterworths, London, 1988, p. 181.
16. N. S. Singer, J. Latella, and S. Yamamoto, European Patent Application 0 323 529, 1987.
17. A. Pegg, W. M. Marrs, and I. C. Dea, Leatherhead Research Report No. 703, 1992.
18. 'Official Methods of Analysis', Association of Official Analytical Chemists, 13th edn, 1980, p. 47.021.
19. M. Dubois, K. A. Gilles, J. K. Hamilton, P. A. Rebers, and F. Smith, *Anal. Chem.*, 1956, **28**, 350.
20. N. Zhuravskaya, E. V. Kiknadze, Yu. A. Antonov, and V. B. Tolstoguzov, *Nahrung*, 1986, **30**, 601.
21. P. Resmini, L. Pellegrino, J. A. Hogenboom, and R. Andreini, *Ital. J. Food Sci.*, 1989, **2**, 51.

Effect of Sodium Caseinate on Pasting and Gelation Properties of Wheat Starch

By C. Marzin, J.-L. Doublier, and J. Lefebvre

LABORATOIRE DE PHYSICO-CHIMIE DES MACROMOLECULES, INRA, BP 527, 44026 NANTES CEDEX 03, FRANCE

1 Introduction

Polysaccharides find extensive applications in the food industry as thickening or gelling agents for controlling the rheological and textural characteristics of food products. The use of blends of polysaccharides allows a widening of the range of applications or an improvement in functional properties, leading to significant cost savings. This arises usually from synergistic effects. The current interpretation of these effects assumes specific interactions between particular segments of the two macromolecules, as in the classical interpretation of the galactomannan + κ-carrageenan or galactomannan + xanthan synergies,[1] the cases which are the most thoroughly studied. However, another hypothesis now has firm experimental support and is gaining increasing favour which is based on the concept of thermodynamic incompatibility between the different macromolecules: each type of macromolecule is excluded from the volume occupied by the other, and this results in a phase separation process in the system.[2] Such a mechanism, which agrees with the great generality of phenomena observed in polysaccharide blends, is also at the origin of the behaviour of most mixed protein + polysaccharide systems.[3,4]

Starch + hydrocolloid mixtures constitute systems more complicated to study and to understand than blends of hydrocolloids in the molecular state. Hydrocolloids have the capacity to modify the properties of starch systems by affecting the phenomena occurring at the granular level during pasting and gelatinization, as well as through their interactions in the aqueous phase with starch macromolecules leached out of the granules. It is well known that addition of a polysaccharide such as guar gum, locust bean gum, xanthan or carrageenan influences strongly the gelatinization and retrogradation of starch, as well as the rheological properties of starch pastes and gels; large increases in viscosity have been described,[5] but visco-elasticity appears to be less dramatically affected.[6]

340

Despite the fact that starch and casein are involved in many food formulations, the investigation of their possible interactions, never mind the potential practical consequences, has hardly been undertaken. The few previously published studies are limited in their scope and methods, and the separate studies have been carried out under quite different experimental conditions;[7,8] they have led, moreover, to conflicting results. Obviously, some more systematic and thorough investigations are needed to assess, analyse and then understand the effects of casein on the behaviour of starch systems.

This work reports on the effect of sodium caseinate on the pasting and gelatinization properties of wheat starch, studied through the modifications of the swelling and solubility patterns and of the rheological behaviour of the systems (flow behaviour and visco-elasticity of the pastes, gelation kinetics and visco-elasticity of the gels). In the range of concentrations used in this work, sodium caseinate does not contribute directly to any appreciable extent to the rheology of the systems; any rheological changes which occur must be an indirect consequence of caseinate–starch interactions.

2 Experimental

The sodium caseinate sample (95.6% protein; 580 mg of calcium per kg) was provided by Kerry Ingredients (Ireland). Wheat starch (27% amylose) was obtained from Roquette Frères (France).

A Brabender Viscograph E instrument was employed using the standard pasting procedure, which comprised three stages: heating the dispersion of starch granules from 30 to 96 °C at a rate of 1.5 °C min^{-1}, then maintaining it for 10 min at 96 °C, and finally cooling it to 80 °C at the rate of 1.5 °C min^{-1}; during the whole cycle the sample was sheared between the pins of the measuring system (rotation speed 75 r.p.m.). These were relatively mild heating rates and shear conditions.

At the end of the pasting process, the samples were immediately transferred into the cup (which was kept at 60 °C) of a coaxial cylinder measuring device of a Contraves Rheomat 120 viscometer. Depending on the consistency of the paste, the measuring device was either a DIN 145 (diameter of the cup 49 mm; diameter and height of the inner cylinder 43 mm and 78 mm) or a DIN 125 (25, 24, and 45 mm). Each sample was submitted to three successive shearing cycles at 60 °C as follows:

1st cycle, the shear rate $\dot{\gamma}$ was increased linearly with time from 0 to 660 s^{-1} in 2 min, and then decreased to 0 s^{-1} at the same rate;
2nd cycle, identical to the first one;
3rd cycle, the shear rate $\dot{\gamma}$ was decreased logarithmically with time from 600 to 0.1 s^{-1} in 2 min (this logarithmic ramp allows a better characterization at low shear rates).

The flow properties were studied at wheat starch concentrations of 5–10 wt%. The sodium caseinate concentration was either 0 or 7.5 wt%.

To study the viscoelasticity of starch pastes and gels, we used a Carri-Med CSL50 constant stress rheometer operated with cone-and-plate geometry (cone angle and diameter, 4° and 5 cm) in the oscillatory mode. At the end of the pasting process, the samples were immediately transferred onto the rheometer plate at 60 °C, the plate was raised, and the sample was covered with a layer of paraffin oil in order to avoid evaporation. Three series of measurements were then performed in succession at a fixed strain of 4% as follows.

1. Mechanical spectrum of the paste: the storage and loss moduli G' and G'' were measured at 60 °C over the frequency range 10^{-2}–10 Hz.
2. Gelation kinetics: after quenching of the sample at 25 °C, the evolution of G' and G'' measured at 1 Hz was monitored for 15 h. At the end of this period, the moduli reached nearly constant values, therefore allowing us to proceed to step 3.
3. Mechanical spectrum of the gel: the moduli G' and G'' of the gelled sample were measured at 25 °C over the frequency range 10^{-2}–10 Hz.

Viscoelasticity measurements were carried out for systems containing 6 wt% wheat starch; when caseinate was added, its concentration was 7.5 wt% in the system.

Swelling and solubility patterns were obtained on 0.5 wt% wheat starch dispersions pasted under the same conditions as above; when caseinate was added, the caseinate concentration was 0.6%, *i.e.* the caseinate:starch ratio (1.2:1) was close to the value used in the visco-elastic measurements (1.25:1). Aliquots of the starch dispersion were withdrawn at 5 °C intervals during the heating period, and then 10, 20 and 30 min after the 96 °C plateau had been reached. After rapid cooling, 8 ml samples were centrifuged for 10 min at 800 *g*. The volume (V_2, ml) of the supernatant and the weight (W, mg) of the sediment were determined. The content of starch material in the supernatant (C_2) was determined by the phenol–sulfuric acid method using water or a solution of caseinate at the appropriate concentration as blanks; with this procedure, protein interference was found to be minimal. The content in protein of the supernatant was obtained from its optical density at 277 nm, corrected for turbidity, using an absorption coefficient of $A^{1\%}{}_{1cm} = 8.6$. The solubility index S (%) and the swelling index G (g g^{-1}) of the starch were then calculated:[9,10]

$$S = 100 V_2 C_2 / C_1 \qquad\qquad G = P/(8C_1 - V_2 C_2) \qquad\qquad (1)$$

where C_1 is the starch concentration in the aliquot determined from its dry matter content and the initial starch/caseinate ratio.

3 Results

Effect of Caseinate on Flow Behaviour of Starch Pastes

Typical flow curves for a 6 wt% wheat starch paste are shown on Figure 1. When submitted to the first shearing cycle, the paste exhibits a highly shear

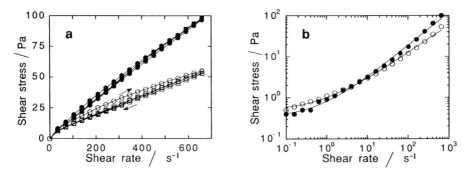

Figure 1   *Flow behaviour at 60 °C of a 6% wheat starch paste with (solid symbols) and without (open symbols) 7.5% added sodium caseinate. Shear stress is plotted against shear rate: (a) first (circles) and second (squares) shearing cycles; (b) third shearing cycle (curves were obtained by fitting eqn (1) to the results)*

thinning behaviour and a hysteresis loop; in the second cycle, the shear thinning character is kept, but the hysteresis loop is no longer observed, indicating that some state of equilibrium is reached after the first cycle (Figure 1a). The third cycle, plotted in double logarithmic co-ordinates, allows then a closer examination of the behaviour at low shear rates; this points to the existence of a yield stress (Figure 1b), a generally observed feature of starch paste flow curves.[11] The third shearing cycle flow curves have therefore been fitted with a Herschel–Bulkeley equation,

$$\sigma = \sigma_0 + K\dot{\gamma}^n \qquad (2)$$

where σ_0 is the yield stress value, K is the consistency index, and n is the power law exponent. Starch concentration does not qualitatively change the flow behaviour of the pastes; its effect on the Herschel-Bulkeley parameters is reported in Figure 2. An increase in starch concentration results in an (exponential) increase in σ_0 and K and a (linear) decrease in n, indicating that the more concentrated the starch dispersion, the more structured the system remains after shearing.

When pasting of a 6% wheat starch dispersion is affected by the presence of caseinate (7.5%), the viscosity of the paste in the $100–600\ \text{s}^{-1}$ shear-rate range is largely enhanced during the two first shearing cycles, but the behaviour appears to be much less shear thinning, and the thixotropic character disappears almost completely (Figure 1a). Figure 1b shows that, in fact, for the third cycle, there is a cross-over between the flow curve for the blend and that for the pure starch. The first one is situated below the second for shear rates lower than $\dot{\gamma}_c$ (with $\dot{\gamma}_c \approx 5\ \text{s}^{-1}$ for the 6% starch, 7.5% caseinate blend), and *vice versa* for $\dot{\gamma} > \dot{\gamma}_c$. The value of $\dot{\gamma}_c$ is higher the higher the starch concentration is; but the effect of caseinate remains qualitatively the same whatever the starch concentration. The presence of caseinate leads to

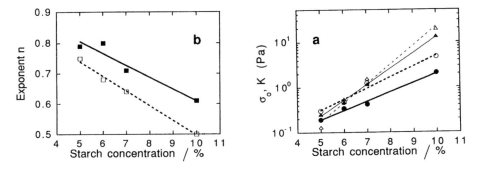

Figure 2   *Variation with starch concentration of (a) the yield stress σ₀ (circles) and the consistency index K (triangles) and (b) the exponent n (squares) for wheat starch pastes prepared with (solid symbols) and without (open symbols) 7.5% added sodium caseinate. The value of parameters σ₀, K and n were obtained by fitting eqn (2) to the results of the third shearing cycle at 60 °C*

a noticeable decrease in σ_0, and in K to a lesser extent, and it causes a moderate increase of the exponent n (Figure 2).

Effect of Caseinate on the Visco-elasticity of Starch Systems

At the end of the pasting procedure, wheat starch pastes examined at 60 °C exhibited a high degree of structure as shown by their gel-like mechanical spectra which displayed an elastic plateau over the experimental frequency range, despite the rather low value of the modulus G' (Figure 3a). The shape of the mechanical spectrum is not changed in the presence of caseinate; its main effect is to shift the G' and G'' curves slightly down and up, respectively, and there seems to be in addition a move of the curves along the frequency axis in the direction of higher frequencies (Figure 3a). These shifts indicate that the presence of caseinate does not change the type of visco-elastic behaviour exhibited by the system, but it does cause a moderate decrease in the magnitude of the elasticity.

After quenching the starch system at 25 °C, G' increases with time following an S-shaped curve (Figure 3b). It is known that this phenomenon is due to the gelation of that fraction of amylose which is leached out of the granules during pasting.[12] The spectrum of the final system (*i.e.* after the phenomenon has levelled off) is of the gel type and it shows G' values much higher—10-fold in the example shown in Figure 3c—than those measured on the paste at 60 °C, as a result of the contribution of amylose gelation in the extra-granular phase. The addition of caseinate accelerates considerably the amylose gelation kinetics but decreases the final plateau value of G' (Figure 3b). Its effect on the mechanical spectrum of the final system is, as far as G' is concerned, the same in direction and magnitude as for the pastes at 60 °C: that is, we see an approximately two-fold decrease over the frequency range (Figure 3c).

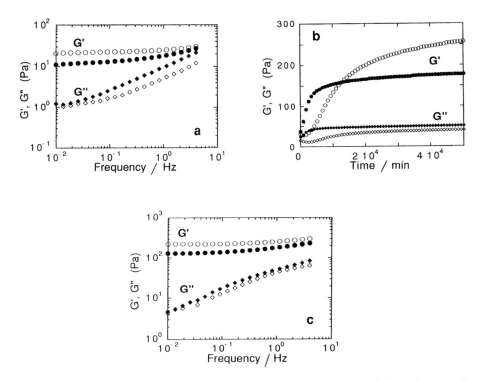

Figure 3   *Visco-elasticity of a 6% wheat starch system with (solid symbols) and without (open symbols) 7.5% sodium caseinate: (a) mechanical spectrum of the paste at 60 °C; (b) gelation kinetics after quenching at 25 °C, monitored by measuring G′ and G″ at 1 Hz; (c) mechanical spectrum at 25 °C of the gel after 15 h of ageing*

Effect of Caseinate on Swelling and Solubilization Patterns of Wheat Starch

The swelling and solubility patterns of wheat starch are shown on Figure 4. The patterns obtained in the presence of caseinate are also reported on the same plots. In the absence of caseinate, the swelling and solubility curves are typical of the temperature dependence of cereal starches. Swelling of starch granules proceeds in two stages, beginning at approx. 55 °C and 75 °C, respectively, and solubilization of granular material into the aqueous phase remains low until a temperature of 80–85 °C is reached. At the end of the pasting process, the swelling and solubility indices have reached the values 15% and 25%, respectively. It is known[11] that the material solubilized is essentially amylose.

Caseinate has a drastic effect on both the swelling and the solubility curves. The second swelling step is almost suppressed and the final swelling index is divided by three. The solubility curve keeps the same shape as in the absence of caseinate, but the values of the solubility index are divided roughly by two. The quantitative results indicate that caseinate is almost entirely (98%) recovered in the supernatant; it does not penetrate the swollen granules.

4 Discussion

By way of simplification, cereal starch pastes can be viewed as suspensions of swollen, deformable particles in a macromolecular solution which forms the continuous aqueous phase.[11] The swollen granules, out of which amylose has been more or less completely leached out, and which have lost most of their inner structure, but nevertheless keep part of their integrity after pasting, constitute the particles; the continuous phase can be considered as an amylose solution. This description of the pastes is well established for 'normal' starches; it is, in particular, consistent with the results of swelling and solubility measurements during pasting and also with microscopic measurements.

The flow behaviour of such systems will be determined principally by the volume fraction Φ of the dispersed phase, the concentration C_s of macromolecules in the continuous phase, and (when Φ or the shear-rate increases) by the deformability and shear resistance of the swollen particles. The first two parameters can be obtained from the results of swelling and solubility determinations:[11]

$$\Phi = [1 - (S/100)]CG \qquad C_s = CS/[100(1 - \Phi)] \qquad (3)$$

If we assume that the values of S and G determined for the wheat starch and caseinate concentrations used in the swelling and solubility experiments (caseinate:starch ratio 1.2:1) remain valid at higher concentrations, the values of Φ and C_s in the pastes submitted to rheological measurement can be calculated, provided the caseinate/starch ratio is (about) the same.

We find using this procedure the values $\Phi = 0.71$ and $C_s = 50$ mg g^{-1} at the end of the pasting process for a 6% starch suspension. In the presence of 7.5% caseinate (caseinate:starch ratio 1.25:1), Φ and C_s drop to 0.28 and 8.9 mg g^{-1}, respectively. This should result in a dramatic decrease in the viscosity of the pastes, but the opposite overall effect is observed. The lowering of the yield stress of the pastes induced by caseinate is consistent with the decrease in Φ. The reduction of the thixotropic character of the wheat starch pastes observed in the presence of caseinate can be qualitatively explained by the reduction in starch granule deformability and mechanical fragility arising from the diminution of their degree of swelling. However, to interpret the consequences of the addition of caseinate on the flow parameters K and n, it has to be considered that some kind of interactions are developed between caseinate and amylose within the continuous phase (the results show that caseinate does not penetrate into swollen wheat starch granules). The visco-elasticity of the pastes at 60 °C is primarily determined by the same parameters, Φ and C_s; in particular, the weak gel character is mainly due to the high value of Φ in the suspension.[6] In the presence of caseinate, the observed shifts in the G' and G''

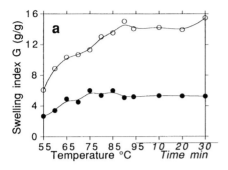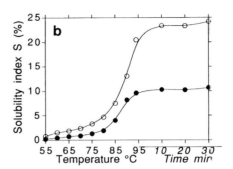

Figure 4   *(a) Swelling pattern and (b) solubility pattern of wheat starch pasted in the absence (open symbols) and in the presence (solid symbols) of sodium caseinate (caseinate:starch ratio: 1.20:1)*

frequency-dependence curves are compatible with the decrease in both Φ and C_s values, but they are much less important than expected, probably due to an antagonistic effect of caseinate–amylose interactions in the continuous phase.

Finally, the effect of caseinate on the gelation kinetics of wheat starch pastes provides a clear demonstration of the presence of a caseinate–amylose interaction. As already mentioned, the phenomenon which is monitored is in fact amylose gelation within the continuous phase. The gelation of amylose in water is known to be induced by a microphase-separation process leading to crystalline junction zones.[13,14] Caseinate apparently accelerates the phase separation process. The decrease in G' values observed can be ascribed to the decrease in amylose concentration C_s in the presence of caseinate. We have observed an accelerating and enhancing effect of caseinate on amylose gelation with model amylose systems (results to be published). We have also shown in previous studies that this kind of phenomenon occurs in the case of polysaccharide biopolymer + starch systems,[6] and is due to an incompatibility between amylose and other polysaccharides resulting in a phase separation within the continuous phase.[15] However, such polysaccharides, unlike case-inate, do not seem to modify starch swelling and solubility, and their effect on the flow behaviour is due mainly to their direct contribution to the viscosity of the aqueous phase;[5] this could explain why the modifications of starch paste and gel visco-elasticity, which they induce, are much smaller than those brought about by caseinate.

5 Conclusions

In the presence of sodium caseinate, the pasting and gelling processes of wheat starch and the rheological properties of the resulting systems undergo profound changes. During pasting, caseinate is concentrated in the continuous phase of the suspension. In this phase, it co-exists with amylose. Because of the incompatibility of caseinate with starch components, swelling and solubility of starch are strongly reduced. Besides, this incompatibility induces a phase separation within the continuous phase. The outcome of these phenomena is a noticeable change in the rheological properties of the systems (pastes and gels) caused by the presence of caseinate.

The interpretation outlined above must necessarily be considered at this stage as a working hypothesis. Since the properties of starch pastes and gels are known to depend largely upon the pasting conditions, other preparation procedures are to be tested.[11] While the overall behaviour can be expected to be the same for other normal cereal starches, a comparison with other types of starches would be enlightening. The structure of wheat starch granules swollen in the presence of caseinate in the continuous phase should be investigated; in addition, the possibility of molecular associations between casein proteins and amylose cannot be ruled out at this point and this has to be explored. Finally, the state of aggregation of casein in the system, which was not controlled in our experiments, could play some role; this point also deserves future attention. Work is currently in progress along these lines.

Acknowledgement

This work is part of a study funded by the EC (contract AIR1 CT92 0245).

References

1. I. C. M. Dea, E. R. Morris, D. A. Rees, E. J. Welsh, H. A. Barnes, and J. Price, *Carbohydr. Res.*, 1977, **57**, 249.
2. J.-L. Doublier, C. Castelain, and J. Lefebvre, in 'Plant Polymeric Carbohydrates', ed. F. Meuser, D. J. Manners, and W. Seibel, Royal Society of Chemistry, Cambridge, 1993, p. 76.
3. V. B. Tolstoguzov, in 'Functional Properties of Food Macromolecules', ed. J. R. Mitchell and D. A. Ledward, Elsevier Applied Science, London, 1986, p. 385.
4. C. Castelain, J. Lefebvre, and J.-L. Doublier, *Food Hydrocolloids*, 1986, **2**, 141.
5. M. Alloncle, J. Lefebvre, G. Llamas, and J.-L. Doublier, *Cereal Chem.*, 1989, **66**, 90.
6. M. Alloncle and J.-L. Doublier, *Food Hydrocolloids*, 1991, **5**, 455.
7. A.-M. Hermansson, 'Protein/starch Complex', US Patent 4 159 982, 1979.
8. J. Lelièvre and J. Husbands, *Starch/Stärke*, 1989, **41**, 236.
9. H. W. Leach, L. D. MacCowen, and T. J. Schoch, *Cereal Chem.*, 1959, **36**, 534.
10. J.-L. Doublier, *Starch/Stärke*, 1981, **33**, 415.
11. J.-L. Doublier, G. Llamas, and M. Le Meur, *Carbohydr. Res.*, 1985, **135**, 257.
12. S. G. Ring, *Stärke*, 1985, **37**, 80.
13. M. J. Miles, V. J. Morris, and S. G. Ring, *Carbohydr. Res.*, 1985, **135**, 257.
14. H. S. Ellis and S. G. Ring, *Carbohydr. Polym.*, 1985, **5**, 201.
15. J.-L. Doublier and G. Llamas, in 'Food Colloids and Polymers: Stability and Mechanical Properties', ed. E. Dickinson and P. Walstra, Special Publication No. 113, The Royal Society of Chemistry, Cambridge, UK, 1993, p. 138.

Colloidal Stability and Sedimentation of Pectin-Stabilized Acid Milk Drinks

By T. P. Kravtchenko, A. Parker, and A. Trespoey

SANOFI BIO INDUSTRIES, RESEARCH AND DEVELOPMENT, LABORATOIRE DE BAUPTE, 50500 CARENTAN, FRANCE

1 Introduction

Acid milk drinks exist all over the world as fruited milk, yoghurt drinks, buttermilk, whey drinks, kefir, *etc.* They are characterized by a pH ranging from 3.6 to 4.5 and by a viscosity close to that of milk. They are usually prepared from yoghurt. Upon slow bacterial acidification of milk, casein micelles become aggregated to form a three-dimensional network of chains and clusters. When examined by electron microscopy,[1,2] casein particle chains look like interconnected 'strings of pearls' where each pearl is a rather spherical casein particle. The manufacture of acid milk drinks requires the destruction of this network by a strong shear treatment such as homogenization. Despite the complexity of the events which occur during the acidification, acid milk drinks can therefore be considered as a suspension of casein particles.

A common fault of such products is 'wheying off', *i.e.* the formation of a clear layer at the top of the drink on storage. Indeed, near the isoelectric point of casein, the van der Waals (and electrostatic) attraction dominate and casein particles aggregate as sedimenting clusters. For more than 30 years high methoxyl pectin has been known[3] to prevent the sedimentation of casein micelles and thus has been used as a stabilizer.

Despite the very widespread commercial interest in acid milk drinks, little has been written about the mode of action of the pectin in this application. The present paper deals with the mechanism by which pectin stabilizes casein particles in acid milk drinks.

2 Material and Methods

The pectin sample was a commercial high methoxyl pectin (AYD31 type) from Sanofi Bio Industries (Paris, France).

Acid milk drinks were prepared from non-fat 'Taillefine' yoghurt made by Gervais-Danone (France). Yoghurt gels were homogenized at 200 bar with a

Mini-Lab 8.30VH homogenizer (Rannie, Denmark) and immediately added to the aqueous pectin solution of the proper concentration in the ratio 2:1. Final acid milk drinks contained 8% of non-fat milk solids. They were stored at 4 °C. The degree of flocculation was determined after 3 weeks storage by measuring the height of the wheying off. Sediment volumes were determined after centrifugation at 13 000 *g* for 30 min using non-heparinated tubes (1.5 × 75 mm) and a haematocrit centrifuge (Hermle). The particle-size distribution was measured by low-angle laser diffraction using a Malvern Mastersizer (45 mm focal length lens, presentation code 0403) after 2 weeks storage at 4 °C. Relative viscosities were measured with an Ubbelhode capillary viscometer (flow time for pure water 90 s). The flow time of the solvent was measured using the supernatant of acid milk drinks after centifugation at 13 000 *g* for 30 min.

3 Mechanism of Stabilization

In an acid milk drink, the volume of clear whey decreases with increasing pectin content[4] up to a threshold value that we have called the critical pectin level (CPL). In other words, the CPL is the minimum amount of a given pectin required to avoid whey separation on storage of the acid milk drink.

Microscopic examination of acid milk drinks[4,5] shows that, in the absence of pectin, casein particles form large aggregates whereas with sufficient pectin, *i.e.*

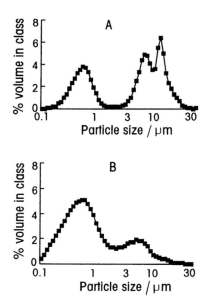

Figure 1   *Particle-size distribution of (A) a non-stabilized acid milk drink and (B) acid milk drink stabilized by the addition of 0.25% AYF31 pectin*

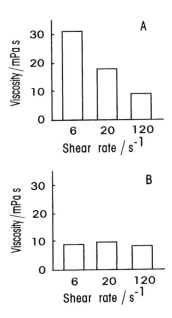

Figure 2   *Flow behaviour of (A) a non-stabilized acid milk drink and (B) an acid milk drink stabilized by the addition of 0.25% AYD31 pectin*

above the CPL, casein particles appear to be well separated and homogeneously distributed. In a previous communication,[6] it was shown that the addition of increasing pectin concentrations up to the CPL provokes several correlated physical effects: the average size of casein particles (Figure 1) decreases and the rheology of the acid milk drink (Figure 2) changes from highly shear-thinning and thixotropic to almost Newtonian. All these experimental observations converge to indicate that pectin expresses its stabilizing effect by eliminating casein aggregation, *i.e.* preventing the formation of flocs that can readily sediment.

Van Hooydonk *et al.*[7] proposed that pectin limits aggregation of casein particles and the subsequent sedimentation of casein aggregates by increasing the viscosity of the whey. The existence of an apparent adsorption isotherm[6] shows that part of the pectin added is actually adsorbed onto the surface of casein particles, indicating that pectin may act as a protective colloid stabilizer rather than a simple thickening agent. According to Glahn[8] and Amberg-Pedersen and Jorgensen,[9] the stabilizing effect of pectin may be due to interparticle repulsion caused by the charge acquired by the casein particles on adsorption of negatively charged pectin. However, it was shown[6] that the zeta potential of casein particles in acid milk drinks containing up to 0.2% of pectin is insufficient to provide the required repulsion force. Moreover, at this pH, the carboxyl groups of pectin are only partially dissociated. It has therefore been suggested[6] that the pectin chains attached to the surface of casein particles 'protect' them from flocculation by a mechanism of steric repulsion. Further evidence for this explanation has been sought by studying the sediment volume of acid milk drinks on centrifugation.

4 Centrifugation Volume of the Sediment

To establish the adsorption isotherm of pectin on casein particles, the free pectin fraction was quantified in the clear supernatant of the acid milk drinks after fast centrifugation. It has been observed (Figure 3) that the amount of sediment increases with increasing pectin content. Usually, the stronger the attractive forces are, the less dense is the sediment: sticky particles associate as loose flocs of high fractal dimension whereas non-interacting particles can slide past each other to form a dense sediment. If the explanation given above for the stabilizing effect of pectin is correct, *i.e.* considering the casein as particles decreasing in adhesiveness with increasing pectin content, the sediment volume would be expected to be larger for non-stabilized acid milk drinks than for stable ones. One very significant difference between the measurements presented here and usual sediment volume measurements is the high centrifugation speed. Under the effect of high centrifugation acceleration, it is likely that any casein network would collapse, so that the effect of flocculation on the sediment volume would no longer be seen. The increase in sediment volume is therefore explained by an increase of the so-called 'voluminosity'[5] of casein particles. In terms of the steric repulsion theory, it is likely that the increase in sediment volume is due to the presence of an adsorbed pectin layer.

Assuming that casein particles are spherical and monodisperse, it is possible to estimate the thickness of the adsorbed layer of pectin as a function of the casein particle radius. The sediment volume can be expressed as $V = kNv$, where k is the volume fraction occupied by the particles, N the number of particles and v the volume of each fundamental particle. In a perfectly packed sediment, the volume fraction occupied by spherical particles is independent of their size and is close to 70%. However, in a series of acid milk drink samples,

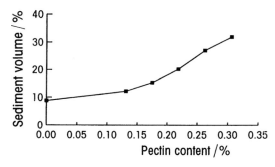

Figure 3   *Increase of sediment volume, on centrifugation at 13 000* **g**, *as a function of the AYD31 pectin content*

the number of fundamental casein particles is kept constant. On the addition of pectin, the increase of sediment volume can therefore only be explained by an increase of the particle volume. Figure 3 shows that the sediment volume is increased by a factor of 2.5 when pectin content increases from zero to the CPL. From the above demonstration, it can be concluded that the volume of the particles has been increased by a factor 2.5, *i.e.* their average radius has been increased by a factor equal to the cube root of 2.5 ($= 1.36$).

In yoghurt made from heated milk, the diameter of casein particles is of the order of 300 nm.[10] At the CPL value, the adsorbed pectin layer may thus be as large as 55 nm. Published values for the pectin molecular size are extremely variable. However, it may be assumed that the number of galacturonic acid monomers per chain is of the order of 1000. The length of each galacturonic acid unit has been determined to be 0.437 nm[11] which makes a total chain length of the order of 450 nm. As pectin chains are rather stiff,[11,12] it seems possible that the adsorbed layer is indeed as thick as 55 nm. Such a thickness is not unreasonable: layers of similar thickness have been reported in the literature for comparable systems. For example, Yokoyama *et al.*[13] measured the adsorbed layer thickness of gum tragacanth on polystyrene latex beads and found a thickness of about 60 nm.

5 Relative Viscosity Measurements

Another approach for measuring the effective adsorbed layer thickness is by viscosity measurements. If we consider acid milk drinks as a dispersion of hard spheres, the relative viscosity η_{rel} can be approximated by the Krieger–Dougherty relation[14]

$$\eta_{rel} = \left(1 - \frac{\phi}{p}\right)^{-[\eta]p}$$

where ϕ is the volume fraction of particles, including their adsorbed layer, $[\eta]$ is the intrinsic viscosity, and p an adjustable parameter.

Figure 4 shows the changes in the relative viscosity of acid milk drinks prepared with increasing amounts of pectin. The viscosity of the dispersing medium is that of the clear supernatant obtained by centrifugation at 13 000 *g*. The flow behaviour of the clear serum is almost Newtonian (results not shown).

The relative viscosity increases from 4.3 to 5.5 cP when increasing the pectin content from 0 to 0.25%. Such an increase in relative viscosity is probably due to an increase of the particle volume fraction,[14] *i.e.* an increase of the hydrodynamic volume of the fundamental particles. Figure 5 shows the increase in volume fraction as calculated by the Krieger–Dougherty formula using 2.5 (the Einstein value) for the intrinsic viscosity and 0.68 (a high shear-rate value) for the p parameter. Values of volume fraction calculated from viscosity data are larger than those found by the centrifugation technique. At low pectin content, this difference may be explained by the stickiness of casein

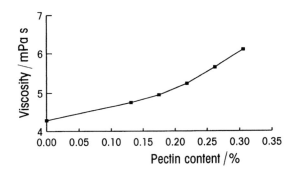

Figure 4   *Increase in relative viscosity of acid milk drinks containing increasing amounts of AYD31 pectin*

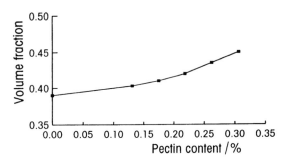

Figure 5   *Increase of the volume fraction occupied by the casein–pectin complexes, as calculated with the Krieger–Dougherty formula, with increasing amounts of AYD31 pectin*

particles: even under the effect of high shear rate, the presence of some aggregates leads to an increase of the viscosity. Above the CPL, casein particles do not stick to one another and the high viscosity cannot be explained by the presence of aggregates. Under such conditions, the difference between the volume fractions estimated by centrifugation and viscometry may be due to some compressibility of the adsorbed pectin layer: under the effect of high acceleration, the pectin layer is compressed, and the hydrodynamic volume of the casein–pectin complexes therefore appears to be smaller than that found with less perturbing methods such as viscosity measurement. This last point suggests that, at ambient hydrostatic pressure, the thickness of the adsorbed pectin layer may be even larger than our earlier estimate of 55 nm. Anyway, the models we have used are probably too simple to describe such complex systems as casein micelles: casein–pectin complexes really cannot be considered as solid spheres.

6 Conclusion

The steep increase of sediment volume with increasing pectin content of acid milk drinks submitted to fast centrifugation, which is a quite unusual behaviour, has been explained on the basis of the theory of polymer adsorption. The explanation reinforces the model which supposes that pectin stabilizes casein particles by steric repulsion rather than by electrostatic repulsion. Pectin adsorbs at the surface of casein particles and forms a thick layer. Stability is attained when the thickness and the density of the pectin layer are large enough to maintain casein particles at a distance large enough to prevent aggregation by the van der Waals attraction.

Not all pectins behave in the same way, for example because of their variable degree of methoxylation. Some become efficient as low dosage whereas others require higher dosage to provide acceptable stability. However, the correlation between the pectin concentration for a sediment volume of 25% and the CPL appears to be general. A novel and rapid application test has therefore been suggested for measuring the stabilizing power of pectins in acid milk drinks. It is based on the centrifugation at 13 000 *g* of a series of standard acid milk drinks containing increasing amounts of pectin. The CPL is the lowest pectin content providing a relative sediment volume of 25%.

References

1. I. Heertje, J. Visser, and P. Smits, *Food Microstruct.*, 1985, **4**, 267.
2. M. Kalab, P. Alla-Wojtas, and B. E. Phipps-Todd, *Food Microstruct.*, 1983, **2**, 51.
3. J. J. Doesburg and L. de Vos, 5th Int. Fruit Juice Congress, Vienna, 1959, p. 32.
4. A. C. M. van Hooydonk, L. Smalbrink, and H. G. Hagedoorn, *Voedingsmiddelen Technol.*, 1982, **17**, 23.
5. P. Walstra, *J. Dairy Res.*, 1979, **46**, 317.
6. A. Parker, T. P. Kravtchenko, and P. Boulenguer, in 'Food Hydrocolloids: Structures, Properties and Functions', ed. K. Nishinari and E. Doi, Plenum, New York, 1993, p. 307.
7. A. C. M. van Hooydonk, L. Smalbrink, H. G. Hagedoorn, and J. C. E. Reitsma, *Voedingsmiddelen Technol.*, 1982, **20**, 25.
8. P. E. Glahn, *Prog. Food Nutr. Sci.*, 1982, **6**, 171.
9. H. C. Amberg-Pedersen and B. B. Jorgensen, *Food Hydrocolloids*, 1991, **5**, 323.
10. M. Kalab, D. B. Emmons, and A. G. Sargant, *Milchwissenschaft*, 1976, **31**, 402.
11. D. A. Rees and A. W. Wight, *J. Chem. Soc. B*, 1971, 1366.
12. I. G. Plashchina, M. G. Semenova, E. E. Braudo, and V. B. Tolstoguzov, *Carbohydr. Polym.*, 1985, **5**, 159.
13. A. Yokoyama, K. R. Srinivasan, and H. S. Fogler, *J. Colloid Interface Sci.*, 1988, **126**, 141.
14. R. J. Hunter, 'Foundations of Colloid Science', Oxford University Press, 1986, vol. 1, p. 538.

Decrease of *In Vitro* Hydrolysis of Soybean Protein by Sodium Carrageenan

By J. Mouécoucou, C. Villaume, H. M. Bau, A. Schwertz, J. P. Nicolas, and L. Méjean

INSERM U.308, 38 RUE LIONNOIS, 54000 NANCY, FRANCE

1 Introduction

Carrageenans are sulfated polysaccharides obtained from marine red algae of the *Rhodophyceae* family. Alginates are linear polysaccharides obtained from marine brown algae of the *Phaeophyceae* family. Both are used in the food industry as gelling, thickening and stabilizing additives.[1]

Technological treatments of foods can produce the formation of cross-linkages between hydrocolloids and food components such as protein, causing a relative protein deficiency. In a previous study, we have observed[2] that the ingestion of heated mixtures of soybean flour and sodium carrageenan leads to a decrease in protein efficiency ratio and to a growth retardation in rats, whereas addition of sodium alginate to soybean flour has no effect. In addition, we have demonstrated[3] that methionine supplementation of heated soybean flour containing carrageenan prevents these effects.

The aim of the present study is to show the possible existence of physico-chemical interactions between hydrocolloids and soybean proteins using viscometry and the influence of hydrocolloids on protein digestibility using a dialysis cell method.

2 Material and Methods

Apparent viscosities of mixtures of soybean flour with 0–3% sodium alginate or sodium carrageenan were measured at 37 °C and pH 6.8 with a Calorifuged Viscosimeter (VT 180 Gebruder, Haake). Samples containing 10% protein were suspended in distilled water and stirred for 1 h before measurements.

In vitro digestion was performed according to the method developed by Savoie and Gauthier[4] with minor modifications.

Pepsin (P 7012), trypsin (Tr 8253), papain (P 3375) and bromelain (B 2252) were purchased from Sigma (Saint Quentin-Fallavier, France). Soybean flour with or without 3% sodium alginate or sodium carrageenan and containing 40 mg of protein nitrogen (N × 6.25) was suspended in 16 ml of 0.1 N HCl (50 ppm thimerosal). The pH was adjusted to 1.8 and the volume adjusted to 19 ml with distilled water. The suspension was treated with 1 ml of pepsin solution in 0.1 N HCl (enzyme:substrate = 1:250 by wt) for 30 min at 37 °C. The mixture was alkalanized (pH = 8.2), the volume was adjusted to 22 ml with distilled water, and the mixture was poured in the dialysis bag (Spectra Por® 6, relative molecular mass cut-off 1000; Bioblock, Strasbourg) of the digestion cell. Trypsin solution (1 ml) in 0.2 M phosphate buffer (pH 8.2) was added. The ratio enzyme:substrate was 1:25 by wt. Digested products were collected after circulation of a 0.2 M phosphate buffer solution at pH 8.2 each hour for 6 h.

Papain and bromelain digestion were allowed to take place without preliminary hydrolysis by pepsin. The mixtures of soybean flour and sodium alginate or sodium carrageenan were suspended in 16 ml of 0.2 M phosphate buffer (pH 6.2) at 37 °C. The pH was adjusted to 6.2 and the volume was made up to 22 ml with distilled water, and each mixture obtained was poured into a dialysis bag. Papain or bromelain (1 ml; enzyme:substrate = 1:250) was added. The digested products were collected as previously described except for the circulating buffer solution which was now phosphate buffer at pH 6.2. Nitrogen content of the digested products was analysed by the Kjeldahl method. In all cases, the nitrogen assay was made on a 20 ml sample of digested products which had been mineralized with 5 ml of 36 N H_2SO_4.

3 Results

As can be seen from Figure 1, the viscosity of soybean flour mixed with increasing levels of carrageenan did not change. On the contrary, the addition of alginate to soybean flour leads to a substantial increase in apparent viscosity.

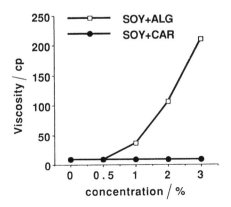

Figure 1 *Apparent viscosity of soybean flour (SOY; 10 wt%) + various concentrations of sodium carrageenan (CAR) or sodium alginate (ALG)*

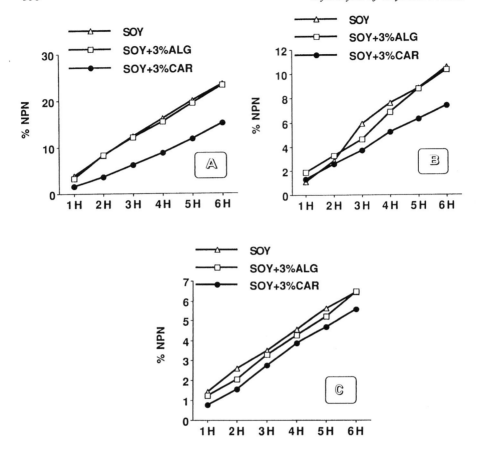

Figure 2   *Effect of (A) trypsin, (B) bromelain and (C) papain on time-dependent nitrogen release from soybean flour (SOY) + 3% sodium carrageenan (CAR) or soybean flour + 3% sodium alginate (ALG) (NPN = non-protein nitrogen)*

Figure 2 shows that the digestion by pepsin + trypsin was very effective on the soybean flour alone and released 23% of non-protein nitrogen. The digestions by bromelain and papain released 10% and 6.5%, respectively. The addition of 3% alginate to soybean flour had no effect on the digestion in the three cases. The addition of 3% carrageenan to soybean flour reduced the nitrogen released by 42% with trypsin, 30% with bromelain, and 14% with papain.

4 Discussion

Our results show that the decrease of protein efficiency ratio and growth retardation previously observed in rats fed on a soybean flour + carrageenan diet cannot be attributed to diet viscosity, but rather to diet digestion. The viscosity of the mixture of soybean flour + sodium carrageenan does not vary significantly with increasing carrageenan level, whereas the viscosity of the mixture of soybean flour + sodium alginate increases with increasing alginate levels. But alginate has no *in vivo* effect, in contrast to carrageenan. However, the results of digestion by trypsin, bromelain and papain, using the dialysis cell, illustrate the existence of a decrease of the nitrogen release with the soybean flour + carrageenan mixture and no change with the soybean flour + alginate mixture. The effectiveness of the proteolysis inhibition caused by carrageenan seems to be related to the effectiveness of the enzymic action on soybean alone. The weak hydrolysis levels observed with bromelain and papain could be due to the lack of a preliminary treatment with pepsin.

The inhibition of proteolysis in the presence of carrageenan has been shown by several authors.[5,6] It occurs whatever the pH, and is due to a binding of carrageenan to the protein. Our results show that the degree of inhibition is related to the percentage of carrageenan added.

The delay in nitrogen release observed *in vitro* when carrageenan is added to the diet agrees with the previous *in vivo* observations.[2,3] Soybean proteins are deficient in methionine, and so the proteolysis decrease caused by added carrageenan could lead to a decrease in the amount of methionine released during the digestion, thus producing an unbalanced diet deficient in methionine.

References

1. J. L. Doublier and J. F. Thibault, 'Additifs et auxiliaires de fabrication dans les industries agro-alimentaires', Tech. Doc., Paris, 1984, p. 305.
2. J. Mouécoucou, C. Villaume, H. M. Bau, J. P. Nicolas, and L. Méjean, *Repr. Nutr. Dév.*, 1991, **31**, 377.
3. J. Mouécoucou, C. Villaume, H. M. Bau, J. P. Nicolas, and L. Méjean, *J. Sci. Food Agric.*, 1992, **60**, 361.
4. L. Savoie, and S. F. Gauthier, *J. Food Sci.*, 1986, **51**, 494.
5. W. Anderson, *J. Pharm. Pharmacol.*, 1961, **13**, 139.
6. N. F. Stanley, *Prog. Food Nutr. Sci.*, 1982, **6**, 161.

Gels and Networks

The Importance of Biopolymers in Structure Engineering

By Anne-Marie Hermansson

SIK, THE SWEDISH INSTITUTE FOR FOOD RESEARCH, PO BOX 5401, S-40229
GOTHENBURG, SWEDEN

1 Introduction

The purpose of many food processes is to create a structure which gives the
product its characteristic properties. A specific food product can be made
using a variety of formulations and processes, which differ in their demand
on the functional ingredient. This is quite evident in the low fat area, where
the fat content is to be reduced without changing the characteristics of the
actual food product. In spite of this, there is a lot to learn about the role of a
functional ingredient and the final product quality. Without such knowledge
it is difficult to develop and optimize processes for the manufacture of
processed food.

Biopolymers make up the backbone structure in many traditional solid and
semi-solid foods such as meat, fish, dairy and cereal products. However, when
we consider the importance of biopolymers in structure engineering, we focus
on the potential of biopolymers to assist in generating new tailor-made
structures with specific properties of the aqueous phase, interfaces, the fat
phase, and the overall behaviour of multiphase systems. Up to now, basic
studies have been made in model systems where one mechanism at a time has
been studied under controlled conditions.

In order to understand fully the functionality of biopolymers in multiphase
food systems, the whole complexity needs to be approached by multivariate
techniques. It is also important to be aware that the behaviour of biopolymers
on several spatial levels has to be taken into account. Today tools are available
to evaluate the contribution from studies of single molecules, and interactions
on the molecular and the supramolecular levels, up to gel networks and phase-
separated systems.[1]

It is not at all possible to cover all aspects of biopolymer functionality in this
short paper. Instead, a few examples will be given in order to illustrate how
biopolymers influence the microstructure of multiphase systems such as water-

in-oil (W/O) emulsions and whipped emulsions of reduced fat content, and how the gel structure and rheological properties of single and mixed gels of some polysaccharides and proteins can be manipulated.

2 Emulsions and Foams

There is increasing interest in controlling the structures of products with reduced fat contents. The possibilities of replacing fat depend on the specific function of the fat, and this varies from one type of product to another. Different approaches have to be taken for the production of low-fat whipped products, water-continuous, and fat-continuous products. In whipped products a crucial function of the fat is to stabilize the air–water interface in the form of crystalline particles, and the fat has to be reduced without loss of stability, product volume, mouthfeel or texture. Low-fat spreads can be fat-continuous, water-continuous, or bicontinuous (*i.e.* having both aqueous and fat continuous phases) and the distribution of phases has an impact on attributes such as appearance, flavour, mouthfeel, texture and microbiological stability.

Effects of Caseins on W/O Emulsions

The aqueous phase is bound to play a more important role in low-fat spreads than in full-fat margarines. Little information is available at present on what properties the aqueous phase should possess in order to obtain the desired sensory characteristics of the final product. There are large variations in both the size and shape distribution of the aqueous phase of commercial products of similar composition.[2] The most commonly used proteins today in low-fat products are the milk proteins, in combination with gelatin or polysaccharides. Milk proteins are known to have a destabilizing effect on W/O emulsions, and it has previously been shown that addition of 1 wt% milk proteins to 3 wt% gelatin results in a considerable increase in the particle size of the aqueous domains (*i.e.* regions) in a spread with 40% fat.[2–4]

In a recent study, the addition of 5 and 7 wt% milk powders and caseinates to 2.5 wt% gelatin in the aqueous phase of a spread with 40 wt% fat was evaluated. The results showed no increase in the size distribution of the aqueous domains when the concentration of skim milk or buttermilk powder was increased from 5 to 7 wt% in the aqueous phase.

Figure 1(a) shows the distribution of the aqueous phase with 7% buttermilk. The domains are spherical and relatively small. A fraction of very small droplets, as well as bigger droplets of *ca.* 10 μm, can be seen. Addition of caseinate results in bigger aqueous domains, which increase with concentration regardless of the type of caseinate (low viscosity, high viscosity, low or high Ca content). Figure 1(b) shows the distribution for an aqueous phase containing 7 wt% caseinate with big domains as well as small droplets. The results illustrate how the distribution of the aqueous phase is influenced by the type of milk protein used. However, the complexity of the multiphase system must

a   b

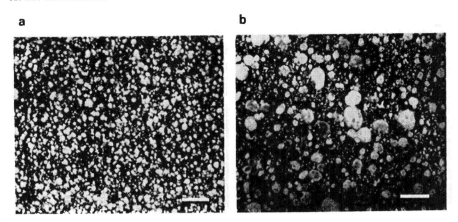

Figure 1   *Light micrograph of a spread containing (a) 7 wt% buttermilk powder + 2.5 wt% gelatin in the aqueous phase; and (b) 7 wt% sodium caseinate + 2.5 wt% gelatin in the aqueous phase (bar = 50 μm)*

also be borne in mind. Recent results obtained by Johansson *et al.*[5] show that milk proteins also affect the behaviour of fat crystals and make them more polar. Variations in processing conditions will, of course, also influence microstructure and product properties.

Electron microscopy reveals differences in the distribution of proteins in the aqueous phase. Figure 2(a) shows a freeze-etched replica of a gelatin droplet.[2] The filamentous gelatin network structure is quite easy to distinguish from any milk protein structure. Figure 2(b) shows part of one of the small droplets in the system containing gelatin + milk powder discussed above. The fracture plane reveals part of the interface as well as the interior of the droplet. Casein micelles can be seen at the interface, and the interior of the droplet is composed of a gelatin network. Thus, there is a distinct phase separation of the two protein components. The same observation was made for small droplets in the gelatin + caseinate systems. However, in the larger aqueous domains, the milk proteins were uniformly distributed and, for example, casein micelles could be seen throughout these aqueous domains. The principles underlying the distribution of biopolymers within the various aqueous domains needs to be better understood, as well as its impact on the quality of fat-continuous products.

Effect of Caseinate on Whipped Emulsions

It is well known that crystalline fat has a stabilizing role in whipped creams, ice-creams and toppings.[6,7] However, the exact mechanism varies depending on the ingredients and the process. In whipped creams the air bubbles are stabilized by adsorption of fat globules, whereas adsorption of big fat crystals gives rise to defective foams with poor whipping properties. This is not the case for the whipped emulsion to be discussed here.

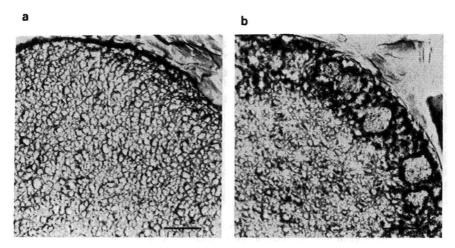

Figure 2   *Part of a droplet (a) containing gelatin; and (b) with 7 wt% skim milk powder + 2.5 wt% gelatin in the aqueous phase (bar = 200 nm)*

A comparison has been made of whipped emulsions with varying ratios of caseinate to emulsifiers (ethoxylated sorbitan ester + sorbitan monostearate) at two fat levels. Figure 3(a) shows an air bubble in the reference sample with 1.5 wt% caseinate, 0.39% emulsifiers and 20 wt% fat (coconut/palm kernel oil). The air bubble is covered with small fat globules partly fused together, as well as fat crystals. This sample gives rise to a stable foam at a fat content of 20%. Figure 3(b) shows part of an air bubble where there is a surplus of caseinate due to removal of 80% of the emulsifiers. It can be seen that the fat globules repel each other and adsorb less well to the interface. This foam is unstable. When a foam is made with a surplus of emulsifiers and the caseinate content is reduced by 50%, there is a strong interaction between the fat globules, as illustrated by Figure 3(c). The globules also seem to interact in the serum phase, which gives rise to a rigid foam. The storage modulus of this foam is three times as high as for the reference foam. When the fat level is reduced to 16%, the ratio of caseinate to emulsifiers has to be reduced in relation to the reference sample in order to maintain the stability of the foam. This example illustrates how the balance between caseinate and emulsifiers controls the structure of a whipped emulsion, and that the balance has to be adjusted according to the fat content. The more caseinate there is, the more the fat globules repel each other.

3 Gels

Gels are responsible for many functional properties, such as water holding, diffusion and rheological properties, and so the gel-forming properties of biopolymers have received a lot of attention. There are several gelation mechanisms for proteins as well as for polysaccharides, and a wide variety of

a   **b**

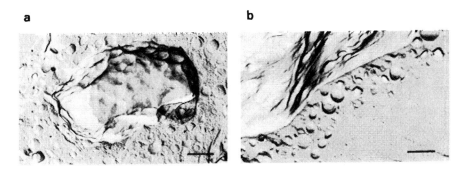

c

Figure 3   *Freeze-etched replicas of air bubbles in whipped emulsions stabilized by caseinate and emulsifiers (ethoxylated sorbitan ester and sorbitan mono-stearate): (a) reference sample; (b) surplus of caseinate in relation to emulsifiers; (c) surplus of emulsifiers in relation to caseinate (bar = 0.5 μm)*

network structures. Even a single biopolymer can be made to form quite different gels by the choice of environmental and processing conditions. Synergistic effects can be obtained by mixing biopolymers, and determining the reasons for these effects is a focus of current research. Electron microscopy provides a valuable tool for the characterization of single biopolymers, supramolecular assemblies, gel networks, and complex multiphase structures [8]

Examples are given below from studies of κ-carrageenan showing how differences can be obtained in the supramolecular network structure, resulting in large variations in rheological properties. Not only factors such as pH and salt concentration, but also factors such as processing conditions, can have a major impact on the gel structure and rheological properties. This is illustrated by the effect of the heating rate on whey protein gels.

Supramolecular Structures of Gel Networks

κ-Carrageenan is a fraction from naturally occurring sulfated galactans of red algae. Gelation occurs on cooling and has been attributed to a two-stage

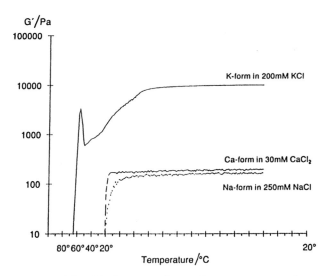

Figure 4   *Storage modulus G' of cationic forms of κ-carrageenan gels as a function of temperature*

reaction involving a coil–helix transition followed by aggregation.[9,10] The coil–helix transition is strongly dependent on the concentration and type of cation.[10] Of the cations commonly associated with κ-carrageenan, potassium is far more efficient for gelation than either sodium or calcium. The potassium form of κ-carrageenan is highly dependent on salt concentration, and strong gels are formed at high KCl concentrations, whereas the weaker gels of the sodium and calcium forms are relatively independent of the sodium and calcium concentrations.[11,12] Figure 4 shows the storage moduli of the three cationic forms of κ-carrageenan: 30 mM Ca^{2+} and 250 mM Na^+ are required for the coil–helix transitions of the calcium and sodium forms, whereas only 7 mM K^+ is required for the potassium form.[10]

In the presence of 50 mM potassium and above, a transient state is found on cooling. The structure of the transient state consists of a fine network, and the junction zones are believed to be double helices. When the temperature is further lowered, the structure becomes unstable, and helices associate into rigid rods. These rods are the building blocks in the formation of a coarse supramolecular structure, where strands are aligned to form the long stiff superstrands typical of the potassium-induced gelation of κ-carrageenan. Figure 5(a) shows the typical stiff superstrands of the potassium form in 100 mM KCl. A very fine structure can be seen in the background as well. The balance between the fine and the coarse structure varies with the KCl concentration and is a reason for the salt dependence of the gel strength.

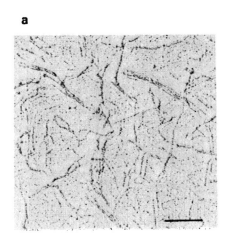

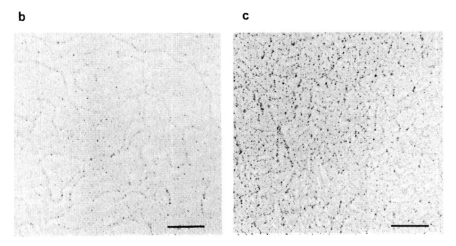

Figure 5   *Images of κ-carrageenan: (a) the potassium form in 100 mM KCl; (b) the sodium form in 250 mM NaCl; and (c) the calcium form in 30 mM CaCl₂; prepared by the mica sandwich technique for TEM (bar = 100 nm)*

The gels formed in the presence of sodium ions are relatively weak and not dependent on the sodium ion concentration to the same extent as the potassium-induced gel. Figure 5(b) shows strands of the network of the sodium form in 250 mM NaCl. This structure is composed of flexible superstrands of a constant thickness, whereas the potassium form has a variable thickness. Similarly to the potassium-induced gels, the network is formed by alignment of at least two double helices or assemblies thereof, which deviate and align with each other.

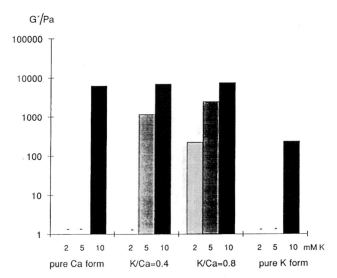

Figure 6   *Storage modulus G' of pure and intermediate forms of κ-carrageenan showing synergistic effects on addition of KCl (from reference 12)*

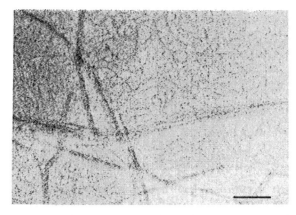

Figure 7   *The structure of an intermediate form of κ-carrageenan with a K/Ca ratio of 0.8 in 5 mM CaCl₂ showing a mixed network structure (bar = 100 nm)*

The pure calcium κ-carrageenan also gives rise to weak gels in the concentration range 30–100 mM. Salting-out effects are observed at higher concentrations. The calcium-induced gel has a different structure from the gels induced by sodium or potassium. Figure 5(c) shows the calcium-induced gel structure. The gel is composed of a very fine network structure without any of the superstrands typical of the sodium or potassium-induced forms of κ-carrageenan. These results illustrate the wide variety of network structures giving rise to large differences in rheological properties that can occur even within one well-defined polysaccharide system.

Mixed gels give rise to interesting properties and are of more importance to foods, which are seldom composed of individual biopolymers. Even single biopolymers can form mixed network structures. As mentioned above, the strongest potassium-induced κ-carrageenan gels were composed of a mixture of a fine- and a coarse-stranded network. Pronounced effects can be obtained with mixtures of potassium and calcium ions, by adding calcium ions to the calcium form, or by making intermediate ionic forms.[12] Depending on the ratio of cations, it is possible to make 1% κ-carrageenan gels with a storage modulus varying from 70 to 43 000 Pa at 20 °C. It is also possible to manipulate the melting temperature for gels of similar strength, *e.g.* 70 to 35 °C. This can be of importance when the melting characteristics of a product have to be considered. Figure 6 shows the effect of KCl addition on the storage modulus of pure and intermediate forms, and Figure 7 shows a corresponding micrograph of an intermediate form. The structure is composed of a mixture of the fine-stranded network typical of the calcium form and the coarse supramolecular structure typical of the potassium form.

Interactions between κ-carrageenan and other biopolymers such as caseins and galactomannans can give rise to synergistic effects with regard to stability and rheological properties. The mechanisms behind these effects are far from understood. It is, for example, not known how the supramolecular structures of carrageenan can be affected by the addition of another biopolymer. As discussed above, quite different structures can be obtained depending on the choice of cation. We are therefore interested in seeing how the different cationic forms of κ-carrageenan interact with other biopolymers.

A recent study has been made of mixed gels of the potassium, sodium and calcium forms of κ-carrageenan + locust bean gum by dynamic visco-elastic

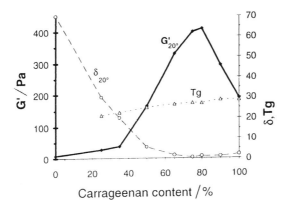

Figure 8   *Gelation temperature T_g, storage modulus s, and phase angle δ of 1 wt% mixed gels of K-κ-carrageenan + galactomannan in 8 mM KCl (from reference 13)*

measurements.[13] Synergistic effects are only observed for mixed gels with K-κ-carrageenan in ≤ 100 mM KCl. Figure 8 shows an example of the synergistic behaviour of K-κ-carrageenan in 8 mM KCl. Gels of K-κ-carrageenan + locust bean gum do not show synergistic effects at high potassium concentration (200 mM), and the sodium and calcium forms of κ-carrageenan do not show synergistic effects with locust bean gum at any concentration. The exact reason for the synergism with potassium but not with the sodium or calcium forms on the supramolecular level is not yet fully elucidated.

In another ongoing project, interactions between κ-carrageenan and caseins are being studied. Published results in this area are contradictory,[14–16] but no studies have previously been made of how casein affects the supramolecular structure of κ-carrageenan. The results obtained so far reveal that the presence of caseinate or casein fractions hampers the association of helices into superstrands regardless of the ionic form of κ-carrageenan, probably by association to the helices. Figure 9 shows micrographs of the coarse super-strands of 0.01 wt% K-κ-carrageenan in 100 mM KCl at pH 7.0 in comparison with the same solution of 0.01 wt% K-κ-carrageenan + 0.005 wt% sodium caseinate. Hardly any superstrands can be seen in the presence of sodium caseinate. This result clearly illustrates that addition of one

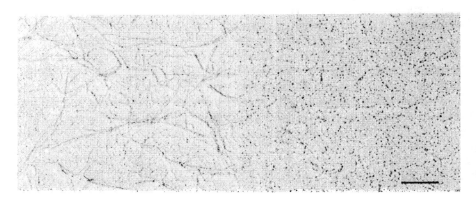

Figure 9   *The supramolecular structures of 0.01 wt% K-κ-carrageenan (left) and 0.01 wt% K-κ-carrageenan + 0.005 wt% sodium caseinate (right) in 100 mM KCl at pH 7 (bar = 100 nm)*

biopolymer can have a major effect on the supramolecular structure of another biopolymer. The nature of the synergistic effect with κ-carrageenan + locust bean gum is less straightforward.

Particulate Gels

Particulate protein systems are important in many applications, especially in the dairy field but also in applications of physically modified protein ingredients and in the development of low-fat products. Particulate protein

a   b

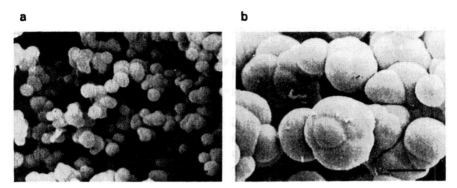

Figure 10 *SEM micrographs of 10 wt% β-lactoglobulin gels at pH 5.3 heated at (a) 12 °C min⁻¹ and (b) 1 °C min⁻¹ (bar = 2 µm)*

a   b

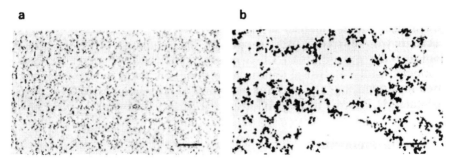

Figure 11 *Light micrographs of 10 wt% β-lactoglobulin gels at pH 5.3 heated at (a) 12 °C min⁻¹ and (b) 1 °C min⁻¹ (bar = 50 µm)*

gels can form at intermediate pH levels where the net charge is low, but this range can be extended by addition of salt or other proteins, or by pre-denaturation of the proteins. Variations in the particle size, strand thickness, pore size and heterogeneity of the gel structure have a major influence on rheological and water-holding, as well as sensory properties.

Relationships between the microstructure, and physical and sensory properties of β-lactoglobulin, as well as mixtures of whey proteins, have been studied in our laboratory.[17-19] The nature of particulate networks is determined by environmental factors such as pH and salt concentration and by processing factors such as shear and heating conditions. The results illustrate that the gelation process itself can offer unique possibilities for controlling the structure, and of generating products with the desired properties. Figures 10(a) and 10(b) show scanning electron microscopy (SEM) micrographs of 10 wt% β-lactoglobulin gels at pH 5.3 formed at two different heating rates. The structures differ both with regard to the particle size and the way the particles have fused together into the network structure. Figures 11(a) and 11(b) show corresponding light micrographs at low magnification. We can see a substantial change in the pore-size distribution due to a change in the heating rate.

The structure parameters contribute in different ways to rheological measurements at small and large deformations. Fracture properties correlate well with the pore-size distribution, large pores being the weakest parts of the gel network at large deformations. The difference in stress at fracture for unnotched samples varies between 25 kPa for the gel formed at the high heating rate and 7 kPa for the gel formed at the low heating rate with large pores. Quite different results are obtained in the small deformation tests. The storage modulus is highest for the gel formed at the lowest heating rate. The visco-elastic properties depend mainly on the composition of the strands for the particular β-lactoglobulin gels. Flexible strands, formed of particles linked together like a string of beads at high heating rates, give rise to a low storage modulus, whereas the stiffer and thicker strands, consisting of particles fused together, have a higher storage modulus.[18]

The next step is to correlate structure parameters to sensory properties. The preliminary results obtained so far suggest that different structure parameters correlate with different sensory properties. Image analysis provides a valuable tool for the quantification of structure parameters. If image analysis is combined with a statistical approach, information about the structure can be used for optimizing processes, for the choice of raw materials, and for product formulation, in order to achieve the desired sensory quality of the product.

Acknowledgements

This review is based on work done in the Biophysics Section at SIK. I would especially like to thank Elvy Jordansson, Maud Langton, Mats Stading, Ewa Eriksson, Annika Altskär and Ina Storm for their contributions.

References

1. 'Physical Techniques for the Study of Food Biopolymers', ed. S. B. Ross-Murphy, Blackie, Glasgow, 1993.
2. A.-M. Hermansson and E. Jordansson, in 'Gums and Stabilisers for the Food Industry', ed. G. O. Phillips, P. A. Williams, and D. J. Wedlock, IRL Press, Oxford, 1992, vol. 6, p. 409.
3. D. P. J. Moran, *Dairy Ind. Int.*, 1990, **55**, 41.
4. A. C. Juriaanse and I. Heertje, *Food Microstruct.*, 1988, **7**, 181.
5. D. Johansson, B. Bergenståhl, and E. Lundgren, *J. Am. Oil Chem. Soc.*, 1994, in the press.
6. B. E. Brooker, *Food Struct.*, 1993, **12**, 115.
7. W. Buchheim and P. Dejmek, in 'Food Emulsions', ed. K. Larsson and S. E. Friberg, Marcel Dekker, New York, 2nd edn, 1990, p. 203.
8. A.-M. Hermansson and M. Langton, in 'Physical Techniques for the Study of Food Biopolymers', ed. S. B. Ross-Murphy, Blackie, Glasgow, 1993, p. 277.

9. V. J. Morris, in 'Functional Properties of Food Macromolecules', ed. J. R. Mitchell and D. A. Ledward, Elsevier Applied Science, London, 1986, p. 121.
10. C. Rochas and M. Rinaudo, *Biopolymers*, 1980, **19**, 1675.
11. A.-M. Hermansson, *Carbohydr. Polym.*, 1989, **10**, 163.
12. A.-M. Hermansson, E. Eriksson, and E. Jordansson, *Carbohydr. Polym.*, 1991, **16**, 297.
13. M. Stading and A.-M. Hermansson, *Carbohydr. Polym.*, 1993, **22**, 49.
14. T. H. M. Snoeren, 'κ-Carrageenan: a Study of its Physicochemical Properties', PhD Thesis, University of Wageningen, Wageningen, The Netherlands, 1976.
15. I. Heertje, *Food Struct.*, 1994, in the press.
16. M. G. Lynch, and D. M. Mulvihill, *Ir. J. Agric. Res.*, 1992, **2**, 209.
17. M. Langton and A.-M. Hermansson, *Food Hydrocolloids*, 1992, **5**, 523.
18. M. Stading, M. Langton, and A.-M. Hermansson, *Food Hydrocolloids*, 1993, **7**, 195.
19. P. Walkenström and A.-M. Hermansson, *Food Hydrocolloids*, 1994, in the press.

Physical Chemistry of Heterogeneous and Mixed Gels

By V. J. Morris and G. J. Brownsey

INSTITUTE OF FOOD RESEARCH, NORWICH LABORATORY, NORWICH RE-
SEARCH PARK, COLNEY, NORWICH NR4 7UA, UK

1 Mixed Polysaccharide Gels

Mixed biopolymer gels are of considerable interest because they provide
realistic models for complex food structures or natural tissues. They also
provide relatively inexpensive methods for manipulating rheology and texture.
Of particular interest with respect to polysaccharide + polysaccharide mixtures
are synergistic interactions which lead to a non-additive enhancement of gel
properties. Likely candidates for such synergistic effects are mixtures of two
polysaccharides showing similar chemical and stereochemical structures. In
this article we discuss two such systems: ι- + κ-carrageenan mixtures, within
which no intermolecular binding appears to occur, and xanthan + carob
mixtures, within which there is mounting evidence for intermolecular binding.

ι- + κ-Carrageenan Gels

Carrageenans are sulfated galactans extracted from a number of species of red
algae.[1] The structures of the polysaccharides extracted from particular species
approximate to one of several idealized structures. By the correct choice of
algae and subsequent processing, it is possible to extract relatively pure
samples of a particular type of carrageenan. Two idealized structures (ι and κ)
are gel-forming carrageenans. The disaccharide repeat unit of ι-carrageenan is
$(1\rightarrow3)\alpha$-D-galactose-6-sulfate$(1\rightarrow4)\beta$-D-3,6-anhydrogalactose-2-sulfate. κ-car-
rageenan has a similar disaccharide repeat unit but the 2-sulfate on the
anhydrogalactose residue is missing. In the condensed state both ι- and κ-
carrageenan form coaxial duplexes comprising parallel right-handed three-fold
helical chains. In ι-carrageenan the pitch of the helical chains is 2.66 nm and
the chains are displaced from each other along their common axis by exactly
half their pitch.[2,3] For κ-carrageenan the helical pitch is 2.5 nm, and the
relative displacement of the chains differs from that observed for
ι-carrageenan.[4,5] The similarities of these structures suggest the possibility of

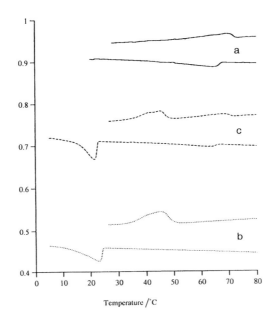

Figure 1   *DSC data for carrageenan gels: (a) 2 wt% aqueous ι-carrageenan gels; (b) 2 wt% aqueous κ-carrageenan gels; (c) 1:1 mixed carrageenan gels (2 wt% total polymer concentration). Heating and cooling rates are 0.5 °C min⁻¹. The vertical axis has arbitrary units*

intermolecular helix formation or association (or co-crystallization) of the double-helical structures.

There are only a few published studies on mixed carrageenan gels.[6-10] Analysis of rheological and optical rotation data by Rochas and co-workers[7] has suggested a synergistic interaction between the two polysaccharides. However, rheological and optical rotation data obtained by other researchers[6,8-10] have been taken to suggest a two-stage gelation process in which the κ- and ι-components gel independently.

In our current studies we are using differential scanning calorimetry (DSC) together with rheological measurements to probe ι- + κ-carrageenan mixed gels. The samples of ι- + κ-carrageenan were purchased from Sigma and used without further purification. Both were mixed salt forms containing sodium, potassium and calcium as counter ions. Carrageenan sols were prepared by dispersing the powdered carrageenan in water or an appropriate salt solution, and then heating the mixture to 85 °C in a sealed tube and maintaining this temperature for approximately 1 h. Mixed sols were prepared by mixing appropriate volumes of hot ι and κ sols and maintaining these samples in sealed tubes at 85 °C for approximately 1 h. In order to prepare gels, the hot sols were poured into appropriate moulds, covered and allowed to cool to room temperature (25 °C). Samples were stored overnight (18 h) before

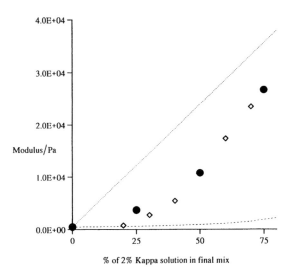

% of 2% Kappa solution in final mix

Figure 2   *Rheological data for mixed carrageenan gels with 2 wt% total polymer concentration: ●, added salt concentration 0.08 M KCl. The data have been compared with the calculated isostress and isostrain boundary regions and equivalent experimental data for pure κ-carrageenan gels containing 0.08 M KCl (◇)*

testing. Rheological measurements were undertaken using an Instron 3250 mechanical spectrometer in the oscillatory mode. Parallel plate geometry was used with the gel mould fixed to the lower plate. Oscillatory studies typically used strains <0.01 operating within the frequency range 0.1–10 Hz. DSC studies were made using a Setaram Micro DSC. All systems were conditioned by holding at 20 °C or 5 °C (depending upon salt level) for 18 h and the maximum temperature attained during the studies was 95 °C. In all cases solvent was used as a reference phase.

DSC data on 2 wt% aqueous carrageenan gels are shown in Figure 1. Heating and cooling curves for ι-carrageenan show a thermo-reversible transition with no evidence of hysteresis. Melting and setting of the gel are coincident with the conformational transition of the molecules attributed to double helix formation.[1] For κ-carrageenan, the transitions associated with setting and melting of the gels are displaced. Setting is accompanied by a sharp transition with melting involving a broader transition at a higher temperature. Once again[11] gel setting is associated with double helix formation, with the higher melting temperature attributed to disruption of helix–helix aggregates. For the mixed gels, two transitions are observed on heating and cooling. This is illustrated in Figure 1 for 50:50 mixed gels. On heating and cooling the two transitions are coincidental with the transitions obtained for the pure components. The high temperature 'ι' transition is thermally reversible and the low temperature 'κ' transition shows the hysteresis effects observed for pure 'κ'. These results confirm the unusual observations of Parker and co-workers[9] that, in the mixed salt forms, the melting and setting of 'ι'-carrageenan occurs at a higher temperature than for 'κ'-carrageenan. This is opposite to the effects observed in the purified salt forms.[12] The observations suggest that the two components associate independently of each other. It is well known that addition of KCl alters the melting and setting of the individual gels. The effect on κ-carrageenan is more pronounced than that on ι-carrageenan. This effect (results not shown) also occurs in the mixed gels, and at 2% total concentrations and with 0.08 M added KCl the two forms melt at the same temperature.

Figure 2 shows rheological data on mixed gels in the presence of added KCl. As observed by other authors[7,10] the modulus of the binary gel lies within the isostress and isostrain boundaries calculated on the basis of the moduli of the pure components (Figure 2). Both Piculell and co-workers[10] and Parker and co-workers[9] suggest that the mixed gels are phase-separated composites. However, at the total polymer concentration of 2%, all of the mixed gels shown in Figure 2 contain sufficient concentrations of both polymers to enable each to form a continuous network. Under such conditions the formation of two interpenetrating networks might be expected. In the similar system of κ-carrageenan + agarose, Zhang and Rochas[13] argued for the formation of an interpenetrating network (IPN) based on the additivity of the Youngs' moduli. In the present case the formation of an IPN would mean that the modulus of the mixture would be dominated by the stiffer κ network. The variation of the modulus should therefore match the appropriate concentration-dependence of the κ component. In Figure 2, data for pure κ gels containing added 0.08 M KCl have been superimposed on the data for the mixed gels. This comparison does seem to favour the IPN model. At volume fractions too low for the minor component to form a gel, localized precipitation of the minor phase within the 'major component' network would be expected. The assertion of independent networks is based on the failure to detect any κ contaminants in our samples (or *vice versa*) by DSC. More detailed analysis[7,14] does suggest that such contaminants may occur in supposedly pure carrageenans. Present data suggests that κ blocks may occur within 'ι' molecules whereas ι contaminants in 'κ' samples are separate molecules and can be removed by KCl precipitation.[7,14] Thus it is possible that some degree of inter-aggregate binding may occur.

The DSC data presented in Figure 1 differ from those cited in support of the currently accepted domain model for carrageenan gelation.[15,16] However, there is evidence that in the studies by Robinson and co-workers[15,16] their ι samples may have been contaminated with quite large levels of κ components.[17] The present studies suggest that gelation and melting of ι-carrageenan gels involves solely formation and melting of double helical junction zones. For κ-carrageenan, gel setting is determined by the coil-to-helix transition accompanied by simultaneous or subsequent aggregation (or crystallization) of helices leading to a higher melting point. These studies support the conclusions of Piculell and co-workers[12] based on optical rotation data.

Xanthan + Carob Gels

Xanthan + carob mixtures are used industrially as thermo-reversible gelling agents. This system is of particular interest as gelation arises on mixing together two 'non-gelling' pure components. The two molecules also possess similarities in chemical structure which could lead to intermolecular binding. Xanthan is an anionic heteropolysaccharide based on a cellulose backbone substituted on alternate glucose residues with a charged trisaccharide sidechain.[18,19] The structure is incompletely substituted with acetate and pyruvate. The sidechain distorts the backbone conformation causing formation of a five-fold helical structure.[20,21] At present it is not possible to distinguish between sterically acceptable single or double helical structures.[20-24] In solution xanthan exhibits a 'helix–coil' transition in which the ordered (helical) structure is favoured by low temperature and/or high ionic strength.[25,26] Carob is a plant galactomannan consisting of a mannan backbone which is incompletely and irregularly substituted with a α-D-galactose unit at the C_6 position.[27] The mannose:galactose ratio is $\approx$ 3.55:1.

Xanthan + carob mixtures form transparent thermo-reversible gels.[27] All current models for gelation invoke an intermolecular association but differ in the detail of the 'mixed junctions' formed.[27-39] There is considerable evidence for intermolecular binding.[27-41] Early models have proposed a binding between the xanthan helix and unsubstituted regions[25,27] or faces[42] of the galactomannan backbone. Mixing experiments[32,33] have suggested that helix denaturation is important for promoting xanthan + carob gelation. Subsequent experimental studies[29,30,39,40] have confirmed this proposal. X-ray fibre diffraction data on mixed gels suggest intermolecular association of the galactomannon with denatured xanthan.[32,33] Qualitative analysis of the data has been taken to suggest co-crystallization of the galactomannan and the denatured xanthan.[35,43] Modelling studies show that disordered xanthan can adopt a two-fold helical structure consistent with such a model.[23]

Recently, it has been reported[36,37] that mixing xanthan and carob at temperatures below the helix–coil transition temperature T_m still produces gelation. This has led to the suggestion that gelation can occur by two mechanisms. If the mixture is prepared at a temperature above T_m, and then cooled, gelation is considered to involve binding of galactomannans to the denatured xanthan. However, if the mixture is prepared at temperatures below T_m and then cooled, the gel is considered to arise from galactomannan binding to the xanthan helix.

This paper describes some further studies on xanthan + carob gels. Samples of xanthan (Keltrol) were purchased from Kelco-AIL. The xanthan was converted into the sodium salt form: 0.1 wt% aqueous samples were clarified by filtration through a 3 μm filter. This sample was ion-exchanged (H^+ Dowex) into the hydrogen form and the solution was neutralized with sodium hydroxide (0.1 M) and then freeze-dried. Xanthan solutions were prepared by dispersing xanthan in water for 0.5 h at 85 °C and then adding NaCl or water to give the required polymer concentration and salt content. The samples were heated for a further 1 h at 85 °C in a vigorously shaken, sealed container. These solutions were passed through a 1.2 μm filter, cooled, left overnight and diluted as required. Carob was purchased from Sigma and used without further purification. Solutions were prepared in the same manner as described for xanthan solutions. Mixed gels were prepared by combining the appropriate xanthan and carob solutions at the required mixing temperature (T_p), holding the samples at this temperature for 1 h in a sealed tube, and then pouring the mixture into appropriate moulds, covering the surface to inhibit evaporation, and allowing the mixture to cool overnight. Gel melting points were measured using the falling ball method. Once gels had set a steel ball was placed on the surface of the gel and the glass tube containing the gel sealed. The gels were heated (approx. 1 °C min^{-1}) and the melting-point (T_g) taken as the temperature at which the ball started to fall through the gel. Rheological studies were made using an Instron 3250 mechanical spectrometer. The mixed gels were examined under sinusoidal oscillation (0.011–10 Hz) at low strain

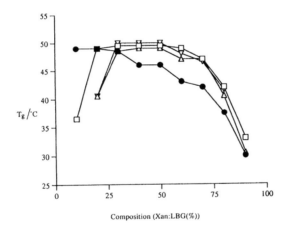

Composition (Xan:LBG(%))

Figure 3   *Melting temperatures* T_g *for xanthan (Xan) + carob (LBG) gels in the presence of different levels of added:* ●, *water;* □, *0.01 M NaCl;* △, *0.02 M NaCl; and* ▽, *0.04 M NaCl*

(<0.5) between flat plates of diameter 40 mm. Gels were examined 24 h after preparation at 20 °C. Optical rotation measurements on xanthan solutions were made with a JASCO DIP-360 polarimeter, using a cell of pathlength 100 mm and a wavelength of 436 nm. Solutions were added to the thermostated cell at 85 °C, held for 1 h to allow air bubbles to clear, cooled over a 3 h period to 20 °C, and held at this temperature overnight. Optical rotation data were then collected on heating.

Figure 3 shows T_g data for mixed gels prepared at 85 °C and then cooled to 20 °C. The values of T_g vary with the xanthan:carob ratio but seem to be independent of ionic strength. For 1:1 mixed gels we obtain $T_g \approx 46$–50 °C, which is in agreement with the observations of Williams and co-workers,[36] who noted a constant value of $T_g \approx 49$ °C irrespective of the level of added salt. Figure 4 shows the variation of T_m – as measured from optical rotation data on xanthan – with the level of added salt. Clearly, by varying T_p and the level of added salt, it is possible to achieve a range of preparation conditions for which $T_p < T_g < T_m$, $T_g < T_p < T_m$ and $T_g < T_m < T_p$. Rheological data obtained

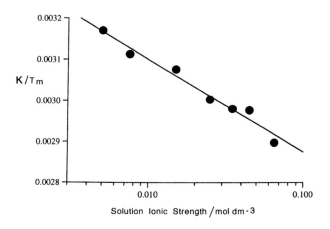

Figure 4   *Dependence of T_m on the level of added NaCl for xanthan + carob gels*

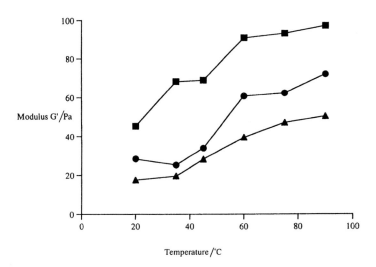

Figure 5   *Effects of T_p and the level of added salt on the shear modulus (G') of 1:1 xanthan + carob gels (total polymer concentration 0.5 wt%) in: ■, water; ●, 0.02 M NaCl; ▲, 0.4 M NaCl*

under these conditions are shown in Figure 5. It can be seen that elastic gels can be formed at all T_p values, even for cold-mixing conditions where $T_p < T_g$. The measured shear modulus G' increases with increasing T_p and decreases with increasing T_m (level of added salt). The most obvious suggestion is that G' is related to the degree of denaturation of the xanthan helix. Recently Morris[54] has argued that the difference in modulus for cold-mixed and hot-mixed gels are not due to the level of denaturation of xanthan helix but rather to the disruption of gel formation caused by the mixing process. Such an effect may contribute to the variation of G' with T_p, but it cannot account for the decrease in G' with added salt, particularly at $T_p > T_g$. Thus, the accumulated experimental data suggest that the galactomannan behaves as a denaturant, shifting the helix–coil transition to allow intermolecular binding with the disordered chain.

2 Heterogeneous Gellan Gels

Studies on gelation of polysaccharides have largely concentrated on the molecular structure of the junction zones within the gels and the mechanisms of gelation. Less effort has been put into investigating the 'larger-scale' network structure of such gels. This article describes recent rheological studies which suggest some degree of heterogeneous structure in certain gellan gels.

Gellan gum is an extracellular polysaccharide, produced commercially by aerobic fermentation of *Pseudomonas elodea*,[44] and marketed as a 'broad spectrum' gelling agent with a wide variety of potential industrial and food applications.[45–47] Gellan is an anionic heteropolysaccharide containing the tetrasaccharide repeat unit[48,49]

$$\rightarrow 3)\beta\text{DGlc}(1 \rightarrow 4)\beta\text{DGlcA}(1 \rightarrow 4)\beta\text{DGlc}(1 \rightarrow 4)\alpha\text{LRha}(1$$

The native product is esterified with both L-glycerate and acetate substituents.[50] Commercial extraction involves subjecting the broth to alkaline conditions which leads to a de-esterified product (Gelrite) in the potassium salt form. Extensive studies have been made on the solid-state conformation of gellan, its conformation in solution, and the mechanisms of gelation. Such studies are referenced in recent articles describing models for gelation.[35,51,52]

Gellan gum can be used to produce gels at low polymer concentration from both aqueous and salt solutions. The gels are thermo-reversible. Gelation is sensitive to the polysaccharide concentration and the nature and concentration of added salt.[46,53–56] At a given polysaccharide concentration, an increase in the level of added salt causes the magnitude of the fracture stress and the modulus to increase to a maximum and then decrease. The salt concentration at which these maxima occur is essentially independent of polysaccharide concentration but sensitive to cation valancy and, to a lesser extent, cation type.[46,53–56] At present there is no satisfactory explanation for the origin of these peaks. The following studies have been made in order to demonstrate and explain these effects.

Commercial samples of Gelrite (Kelco-AIL) were used without further purification. The gum was dispersed in water in sealed tubes and heated to 110 °C using an oil bath. Sols were prepared by thoroughly mixing appropriate volumes of hot (90 °C) gellan samples with hot (90 °C) salt solutions. Gels were prepared by pouring hot sols into appropriate moulds, covering to prevent solvent loss, and allowing to cool to room temperature. Gels were tested approximately 24 h after preparation. Conventional compression and relaxation tests were carried out using an Instron 1122. Small deformation oscillatory studies were made using an Instron 3250 at a fixed frequency (2 Hz), with parallel plates (diameter 40 mm) and a low strain sweep (<0.01).

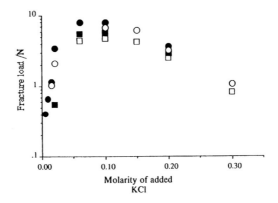

Figure 6   *Fracture load of gellan gum gels as a function of added KCl. Polymer concentrations:* □, *0.6 wt%;* ■, *0.7 wt%;* ○, *0.8 wt%;* ●, *0.9 wt%. The fracture load is proportional to the fracture stress*

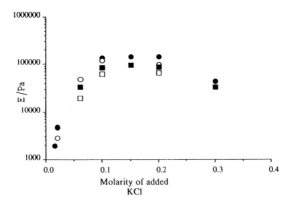

Figure 7   *Young's modulus determined at constant strain (0.15)versus level of added KCl. Polymer concentrations:* □, *0.9 wt%;* ■, *1.1 wt%;* ○, *1.3 wt%;* ●, *1.5 wt%*

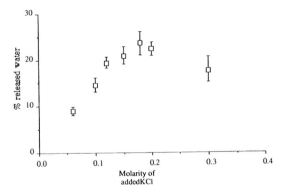

Figure 8   *Measured water loss from 0.8 wt% gellan gels (strain 0.09) as a function of added KCl concentration. Water loss was calculated as a percentage of the total water content of the gel*

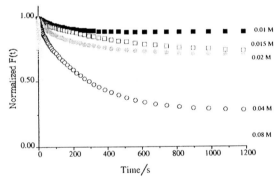

Figure 9   *Normalized stress relaxation data for 0.8 wt% gellan gels as a function of KCl. Data are expressed as the normalized force F(t). Cylindrical samples of 13 mm diameter were compressed (cross-head speed 1 mm min^{-1}) with an initial force of 0.2 N*

Figure 6 shows measured fracture stress as a function of added KCl for different polysaccharide concentrations. It was difficult to locate a linear region at small strain for the compression studies. Thus approximate values of Young's modulus E were calculated for a strain of 0.15. Figure 7 shows the variation of E with added salt for different polymer concentrations. Both the fracture stress and the modulus show a maximum at approximately 0.1 M KCl. Visual inspection of the gels reveals an increase in turbidity with increasing levels of added salt. Compression of the gels causes release of fluid to a maximum extent at around 0.1 M KCl (Figure 8).

Stress relaxation data (Figure 9) indicate a shrinkage of the gel which is accompanied by loss of fluid. At low salt levels the gel starts to approach the ideal of an elastic body, but there is more pronounced relaxation behaviour

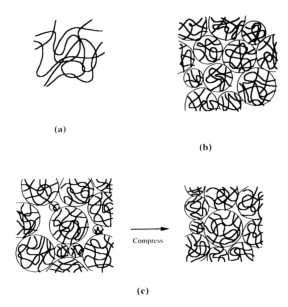

(a)

(b)

(c)

Figure 10   *Schematic model of gellan gum gels: (a) mainly elastic network at low added salt levels; (b) microgel structure at higher salt levels; (c) shear-hardening arising from denser packing of microgel particles upon compression*

Table 1   *Measurement of storage modulus G′ as a function of applied normal force for 0.8% gellan in 0.06 M KCl*

Applied force	Relaxed force	G′
N	N	kPa
0.04	nm[a]	2.3(5)
0.2	0.08	2.6(0)
1.0	0.2	2.9(6)
1.0	0.4	3.3(7)
1.0	0.6	3.8(4)
1.0	0.6	4.1(8)
1.0	0.8	5.1(0)
1.0	0.8	5.5(0)

[a] Not measured

with increasing levels of added salt. Table 1 shows the effect of normal force upon measurements of storage shear modulus $G′$ using dynamic oscillatory measurements. These results suggest that increasing the normal force, or successive application of a given normal force, reduces the level of stress and increases $G′$. Therefore the gels are shear hardening.

Two competing processes are therefore occurring in gellan gels. Addition of salt leads to an increase in G' presumably due to an increase in the number density of cross-links. However, increasing the ionic strength also seems to introduce a 'viscous-like' behaviour. Competition between these two mechanisms could account for the peaks in the fracture strength and moduli values. A possible explanation for the viscous component of the behaviour could lie in a decreasing solubility of the polymer with increasing salt content. The presence of incompletely dispersed aggregates would lead to the formation of microgel particles with the number density of such microgel particles increasing with increasing salt content (Figure 10a and 10b). Stress relaxation, shear-hardening and water loss from such systems would be envisaged as arising from the disruption of weaker inter-microgel contacts and a repacking of the microgel particles (Figure 10c). Such a heterogeneous structure provides a possible explanation for the marked flavour-releasing properties of gellan gum gels. Biting will tend to concentrate stress at weaker inter-microgel linkages leading to fracture and collapse of the gel into small microgel particles with a large surface area favourable for release of flavour components. On such a model, increased heterogeneity should favour improved flavour release.

Acknowledgements

Both authors wish to thank M. J. Ridout and A. Tsiami for providing research results prior to publication.

References

1. T. Painter, in 'The Polysaccharides', ed. G. O. Aspinall, Academic Press, New York, 1983, vol. 2, p. 195.
2. N. S. Anderson, J. W. Campbell, M. M. Harding, D. A. Rees, and J. W. B. Samuel, *J. Mol. Biol.*, 1969, **45**, 85.
3. S. Arnott, W. E. Scott, D. A. Rees, and G. C. McNab, *J. Mol. Biol.*, 1974, **90**, 253.
4. R. P. Millane, R. Chandrasekaran, S. Arnott, and I. C. M. Dea, *Carbohydr. Res.*, 1988, **182**, 1.
5. P. Cairns, E. D. T. Atkins, M. J. Miles, and V. J. Morris, *Int. J. Biol. Macromol.*, 1991, **13**, 65.
6. O. Christensen and J. Trudsoe, *J. Texture Stud.*, 1980, **11**, 137.
7. C. Rochas, M. Rinaudo, and S. Landry, *Carbohydr. Polym.*, 1989, **10**, 115.
8. L. Piculell, in 'Gums and Stabilisers for the Food Industry', ed. G. O. Phillips, D. J. Wedlock, and P. A. Williams, IRL Press, Oxford, 1993, vol. 6, p. 155.
9. A. Parker, G. Brigand, C. Miniou, A. Trespoey, and P. Vallée, *Carbohydr. Polym.*, 1993, **20**, 253.
10. L. Piculell, S. Nilsson, and P. Muhrbeck, *Carbohydr. Polym.*, 1992, **18**, 199.
11. D. A. Rees, E. R. Morris, D. Thom, and J. K. Madden, in 'The Polysaccharides', ed. G. O. Aspinall, Academic Press, New York, 1982, p. 195.
12. L. Piculell, C. Hakansson, and S. Nilsson, *Int. J. Biol. Macromol.*, 1987, **9**, 297.
13. J. Zhang and C. Rochas, *Carbohydr. Polym.*, 1990, **13**, 257.
14. C. Bellion, G. K. Hamer, and W. Yaphe, *Proc. Int. Seaweed Symp.*, 1981, **10**, 379.

15. G. Robinson, E. R. Morris, and D. A. Rees, *J. Chem. Soc., Chem. Commun.*, 1980, 152.
16. E. R. Morris, D. A. Rees, and G. Robinson, *J. Mol. Biol.*, 1980, **138**, 349.
17. I. T. Norton, D. M. Goodall, E. R. Morris, and D. A. Rees, *J. Chem. Soc., Faraday Trans.* 1, 1983, **79**, 2501.
18. P. E. Jansson, L. Kenne, and B. Lindberg, *Carbohydr. Res.*, 1975, **45**, 275.
19. L. D. Melton, L. Mindt, D. A. Rees, and G. R. Sanderson, *Carbohydr. Res.*, 1976, **46**, 245.
20. R. Moorhouse, M. D. Walkinshaw, and S. Arnott, in 'Extracellular Microbial Polysaccharides', ed. P. A. Sandford and A. Laskin, ACS Symp. Ser. vol. 45, American Chemical Society, Washington, DC, 1977, p. 90.
21. K. Okuyama, S. Arnott, R. Moorhouse, M. D. Walkinshaw, E. D. T. Atkins, and C. H. Wolf-Ullish, in 'Fiber Diffraction Methods', ed. A. D. French and K. D. Gardener, ACS Symp. Ser. vol. 141, American Chemical Society, Washington, DC, 1980, p. 411.
22. R. P. Millane and T. V. Narasaiah, *Carbohydr. Polym.*, 1990, **12**, 315.
23. R. P. Millane and B. Wang, *Carbohydr. Polym.*, 1990, **13**, 57.
24. R. P. Millane and B. Wang, in 'Gums and Stabilisers for the Food Industry', ed. G. O. Phillips, D. J. Wedlock, and P. A. Williams, IRL Press, Oxford, 1992, vol. 6, p. 541.
25. E. R. Morris, D. A. Rees, G. Young, M. D. Walkinshaw, and A. Darke, *J. Mol. Biol.*, 1977, **110**, 1.
26. I. T. Norton, D. M. Goodall, S. A. Fangou, E. R. Morris, and D. A. Rees, *J. Mol. Biol.*, 1984, **175**, 241.
27. I. C. M. Dea and A. Morrison, *Adv. Carbohydr. Chem. Biochem.*, 1975, **31**, 241.
28. I. C. M. Dea, E. R. Morris, E. J. Welsh, H. A. Barnes, and J. Price, *Carbohydr. Res.*, 1977, **57**, 249.
29. N. W. H. Cheetham, B. V. McCleary, G. Teng, and Maryanto, *Carbohydr. Polym.*, 1986, **6**, 257.
30. N. W. M. Cheetham and E. N. M. Mashimba, *Carbohydr. Polym.*, 1991, **14**, 17.
31. M. Tako, A. Asato, and S. Nakamura, *Agric. Biol. Chem.*, 1984, **48**, 2995.
32. P. Cairns, M. J. Miles, and V. J. Morris, *Nature (London)*, 1986, **322**, 89.
33. P. Cairns, M. J. Miles, and G. J. Brownsey, *Carbohydr. Res.*, 1987, **160**, 411.
34. E. R. Morris, in 'Food Gels', ed. P. Harris, Elsevier Applied Science, London, 1992, p. 291.
35. V. J. Morris, *Food Biotechnol.*, 1990, **4**, 45.
36. P. A. Williams, D. H. Day, M. J. Langdon, G. O. Philips, and K. Nishinari, *Food Hydrocolloids*, 1991, **4**, 489.
37. R. O. Mannion, C. D. Melia, B. Launay, B. Culvelier, S. E. Hill, S. E. Harding, and J. R. Mitchell, *Carbohydr. Polym.*, 1992, **19**, 91.
38. D. F. Zhang, M. J. Ridout, G. J. Brownsey, and V. J. Morris, *Carbohydr. Polym.*, 1993, **21**, 53.
39. N. W. M. Cheetham and E. N. M. Mashimba, *Carbohydr. Polym.*, 1988, **9**, 195.
40. B. V. McCleary and H. Neukom, *Prog. Food Nutr. Sci.*, 1982, **6**, 109.
41. N. W. M. Cheetham and A. Punruckrong, *Carbohydr. Polym.*, 1989, **10**, 129.
42. B. V. McCleary, A. H. Clark, I. C. M. Dea, and D. A. Rees, *Carbohydr. Res.*, 1979, **71**, 205.
43. V. J. Morris, in 'Food Polymers, Gels and Colloids', ed. E. Dickinson, Special Publication No. 82, The Royal Society of Chemistry, Cambridge, UK, 1991, p. 310.

44. K. S. Kang, G. T. Veeder, P. J. Mirrasoul, T. Kanecko, and I. W. Cottrell, *Appl. Environ. Microbiol.*, 1982, **43**, 1086.

45. G. T. Colegrave, *Ind. Eng. Chem., Prod. Res. Dev.*, 1983, **22**, 456.

46. G. R. Sanderson and R. C. Clark, *Food Technol.*, 1983, **37(4)**, 63.

47. P. A. Sandford, I. W. Cottrell, and D. J. Pettit, *Pure Appl. Chem.*, 1984, **56**, 879.

48. M. A. O'Neill, R. R. Selvandran, and V. J. Morris, *Carbohydr. Res.*, 1983, **124**, 123.

49. P. E. Jansson, B. Lindberg, and P. A. Sandford, *Carbohydr. Res.*, 1983, **124**, 135.

50. M. S. Kuo, A. J. Mort, and A. Dell, *Carbohydr. Res.*, 1986, **156**, 173.

51. V. J. Morris, *Agro Food Industry, Hi-Tech*, 1992, **3**, 3.

52. G. Robinson, C. E. Manning, and E. R. Morris, in 'Food Polymers, Gels and Colloids', ed. E. Dickinson, Special Publication No. 82, The Royal Society of Chemistry, Cambridge, UK, 1991, p. 22.

53. P. T. Attwool, PhD Thesis, University of Bristol, 1987.

54. V. J. Morris, in 'Food Biotechnology', ed. R. D. King and P. S. J. Cheetham, Elsevier Applied Science, London, 1986, p. 193.

55. H. Moritaka, H. Fukuba, K. Kumeno, N. Nakahama, and K. Nishinari, *Food Hydrocolloids*, 1991, **4**, 495.

56. M. M. A. K. Nussinovitch, M. D. Normand, and M. Peleg, *J. Texture Stud.*, 1991, **21**, 37.

Investigation of Sol–Gel Transitions of β-Lactoglobulin by Rheological and Small-angle Neutron Scattering Measurements

By Denis Renard, Monique A.V. Axelos, and Jacques Lefebvre

LABORATOIRE DE PHYSICO-CHIMIE DES MACROMOLÉCULES, INRA, BP 527, 44026 NANTES CEDEX 03, FRANCE

1 Introduction

Gelation of a globular protein, such as β-lactoglobulin (β-lg), is the result of an aggregation process, which is generally triggered by a conformational change of the protein induced by a modification of solvent conditions, usually a rise in temperature (where it is an example of a heat-set gel). The structure and properties of the resulting gel depend on the mode and kinetics of the aggregation process.

As protein molecules bear a net charge, except at their isoelectric point, interparticle electrostatic repulsions tend to oppose aggregation. Consequently, pH and ionic strength would be expected to affect the formation and the characteristics of globular protein gels. It has been shown[1,2] that gels from globular proteins are clear and present a fine-stranded structure when obtained under conditions of strong electrostatic repulsion, but are opaque with a coarse lumpy structure under conditions of weak electrostatic repulsion. The characterization of these gel structures, and particularly those formed from β-lg, has already been investigated by rheological and microscopical studies.[3,4] However, further investigations are needed on the sol–gel transition of β-lg, and on how controlling factors such as pH and ionic strength affect it, in order to be able to understand the functionality and to be able to predict the behaviour of this type of colloidal system. We have previously established the sol–gel transition diagrams of β-lg and have related the critical gelation concentration to the intensity and range of electrostatic interactions using a simple theoretical model.[5] We now investigate the sol–gel transition of this protein by dynamic rheological measurements and small-angle neutron scattering (SANS). The aim of this study is to understand the structure of the system at different scales throughout the transition. We try to relate the concentration dependence of the storage modulus to the fractal dimensionality of the system as proposed by the scaling theories of fractal networks[6,7] and compare these values with those obtained by SANS.

2 Experimental

The β-lg sample (three-times crystallized and lyophilized) was purchased from Sigma and was a mixture of genetic variants A and B. The protein was dissolved in sodium chloride solution and the pH was adjusted to 2, 7 or 9 with HCl or NaOH solution (0.1 or 1 M). The solutions were stirred overnight and undissolved material was removed by centrifugation at 16 000 g for 40 min. Protein concentrations were determined as already described.[5]

For the SANS experiments, β-lg was dissolved in D_2O (99.8% pure) containing 0.03 or 0.1 M NaCl, adjusted to the desired pD value with DCl or NaOD, and filtered through 0.02 μm Anotop filters (only for the pH 9 solution). The solutions were heated for 1 h at 80 °C in an ethylene glycol bath surrounding the quartz cuvette (1 or 2 mm path length) and then quenched at room temperature. All the measurements were performed at room temperature.

The instrument used for the SANS experiments was the PACE small-angle spectrometer at the ORPHEE reactor (CEA-CEN Saclay). The spectra were recorded for each sample using two different spectrometer configurations (changing the incident wavelength λ and the distance of the sample to the detector). Two partially overlapping ranges for the wave-vector $q = (4\pi/\lambda)\sin(\theta/2)$, θ being the scattering angle, were thus covered: 4.75×10^{-2}– 5.05×10^{-1} nm^{-1} and 2.5×10^{-1}–2.64 nm^{-1}. The data were normalized for transmission and sample path length and divided by the water spectrum. In order to obtain an absolute intensity scale, we used the absolute value of the water intensity in units of cross-section (cm^{-1}). The scattering intensity $I(q)$ was then expressed in molecular mass (g mol^{-1}) for a neutron contrast $\Delta\rho^2$ corresponding to 0% hydrogen–deuterium exchange between protein and solvent:

$$I(q) = A\,P(q)\,S(q) \qquad or \qquad (N_a d^2 / \Delta\rho^2)\,I(q)\,/\,C = M\,P(q)\,S(q) \quad (1)$$

$P(q)$ is the 'form factor' of the macromolecule, and $S(q)$ is the 'structure factor' which accounts for intermolecular interactions. The quantity A is a factor which includes the concentration C, the mass M, the density d, Avogadro's number N_a and the neutron contrast $\Delta\rho^2$. The contrast $\Delta\rho^2$ was calculated for 0% hydrogen–deuterium exchange between protein and solvent. Since the effective amount of exchange was not determined, absolute values of the mass of the scattering objects could not be obtained.

Small-amplitude oscillatory shear experiments were performed at 80 and 20 °C with a Carri-Med CS 50 rheometer using the cone-and-plate geometry (cone diameter 6 cm; angle 3.58°). For gel-cure experiments, measurements were performed at 1 Hz for 1 h at 80 °C, the strain amplitude being maintained

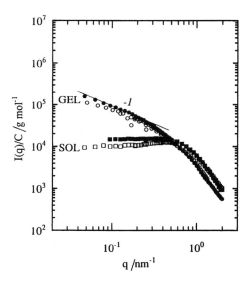

Figure 1   *Specific neutron scattering intensity* I(q)/C *as a function of the wave-vector*
q *for β-lg solutions and gels at pH 2: 2 wt% native β-lg solution in 0.03 M*
NaCl (□) and 0.1 M NaCl (■); 6 wt% β-lg gel in 0.03 M NaCl (○) and
0.1 M NaCl (●). Gels were obtained from protein solutions heated for 1 h at
80 °C and quenched to 20 °C

at 0.01. After quenching at 20 °C, the characteristics of the final system were
described by the mechanical spectrum recorded between 10^{-3} and 10 Hz at
0.03 strain amplitude.

3 Results and Discussion

SANS of β-lg Solutions and Gels at pH 2

The scattering curve of the native β-lg solution in 0.1 M NaCl (Figure 1) is
typical of a dispersion of non-interacting particles, and it shows the usual
Guinier and Debye regions, at q values lower and higher respectively than
5×10^{-1} nm^{-1}. But in 0.03 M NaCl, a correlation peak appears at
$q \approx 5 \times 10^{-1}$ nm^{-1} (Figure 1) which suggests an arrangement of protein
molecules in the solution, induced by the intermolecular electrostatic repulsions, which are only partially screened at low ionic strength. At higher q
values, the scattering curves are superimposed, indicating that the shape of the
molecule is unaffected by the ionic strength.

For gels, the Guinier plateau is displaced to the left to a position out of the
experimental q range (Figure 1), as is expected since the structural units are
now much larger objects; the observation window here corresponds to the
Debye region, which informs us about the structure of the aggregates. Up to

$q \approx 4 \times 10^{-1}$ nm^{-1}, *i.e.* in the observation scale range 300–15 nm, the neutron scattering intensity follows the power law $I(q)/C \propto q^{-1}$; such behaviour means that the structure of the aggregates in the gel is fractal, and the value of the exponent gives the fractal dimension $D = 1$. The gel appears therefore to be formed of strings of beads, with relatively long distances between the cross-links perhaps due to strong electrostatic repulsions at local scale between primary particles. The ionic strength does not affect the structure of the aggregates in the gel: the scattering curves obtained at the two salt concentrations superimpose over the whole q range (Figure 1).

The scattering spectrum obtained for the gel at high ionic strength was fitted accurately by a model of a linear aggregate composed of 20 spheres each of 1.75 nm radius. The fractal structure of $D = 1$ for β-lg gels at pH 2 is thus confirmed; the primary particle appears to be very close in size to the β-lg monomer. This type of structure was also observed by microscopy for β-lg gels at pH 3.5 in water by Langton and Hermansson,[8] who found that the strands had a thickness in the range 4–6 nm, corresponding to one or two β-lg monomers. This is in accordance with the homogeneous and transparent macroscopic appearance of the gels irrespective of the ionic strength.

SANS of β-lg Solutions and Gels at pH 7

Figure 2 illustrates the spectra obtained at pH 7. The scattering intensities for native β-lg solutions are the same at the two salt concentrations over all the

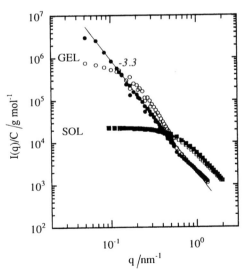

Figure 2 *Specific neutron scattering intensity* $I(q)/C$ *as a function of the wave-vector* q *for* β-lg *solutions and gels at pH 7: 2 wt% native* β-lg *solution in 0.03 M NaCl (□) and 0.1 M NaCl (■); 6 wt%* β-lg *gel in 0.03 M NaCl (○) and 0.1 M NaCl (●). Gels were obtained from protein solutions heated for 1 h at 80 °C and quenched to 20 °C*

observation scale and they do not display any correlation peak: because the net charge of the protein is much lower at pH 7 than at pH 2, 0.03 M NaCl suffices to screen out completely the double-layer interaction.

The specific scattering intensities of the gels are much higher in the low-q region than those of the solutions, indicating the formation of large, compact structures in the gels. In this region, the curve for 0.1 M NaCl is higher than the one for 0.03 M NaCl, pointing out an effect of the ionic strength on the structure of the aggregates although the intermolecular electrostatic interaction is screened out. The curve obtained in 0.1 M NaCl shows clear evidence of a power-law region, with an exponent of 3.3. Therefore, no conclusion can be drawn about the fractal or non-fractal character of the structure of the aggregates (the exponent being higher than 3, the slope cannot be considered as a volume fractal dimension). The results are in agreement with the macroscopic appearance of the gels; homogeneous with a faint opalescence at low ionic strength, and inhomogeneous and opaque at the higher ionic strength. For $q > 4 \times 10^{-1}$ nm^{-1}, the gel spectra superimpose, demonstrating that the type of structure in the range 5–16 nm is the same irrespective of the ionic strength. Moreover, the primary particle is different from the β-lg monomer and it seems to be composed of a dense, compact aggregate.

SANS of β-lg Solutions and Gels at pH 9

The scattering spectra of native β-lg solutions at pH 9 (Figure 3) are very similar to those obtained at pH 2. In particular, a correlation peak is visible in

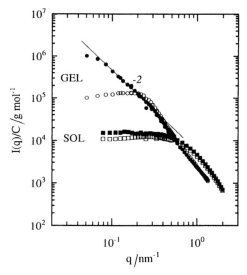

Figure 3   *Specific neutron scattering intensity* $I(q)/C$ *as a function of the wave-vector* q *for* β-lg *solutions and gels at pH 9: 2 wt% native* β-lg *solution in 0.03 M NaCl (☐) and 0.1 M NaCl (■); 6 wt%* β-lg *gel in 0.03 M NaCl (○) and 0.1 M NaCl (●). Gels were obtained by heating protein solution for 1 h at 80 °C and quenching to 20 °C*

0.03 M NaCl but it disappears at 0.1 M NaCl; it is centred at a lower q value (approx. 0.25 nm^{-1}) at pH 9 than at pH 2 (approx. 0.5 nm^{-1}) for the same protein concentration (2 wt%), as can be expected, since the absolute value of the protein net charge is lower at pH 9 than at pH 2 (12.3 instead of 20).

The ionic strength seems to affect the structure of the aggregates in the gels in a way rather similar to that observed at pH 7, as shown by the curves of Figure 3 for $q < 0.2$ nm^{-1}. But at this pH, the curve obtained in 0.1 M NaCl seems to follow in the $0.08 < q < 0.3$ nm^{-1} region a power law with an exponent 2. However, the range of q over which the power law holds is too narrow to be affirmative about the fractal character of the structure of the aggregates. Compact, spherical aggregates with diameters in the range of 0.1 to 2 μm were also observed by scanning electron microscopy.[9] In the case of the gel obtained at pH 9 and 0.03 M NaCl, a correlation peak can be observed with its maximum situated at approx. 0.2 nm^{-1}, showing that a preferential organization due to electrostatic interparticle repulsions, as occurs in the corresponding initial solution, is retained in the gel state.

Visco-elasticity of β-lg Gels

The visco-elastic behaviour of gels is related to their structure. In the particular case of fractal, colloidal networks, the models of Bremer *et al.*[6] and Shih *et al.*[7], which consider the network as a collection of fractal flocs closely packed through the sample, allows the fractal dimension D of the aggregates to be extracted from the exponent A' of the power law relating the elastic modulus G to the volume fraction Φ of the particles, *i.e.*

$$G = K \, \Phi^{A'} \tag{2}$$

where K is the elastic constant of the individual aggregates. The model of Bremer assumes K to be independent of the aggregate size and gives

$$A' = B / (3 - D) \tag{3}$$

with $B = 2$ or $B = 3$ for low and high volume fractions, respectively. In the model of Shih, K is considered to be inversely proportional to the size of the aggregates and determined by their effective backbone. This backbone has also a fractal structure, characterized by a second fractal dimension x which is related to the strain limit γ_0 above which strain and stress amplitudes are no longer proportional:

$$\gamma_0 \propto \Phi^{-(1+x)/(3-D)} \tag{4}$$

The parameter A' depends on both D and x:

$$A' = (3 + x) / (3 - D) \tag{5}$$

The values of A' are hence higher than in the model of Bremer.

When the visco-elasticity of the network is characterized by dynamic measurements, as it is generally for gels, the equilibrium value (for a

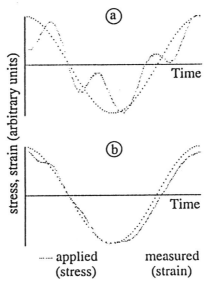

Figure 4   *Typical examples of strain response to an applied sinusoidal shear stress during the gelation at 80 °C of (a) a β-lg solution and (b) the resulting gel quenched to 20 °C after 1 h*

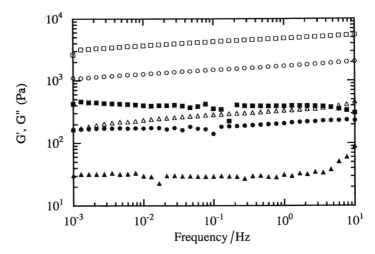

Figure 5   *Mechanical spectra at 20 °C of 6 wt% β-lg gels in 0.1 M NaCl at pH 2 (○, G′; ●, G″), pH 7 (□, G′; ■, G″) and pH 9 (△, G′; ▲, G″). (Gels were obtained by heating protein solution for 1 h at 80 °C and quenching to 20 °C)*

permanently cross-linked gel) or the rubbery plateau value (for a transient network) of the storage modulus G' is taken to be equal to G. Two problems are too often overlooked: the response to the sinusoidal oscillation must be itself sinusoidal (linear visco-elastic behaviour) in order that the instrumental

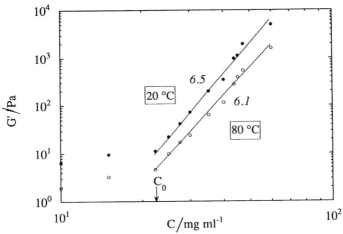

Figure 6 *Dependence of G' on β-lg concentration for gels at pH 7 in 0.1 M NaCl: ○, gels formed after heating β-lg solutions for 1 h at 80 °C; ●, same gels after quenching to 20 °C. G' is measured at 1 Hz. The slope values are given in italics. C_0 is the limiting gelation concentration determined as explained in reference 5*

data can be converted to values of the storage (G') and loss (G'') moduli according to the classical procedure; and G' has to be determined at a frequency at which its value is really representative of the height of the elastic plateau. The first condition is more stringent than the usual criterion of linearity which is the proportionality between stress and strain amplitudes, as illustrated in Figure 4 for a β-lg gel. Although the amplitude of the applied stress is within the proportionality region, the strain is clearly non-sinusoidal, harmonics being superimposed to the fundamental wave. The visco-elastic behaviour of β-lg gels appear to be non-linear in this respect even at low stress and strain amplitudes and whatever the protein concentration. Consequently, the moduli determined are only apparent ones; however, since we have $G' \gg G''$ even from the beginning of the gelation kinetics, the apparent value of G' can be taken as a reasonable approximation to the actual storage modulus. As for the second condition, Figure 5 shows that the dependence of G' on the frequency is slight within the experimental window; therefore, the apparent storage modulus G' determined at 1 Hz can be taken as a measure of the height of the elastic plateau of the gels.

We have studied the variation of G' with protein concentration for gels obtained in the presence of 0.1 M NaCl at the three pH values. An example is given in Figure 6. In all cases, a power law is observed for concentrations above the limiting gelation concentration C_0 determined by visual inspection as described in a previous paper.[5] From the results, we calculate the fractal dimensions of β-lg gels using eqns (3) and (4), and they are compared with those obtained by SANS in Table 1. We have also determined γ_0 for these gels. The values of D obtained from the rheological behaviour are comparable to

Table 1   *Fractal dimensionality* D *of* β-lg *gels at 20 °C calculated from rheology (models of Bremer et al.[6] and of Shih et al.[7]) and SANS measurements (6 wt%* β-lg *in 0.1 M NaCl, heated for 1 h at 80 °C)*

pH	A'^{a}	x^{b}	D^{c} (Bremer)	D^{d} (Shih)	D (SANS)
2	4.2	1.41	2.29 ± 0.02	1.95 ± 0.05	1.0 ± 0.10
7	6.5	0.62	2.54 ± 0.04	2.44 ± 0.02	–
9	8.3	0.54	2.64 ± 0.05	2.57 ± 0.06	2.0 ± 0.08

[a] Equation (2); [b] equation (4); [c] equation (3); [d] equation (5).

those found in a similar way by others for β-lg[9] or whey protein[10,11] gels, but are substantially higher than those determined by SANS. Moreover, except at pH 2, they are much larger than those predicted by computer simulation and found experimentally for diluted aggregating systems for diffusion-limited ($D = 1.7$–1.8) or reaction-limited ($D = 2.1$) cluster–cluster aggregation.[12] Three arguments can be put forward to explain the discrepancies:

(i)   The first hypothesis derives from the theoretical approach used to determine the fractal dimensionality D: a change in the average size of the primary β-lg particles may lead to a difference in the fractal dimensions obtained. The rheological measurements would give a higher fractal dimensionality if the size of the primary particle increases in the growing aggregate. This hypothesis is partly confirmed by SANS results since the primary particles we observed at pH 7 and 9 were dense, compact aggregates; an 'octamer of an octamer' was proposed as the primary particle in β-lg gels formed at pH 7 in 0.1 M NaCl.[13] But the argument fails when the results at pH 2 are considered, because the SANS results give strong evidence that the primary particle is then the β-lg monomer.

(ii)   Another explanation could be that the gels, under the conditions studied, and especially in the high concentration range, are not fractals. The scaling behaviour of G' *versus* protein concentration could be fortuitous, the relation being verified for no more than one decade of protein concentration and only above a certain concentration C_0.

(iii) Finally, the difference in the scale of observation for the two techniques, SANS and rheology, could explain the discrepancy obtained for the effective fractal dimension values. The SANS technique takes into account the interparticle correlations whereas rheology measures a consequence of the connectivity.

4 Conclusion

We have presented evidence that the difference in structure and rheological properties of β-lg gels as a function of pH and ionic strength are mainly governed by double-layer repulsions. The existence of fractal structures for the aggregates formed by heating β-lg solutions at pH 2, independently of the ionic strength, is clearly demonstrated; the aggregates are linear, organized as a 'string of beads' with the β-lg monomer as the primary particle. The existence of fractal structures at pH 7 and 9 remains to be confirmed by scattering measurements over a larger q-range and on more dilute systems. The primary particle of the clusters at these pH values appears to be a dense, compact structure much larger than the monomer.

Fractal dimensions calculated from rheological measurements are systematically higher than those obtained by SANS. This raises the difficulty of using the theoretical approach of fractal analysis in a concentrated medium.

References

1. A. H. Clark and S. B. Ross-Murphy, *Adv. Polym. Sci.*, 1987, **83**, 57.
2. A.-M. Hermansson, in 'Food Structure – Its Creation and Evaluation', Butterworths, London, 1988, p. 25.
3. M. Stading and A.-M. Hermansson, *Food Hydrocolloids*, 1990, **4**, 121.
4. M. Paulsson, P. Dejmek, and T. van Vliet, *J. Dairy Sci.*, 1990, **73**, 45.
5. D. Renard and J. Lefebvre, *Int. J. Biol. Macromol.*, 1992, **14**, 287.
6. L. G. B. Bremer, T. van Vliet, and P. Walstra, *J. Chem. Soc., Faraday Trans.*, 1989, **85**, 3359.
7. W.-H. Shih, W. Y. Shih, S.-I. Kim, J. Lui, and I. A. Aksay, *Phys. Rev. A.*, 1990, **42**, 4772.
8. M. Langton and A.-M. Hermansson, *Food Hydrocolloids*, 1992, **5**, 523.
9. M. Stading, M. Langton, and A.-M. Hermansson, *Food Hydrocolloids*, 1993, **7**, 195.
10. P. B. Fernandes, *Food Hydrocolloids*, 1994, **8**, 277.
11. R. Vreeker, L. L. Hoekstra, D. C. den Boer, and W. G. M. Agterof, *Food Hydrocolloids*, 1992, **6**, 423.
12. M. Kolb, R. Jullien, and R. Botet, *Phys. Rev. Lett.*, 1983, **51**, 1123.
13. J.-C. Gimel, D. Durand, and T. Nicolai, *Macromolecules*, 1994, **27**, 583.

High Pressure Gelation of Fish Myofibrillar Proteins

By Anne Carlez, Javier Borderias,[*] Eliane Dumay, and Jean-Claude Cheftel

UNITÉ DE BIOCHIMIE ET TECHNOLOGIE ALIMENTAIRES, CENTRE DE GÉNIE ET TECHNOLOGIE ALIMENTAIRES, UNIVERSITÉ DE MONTPELLIER II, 34095 MONTPELLIER CEDEX 5, FRANCE

1 Introduction

Derived from other industrial fields (ceramics, superalloys, artificial diamonds), the use of high pressure (100–500 MPa) appears to have numerous potential applications in food processing. Commercial products are already available on the Japanese market: fruit juices and sauces, jams and jellies, sweetened fruits for ice-cream, acid dairy desserts, raw beef and ham, and some seafoods. High pressure does inactivate non-sporulated microorganisms and some enzymes, without altering the taste, flavour or nutrient content of foods. High pressure may also induce protein denaturation and facilitate the texturization of foods.[1,2] Experiments have already been carried out to assess the effectiveness of high pressure in promoting the formation of gels from fish muscle or surimi.[3,4] However, it has not yet been attempted to investigate the mechanisms involved in such protein gelation at high pressure and low temperature.

The object of the present study was to find the optimum pressure conditions for the formation of protein gels from bream surimi, and to compare the mechanisms of pressure gelation with those of heat gelation as applied to fish myofibrillar proteins.

2 Materials and Methods

Threadfin bream (*Nemipterus tambuloides*) surimi from Thailand (with 76.8 ± 0.2 wt% water, 15 wt% protein, 4 wt% sucrose and 4 wt% sorbitol) was stored at $-20\,°C$ for about 14 months and thawed for 2 h at room temperature before use. Mixing was done as described by Carlez *et al.*[5] to

[*] Present address: Instituto del Frio, Ciudad Universitaria, 28040 Madrid, Spain

reach a final moisture content of 78% and a final NaCl content of 0, 1 or 2% on a total weight basis. The following additives were used: $CaCl_2$ (Prolabo, Paris), mercaptoethanol (Merck, Darmstadt), ethylenediaminetetraacetic acid (EDTA) and sodium dodecyl sulfate (SDS) (both from Sigma, St. Louis, MO, USA). The final mix was introduced into polyvinylidene chloride tubing (Krehalon brand, Deventer, Netherlands) and subjected to pressure- or heat-processing (90 °C for 30 min). In some cases, 'setting' (pre-incubation) was carried out at 4 °C for 24 h or at 37 °C for 30 min. Cooking was carried out in a waterbath at 90 °C for 30 min. Pressure processing was performed in water at 5–10 °C for 15 min in a 1 litre pressure unit (ACB, Nantes), as described by Carlez *et al.*[6] Gels were stored at 4 °C for 18–24 h before texture analysis. Some gels were also stored at 4 °C for up to 12 days.

The following texture measurements were carried out: yield force at gel rupture, F; yield deformation at gel rupture, d; 'gel strength', $F \times d$; and force at a penetration of 6 mm, F_6 (indicative of rigidity), as described by Carlez *et al.*[5] A few measurements were also carried out by compression–relaxation to determine $(1 - A)$, the index of gel elasticity. We define $A = (F_0 - F_{10}) / F_0$, where F_0 and F_{10} are the forces exerted at relaxation times 0 and 10 min respectively. Gel character was also assessed using the folding test (FT) on a 1 to 5 scale. The maximum score was obtained when no cracks appeared on the folds when the slice was double-folded (quarter-circle) and pressed between the fingers. In the case of some gels, non-destructuve oscillatory rheology was carried out as described by Carlez *et al.*[5] The storage modulus G', the loss modulus G'', and tan δ (δ = angle of phase displacement = G''/G') of gel slices (2–3 mm thickness) were determined.

The L^*, a^*, b^* colour values of gels were determined by reflectance using a Minolta Chromameter II (Tokyo, Japan). The incident light corresponded to the 'C illuminant'.

3 Results and Discussion

Texture and Colour Characteristics of Fish Protein Gels

In the absence of NaCl, thermal processing gave no gelation, while processing at 300 or 450 MPa gave very weak gels of low gel strength, measured by the low yield force and low yield deformation (FT = 2; $F = 102$ g; $d = 4.8$ mm or FT = 2; $F = 120$ g; $d = 5.2$ mm). Gels with good characteristics (FT = 5) were formed at either 1 or 2 wt% NaCl. The formation of fish protein gels was adequate only when proteins were partly solubilized by kneading surimi in the presence of salt: 1 wt% NaCl was selected for the rest of the experiments.

As shown in Figure 1, no gels were formed after processing bream surimi at 150 MPa, even in the presence of 1 wt% NaCl. This confirms the data of Yamamoto[7] who did not observe any pressure-induced gelation at 140 MPa (20 °C, 5–30 min) of suspensions of myosin filaments with 1–5 mg of protein per ml of 0.1 M KCl. Results shown in Figure 1 indicate a higher gel strength

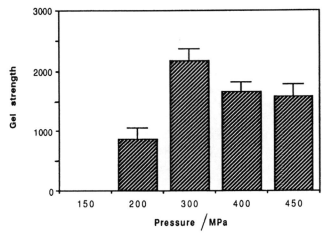

Figure 1   *Gel strength (F × d) of gels obtained by pressure processing (150 to 450 MPa, 5–10 °C, 15 min) of a bream surimi + NaCl mixture (1 wt% NaCl)*

of samples processed at 300 MPa than of those treated at 200, 400 or 450 MPa. This was due to an increase in both yield force and deformation. No difference in rigidity (force at 6 mm penetration) was observed. Gels obtained at 450 MPa were whiter than those obtained at 300 MPa ($L^* = 65.0$ as compared with $L^* = 67.2$). A pressure level of 300 MPa was therefore selected for subsequent experiments. Shoji et al.[8] also observed gelation of Alaska pollack surimi at or above 200 MPa, the firmest gel being formed at 300 MPa (as compared with 200, 400 or 500 MPa) and the whitest at 500 MPa.

When pressure-induced gels were compared with heat-induced gels (Figure 2), the yield force and rigidity of the former were observed to be lower ($F = 203 \pm 21$ g, $F_6 = 82 \pm 8$ g as compared with $F = 309 \pm 38$ g, $F_6 = 191 \pm 10$ g), while the yield deformation was higher ($d = 13.7 \pm 1.1$ mm as against 9.8 ± 0.9 mm). This appeared to indicate that pressure-induced gels were less firm and more pliable. The index of elasticity $(1 - A)$ was slightly lower in the case of pressure-induced gels (0.497 ± 0.006 as compared with 0.531 ± 0.019). Pressure-induced gels were also more transparent, less white and less yellow than heat-induced gels ($L^* = 68.4 \pm 1.9$, $b^* = -0.7 \pm 0.8$ as against $L^* = 76.2 \pm 2.1$, $b^* = 4.0 \pm 0.7$) (Figure 2). Shoji et al.[8] observed that the yield deformation of pressure-induced gels was higher than that of heat-induced gels (30 °C for 60 min + 90 °C for 30 min). In addition, they found pressure-induced gels to be more transparent and less white than heat-induced gels. Okamoto et al.[9] also observed that pressure-induced gels from various proteins were more glossy, more dense and smoother than their heat-induced counterparts.

Pressure processing (300–450 MPa, 5 °C, 15 min) followed by heat processing (90 °C, 30 min) resulted in higher yield force, gel strength and rigidity, but lower yield deformation ($F = 342 \pm 54$ g, $d = 7.7 \pm 0.8$ mm,

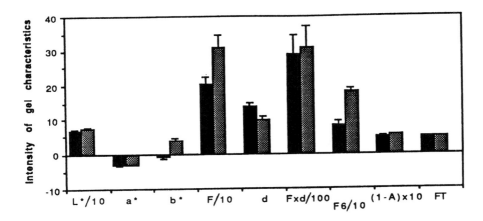

Figure 2   *Colour and texture characteristics of pressure-induced gels (■) (300 MPa, 5–10 °C, 15 min) and of heat-induced gels (▒) (90 °C, 30 min) from bream surimi (1 wt% NaCl). Results are means of nine measurements made on different gel samples*

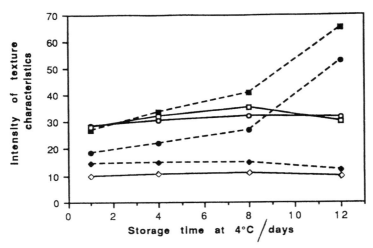

Figure 3   *Changes in the texture characteristics of pressure-induced or heat-induced gels from bream surimi as a function of storage time at 4 °C after gel formation: - - -, pressure-induced gels (300 MPa, 5–10 °C, 15 min); ―――――, heat-induced gels (90 °C, 30 min); ●, ○, yield force × 10⁻¹(g); ◆, ◇, yield deformation (mm); ■, □, gel strength F × d × 10⁻². The NaCl content of all gels is 1 wt%. Results are means of three measurements on the same gel sample*

$F_6 = 294 \pm 55$ g), as compared with pressure processing alone, indicating a very firm and poorly pliable gel. The score with the folding test was reduced from 5 to 4 or 2, clearly indicating less cohesive and/or elastic gels.

The characteristics of pressure- and heat-induced fish protein gels were compared, with or without a prior setting step at 37 °C for 30 min or at 5 °C for 24 h. For pressure-induced gels, prior setting did not modify the texture characteristics. For heat-induced gels, prior setting, expecially at 37 °C for 30 min, did appear slightly to increase both the yield force and the yield deformation (results not shown). The different responses to setting of pressure- and heat-induced fish protein gels were probably due to differences in the mechanisms of gel formation under pressure or by heating.

The changes in the texture characteristics of pressure- and heat-induced gels upon gel storage at 4 °C are shown in Figure 3. It can be seen that the yield force and the yield deformation remained constant for heat-induced gels throughout the 12 days of chilled storage. In contrast, these characteristics increased in the case of pressure-induced gels, especially the yield force between 8 and 12 days. The same phenomenon was reported by Shoji *et al.*[10,11] who noted an increased yield strength and deformation during storage of pressure-induced gels of Alaska pollack surimi (2.5 wt% NaCl), and recommended that pressure-induced gels be subsequently stabilized by heat processing, although this may cause a slight decrease in yield deformation and in transparency. No changes in colour characteristics were observed during the 12 days of chilled storage of pressure- or heat-induced fish protein gels.

Visco-elasticity Characteristics of Pressure- and Heat-induced Gels

Rheological measurements were carried out in the oscillatory mode on pressure-induced, heat-induced and pressure-plus-heat-induced gels of bream surimi in the presence of 1 wt% NaCl. Pressure-induced gels were significantly different from heat-induced gels, with G' and G'' values 2.4 to 3.3 times lower ($G' = 19\,760 \pm 2380$, $G'' = 4341 \pm 622$ *versus* $G' = 54\,176 \pm 4986$, $G'' = 10\,305 \pm 988$). The higher $\tan \delta$ value of pressure-induced gels ($0.220 \pm 6.5 \times 10^{-3}$ *versus* $0.190 \pm 2.6 \times 10^{-3}$) clearly indicated a more 'liquid-like' and less 'solid-like' behaviour. The heat-induced and pressure-plus-heat-induced gels displayed the same solid/liquid behaviour, although both G' and G'' were significantly higher in the case of gels that had been subjected to pressure before heat processing (results not shown). These results were very similar to those obtained by Ishikawa *et al.*[12,13] with sardine, pollack and skipjack surimi (in the presence of 1.5 wt% NaCl) processed at $50-400$ MPa for 20 min, or at 85 °C for 15 min, or pressure-plus-heat-processed. In all cases, gels induced by heat or by pressure and heat were similar, whereas pressure-induced gels had significantly lower G' and G'' values and higher $\tan \delta$ values. The molecular characteristics of the gel network responsible for these differences in rheological behaviour are not known.

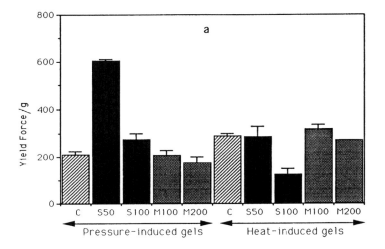

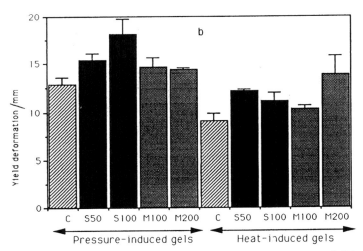

Figure 4   *Effect of addition of SDS or MSH on (a) the yield force and (b) the yield deformation of pressure-induced (300 MPa, 5–10 °C, 15 min) or heat-induced (90 °C, 30 min) gels from bream surimi: C, control; S50, SDS (50 mM); S100, SDS (100 mM); M100, MSH (100 mM); M200, MSH (200 mM). The NaCl content of all gels is 1 wt%*

Influence of Various Chemical Additives Present during Gel Formation

In pressure-induced gels (300 MPa), a low SDS concentration (50 mM) was found to cause an increase in gel strength (both yield force and deformation were raised) and rigidity, while it increased the yield deformation and decreased the rigidity of heat-induced gels (Figures 4a and 4b). At a higher SDS concentration (100 mM), gel strength and rigidity decreased, especially in the case of heat-induced gels, whereas gel strength and rigidity were lower than in the control sample without SDS, indicating softer gels (*F* decreased but *d* increased). All gels obtained with SDS were less uniform (with more aggregation) and contained many air bubbles. These differences (in spite of identical pre-treatment with SDS) supported the hypothesis that protein–protein interactions are different in pressure- and heat-induced gels. In the case of heat-induced gels from sardine surimi, Roussel and Cheftel[14] have suggested that the presence of small amounts of SDS may cause slight protein unfolding and improved solubilization, thus enhancing the formation of a more regular and elastic gel network. Larger amounts of SDS would tend, however, to induce excessive unfolding, prevent hydrophobic interactions, and so affect gelation in a detrimental manner. The score in the folding test was not influenced by the presence of SDS. This score thus appeared to be a much less sensitive indicator of texture changes than the yield force or the yield deformation of the gel. The L^* colour parameter (luminance) of protein gels was found to decrease in the presence of SDS.

Texture characteristics of pressure-induced gels were slightly affected by addition of mercaptoethanol (MSH) (decreased rigidity, slight increase in gel deformation). The yield deformation and gel strength of heat-induced gels slightly increased (Figures 4a and 4b), at both concentrations of MSH. Rigidity decreased, indicating softer gels. All gels (heat and pressure) prepared with MSH were more uniform (less air bubbles), brighter, and smoother. Roussel and Cheftel[14] interpreted the influence of this reducing agent on the texture of kamaboko gels by suggesting that low concentrations of MSH may enhance SH–SS interchange reactions and the formation of intermolecular disulfide bonds. Previously, Jiang *et al.*[15] had observed that the gel-forming ability (particularly the gel strength) of cod or mackerel surimi was improved by the addition of MSH (6.4 to 12.8 mM) or other reducing agents. The score in the folding test for pressure- or heat-induced gels was not influenced by the presence of MSH at both concentrations tested. The luminance (L^* value) of pressure-induced gels slightly increased in the presence of 100 or 200 mM MSH.

In pressure-induced gels, the presence of 15 or 30 mM $CaCl_2$ caused a marked enhancement, with higher yield force, yield deformation and gel strength values, and slightly higher rigidity values (Figures 5a and 5b). In heat-induced gels, $CaCl_2$ increased the brittleness (more aggregation), as judged by the low folding score of 3 at 15 mM $CaCl_2$ and of 2 at 30 mM. The yield force, gel strength, and rigidity increased slightly, while the yield deformation was hardly affected. The colour parameters of the pressure-induced gels were hardly affected by the presence of $CaCl_2$. In the case of heat-induced gels, however, the L^* value increased with $CaCl_2$ concentration, perhaps because the gel had a more aggregated structure.

In the case of addition of EDTA, the score in the folding test of pressure-induced gels remained high, but the yield deformation, yield force and gel

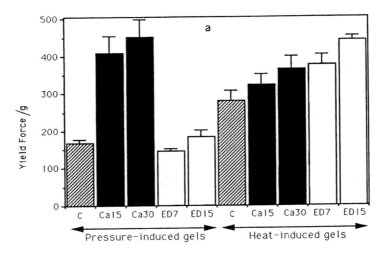

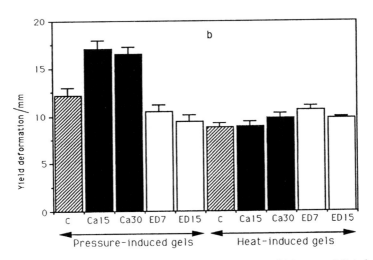

Figure 5   *Effect of addition of CaCl$_2$ or EDTA on (a) the yield force and (b) the yield deformation of pressure-induced (300 MPa, 5–10 °C, 15 min) or heat-induced (90 °C, 30 min) gels from bream surimi: C, control; Ca15, CaCl$_2$ (15 mM); Ca30, CaCl$_2$ (30 mM); ED7, EDTA (7 mM); ED15, EDTA (15 mM). The NaCl content of all gels is 1 wt%*

strength were decreased by the presence of EDTA (Figures 5a and 5b). In the case of heat-induced gels, the folding score decreased from 5 at 7 mM EDTA to 4–5 at 15 mM. The yield force, gel strength and rigidity increased, especially at 15 mM EDTA. The yield deformation was little changed. This apparent discrepancy between the effect of EDTA on the folding score and on instrumental texture is probably due to the more aggregated structure. Roussel and Cheftel[14] tested the effects of EDTA on heat-induced kamaboko gels from sardine surimi, and observed considerable decreases in folding score and gel strength over the same EDTA concentrations: they noted an increase in gel rigidity that might be explained by an artifact due to the granular (aggregated) texture of gels.

The different responses to $CaCl_2$ and to EDTA of pressure- and of heat-induced gels of bream surimi are not easily interpreted. In the case of pressure-induced gels, it is clear that the addition of calcium ions enhances all measured texture characteristics, while the complexation of calcium ions with EDTA decreases these same characteristics. Protein–calcium–protein electrostatic interactions therefore appear to play an important role in the gel network of pressurized myofibrillar proteins. The fact that addition of either $CaCl_2$ or EDTA increases the brittleness (lowers the folding score) and the yield force of heat-induced gels does not permit straightforward interpretation. The L^* value (luminance) of pressure- and of heat-induced gels slightly increased with increasing concentrations of $CaCl_2$ or of EDTA.

4 Conclusions

Pressure processing of surimi + NaCl mixes at low temperature (5–10 °C) resulted in protein gels with texture and colour characteristics markedly different from those of heat-induced gels. With pressure or heat conditions giving almost identical gel strength, pressure-induced gels displayed a smaller yield force and a higher yield deformation than heat-induced gels. Oscillatory rheometry further indicated that pressure-induced gels had a more liquid-like behaviour than heat-induced gels. The former were also more continuous, smoother, more transparent, less white, less yellow, and contained fewer air bubbles than heat-induced gels.

The study of the influence of chemical additives pointed to a different gelling behaviour of heat-induced and of pressure-induced gels. A given additive often induced different texture modifications, depending on the type of gel. This strongly suggests that the mechanisms of gelation and the proportion of hydrogen bonds, hydrophobic interactions, electrostatic interactions, and disulfide bonds are different in the two types of gels. It is difficult to be more specific, however. The presence of sucrose and sorbitol in bream surimi may also influence the mechanisms of protein unfolding and aggregation. Preliminary experiments using differential scanning calorimetry or electrophoresis of solubilized protein constituents have indicated that pressure (300 MPa, 5–10 °C, 15 min) causes less protein unfolding and less aggregation of myosin heavy chains than does heating (90 °C, 30 min). The gel network of pressure-induced gels thus appears to be less dependent on myosin cross-linking, and more sensitive to the presence of calcium ions, than that of heat-induced gels.

From a practical standpoint, the texture characteristics of pressure-induced gels may be of interest for the preparation of some seafood analogues and/or prepared dishes. Other known benefits are the inactivation of vegetative microbial cells, especially of gram negative organisms, and the possibility of moulding gels into given shapes.

Acknowledgements

This work was supported by a grant from the French Ministry of Research and Technology, Programme Aliment 2002, No. 91 G 0574. One of us (J.B.) benefited from a grant from Consejo Superior de Investigaciones Cientificas (Madrid) during his stay in Montpellier.

References

1. J.-C. Cheftel, in 'High Pressure and Biotechnology', ed. C. Balny, R. Hayashi, K. Heremans, and P. Masson, Colloque INSERM/John Libbey Eurotext, London, 1992, p. 195.
2. R. Hayaski, in 'High Pressure and Biotechnology', ed. C. Balny, R. Hayashi, K. Heremans, and P. Masson, Colloque INSERM/John Libbey Eurtext, London, 1992, p. 185.
3. M. Okamoto, Y. Kawamura, and R. Hayashi, *Agric. Biol. Chem.*, 1990, **54**, 183.
4. T. Ohshima, H. Ushio, and C. Koizumi, *Trends Food Sci. Technol.*, 1993, **4**, 370.
5. A. Carlez, J. Borderias, E. Dumay, and J.-C. Cheftel, submitted for publication.
6. A. Carlez, J. P. Rosec, N. Richard, and J.-C. Cheftel, *Lebensm.-Wiss. Technol.*, 1993, **26**, 357.
7. K. Yamamoto, in 'High Pressure Science for Food', ed. R. Hayaski, San-Ei Publishing, Kyoto, 1991, p. 285.
8. T. Shoji, H. Saeki, A. Wakameda, and M. Nakamura, in 'High Pressure Science for Food', ed. R. Hayashi, San-Ei Publishing, Kyoto, 1991, p. 300.
9. M. Okamoto, T. Deuchi, and R. Hayashi, in 'Use of High Pressure in Food', ed. R. Hayashi, San-Ei Publishing, Kyoto, 1989, p. 89.
10. T. Shoji, H. Saeki, A. Wakameda, M. Nakamura, and M. Nonaka, *Nippon Suisan Gakkaishi*, 1992, **58**, 329.
11. T. Shoji, H. Saeki, A. Wakameda, M. Nakamura, and M. Nonaka, *Nippon Suisan Gakkaishi*, 1992, **58**, 2055.
12. M. Ishikawa, K. Sakai, T. Yamaguchi, and S. Rachi, in 'High Pressure Science for Food', ed. R. Hayashi, San-Ei Publishing, Kyoto, 1991, p. 309.
13. M. Ishikawa, K. Sakai, T. Yamaguchi, and S. Rachi, in 'High Pressure Bioscience and Food Science', ed. R. Hayashi, San-Ei Publishing, Kyoto, 1993, p. 184.
14. H. Roussel and J.-C. Cheftel, *Int. J. Food Sci. Technol.*, 1990, **25**, 260.
15. S. T. Jiang, C. C. Lan, and C. Y. Tsao, *J. Food Sci.*, 1986, **51**, 310.

Gelation of Protein Solutions and Emulsions by Transglutaminase

By Y. Matsumura, Y. Chanyongvorakul, T. Mori, and
M. Motoki[1]

RESEARCH INSTITUTE FOR FOOD SCIENCE, KYOTO UNIVERSITY,
GOKASHO, UJI, KYOTO 611, JAPAN
[1]FOOD RESEARCH AND DEVELOPMENT LABORATORIES, AJINOMOTO CO.,
INC, 1-1, SUZUKI-CHO, KAWASAKI-KU, KAWASAKI-SHI 210, JAPAN

1 Introduction

Transglutaminase (TGase) is an enzyme which is capable of catalysing the
acyl-transfer reaction between peptide-bound glutamine residues and a variety
of primary amines. When protein-bound lysyl residues act as acyl receptors,
intra- and intermolecular ε-(γ-glutamyl)–lysine bonds are formed by the
enzyme reaction. Therefore, TGase can be used to mediate the cross-linking of
food proteins in order to improve their rheological properties.

There has been a number of reports concerning the TGase-catalysed
polymerization and gelation of food proteins.[1-6] However, the studies have
been restricted by insufficient supply of TGase, which is usually extracted from
the blood plasma or livers of mammals. Ando et al.[7] have screened the
microorganism (*Streptoverticillium*) that produces the Ca^{2+}-independent
TGase. This Ca^{2+}-independent activity of the enzyme is of a great advantage
for food applications because the physical states of many food systems are
affected by the presence of Ca^{2+}. We report here our recent results on the
application of this new microbial TGase to food model systems.

2 Enhanced Susceptibility to Transglutaminase of α-Lactalbumin Subjected to Partial Denaturation

While it is well known that caseins are good substrates of TGase,[1,3] TGase
generally cannot act on globular proteins[8,9] such as bovine serum albumin
(BSA), α-lactalbumin, β-lactoglobulin, ovalbumin, *etc.* However, the incorporation of histamine into BSA and ovalbumin by TGase reaction does occur
when lysyl residues of both proteins are chemically modified.[8] Larré et al.[9]
have also pointed out that citraconylation significantly increases the suscept-

ibility of legumin to TGase reaction. These results suggest that the chemical modification opens up globular proteins to TGase. Furthermore, it has been shown that TGase can polymerize globular proteins in the presence of reducing reagents such as dithiothreitol.[4-6,9] It is thought that the reducing agents induce a conformational change of the globular proteins and exposure of the reactive sites by disruption of intramolecular disulfide bonds. These results overall indicate that the conformational change of substrate proteins enhances their susceptibility to the TGase reaction. Chemical modification of proteins, however, even if it is irreversible, is not good for application to real food systems. It is also undesirable for relatively high concentrations of reducing agents to be present in foods. Therefore, it would be interesting to investigate whether the conformational change induced by more physical treatments such as heating, interfacial denaturation, *etc.*, is also applicable to enhance the susceptibility to the TGase reaction.

α-Lactalbumin is one of the major components of milk whey protein. As described above,[8,9] TGase cannot act on α-lactalbumin in the native state. However, it is well known that α-lactalbumin assumes structures that are characteristic of the 'molten globule state' under a variety of solution conditions.[10,11] The molten globule state is characterized as the compact globular molecule with native-like secondary structure and with disordered (unfolded) tertiary structure.[10,11] The molten globule state is acknowledged as an intermediate between the native and the completely unfolded states. To investigate the relation between the conformation of the substrate protein and the susceptibility to the TGase-catalysing cross-linking reaction, the reactivities of α-lactalbumin in the native state and in the molten globule state are compared.

α-Lactalbumin is a calcium-binding protein. Removal of bound Ca^{2+} destabilizes the native structure and induces a transition from the native state to the molten globule state. The thermal transition from the native state to the denatured state of holo-α-lactalbumin occurs at about 60 °C. Apo-α-lactalbumin, however, transforms to the molten globule state by thermal denaturation at 40 °C.[12] Therefore, holo- and apo-α-lactalbumin take on the native and molten globule states, respectively, at 40 °C.

The TGase reaction at 40 °C for holo-α-lactalbumin (native state) and apo-α-lactalbumin (molten globule state) was followed by the determination of the ammonia released from the proteins. Approximately 1.4 mol of NH_4^+ per mol of protein was released from α-lactalbumin in the molten globule state over 4 h, although less than 0.2 mol of NH_4^+ per mol of protein was released in the native state (results not shown). This indicates that the TGase reaction proceeds much faster when α-lactalbumin is in the molten globule state.

To detect the presence of intermolecular cross-linking, the change in molecular size of α-lactalbumin after TGase reaction for 4 h was analysed by sodium dodecyl sulfate–polyacrylamide gel electrophoresis (SDS/PAGE) in the presence of 2-mercaptoethanol (2-ME). The results are shown in Figure 1. The addition of 2-ME excluded the possibility of the presence of polymers formed through intermolecular disulfide bonds. Gels (1) and (2) show

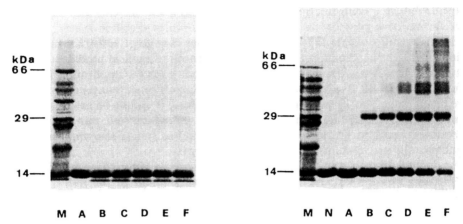

Figure 1   *SDS/PAGE of α-lactalbumin incubated with TGase for various times. The concentration of α-lactalbumin was 0.5 wt% and the enzyme:substrate ratio was 1:50 by weight. α-Lactalbumin was in (1) the native state and (2) the molten globule state, respectively. M, molecular mass markers; N, no enzyme; A, B, C, D, E, and F, incubation for 0, 15, 30, 60, 120, and 240 min, respectively*

respectively the results for holo-α-lactalbumin and apo-α-lactalbumin. We see that, when the native molecule was the substrate, TGase-catalysed polymerization did not significantly occur. In contrast, the amount of polymerized α-lactalbumin and the sizes of oligomers were found to increase with time in the case of the molten globule state. Time-dependent decreases in the content of the α-lactalbumin monomer were densitometrically determined. While less than 20% of monomers were lost after 4 h in the case of the native state, more than 80% of monomers were involved in the formation of polymers when α-lactalbumin in the molten globule state was used as the substrate (results not shown).

The high susceptibility of α-lactalbumin in the molten globule state shows that lysyl and glutaminyl residues of α-lactalbumin in this state are accessible to TGase. However, it is thought that either lysyl or glutaminyl, or both of these types of residues, are inaccessible to TGase in the native state. Therefore, we have investigated whether the transformation of α-lactalbumin from the native state to the molten globule state causes a change in the reactivity of lysyl and glutaminyl residues toward TGase. The reaction of lysyl and glutaminyl residues was estimated by the incorporation of carbobenzoxy (CBZ)-Gln-Gly and monodansylcadaverine by the TGase reaction, respectively. There was no significant difference in the amount of incorporated CBZ-Gln-Gly into lysyl residues between the native state and the molten globule state. The extent of incorporation of monodansylcadaverine which was determined fluorimetrically, was however, very much larger in the molten globule state than in the native state. These resul:s indicated that the transformation of α-lactalbumin from the native state to the molten globule state unmasked the glutaminyl residues and increased the reactivity towards TGase.

To identify the site of monodansylcadaverine incorporation by TGase in α-lactalbumin, dansylcadaverine-modified α-lactalbumin was subjected to digestion by chymotrypsin, and the resultant peptides were separated by reversed-phase HPLC (results not shown). The fluorescence-labelled peptide was analysed by an amino acid analyser and a gas-phase amino acid sequencer. The results showed that the incorporation site was Gln[54] in α-lactalbumin. On the three-dimensional structure of α-lactalbumin (not bovine but baboon) reported by Acharya *et al.*,[13] it was shown that this Gln[54] site was involved in a type I turn structure and exposed to the solvent in the native state. (If it is true that the secondary structure of α-lactalbumin in the molten globule state is native-like,[10,11] this site should be also situated in the region of the type I structure in case of the molten globule state). Coussons *et al.*[14] have compared the amino acid sequences around known sites of TGase reactivity in various proteins. They could not point out particularly striking patterns in sequences, but most of glutamine residues were found to be in regions which were predicted to be reverse turns. Our results are in agreement with those findings. However, it should be noted that Gln[54] in α-lactalbumin could not be attacked by TGase in the case of the native state, although this glutamine residue was in the reverse turn. This means that not only the secondary structure but also the tertiary structure or a local structural perturbation played an important role in the enhancement of susceptibility of proteins to TGase reaction.

We also found that interfacial denaturation of globular proteins at the oil–water interface in emulsions enhanced their susceptibility to TGase reaction (results not shown).

3 Gelation of Solutions and Emulsions containing Bean Globulins

Glycinin and legumin are the major storage proteins of soy bean and broad bean, respectively. These proteins are classified as 11S globulin which is widely found in the protein fractions of various seeds. Although seed storage proteins are a rather poor substrate for TGase in comparison with casein, polymerization and gelation of glycinin by TGase have been reported previously.[2,3,15] We have demonstrated[16] that gelation of legumin by TGase proceeds in a similar manner to that of glycinin. While acidic subunits were involved in polymerization during TGase reaction of both bean globulins, basic subunits were rather resistant to attack by TGase. Kang *et al.*[15] succeeded in enhancing the susceptibility of basic subumits of glycinin to TGase reaction by heat treatment.

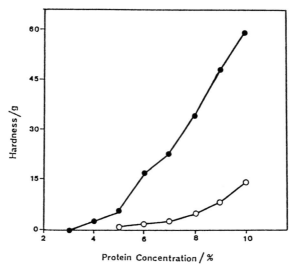

Figure 2   *Hardness of legumin gels formed by TGase reaction and heat treatment at various protein concentrations. Gels were formed in stainless steel moulds (6 mm diameter, 5 mm height) and subjected to compression tests without being removed from the moulds (20 °C). The gels were compressed to 60% of their original heights. Hardness (g) was evaluated from the load value at this deformation point: ●, gels formed by transglutaminase reaction; ○, gels formed by heating at 100 ° C for 1 h*

From the fundamental and/or commercial viewpoints, it has been interesting to compare physical properties of gels formed by heat treatment with those of gels formed by TGase reaction. From Figure 2, it can be seen that the minimum protein concentrations for gelling of legumin were 3 wt% and 5 wt% for TGase-catalysed gelation and thermal gelation, respectively. The values of 'hardness' were also larger for gels formed by the TGase reaction as compared with heat-induced gels at the same protein concentrations. These results indicated that a lower protein concentration was necessary for TGase-catalysed gelation, and that the gel formed was stiffer than the former gel. In the case of glycinin gels, a similar tendency was observed.

It has been suggested that endogenous TGase induces the setting of fish meat sol to a gel at room temperature.[17,18] The unique elasticity of the setting gel of fish meat is thought to be due to the formation of ε-(γ-glutamyl)–lysyl cross-links in gels. Therefore, we have investigated whether such elasticity is found in two globulin gels formed by the TGase reaction. Compression–decompression tests and creep analyses have revealed that the globulin gels formed by the TGase reaction show more elastic properties in comparison with the heat-induced gels (results not shown).

The relation between the microstructure and physical properties of gels was also investigated. Figure 3 shows the scanning electron microscopy (SEM) images of legumin gels. Micrographs (A) and (B) show the gels formed by heating and TGase reaction, respectively. It can be seen that the network structure has been well developed in the gel formed by TGase (Figure 3B). Thick strands (approximately 50 nm in width) formed the network structure, and clusters more than 100 nm in width or length were clearly observed. In the case of the heat-induced gel (Figure 3A), however, small particles less than 30 nm in diameter were seen. The surfaces of particles were smooth compared with those of the gel formed by TGase. Since particulates were distributed non-uniformly and associated irregularly, we could not see the developed network structure in the micrograph.

A

B

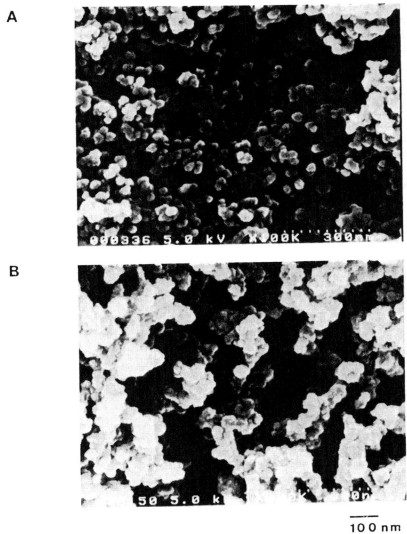

100 nm

Figure 3   *SEM images of legumin gels at a magnification of 100 000 ×. The protein concentration was 12.5 wt%. A, gels formed by heat treatment; B, gels formed by transglutaminase reaction*

The details of the microstructure were correlated with data on rheological properties. The TGase-catalysed gel had a developed network structure formed by thick strands. This may have contributed to the elastic and stiff behaviour of the TGase-catalysed gel. However, with the heat-induced gel, relatively irregular aggregation of small particles may have been responsible for the less elastic and stiff properties of the gel.

Not only protein solutions, but also emulsion-stabilized proteins, could be gelled by TGase.[19] We have previously reported[20] that emulsions made with soya oil and glycinin were gelled by TGase to form emulsion gels. In this report,[20] we showed, using small deformation mechanical tests, that the viscoelastic properties of emulsion gels are affected by the oil droplet size and the presence of a low molecular mass surfactant.

4 Conclusions

Our results have shown the possibility that TGase can generally act on proteins which are partially denatured by physical means (for instance by heat treatment). It has also been demonstrated that the gel formed by the TGase reaction has a different microstructure and physical properties from those of the gel formed by heat treatment. Proteins play important roles in various food systems such as gels, emulsions, foams, *etc.* Although there have been numerous approaches to modify the functional properties of proteins by heat treatment, there is a continuous desire to use proteins more effectively and under more demanding conditions. TGase treatment could be one of the most useful methods to improve the functional properties of proteins, and give a desirable texture to various foods containing proteins.

References

1. K. Ikura, T. Kometani, M. Yoshikawa, R. Sasaki, and H. Chiba, *Agric. Biol. Chem.*, 1980, **44**, 1567.
2. K. Ikura, T. Kometani, R. Sasaki, and H. Chiba, *Agric. Biol. Chem.*, 1980, **44**, 2979.
3. N. Nio, M. Motoki, and K. Takinami, *Agric. Biol. Chem.*, 1985, **49**, 2283.
4. S.-Y. Tanimoto and J. E. Kinsella, *J. Agric. Food Chem.*, 1988, **36**, 281.
5. F. Traoré and J.-C. Meunier, *J. Agric. Food Chem.*, 1992, **40**, 399.
6. R. Aboumahmoud and P. Savello, *J. Dairy Sci.*, 1989, **73**, 256.
7. H. Ando, M. Adachi, K. Umeda, A. Matsuura, M. Nonaka, R. Uchio, H. Tanaka, and M. Motoki, *Agric. Biol. Chem.*, 1989, **53**, 2613.
8. K. Ikura, M. Goto, M. Yoshikawa, R. Sasaki, and H. Chiba, *Agric. Biol. Chem.*, 1984, **48**, 2347.
9. C. Larré, Z. M. Kedzior, M. G. Chenu, G. Viroben, and J. Gueguen, *J. Agric. Food Chem.*, 1992, **40**, 1121.
10. O. B. Ptitsyn, *J. Protein Chem.*, 1987, **6**, 273.
11. K. Kuwajima, *Proteins: Struct. Funct. Genet.*, 1989, **6**, 87.

12. D. A. Dolgikh, L. V. Abaturov, I. A. Bolotina, E. V. Brazhnikov, V. E. Bychkova, R. I. Gilmanshin, Yu. O. Levedev, G. Semisotonov, E. I. Tiktopulo, and O. B. Ptitsyn, *Eur. Biophys. J.*, 1985, **13**, 109.
13. K. R. Acharya, D. I. Stuart, N. P. C. Walker, M. Lewis, and D. C. Phillips, *J. Mol. Biol.*, 1989, **208**, 99.
14. P. J. Coussons, S. M. Kelley, N. C. Price, C. M. Johnson, B. Smith, and L. Sawyer, *Biochem. J.*, 1991, **273**, 73.
15. I. J. Kang, Y. Matsumura, K. Ikura, M. Motoki, H. Sakamoto, and T. Mori, *J. Agric. Food Chem.*, 1994, **42**, 159.
16. Y. Chanyongvorakul, Y. Matsumura, H. Sakamoto, M. Motoki, K. Ikura, and T. Mori, *Biosci. Biotech. Biochem.*, 1994, **58**, 864.
17. N. Seki, H. Uno, N. H. Lee, I. Kimura, K. Toyota, T. Fujita, and K. Arai, *Nippon Suisan Gakkaishi*, 1990, **56**, 132.
18. Y. Tsukamasa, K. Sato, Y. Shimizu, C. Imai, M. Sugiyama, Y. Minegishi, and M. Kuwabata, *J. Food Sci.*, 1993, **58**, 785.
19. N. Nio, M. Motoki, and K. Takinami, *Agric. Biol. Chem.*, 1986, **50**, 1409.
20. Y. Matsumura, I. J. Kang, H. Sakamoto, M. Motoki, and T. Mori, *Food Hydrocolloids*, 1993, **7**, 227.

Sintering of Fat Crystal Networks in Oils

By Dorota Johansson, Björn Bergenståhl, and Eva Lundgren

INSTITUTE FOR SURFACE CHEMISTRY, PO BOX 5607, S-114 86 STOCKHOLM, SWEDEN

1 Introduction

Foods like margarine, butter, and chocolate consist of semi-solid fat, in a continuous phase, which determines the product properties such as consistency and stability. Semi-solid fat consists of colloidal fat crystals, with high melting points, dispersed in a liquid oil with a low melting point. Most natural fats demonstrate a broad spectrum of melting points such that small to moderate variations in temperature may start melting or post-crystallization processes of some components. Extensive changes in product character and quality occur as a result. These changes may be especially critical for low-fat foods where the amount of fat and/or saturated (solid) fat is close to the necessary limit for creating proper texture and stability.

Crystals in semi-solid fat form a network due to mutual adhesion in which adhesion strength determines structure and consistency (Figures 1 and 2). Dispersion (van der Waals) forces are not strong enough to account for the

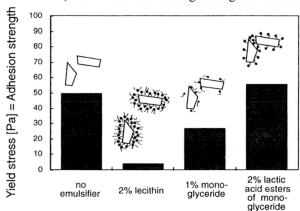

Figure 1   *Influence of emulsifiers on yield stress of 25 wt% fat crystals (tristearin β) in soybean oil. Emulsifiers formed different adsorbed layers on the crystals as schematically indicated*

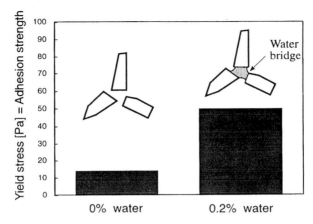

Figure 2   *Yield stress of 25 wt% fat crystals (tristearin β) in soybean oil with and without water*

network strength observed experimentally.[1] Therefore, other adhesive forces, *e.g.* water bridges, must occur.[2] Formation of water bridges results in an increase in the yield value of semi-solid fat upon the addition of small amounts of water (Figure 1). Additionally, oil-soluble emulsifiers which adsorb to fat crystals in oil[3] influence the adhesion forces.[1] For example, unsaturated monoglyceride or lecithin lowers the network strength (Figure 2) due to steric interaction. Lecithin is most effective in this sense. Other emulsifiers, such as esters of monoglycerides, give additional adhesion due to polar interactions. One source of strong adhesion between fat crystals in oils is the presence of solid bridges, often referred to as primary bonds.[4-6] Their process of formation has been referred to as sintering.[7,8] No extensive studies of this phenomenon have been performed. The aim of this work is to demonstrate control of sintering processes in well defined dispersions of fat through variation of temperature and composition.

2 Materials and Methods

In order to reproduce the broad melting-temperature range of real products, three types of fat were mixed together into model samples: (1) high-melting crystalline fats—a β'-stable palm stearin (a fraction of palm oil) with the melting point range 57–59 °C, and a β-stable tristearin with the melting point 72 °C; (2) low-melting refined oils (soybean and rapeseed) with melting points below 0 °C; (3) intermediate-melting fats (sintering elements)—β'-stable palm kernel fat with the melting point range approx. 20–36 °C, and a β-stable fraction of rapeseed fat, lobra 34, with the melting point range approx. 10–35 °C. Properties of these fats have been discussed in our previous work.[9]

Intermediate fats, if present, were initially added to the oil in amounts under their solubility limit. Afterwards, dry crystals of high-melting fat recrystallized from acetone were mixed into the oil. The concentration of crystals was relatively low (3–10%) compared with the composition of fat in margarine or chocolate. This fat level was chosen according to the limitations of the experimental methods. Samples were prepared at room temperature (approx. 20 °C) and then lowered to 10 °C at a controlled cooling rate (strictly speaking a controlled time for the temperature gradient 20 to 10 °C), at which the sedimentation or rheological measurements were performed. Some rheological samples were crystallized directly in oil (with melting of crystals at 60 °C for 1 h, followed by lowering of temperature to 10 °C to achieve crystallization). After the first measurement at 10 °C, all rheological samples were conditioned at 40 °C for 24 h prior to the second measurements at 10 °C.

Sediment volumes of dispersions are very sensitive to interactions between particles: stronger adhesion gives higher sediment volumes.[10] Since a solid bridge is considered to form a very strong adhesive interaction, the sedimentation method is very useful for studying sintering of fat crystals in post-crystallization processes. In order to interpret the results in terms of adhesion and sintering, dispersion parameters (such as crystal size, shape, distribution, and concentration) and other parameters (such as temperature, humidity, pressure, and oil viscosity) have been kept constant.

Rheological properties are sensitive to interactions in dispersions, much in the same way as are sedimentation properties. Rheological measurements have been performed on a Controlled Stress Rheometer (Bohlin, Sweden). The Bingham yield stress was evaluated since it is related to the adhesion strength (depth of energy minimum) between the particles.

3 Sintering by Lowering the Temperature

The sediment volume of 5 wt% palm stearin β' crystals in soybean oil as a function of concentration of intermediate melting fats in the oil is presented in Figure 3. Sample temperature during sedimentation was 10 °C, the temperature being lowered from 20 °C prior to sedimentation during the gradient time of approx. 8 min. The sediment volume of samples containing β'-stable palm kernel fat increased compared with the reference sample (without intermediate melting fat). A maximum increase of *ca.* 40% was attained at a concentration of 0.5 wt%. A slightly lower increase (up to 25%) and no maximum was observed with β-stable lobra 34.

The sediment volume of 5 wt% palm stearin β' crystals in soybean oil as a function of gradient time (20 to 10 °C) during cooling is presented in Figure 4. The oil contained 0.5 wt% intermediate-melting fat. The sediment volume has been expressed relative to the reference sample (without intermediate melting fat). For β'-stable palm kernel fat, an increase in sediment volume was observed for all gradient times. The degree of increase was more than 40% for rapid cooling (short gradient times), and only approx. 15% for the slower rate of cooling. In the case of β-stable lobra 34, an increase of approx. 15% was observed for rapid cooling, and no increase was observed for slow cooling.

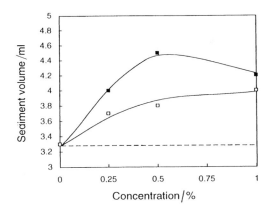

Figure 3   *Sediment volume of 5 wt% high-melting fat crystals (palm stearin β') in soybean oil as a function of concentration of intermediate melting fats: ■, palm kernel; □, lobra 34. The volume of the reference sample (without intermediate fat) is indicated as a broken line. The samples were prepared at 20 °C, rapidly cooled to 10 °C (gradient time approx. 8 min) and left to settle over a period of weeks*

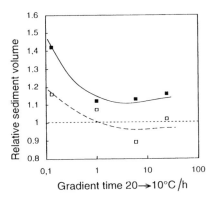

Figure 4   *Sediment volume of 5 wt% high-melting fat crystals (palm stearin β') in soybean oil as a function of cooling rate (expressed as gradient time) from 20 to 10 °C. The samples contained 0.5 wt% intermediate-melting fats: ■, palm kernel fat; □, lobra 34 fat. The sediment volume is presented relative to sediment volume of the reference sample (without intermediate fat)*

The sediment volume dependence on the rate of cooling was found to change when the polymorphic form of high-melting fat was changed from β' (palm stearin, Figure 4) to β (tristearin, Figure 5). For tristearin β, an increase in sediment volume was observed with β-stable lobra 34. The degree of increase was enhanced by very slow cooling rates (very long gradient times). No increase was observed with β'-stable palm kernel fat unless very long gradient times were applied during cooling.

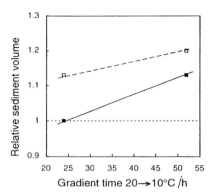

Figure 5   *Sediment volume of 3 wt% high-melting fat crystals (tristearin β) in soybean oil as a function of cooling rate (expressed as gradient time) from 20 to 10 °C. The samples contained 1 wt% intermediate-melting fats:* ■, *palm kernel;* □, *lobra 34. The sediment volume is presented relative to sediment volume of the reference sample (without intermediate fat)*

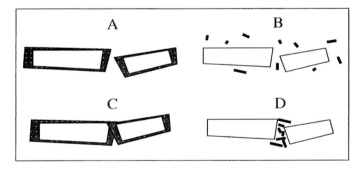

Figure 6   *Schematic picture of competitive phenomena occurring during cooling of samples containing intermediate-melting fats from 20 to 10 °C. (A) Crystal growth; (B) nucleation of new crystals; (C) formation of solid bridges between fat crystals; (D) formation of bridges of flocculated nuclei*

The increase in sediment volume observed in Figures 3–5 was intimately associated with the content of intermediate-melting fat and with the cooling of the samples. Therefore, crystallization of intermediate fat must be involved. Various possible schemes of post-crystallization processes are presented in Figure 6. The schemes are: (A) crystal growth, (B) nucleation of new intermediate-melting crystals, (C) sintering of crystals by intermediate-melting fat bridges, and (D) bridging of crystals by flocculated nuclei of intermediate-melting fat.

Since the concentration of intermediate fat was very low, the amount of fat capable of crystallizing was also very low. In this case, crystal growth (Figure 6A) or formation of new nuclei (Figure 6B) was very limited and therefore could not be responsible for the experimentally observed increase in sediment volume. Bridging according to Figure 6C or 6D must have occurred since the formation of even a limited number of bridges could give rise to additional adhesive forces and a higher sediment volume in the system.

The dependence of sediment volume on cooling rate and polymorphism of intermediate fat was therefore reversed in β' and β fat crystal dispersions (Figures 4 and 5). The β'-stable palm kernel fat bridged β' crystals of palm stearin, and β-stable lobra 34 bridged β crystals of tristearin. In the case of very rapid cooling, lobra 34 crystallized in the undercooled β' form and was able to bridge palm stearin β' crystals. For very slow cooling, palm kernel fat crystallized in the β form and bridged tristearin β crystals. Thus, β' crystals may have been sintered by β' fat bridges, which is favoured by rapid cooling. Also β crystals may have been sintered by β fat bridges, which is favoured by slow cooling. The necessity of identical polymorphic forms of the crystal and the bridge to produce sintering leads to the conclusions that solid bridges (Figure 6C) rather than bridges formed by small crystal nuclei (Figure 6D) were formed.

The maximum degree of sintering for palm kernel fat, having rapid crystallization kinetics, most likely occurs due to competition between bridge formation/crystal growth and the formation of new nuclei at concentrations $\geqslant$ 1 wt%. No such maximum occurred for lobra 34, which has slow crystallization kinetics.

4 Temporary Elevation of Temperature

The firmness (Bingham yield stress) of 10 wt% palm stearin β' crystals at 10 °C was found to increase extensively during storage for 24 h at the elevated

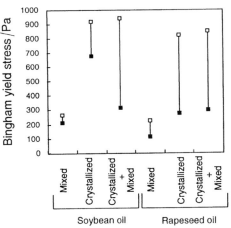

Figure 7   *Bingham yield stresses of 10 wt% palm stearin β' crystals in soybean and rapeseed oil. The dispersions were prepared in three different ways: mixing of dry crystals (recrystallized in acetone) into the oils; recrystallization in the oils; or recrystallization with subsequent mixing. The measurements were performed at 10 °C, directly after sample preparation at room temperature (■), and after their storage for 24 h at 40 °C (□)*

temperature (40 °C), as presented in Figure 7. This increase occurred both in soybean oil and in rapeseed oil and was independent of sample preparation (mixing of dry samples into the oils, recrystallization in oils, or recrystallization with subsequent mixing). The degree of increase was much higher for samples

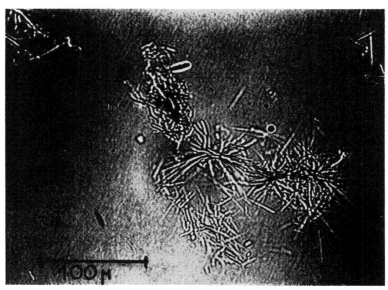

Figure 8   *Light microscope photographs of 5 wt% palm stearin β′ crystals in soybean oil at room temperature. (A) Samples stored at room temperature or below; (B) samples stored for 24 h at 40 °C and afterward at room temperature or below*

recrystallized in oil, where the crystals were expected to be very small compared with the mixed samples.

Microscopic photographs of a sample of 10 wt% palm stearin β' before and after warm storage are presented in Figures 8A and 8B, respectively. The photographs indicate that crystals grew during warming. Crystal growth alone was expected to lead to a decrease in the yield stress, and not the opposite as was observed.[13] Warm storage led to increased crystal adhesion and subsequent flocculation as a result. Warm storage also led to partial melting of crystal surfaces and most likely caused sintering processes during subsequent cooling. Star or brush-like flocs were easily observed in Figure 8B. Their image was different from that of typical β crystal sperulites in oils,[14,15] suggesting that the polymorphic transformation did not occur extensively during warm storage.

Temporary increases in temperature of a semi-solid fat below its melting point leads to fat crystal growth. Strongly increased firmness is observed at elevated temperatures due to flocculation caused by increased adhesion forces. Partial melting of fat most likely leads to sintering during subsequent cooling.

References

1. D. Johansson and B. Bergenståhl, *J. Am. Oil Chem. Soc.*, 1992, **69**, 718.
2. D. Johansson and B. Bergenståhl, *J. Am. Oil Chem. Soc.*, 1992, **69**, 728.
3. D. Johansson and B. Bergenståhl, *J. Am. Oil Chem. Soc.*, 1992, **69**, 705.
4. M. Van den Tempel, *J. Colloid Sci.*, 1961, **16**, 284.
5. J. M. Deman and A. M. Beers, *J. Texture Stud.*, 1987, **18**, 303.
6. D. Precht, in 'Crystalization and Polymorphism of Fats and Fatty Acids', ed. N. Garti and K. Sato, Marcel Dekker, New York, 1988, p. 305.
7. P. Walstra, in 'Food Structure and Behaviour', ed. J. M. V. Blanshard and P. Lillford, Academic Press, New York, 1987, p. 67.
8. I. Heertje, *Food Struct.* 1993, **12**, 77.
9. D. Johansson and B. Bergenståhl, *J. Am. Oil Chem. Soc.*, 1994, in the press.
10. F. M. Tiller and Z. Khatib, *J. Colloid Interface Sci.*, 1984, **100**, 55.
11. T. J. Gillespie, *J. Colloid Sci.*, 1960, **15**, 219.
12. Th. F. Tadros, *Chem. Ind.*, 1985, **7**, 210.
13. Th. F. Tadros, *Langmuir*, 1990, **6**, 28.
14. A. Naguib-Mostafa, A. K. Smith, and J. M. deMan, *J. Am. Oil Chem. Soc.*, 1985, **62**, 760.
15. A. C. Juriaanse and I. Heertje, *Food Microstruct.*, 1988, **7**, 181.

Thermal Gelation of Sunflower Proteins

By A. C. Sánchez and J. Burgos

UNIVERSITY OF ZARAGOZA, FOOD TECHNOLOGY DEPARTMENT, FACULTY OF VETERINARY SCIENCE, MIGUEL SERVET 177, 50013 ZARAGOZA, SPAIN

1 Introduction

A considerable number of foods are protein gel systems. Gelation is a functional property of proteins which can be a determining factor in the potential use of a protein-rich material in food.[1] Gelation implies the formation of a full three-dimensional network of asymmetric molecules through the entire system. This requires intermolecular cross-linking involving segments of two or more polymer molecules, usually in well defined structures called junction zones. Network formation during protein gelation is derived from a complex range of intermolecular interactions, mainly hydrophobic and electrostatic interactions, hydrogen bonds and disulfide bridges. There are several approaches for making protein gels which are in use in the food industry; but usually protein gel formation requires heating to produce protein unfolding (and in some cases subunit dissociation) prior to aggregation.

Gelation of proteins induced by heat has given rise to considerable research work and some interesting reviews.[1-3] Contrary to what occurs with many heat-induced polysacharides gels, cooling after transition to the disordered state is not needed for protein gel setting, except for a few special cases, such as gelatin. Heat-induced protein gels are usually thermally irreversible,[1,4] although liquefaction above 100 °C and 'melting' behaviour are not unknown.[5,6]

Sunflower is a widespread protein-rich crop lacking toxicants and anti-nutritive factors other than phytate. Its protein is reputed to be only slightly deficient in lysine and threonine.[7] However, sunflower protein isolates find only very marginal use in the food industry due in part to problems associated with their chlorogenic acid contents, although a number of processes to obtain very low phytate and chlorogenic acid protein isolates are now available.[8-10] The functional properties of sunflower proteins have so far been little investigated. The aim of this work is to explore the possibilities of inducing gelation in low phytate and chlorogenic acid sunflower protein preparations and to characterize the rheological behaviour of the gels which may be formed.

2 Materials and Methods

Low-temperature defatted and desolvated sunflower meal, prepared from mechanically dehulled seeds, was provided by Gerdoc (Pessac, France). Globulins were extracted by the method of Raymond *et al.*[11] Low phytate and chlorogenic acid isolates were prepared by extraction of the protein from the sunflower meal, at pH 9, with Na_2SO_3 (0.25 wt% aqueous solution) followed by isoelectric precipitation at pH 4.5 and five successive extractions of the precipitated protein with 4 volumes of cold ($-15\ °C$) and acidified (8 wt% 0.005 M HCl) acetone to remove chlorogenic acid.

Gelation was monitored by dynamic rheological measurements using a Bohlin CS/ETO Rheometer working in the oscillatory mode. Test conditions were established as follows: frequency 0.1 Hz; maximum strain 0.02; plate and cone (4°) geometry; sample covered with silicone oil to avoid drying. Gelation time and gelation temperature were defined as the time and temperature after which the angle phase remained under 45°. Unless otherwise stated, gels were formed after heating in the interval 60–98 °C at a rate of 1 °C min^{-1}, followed by holding at 98 °C for 60 min, and the rheological measurements were made at 98 °C.

Trypsinization was performed at 10 wt% protein concentration, 27–28 °C and pH 8.1 with a trypsin:protein ratio of 75 units g^{-1}; proteolysis was stopped by aprotinin (4 units μg^{-1} of trypsin). The degree of trypsinization reached (expressed as the increase in 12% trichloroacetic acid-soluble amine nitrogen) in the partial trypsinization was 147 μmol g^{-1}.

Calorimetric determinations were performed in a Dupont 2010 calorimeter equipped with a pressure cell.

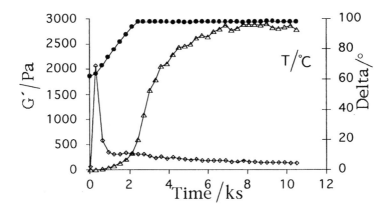

Figure 1   *Storage modulus G' (△) and phase angle (delta, ◇) as a function of time during heating and holding at 98 °C of a 10 wt% solution of partially trypsinized protein isolate in 0.05 M Tris buffer, pH 8. Temperature profile (T, ●) is also shown*

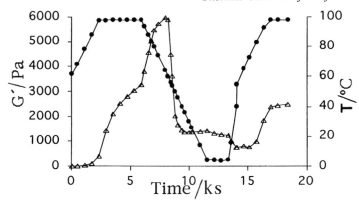

Figure 2   *Changes in temperature* T *(●) and storage modulus* G' *(△) of a 10 wt % solution of a partially trypsinized protein isolate in 0.05 M Tris buffer, pH 8, during heating, holding, cooling and reheating*

3 Results and Discussion

Neither the protein isolates nor the purified globulins could be gelled just by heat treatment (30 min at 100 °C or 110 °C). Gelation could not be obtained either by controlled acidification (dialysis against phosphate buffer, pH 5) or by heating after calcium enrichment (5–30 mM), or in the presence of mercaptoethanol (20 or 70 mM), propylene glycol (5–20%) or sodium chloride (0.5–3%), although some of these treatments induced protein precipitation.

When heated up to 98 °C after partial trypsinization, however, both native globulins and protein isolates gelled at a temperature dependent on pH and protein concentration. The storage modulus growth reached a plateau after about 1 h of holding at 98 °C (Figure 1). If the gel was then cooled, the storage modulus grew further at the beginning of the cooling period, but started to decrease when the gelation temperature was reached. However, the gel did not disintegrate even after 30 min at 4 °C. If the gel was reheated, the modulus, after an initial further drop, began to increase again when the gelation temperature was attained, but it never reached the first plateau value (Figure 2).

Although thermo-reversibility is currently considered of limited value as a means of clarifying cross-linking behaviour,[3] it clearly points to physical interactions – mainly hydrophobic – rather than covalent bonds playing the dominant role in maintaining the network.

The mechanical spectrum is frequently used to determine the similarity between a sample gel and a covalent one, since a covalent gel is frequency independent but a physical one is not. The sunflower protein gels studied here are relatively frequency independent in the frequency interval 1–30 Hz. The slope of the log–log plot of storage modulus *versus* frequency is equal to 0.009 in the frequency range 1–10 Hz and 0.0079 between 10 and 30 Hz.

Partially trypsinized sunflower protein gelation is strongly pH dependent: below pH 7 protein aggregates are formed, but over pH 9 the gels are very weak. The storage modulus reaches a maximum value at pH 8. Gelation time and temperature at constant protein concentration increase with pH in the range 7–11. The increase in gelation time with pH and the low maximum storage modulus at high pH are attributable to increased electrostatic repulsion. Mature gel storage modulus, gelation temperature and gelation time are also protein concentration dependent. The gel storage modulus is not related by a single power law to protein concentration. The critical concentration at pH 8 is estimated to be lower than 1.7 wt%. Gelation temperature, under our experimental conditions, is inversely proportional to the logarithm of the protein concentration. Plotting log (1/gelation time) *versus* log (concentration), a straight line of slope unity is obtained. According to the Oakenfull kinetic model of gelation,[12–14] this figure would correspond to the number of polymer chains participating in the formation of a junction zone. Since junction zone formation requires intermolecular cross-linking, this type of initial reaction rate/concentration dependence is taken as revealing that the rate limiting step of the gelation is a first-order reaction.

It is commonly assumed that a protein denatures before it aggregates and forms a gel. However, the gelation temperature is well below the denaturation temperature of the partially trypsinized protein isolate at the pH (8) of the gelation experiments. Although differential scanning calorimetry reveals that the acetone extraction step denatures sunflower globulins, the experiments performed with native globulins subjected to the same experimental conditions reveal that gelation can occur about 20 °C below the denaturation temperature. That protein gel formation, under certain conditions, may precede denaturation has been also demonstrated by Stading and Hermansson[15] for β-lactoglobulin.

Acknowledgements

The financial support of the Spanish CICYT (ALI 91–035) and Ministry of Education (predoctoral fellowship for A.C.S.) is gratefully acknowledged.

References

1. J. E. Kinsella, *CRC Crit. Rev. Food Sci. Nutr.*, 1976, **7**, 219.
2. A. H. Clark and C. D. Lee-Tuffnell, in 'Functional Properties of Food Macromolecules', ed. J. R. Mitchell and D. A. Ledward, Elsevier Applied Science, London, 1986, p. 203.
3. D. A. Ledward, in 'Functional Properties of Food Macromolecules', ed. J. R. Mitchell and D. A. Ledward, Elsevier Applied Science, London, 1986, p. 171.
4. A. H. Clark, in 'Food Polymer, Gels and Colloids', ed. E. Dickinson, Special Publication No. 82, The Royal Society of Chemistry, Cambridge, UK, 1991, p. 322.
5. T. M. Bikbov, V. Grinberg, H. Schmandke, T. S. Chaika, I. A. Vaintraub, and V. B. Tolstoguzov, *Colloid Polym. Sci.*, 1991, **259**, 536.
6. K. D. Navin Kumar, P. K. Nandi, and M. S. Narasinga Rao, *Int. J. Pept. Protein Res.*, 1980, **15**, 67.

7. B. Gassman, *Nahrung*, 1983, **27**, 351.
8. S. Gheyasuddin, C. M. Cater, and K. F. Matil, *Food Technol.*, 1970, **24**, 242.
9. M. Saeed and M. Cheryan, *J. Food Sci.*, 1988, **53**, 1127.
10. G. Vermeersch, *Rev. Fr. Corps Gras.*, 1987, **34**, 333.
11. J. Raymond, J. L. Azanza, and M. Fotso, *J. Chromatogr.*, 1981, **212**, 199.
12. D. G. Oakenfull and A. Scott, in 'Gums and Stabilisers for the Food Industry', ed. G. O. Phillips, D. J. Wedlock, and P. A. Williams, Elsevier Applied Science, London, 1986, vol. 3, p. 465.
13. D. G. Oakenfull and V. J. Morris, *Chem. Ind.*, 1987, 201.
14. D. G. Oakenfull and A. Scott, in 'Gums and Stabilisers for the Food Industry', ed. G. O. Phillips, D. J. Wedlock, and P. A. Williams, IRL Press, Oxford, 1988, vol. 4, p. 127.
15. M. Stading and A. M. Hermansson, *Food Hydrocolloids*, 1990, **4**, 121.

Binding of Calcium Ions by Pectins and Relationship to Gelation

By Catherine Garnier, Monique A. V. Axelos, and Jean-François Thibault

INSTITUT NATIONAL DE LA RECHERCHE AGRONOMIQUE, RUE DE LA GÉRAUDIÈRE, BP 1627, 44316 NANTES CEDEX 03, FRANCE

1 Introduction

Pectins are important structural polysaccharides of plant cell walls and are of considerable interest as gelling agents in the food industry.[1] The pectin polymers consist mainly of linearly connected α-1→4 D-galacturonic acid residues and their methyl esters, interrupted by some 1→2 linked rhamnose residues.[2] The ability of pectins with a degree of methylation (DM) below 50% to form gels in the presence of such divalent ions as calcium is markedly affected by different parameters, including the DM, the charge distribution along the backbone, the average molecular mass of the sample, the ionic strength, the pH, the temperature, and the presence of co-solutes.[3] The scope of this paper is to describe the effects of the DM and ionic strength on the gelation of pectin + calcium systems. Therefore, sol–gel–syneresis phase diagrams have been established and the binding of calcium ions by pectins has been followed by potentiometry at various polymer and calcium concentrations. Finally, rheological experiments were performed in order to determine the gel point.

2 Materials and Methods

Two pectin samples were used: sample A30 was a commercial apple pectin (batch 618F, Unipectine, France), and sample C48 was obtained by acid de-esterification of a commercial citrus pectin (Copenhagen Pectin Factory, Denmark) as described previously.[4] The galacturonic acid contents were 76.8% and 82.6% and the DM values were 28% and 48% for samples A30 and C48, respectively. The intrinsic viscosities $[\eta]$ were determined in 0.1 M NaCl and at 20 °C and were 0.282 1 g^{-1} and 0.536 1 g^{-1} for samples A30 and C48, respectively. For a salt-free solution of pectin, the variation of the intrinsic viscosity with the ionic strength brought about by the polymer was also taken into account.[4] The weight average molecular masses $\bar{M}_w$ determined by HPSEC–MALLS were 175 700 and 160 300 for samples A30 and C48, respectively.

Pectin + calcium systems were prepared by mixing pectin solutions with calcium chloride solutions at 70 °C for 3 min. The phase diagram (sol–gel–syneresis) was established as calcium concentration *versus* reduced polymer concentration by tilting the test tubes after standing for 48 h at 20 °C. When the meniscus could not be seen to deform under its own weight, we said that the system had gelled. Syneresis was detected by the presence of free solvent at the gel surface.

The selectivity and co-operativity of the binding of calcium by pectins were studied by potentiometry, using a calcium-specific electrode (F2112 Ca, Radiometer) and a saturated calomel electrode as reference.[5] The calcium activity was determined in pectin + calcium systems in the sol and gel states. Calibration curves were obtained using standard $CaCl_2$ solutions, the concentrations being measured through conductimetric determinations with silver nitrate.

Rheological experiments were carried out on pectin gels with a Carri-Med CS50 dynamic controlled stress rheometer, in oscillatory shear, with a cone–plate device (diameter 5 cm, cone angle 4°). The gels were cured *in situ* at 20 °C. After 24 h a mechanical spectrum (G' and G'', the storage and the loss moduli, *versus* the frequency) was recorded, which gave access to the equilibrium elastic modulus at zero frequency (G'_0). A low deformation of 0.04 was maintained whatever the frequency range explored between 10^{-3} and 5 Hz.

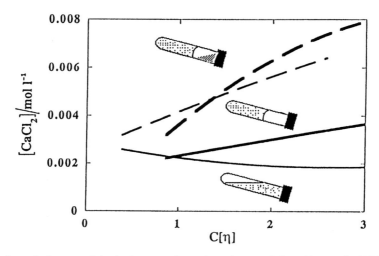

Figure 1   *Influence of the ionic strength on the gelation of the A30 sample (20 °C, pH 7). Thin lines (solid and broken) 0.1 M NaCl; thick lines (solid and broken) salt-free solution*

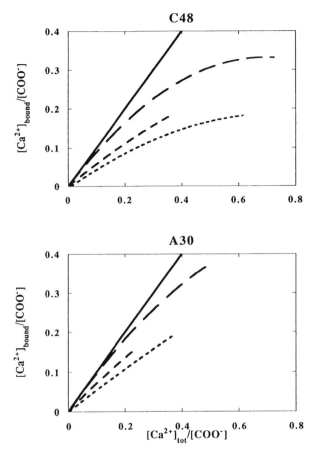

Figure 2   *Influence of the ionic strength and the polymer concentration* C_p *on the calcium binding by pectin chains at pH 7.4 and 20 °C:* ⎯⎯⎯⎯*, total binding of the added calcium; – – –, 0.1 M NaCl,* $C_p = 1.3 \times 10^{-2}$ *eq* l^{-1}; ⎯⎯⎯ ⎯⎯⎯*, water,* $C_p = 1.3 \times 10^{-2}$ *and* 2.8×10^{-2} *eq* l^{-1}; *- - -, 0.1 M NaCl,* $C_p = 2.8 \times 10^{-2}$ *eq* l^{-1}

3 Results and Discussion

The phase diagram of sample A30 at 20 °C and pH 7 in salt-free solution and in 0.1 M NaCl is shown in Figure 1. The reduced polymer concentration $C[\eta]$ was used as abscissa in order to compare the gelation properties under different conditions. Indeed, $C[\eta]$ corresponds to the volume filled in solution by the polymer. Three states can be seen: solution at low calcium concentrations, gel at higher calcium concentrations, and syneresis for the highest calcium concentrations. In 0.1 M NaCl, the gelation is generally possible before the overlapping of polymer chains, defined as $C[\eta] \approx 0.8$ by Graessley,[6]

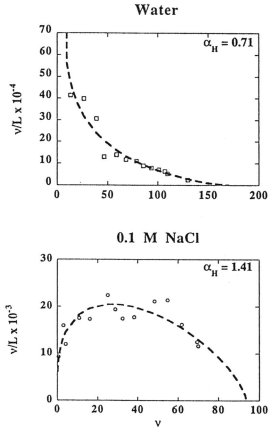

Figure 3   *Scatchard plots of the potentiometric data and Hill fits obtained at* $[COO^-] \approx 1.3 \times 10^{-2}$ *eq* l^{-1} *for sample C48*

because the electrostatic repulsions between negatively charged chains are screened by sodium ions. In salt-free solutions, calcium ions have first to screen the charges before acting as cross-linking agents. By this method, it was also shown that a decrease in the degree of methylation lowers the amount of $CaCl_2$ required to obtain gelation in the systems.

The influence of the ionic strength, the polymer concentration and the DM value on the binding isotherms of calcium ions are shown in Figure 2. The affinity of calcium ions for pectins decreases when the ionic strength increases, whatever the values of DM or the polymer concentration. In water, the calcium binding is independent of the polymer concentration. In 0.1 M NaCl, the binding of calcium ions increases with the polymer concentration, for both values of DM. In the presence of added salt as well as in salt-free solution, calcium binding increases when the DM value of the sample decreases. In order to obtain information on the co-operativity of the binding of calcium

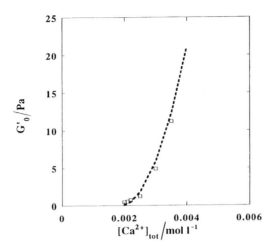

Figure 4   *Evolution of the equilibrium elastic modulus G'_0 with the amount of calcium chloride $[Ca^{2+}]_{tot}$ for sample C48 at $[COO^-] \approx 1.3 \times 10^{-2}$ eq l^{-1} (20 °C, pH 7.4); $[Ca^{2+}]_c \approx 0.0019$ mol l^{-1}*

ions to pectins, the binding isotherm data of sample C48 are plotted in terms of Scatchard plots, v/L *versus* v, where v is the amount of calcium bound by the macromolecule and L is the concentration of free calcium ions (Figure 3). A fit of the Scatchard plots was obtained by the use of the semi-empirical Hill equation

$$\ln L = -(1/\alpha_H) \ln[n_H/\nu) - 1] + \ln K_H \qquad (1)$$

where n_H is the number of sites per macromolecule able to bind cations, K_H is the apparent dissociation constant for the interacting sites and α_H is the Hill constant. When $\alpha_H > 1$, the binding is co-operative; when $\alpha_H = 1$, the binding is non-co-operative; and $\alpha_H < 1$ indicates that the binding is anti-co-operative.[7] In water, the concave curvature and the value of α_H obtained (0.71) indicate anti-co-operative interactions between the polymer and calcium ions[8] and this can be ascribed to a typical polyelectrolyte behaviour. In 0.1 M NaCl, the curve is convex-shaped and we get $\alpha_H = 1.41$, which indicates a co-operative binding of calcium ions by pectin.[9]

Rheological measurements show that G'_0 increases with the amount of calcium in the system. Above the sol–gel transition and close to the gel point, G'_0 varies with the total calcium concentration $[Ca^{2+}]_{tot}$ as

$$G'_0 = 17.3 \left| \frac{[Ca^{2+}]_{tot} - [Ca^{2+}]_c}{[Ca^{2+}]_c} \right|^{1.98} \qquad (2)$$

where $[Ca^{2+}]_c$, the value of $[Ca^{2+}]_{tot}$ at the gel point, is equal to 0.0019 mol l^{-1}. The value of the exponent (1.98) agrees well with results already published (1.93),[10] which confirms that the percolation model can accurately describe the properties of pectin + calcium gels at the sol–gel transition. The linear relation between bound and total calcium in water justifies the use of $[Ca^{2+}]_{tot}$ as a control parameter of the cross-linking reaction.

References

1. D. B. Nelson, C. J. B. Smit, and R. R. Wiles, in 'Food Colloids', ed. H. D. Graham, Avi, Westport, 1977, p. 418.
2. D. Darvill, P. Albersheim, M. MacNeil, J. Lau, W. York, T. Stevenson, J. Thomas, S. Doares, D. Gollin, P. Chelf, and K. Davis, *J. Cell Sci. Suppl.*, 1985, **2**, 203.
3. M. A. V. Axelos and J.-F. Thibault, in 'The Chemistry and Technology of Pectin', ed. R. H. Walter, Academic Press, New York, 1991, p. 109.
4. C. Garnier, M. A. V. Axelos, and J.-F. Thibault, *Carbohydr. Res.*, 1993, **240**, 219.
5. C. Garnier, M. A. V. Axelos, and J.-F. Thibault, *Carbohydr. Res.*, 1994, **256**, 71.
6. W. W. Graessley, *Polymer*, 1980, **21**, 258.
7. A. Lips, A. H. Clark, N. Cutler, and D. Durand, *Food Hydrocolloids*, 1991, **5**, 87.
8. J. Mattai and J. C. T. Kwak, *Biochim. Biophys. Acta*, 1981, **677**, 303.
9. C. R. Cantor and P. R. Schimmel, 'Biophysical Chemistry. Part III: Behaviour of Biological Macromolecules', Freeman, San Francisco, 1980, p. 849.
10. M. A. V. Axelos and M. Kolb, *Phys. Rev. Lett.*, 1990, **64**, 1457.

Heat-induced Denaturation and Aggregation of β-Lactoglobulin: Influence of Sodium Chloride

By M. Verheul, S. P. F. M. Roefs, and C. G. de Kruif

NETHERLANDS INSTITUTE FOR DAIRY RESEARCH (NIZO), PO BOX 20, 6710 BA EDE, THE NETHERLANDS

1 Introduction

β-Lactoglobulin (β-lg) is the major whey protein in bovine milk. It is a globular protein with a diameter of approximately 3.0 nm. The protein contains 162 amino acid residues and has a molecular mass of approximately 18 300 Da. At room temperature β-lg exists as a dimer in water at a pH between 5.5 and 7.5. The monomeric units each contain two intramolecular disulfide bonds and one thiol group.[1]

On heating a β-lg solution several intramolecular and intermolecular changes take place. Raising the temperature to 30–50 °C induces the dissociation of the dimer into its monomeric units, and at temperatures above approximately 60 °C the monomers undergo a denaturation step in which they unfold.[2] The denaturation can be followed by an irreversible aggregation reaction so that the whole process becomes irreversible.[3] Although many studies have been performed on the heat-induced denaturation and aggregation of β-lg, the precise mechanism by which the reactions take place remains unclear. The changes that take place on heating β-lg are influenced by many factors like electrostatic and hydrophobic interactions, hydrogen bonding and disulfide cross-linking.[4]

On heating β-lg in water at neutral pH in the temperature range 60–70 °C, a transparent dispersion or gel is formed that contains polymeric protein particles.[5] Under these conditions the heat-induced denaturation and aggregation of β-lg can be quantitatively described by a model which has recently been developed in our laboratory. This model predicts the formation of protein polymer particles with a certain average length and is based on intermolecular exchange reactions between intramolecular disulfide bonds and exposed reactive thiol groups.[6] Adding NaCl to the β-lg solution prior to heating results in the formation of a turbid dispersion or gel which means that larger

437

aggregates are present. The formation of these larger aggregates may be caused by physical aggregation in addition to the disulfide exchange reactions.

In this paper a systematic study of the influence of NaCl on the heat-induced denaturation and aggregation of β-lg is presented. The kinetics of this denaturation/aggregation process have been followed by measuring the concentration decrease of native β-lg as a function of heating time. The formation of the protein aggregates was monitored *in situ* by dynamic light scattering.

2 Generalized Model for Heat-induced Denaturation and Aggregation

The denaturation and aggregation of β-lg is a very complex process in which many steps are involved, and many factors can influence the different steps. For the heat-induced denaturation and aggregation of β-lg in NaCl solutions, we propose a simplified reaction scheme consisting of two consecutive reactions – a denaturation step and then an aggregation reaction:

$$B \rightleftharpoons B^* \qquad \text{denaturation}$$
$$B^* + B^* \rightarrow B_2^* \rightarrow \cdots \rightarrow B_n \qquad \text{aggregation}$$

The first denaturation step in principle is an equilibrium reaction between the native state and the partly unfolded state of the protein. Upon unfolding, the thiol group is exposed and becomes reactive.[6] The unfolding, or rather the conformational change, is considered to be a first-order reaction although it may involve a number of consecutive changes in the molecule. After denaturation, several irreversible aggregation reactions (step 2) can take place. All the aggregation steps are bimolecular and are therefore second-order reactions. In step 2 the formation of protein particles takes place *via* exchange reactions between free reactive thiol groups and disulfide bonds (chemical aggregation) or physical aggregation of unfolded protein molecules, or a combination of chemical and physical aggregation. The reaction order of the overall reaction is expected to be somewhere between 1 and 2 depending on the ratio of the reaction rates of the different steps.

3 Materials and Methods

Purified β-lg was prepared at NIZO from cheese whey, basically following the procedure of Maubois.[7] The sample contained 91% β-lg, a few percent of α-lactalbumin, less than 1% salt and approximately 3% water. NaCl was obtained from Merck (analytical grade). Solutions were made using double distilled water. The β-lg was dissolved in 0–500 mM NaCl solution with the β-lg concentration in the range 2–30 g l^{-1}.

The reaction kinetics of the heat-induced denaturation and aggregation were followed by heating filtered β-lg solutions (0.2 μm non-protein-adsorbing filter) in test tubes at 68.5 °C for different time periods. The tubes were cooled in ice water, the pH was adjusted to 4.7, and the denatured/aggregated proteins were sedimented by centrifugation for 30 min at 20 000 *g*. The native β-lg concentration in the supernatant was determined by high-performance gel permeation chromatography.[8]

The average size of the aggregates formed during heating at 68.5 °C was measured *in situ* using a dynamic light-scattering set-up (90° configuration, wave vector 0.0186 nm^{-1}). Both the time-averaged scattering intensity and the effective Stokes–Einstein particle diameter were evaluated as a function of time. The β-lg solutions were double filtered (0.1 μm non-protein-adsorbing filters) before use. The calculated Stokes–Einstein particle diameter (cumulant method) was only an apparent diameter since the protein particles formed are regarded as homodisperse hard spheres and no corrections were made for polydispersity and concentration effects.

4 Results and Discussion

The effect of NaCl on the kinetics of the heat-induced denaturation and aggregation of β-lg is depicted in Figure 1 for an initial β-lg concentration of 10 g l^{-1}. The NaCl concentration has an enormous effect on the denaturation/aggregation reaction at 68.5 °C. Up to 20 mM NaCl the conversion rate of native β-lg is increased and a further increase in the NaCl concentration induces a gradual decrease in the conversion rate of β-lg. The experimental points are fitted according to different reaction orders using the general equation for reaction kinetics:[8]

$$-\frac{dC}{dt} = k_n\,C^n \tag{1}$$

where *t* is the heating time (s), *C* is the concentration (g l^{-1}), k_n is the reaction rate constant (g^{1-n} l^{n-1} s^{-1}) and *n* is the order of the reaction. The order of the best fit through the experimental points shifts from 1.5 or higher to 1.0 with increasing NaCl concentration. In Figure 1 the drawn lines are calculated according to a reaction order of 1.5 and the broken lines with reaction order 1.0. The experimental reaction orders are in agreement with the reaction scheme proposed in Section 2 where the subsequent denaturation and aggregation reactions are, respectively, first and second order reactions. The shift to lower order with increasing NaCl concentration indicates that the first step in the reaction scheme, the denaturation or unfolding step, becomes more and more rate limiting with increasing NaCl concentration. This indicates that the denaturation reaction rate is decreased by the salt and/or the aggregation reaction rate is increased. As the overall reaction rate (Figure 1) strongly decreases at high NaCl concentrations, the shift in reaction order cannot be caused by only an increase in aggregation rate. Consequently, we can infer that at least the denaturation reaction is decreased by the addition of NaCl. The increase in the overall conversion rate of native β-lg at low NaCl concentration also points to an increase in the aggregation rate with increasing NaCl concentration. The decreasing

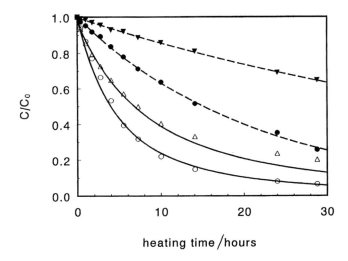

Figure 1   *Relative native β-lg concentration as a function of heating time at 68.5 °C.*
Initial β-lg concentration is $C_0 = 10$ g l^{-1}. NaCl concentrations: △, 0 mM;
○, 20 mM; ●, 200 mM; ▼, 500 mmM. Fitted curves with reaction orders:
————, 1.5; - - -, 1.0

denaturation rate strongly exceeds the increasing aggregation rate at high
NaCl concentration. The decrease in the unfolding rate can be explained by
the stabilizing effect of NaCl on the structure of native β-lg. Salts like NaCl
can have a stabilizing effect on the native structure by increasing the degree
of hydration of the protein.[9] The increase in aggregation rate induced by

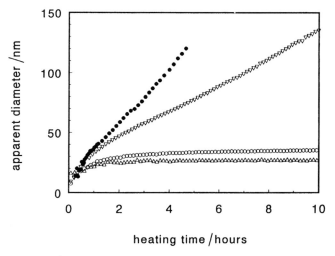

Figure 2   *Apparent diameter against heating time at 68.5 °C for β-lg solutions of 10 g*
l^{-1}. NaCl concentrations: △, 0 mM; ○, 20 mM; ▽, 100 mM; ●, 200 mM

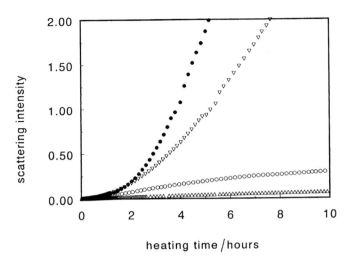

Figure 3 *Scattering intensity (arbitrary units) as a function of heating time at 68.5 °C for β-lg solutions of 10 g l⁻¹. NaCl concentrations: △, 0 mM; ○, 20 mM, ▽, 100 mM; ●, 200 mM*

NaCl is explained by an increased screening of the charge of the protein molecules.[9,10]

In Figures 2 and 3 the apparent diameter and the scattering intensity at 90^c are shown; these were measured *in situ* at 68.5 °C for β-lg samples with an initial concentration of 10 g l⁻¹. In Figure 2 it is seen that, without added salt, the size of the protein particles initially grows rapidly and after a short time reaches a more or less constant particle size of 25–30 nm. In water β-lg polymer particles are formed,[6] and comparison with Figure 1 indicates that the polymer particles do not grow in size but in number concentration upon heating. If NaCl is added prior to heating, the apparent diameter of the protein aggregates no longer remains constant but continues to grow with time. The particle size strongly increases with NaCl concentration, probably due to a decreased solubility of the protein particles in water (the 'salting-out' effect).[10]

The scattering intensity (Figure 3) is even more affected by the NaCl concentration than is the particle size. To a first-order approximation (small particles, low particle concentration, small wave-vector), it is given by[11]

$$I \sim C M \tag{2}$$

where I is the scattering intensity, C is the weight concentration of the particles (g l⁻¹) and M is the molecular mass of the particles (g mol⁻¹). On heating a β-lg solution the scattering intensity will increase because of the growing particle size and the growing particle concentration (see Figure 3). Without NaCl it increases only very gradually, whereas with NaCl two stages can be discerned: a slow initial increase in intensity and a subsequent steep increase. This observation can be explained by considering the aggregation as a two-step process, with, in the first step, the formation of relatively small aggregates as in the absence of salt and in the second step the aggregation of these smaller aggregates into larger entities which can be regarded as a secondary aggregation.

This two-step aggregation process was investigated in more detail by heating at 68.5 °C solutions with four different β-lg concentrations each dissolved in 100 mM NaCl. For each β-lg concentration the onset time for secondary aggregation has been determined. This onset time is defined as the point of intersection between the initial slope of the scattering intensity *versus* time curve and the slope of the secondary stage of the scattering intensity as indicated in Figure 4. The onset time of the secondary aggregation increases with decreasing initial protein concentration. If aggregation is subject to the above described two-step process it can be assumed that the secondary step is induced at a critical concentration of primary aggregates. Then, for different initial β-lg concentrations, the same critical concentration should be found. At 100 mM NaCl the overall order of the denaturation/aggregation process is close to unity (see Figure 1) and therefore the initial rate of the formation of primary particles is proportional to the initial β-lg concentration. This means that the time required to reach the critical concentration of primary particles is inversely proportional to the initial β-lg concentration and, likewise, the onset time of the secondary aggregation should also be inversely proportional to the initial β-lg concentration. As indicated in Figure 5 a linear dependence between the onset time of the secondary aggregation and the reciprocal of the initial β-lg concentration is found. This confirms that the heat-induced aggregation of β-lg in NaCl solutions can be regarded as a two-step aggregation process, in which β-lg monomers are first transformed into

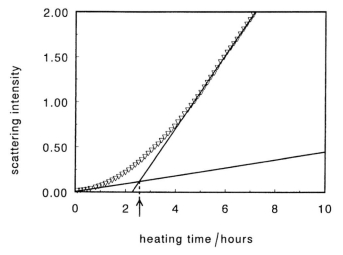

Figure 4   *Scattering intensity (arbitrary units) as a function of heating time at 68.5 °C for a β-lg solution of 10 g l⁻¹ containing 100 mM NaCl. Arrow indicates definition of the onset time of secondary aggregation*

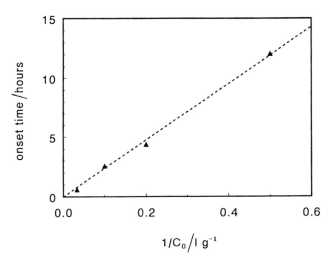

Figure 5   *Onset time of secondary aggregation (see Figure 4) as a function of the reciprocal of the initial β-lg concentration C_0 at a heating temperature of 68.5 °C. NaCl concentration is 100 mM*

primary aggregates and subsequently these primary particles aggregate further to form larger secondary aggregates.

5 Conclusions

The heat-induced denaturation and aggregation of β-lg is a complex process in which intra- and intermolecular changes and reactions occur. Both intra- and intermolecular processes are very much affected by the concentration of NaCl. The overall reaction order shifts from 1.5 to 1.0 when the NaCl concentration is increased from 0 mM to 500 mM, which is in accordance with the proposed simplified reaction scheme. The denaturation or the unfolding rate of β-lg is decreased by NaCl probably due to a stabilizing effect on the native structure. Added NaCl seems to increase the rate of intermolecular aggregation by screening the charge of the protein and to induce the formation of larger aggregates by decreasing the solubility of the protein in water. The heat-induced aggregation of β-lg in NaCl solutions proceeds in a two-step process in which the second step starts at a critical concentration of primary particles.

Acknowledgements

The authors thank Eric Driessen for performing some of the experiments. This research was financially supported by the Ministry of Economic Affairs through the programme IOP-Industrial Proteins, by Friesland Frico Domo/ Dairy Foods, Beilen, and by Coberco Research, Deventer.

References

1. D. M. Mulvihill and M. Donovan, *Ir. J. Food Sci. Technol.*, 1987, **11**, 43.
2. C. Georges, S. Guinand, and J. Tonnelat, *Biochim. Biophys. Acta*, 1962, **59**, 737.
3. M. Paulsson, 'Thermal Denaturation and Gelation of Whey Proteins and their Adsorption at the Air–Water Interface', PhD dissertation, University of Lund, Sweden, 1990.
4. D. M. Mulvihill, D. Rector, and J. E. Kinsella, *Food Hydrocolloids*, 1990, **4**, 267.
5. M. Langton and A.-M. Hermansson, *Food Hydrocolloids*, 1992, **5**, 523.
6. S. P. F. M. Roefs and C. G. de Kruif, *Eur. J.Biochem.*, in the press.
7. J. L. Maubois, *Bull. IDF*, 1979, **212**, 154.
8. J. N. de Wit, *J. Dairy Sci.*, 1990, **73**, 3602.
9. T. Arakawa and S. N. Timasheff, *Biochemistry*, 1987, **26**, 5147.
10. M. M. Kristjánsson and J. E. Kinsella, *Adv. Food Nutr. Res.*, 1991, **35**, 237.
11. H. C. van de Hulst, 'Light Scattering by Small Particles', Dover Publications, New York, 1981.

Rheological and Mechanical Properties

Mechanical Properties of Concentrated Food Gels

By Ton van Vliet

DEPARTMENT OF FOOD SCIENCE, WAGENINGEN AGRICULTURAL UNIVER-
SITY, PO BOX 8129, 6700 EV WAGENINGEN, THE NETHERLANDS

1 Introduction

The two main reasons for studying the mechanical properties of food gels are: (i) they form important quality characteristics of many food products with respect to handling properties and eating quality; and (ii) they may yield information on the structure of the product. The latter is the motive of much scientific research in this area. It mostly concerns experiments at such *small deformations* that the structure of the material is not affected by the experiment. Most theories, relating the structure of a gel to its mechanical properties, apply only to this kind of experiment. However, food technologists involved in product development or quality assurance are interested in the mechanical properties of the gel in relation to shaping, stand-up, handling, cutting/slicing, or eating characteristics. Most of these properties relate to *large deformation* behaviour, including fracture characteristics at various deformation rates, and only to a lesser extent to those determined at small deformations (see Table 1). For concentrated gels, fracture properties involved in, for instance, mastication or slicing may comprise not only well known characteristics such as fracture stress, strain and energy, but also lesser known characteristics such as notch sensitivity, crack propagation rate and the difference between the critical stresses for initiation and propagation of cracks.[1-3]

A study of these practically important properties requires that (i) the physical parameters are measured and the relevant time-scales are established; (ii) experiments are set up in accordance with the outcome of (i); (iii) theory is developed for the behaviour at large deformations and for fracture and yielding phenomena, starting at the relevant scale of the structure (*e.g.*, for the yielding behaviour of a fat, the size and geometrical structure of and the interaction forces between clusters of aggregated fat crystals, and not of the fat molecules); and (iv) finally, if at all possible, an explanation is sought at the molecular level.

Table 1   *List of the main practical and relevant mechanical properties of gels.*
'Time' relates to the time period that a stress is exerted on a product

Property desired	Relevant parameter	Relevant conditions
Thickness	Apparent viscosity after yielding	Stress/strain rate
Firmness	Modulus, fracture stress, yield stress	Time, strain, strain rate
Stand-up	Yield stress, (apparent viscosity)	Time, stress
Shaping	Yield stress + restoration time on setting	Strain rate, stress
Immobilization of solvent	Permeability, tan δ, modulus, yield stress	Stress, time
Handling, slicing	Fracture stress and strain, work of fracture	Strain rate
Eating characteristics	Yield or fracture properties, modulus at large deformation	Strain rate

Several basic types of food gels can be distinguished. Extreme types are gels of cross-linked flexible macromolecules (*e.g.* gelatin gels) and gels of hard particles (*e.g.* margarine). In between, we have, for instance, gels made of cross-linked relatively stiff macromolecules (most polysaccharide gels) and gels of deformable particles (*e.g.* milk gels). By adjusting the concentration, gels can be made in each of the mentioned classes that all have about the same small deformation properties: however, their large deformation properties will usually vary greatly, illustrating that small and large deformation properties need not be related.

In this paper, the possibilities and impossibilities of extrapolating mechanical properties determined at small deformations to the (mostly much more relevant) mechanical properties at large deformations (Table 1) will be illustrated. As an example of the basic difficulty or impossibility of obtaining a simple relation between 'fundamental' mechanical properties and usage properties, a rather speculative analysis will be made of the fundamental mechanical properties required to characterize fully the crispy behaviour of a food.

2 Relation between Small and Large Deformation Properties

Parameters often used to describe the rheological behaviour of gels are the shear or Young's moduli determined in the so-called linear region (*i.e.* the strain region where the stress required to deform a material is proportional to the strain). For visco-elastic gels, the storage modulus G' (a measure of the energy stored during a periodic application of stress) and the loss modulus G'' (a measure of the energy dissipated), and their ratio tan $\delta = G''/G'$ (called the loss tangent), as determined by dynamic mechanical spectroscopy, give a good characterization of the small deformation properties.[4] The timescale of the experiment is the time during which a stress of a certain size and direction is

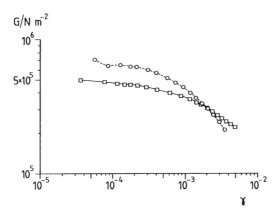

Figure 1   *The shear modulus G of short doughs[5] differing in fat content as a function of the applied strain: (○) standard dough (24% fat), (□) low fat dough (9.9% fat)*

applied to the gel. In a dynamic experiment it is equivalent to the reciprocal frequency ω^{-1} of the sinusoidally oscillating stress or strain.

For most foods the ratio between the stress and strain is not constant at large deformations, *i.e.* the stress–strain relation becomes non-linear. The deviation from linearity is a product-dependent characteristic. In this non-linear region, products can still be characterized by a modulus, but its magnitude will depend on the strain applied and often also on the strain history. It is this difference in non-linear characteristics that makes margarine and gelatin gels so different. The same applies to standard and low-fat dough, whose shear moduli, as a function of shear strain, are shown in Figure 1. The low-fat dough (fat content 9.9%) has a lower modulus than the standard dough (24% fat) at small deformations, but the reverse is true at large deformations. This higher resistance to deformation at large strains means that depositing the low-fat dough on the baking plate is more difficult than depositing the standard dough.

Characterization of the fracture behaviour of food is more complicated than that of the rheological properties.[2] The parameters commonly used are the fracture stress and strain, and sometimes also the work of fracture, but others may also play a part.[1-3] Only in certain cases is there a direct relation between the rheological properties determined at small deformations and the parameters describing fracture. For an ideally elastic material exhibiting brittle fracture, a direct relation may be expected between the modulus and the fracture stress as long as the fracture strain does not change. In the case of visco-elastic materials, a direct relation can sometimes be observed between the loss tangent and the dependence of the strain at fracture on the deformation rate.[3] However, more often than not, there is no simple relation between small deformation, large deformation and the fracture properties. The prime reason for this is the inhomogeneity of most materials, as will be illustrated below.

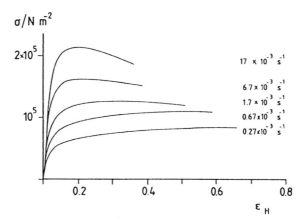

Figure 2   *Stress* versus *Hencky strain [ln* (H_t/H_0) *where* H_0 *is the initial height and* H_t
*is the actual height of the test piece] for 12% glycerol lacto-palmitate gels in
a 2% sodium caseinate solution at various strain rates* $\dot{\varepsilon}$ *(indicated).
Extensive yielding occurs at a strain of ca. 0.05, but yielding starts earlier.
The test pieces were found to show macroscopic fracture at the end of the
curves (from reference 8)*

The starting point for theoretical analysis of the fracture and yielding of materials is that these processes are initiated by the growth of defects (weak spots) which cause stresses to be locally higher.[1,6,7] Important factors are the shape and size of the defects as well as other material properties. Fracture or yielding of a material starts if somewhere within it the local stress exceeds the breaking stress of the bonds between the structural elements responsible for the solid-like character of the material. For yielding, this has to occur at many places throughout (a considerable part of) the material at roughly the same time.[1] Fracture occurs when a small crack grows to a size comparable with the size of the material. A small crack may grow spontaneously (called fracture propagation) if the growth causes more strain energy to become available than the energy required to form two new surfaces.[1,6] The net strain energy available for crack propagation strongly depends on the energy-dissipating mechanisms that occur during deformation of the material and during the transport of the strain energy towards the crack tip.[3] Important energy dissipating mechanism are (i) viscous flow in visco-elastic materials, and (ii) friction between different structural elements of a material due to its inhomogeneous deformation at large strains. The first mechanism will cause the fracture strain to be lower at faster rates of deformation (strain rates), whereas the reverse may be the case if extensive energy dissipation caused by friction occurs.[3] Experiments in the linear region may yield some information on visco-elasticity, but not on the contribution of friction processes to large deformation behaviour. The same applies for flow processes occurring after yielding of (part of) a material, as observed for glycerol lacto-palmitate gels in caseinate solution (Figure 2).

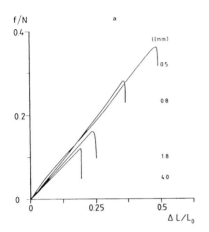 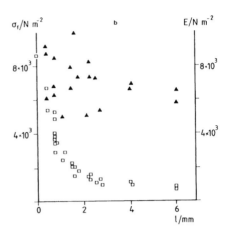

Figure 3   *(a) Force f versus extension ΔL curves of 10% potato starch gels to which a notch of depth l had been applied (value indicated); L_0 is the initial length of the test peice. (b) Fracture stress σ_f (□) and Young's modulus E (▲) as a function of the depth of the applied notch l in tension for similar gels at 20 °C (initial strain rate $2.11 \times 10^{-3}\ s^{-1}$)*

Role of Large Defects

Fracture properties are strongly affected by the presence of large inhomogeneities.[1,2,6,7] In so-called notch-insensitive materials, the overall fracture stress σ_f decreases in proportion to the relative decrease in cross-section of the test piece due to the notch. In a notch-sensitive material, the decrease is much stronger. (For an example see Figure 3b.) However, the shape of the curves of force *versus* relative deformation is roughly independent of the notch length (Figure 3a). Moreover, there is no clear effect of notch length on the value of the Young's modulus E within the accuracy of the experiment (error ± 12%). The effect of a large inhomogeneity on fracture properties is thus much more pronounced than that on a small deformation property, such as Young's modulus.

In linearly elastic materials, the following relation between the overall fracture stress and the length of a notch applies, both for fracture initiation and for propagation:[6,7]

$$\sigma_f \propto \sigma_1\, c/\sqrt{l} \qquad (1)$$

In eqn (1), σ_1 is the local stress at the point at which fracture occurs, and l is the notch length or half the crack length. The quantity c is a constant which is of the order of the square root of the inherent defect length of the material in

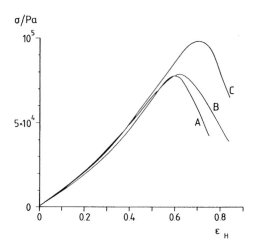

Figure 4   *Stress σ versus Hencky strain ε_H in uniaxial compression for 30% wheat starch gels: (A) no addition; (B) 4.2 vol% glass spheres (diameter ≈90 μm) added; (C) 4.2 vol% glass flints (diameter ≈0.5 μm) added[14]*

the case of fracture initiation, and which depends on the ratio of the energy required for forming two new surfaces to the elastically stored energy for fracture propagation. The observed data, for σ_f as a function of notch length shown in Figure 3b, are in accordance with eqn 1. Because the inherent defect length for starch gels is of the order of 0.1 mm, a notch of length $l = 1$ mm already has a large effect on σ_f.

In the case of a small deformation, the following relation can be deduced between the modulus G and the properties of the gel network:[9,10]

$$G = NC \, (\mathrm{d}^2 A / \mathrm{d} x^2) \qquad (2)$$

In eqn (2), N is defined as the effective strand length per unit volume, C is a characteristic length scale, determined by the geometry of the network, which relates the local deformation to the macroscopic strain, and $\mathrm{d}A$ is the change in free energy due to a change $\mathrm{d}x$ in distance between the cross-links. The presence of a notch will make a certain number of elastically effective strands ineffective, and so will lower G. The extent of the effect will depend on the relative proportion of strands that become ineffective. The results presented in Figure 3 were obtained in a tensile test. The size of the specimen was $40 \times 10 \times 10$ mm.[11] After applying a notch of length l about two triangular volumes, each of a size of approximately $\frac{1}{2}l^2$ times the specimen thickness, the specimen becomes stress free.[1,6] For the results presented in Figure 3, this implies a decrease in modulus E by 4 or 9% due to a notch of 4 or 6 mm, respectively. (Note that 6 mm is 60% of the thickness of the test piece.)

Effect of Dispersed Particles

The effect of particles on large deformation properties is much more difficult to predict than that on small deformation properties, and often it seems that there is no relation between the two.[12] An example is given in Figure 4 where stress *versus* strain curves are given for pure 30% wheat starch gels and for gels containing 4.2 vol% glass particles of average size *ca.* 0.5 μm or 90 μm. The large particles are relatively smooth spheres, whereas the smaller ones are flints. In the filled gels the value of E is approximately 10% higher, in accordance with theory.[13] For the glass spheres the values of σ_f and ε_f remain about the same, but for the flints a clear increase is found (Figure 4). Addition of the flints results in an increase of the specific fracture energy, as determined by cutting experiments, of approximately 40–60%, whereas for the glass spheres this increase is by approximately 10–25%. These higher specific fracture energies probably stem from greater energy dissipation due to friction at large deformations. Such an increase would result in larger values of σ_f and ε_f.[3] The 90 μm glass spheres are (on average) only somewhat smaller than the inherent defect length (*ca.* 140 μm) in the starch gels, and so they may give rise to a slightly stronger local stress concentration. The last effect probably compensates for the effect of the higher friction. Besides, the friction term is probably larger in the presence of the small glass particles with the inherently larger surface area.[14]

3 Relation between Sensory Terms and Fundamental Parameters

In sensory assessment a combination of rheological parameters is always perceived. This makes it difficult to predict sensory assessments by consumers from results of measurements yielding fundamental rheological parameters. Therefore an empirical method, by which mostly the effect of a combination of fundamental parameters is determined, may work better. Moreover, in research, not always the right parameters are determined. Below, we give a partly speculative, theoretical analysis of fundamental, experimentally accessible, characteristics required to obtain crispy behaviour, in order to illustrate the difficulties encountered when trying to predict the handling or eating properties of a food.

In general, a crispy food is dry and difficult to deform, but it breaks easily producing a sharp sound. The acoustic property is an essential characteristic and is sometimes even used to characterize crispy behaviour.[15] The characteristic 'not easy to deform' implies a (rather) high modulus E. 'Breaking easily' implies that the fracture stress and fracture energy are rather low. A low fracture energy for a product having a high E implies that the fracture strain must be small. Moreover, the energy dissipation must be small, and so the material has to behave elastically up to the fracture strain, and the energy dissipation due to friction between the various structural components must be small. To have a low fracture stress in a material with a high modulus, the material has to contain rather large defects and be notch sensitive. The high stress concentration will lead to a large reduction in fracture stress and to a reduction in fracture strain.[16] To generate audible noise, the crack has to proceed at such a high speed that shock waves with the speed of sound are formed. This would imply that energy consumption during crack propagation is small and involves only the required surface free energy (specific fracture energy). Dissipation of the strain energy (becoming available from stress relaxation on further crack growth) during its transport to the crack tip would immediately cause a decrease in crack speed and thereby result in sound of a much lower pitch or even no audible sound at all. The above properties would also seem to imply that crack propagation must closely follow crack initiation.

Fundamental studies to investigate these mechanical characteristics should involve tension or bending tests in which (i) E, σ_f, ε_f and fracture energies are determined as a function of strain rate, and (ii) the effect of artificial notches on σ_f and ε_f is studied.[2] From the results the inherent defect length should be determined. Moreover, fracturing should be filmed, using a high-speed camera to observe fracture initiation and propagation and to determine crack speeds. Besides, although not a mechanical characteristic, the sound produced should be registered. In fact, the last measurement could probably replace the determination of crack speed. For quality control, or for studies meant to obtain a rough indication of crispness quality, a simple bending test combined with sound analysis may be the best approach.

4 Conclusion

For most foods there is no direct relation between the small and the large deformation properties, including fracture and yielding behaviour. An important factor is the inhomogeneous structure. This, in combination with the fact that most practically relevant properties of food gels are based on a combination of several fundamental mechanical parameters, makes a prediction of these properties a complicated, albeit feasible, task.

Acknowledgement

The author gratefully acknowledges valuable discussions with Dr H. Luyten and Professor P. Walstra.

References

1. T. van Vliet, H. Luyten, and P. Walstra, in 'Food Polymers, Gels and Colloids', ed. E. Dickinson, Special Publication No. 82, The Royal Society of Chemistry, Cambridge, UK, 1991, p. 392.
2. H. Luyten, T. van Vliet, and P. Walstra, *J. Texture Stud.*, 1992, **23**, 245.
3. T. van Vliet, H. Luyten, and P. Walstra, in 'Food Colloids and Polymers: Stability and Mechanical Properties', ed. E. Dickinson and P. Walstra, Special Publication No. 113, The Royal Society of Chemistry, Cambridge, UK, 1993, p. 175.

4. J. D. Ferry, 'Visco-elastic Properties of Polymers', Wiley, New York, 1980.
5. A. Baltsavias, A. Jurgens, and T. van Vliet, to be published.
6. A. G. Atkins and Y.-M. Mai, 'Elastic and Plastic Fracture', Ellis Horwood, Chichester, 1985.
7. H. L. Ewalds and J. H. R. Wanhill, 'Fracture Mechanics', Delftse Uitgevers Maatschappij, Delft, The Netherlands, 1984.
8. J. M. M. Westerbeek, PhD Thesis, Wageningen Agricultural University, The Netherlands, 1989.
9. T. van Vliet and P. Walstra, *Neth. Milk Dairy J.*, 1985, **39**, 115.
10. L. G. B. Bremer and T. van Vliet, *Rheol. Acta*, 1991, **30**, 98.
11. H. Luyten, M. G. Ramaker, and T. van Vliet, in 'Gums and Stabilisers for the Food Industry', ed. G. O. Phillips, D. J. Wedlock, and P. A. Williams, IRL Press, Oxford, 1992, vol. 6, p. 101.
12. H. Luyten and T. van Vliet, in 'Rheology of Food, Pharmaceutical and Biological Materials with General Rheology', ed. R. E. Carter, Elsevier Applied Science, London, 1990, p. 43.
13. T. van Vliet, *Colloid Polym. Sci.*, 1988, **266**, 518.
14. K. Grolle, H. Luyten, and T. van Vliet, to be published.
15. Z. M. Vickers and M. C. Bourne, *J. Food Sci.*, 1976, **41**, 1158.
16. P. Purslow, *J. Mater. Sci.*, 1991, **26**, 4468.

Scaling Behaviour of Shear Moduli during the Formation of Rennet Milk Gels

By David S. Horne

HANNAH RESEARCH INSTITUTE, AYR KA6 5HL, UK

1 Introduction

Curd or gel formation following the inoculation of milk with proteolytic enzymes (commonly known as rennet) forms an early stage in the manufacture of cheese. The rennet hydrolyses a specific bond of κ-casein which destabilizes the colloidal casein micelles. It is generally accepted today that κ-casein forms a stabilizing coat on the surface of the casein micelle, the colloidal aggregate of the casein proteins of milk.[1,2] Upon its removal by enzymic hydrolysis to a soluble glycomacropeptide and a larger hydrophobic N-terminal peptide which remains with the micelle, the micelles become essentially insoluble and in quiescent conditions they aggregate to form a gel or curd.

In most research devoted to this topic, there has been a tendency to concentrate on the mechanism of the initial stages, the destabilization of the casein micelle, and the beginnings of aggregation,[3–6] and to neglect the development of firmness, of mechanical strength in the gel, a stage which is regarded as crucial in the manufacture of high-quality products. In those studies devoted to development of gel strength, a wide range of measuring devices have been employed. Unfortunately, in the majority of cases, these have provided results only in the form of arbitrary instrument units, reducing the analysis of the data to a largely phenomenological description of what is taking place. More recently, however, several groups have begun to employ rheological techniques with gel visco-elastic properties quantified in absolute terms.[7–12]

In this paper, the kinetics of the formation of rennet milk gels are monitored using dynamic oscillatory rheometry. An analysis of the resulting asymmetric sigmoidal curves shows that they can be reduced to a single master curve by a judicious choice of scaling factors for both the time and gel strength axes.

2 Materials and Methods

All measurements were made on a Bohlin VOR Rheometer (Bohlin Reologi, Lund, Sweden) operating in sinusoidal oscillatory mode with a Couette-type bob and cup system (bob diameter 25 mm). The angular oscillation of the cup was servo-driven and the bob was suspended from a calibrated torsion element, whose angular deflection was sensed by a differential transformer. After correction for resonance effects and torsion element compliance, phase angles and dynamic shear moduli were calculated via fast Fourier transform on the dedicated PC running the rheometer.

Fresh raw milk was collected from individual cows in the Hannah Research Institute herd at their morning milking and skimmed by centrifugation at 4000 *g* to remove fat. The milk was equilibrated at the reaction temperature of 30 °C in a thermostatted bath for at least 30 min before enzyme addition. The enzyme used in all our experiments was a cloned chymosin (Maxiren 15, Gist-Brocades), diluted 0.3 ml to 10 ml with 120 μl of this diluted solution being added to 20 ml of milk to initiate the reaction. The inoculated milk sample was stirred for 2 min to ensure thorough dispersion and mixing, before being transferred to the thermostatted cup of the rheometer, where measurements of shear moduli at 30 s intervals over several hours were begun, generally 4 or 5 min after initiation of the reaction.

3 Results and Discussion

The experiments reported here have studied the complete maturing process of gel formation. If the oscillation frequency is sufficiently high, the increased collision rate above the diffusional background may promote gelation and increase the rate of firming. Conversely, excessive strain may break bonds

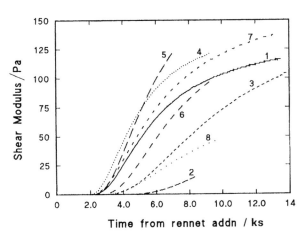

Figure 1 *The range of gel firming kinetics encountered in the milks of eight individual cows following inoculation of the milks with chymosin. Plots show the complex elastic modulus as a function of time after enzyme addition. All reactions were carried out at the natural pH (6.6) of the milk at 30 °C, adding 120 μl of diluted enzyme to a 20 ml sample of the milk. The numbers on the curves assist in identifying the data in Figures 2 and 3*

already formed. Any rheological measurement on a gelling system is therefore a compromise. For all our measurements we have chosen an angular frequency of oscillation of 0.1 Hz and a shear strain of 2%. Although the rheometer calculates the phase angle at which measured stress lags behind applied strain and allows derivation of the loss and storage components of the elastic modulus, only the behaviour of the complex modulus $G^* = (|G'|^2 + |G''|^2)^{\frac{1}{2}}$ of the rennet gels is considered in this paper.

Typical results for the renneting behaviour of the milks drawn from a series of individual cows are shown in Figure 1. Each curve of complex modulus as a function of time after enzyme addition shows a lag period where no elastic modulus is detectable. Instrument noise during this period amounts to less than 100 mPa. The presence of a gel is then signalled by detection of a value above the noise level, followed by a rapidly accelerating increase in complex modulus. The time at which this detectable signal is perceived is designated the gel time t_g.

As is apparent from Figure 1, the gel time varies from milk sample to milk sample, as does the rate of increase in shear modulus (also known as the rate of firming) and the ultimate gel strengths of the curds. Our aim is to designate the important parameters which characterize these kinetic curves so that their response to variations in milk composition and processing treatments can be mapped and thus a greater understanding of cheese technology can be obtained. Without suggesting that the same mechanism is being followed here, Horne and Dalgleish[12] have found that the curves of increase of turbidity in the calcium-induced precipitation of α_{s1}-casein, the individual casein protein, also showed a lag phase before aggregation followed. Furthermore, the kinetic aggregation profiles obtained at the same protein concentration with different calcium levels could be reduced to a single plot when the time axis for each run was scaled by dividing by the coagulation time for that particular run. It was

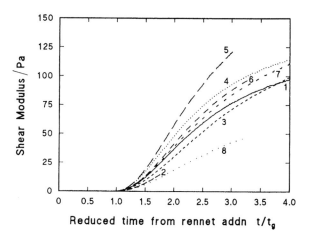

Figure 2   *The elastic modulus data of Figure 1 replotted as a function of reduced time*
t/t_g, where t_g is the gelation time for each particular run, as defined in the text

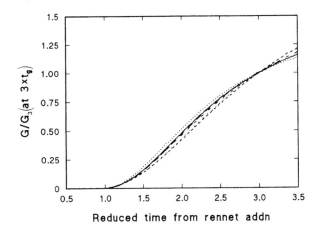

Figure 3   *The data of Figure 2 with elastic modulus G for each curve normalized to the value of the modulus G_3 attained at the reduced time $t/t_g = 3$*

subsequently demonstrated that the calcium-induced aggregation of α_{s1}-casein follows the form of a nucleation and growth mechanism.[13] Although the mechanism by which the aggregating particles are produced is different, *i.e.* although the micelles are destabilized by enzymic hydrolysis in the production of rennet gels, we have replotted in Figure 2 the gelation profiles of Figure 1 as a function of reduced time t/t_g, with t_g varying for each individual curve. The outcome of this exercise is not a single curve as was obtained for the α_{s1}-casein series but a series of curves now apparently ordered according to ultimate gel strength, the value of the elastic modulus at infinite time. This suggests, in itself, that the vertical (elastic modulus) axis should be normalized against this parameter. Determining the asymptotic value of the elastic modulus, G_∞, is not without problems, however, and so we choose to normalize these reduced time plots to the value of the elastic modulus at a particular value of the reduced time, at $t^* = t/t_g = 3$. The result is that the separate curves of Figures 1 and 2 reduce to a common universal curve as shown in Figure 3. The curves profiling the development of gel strength in a rennet curd are therefore demonstrating a simple scaling behaviour.

The presence of scaling behaviour has several implications. It tells us that the mathematical expression describing the time-dependent growth of gel strength can be factorized into the produce of two functions, *i.e.* it can be expressed as

$$G = G_\infty f(t/t_g) \tag{1}$$

One part of eqn (1) is a simple constant, G_∞, the asymptotic value of the shear modulus for the particular reaction. The second function is more complex, but it is nevertheless dependent on the single reduced time variable, t/t_g, and all data sets are taken to follow the same function. With this in mind, we can note that the value of the elastic modulus G at some defined value of the reduced time, say $t^* = t/t_g = 3$, will always be the same fraction of G_∞, since defining the value of t/t_g fixes the value of $f(t/t_g)$. Thus there is no absolute necessity to attempt to determine the value of G_∞. Between milks, comparison of the value of G at a reaction time of $3t_g$ will provide the same information.

The scaling behaviour also clearly demonstrates that selecting the rate of firming as a parameter to characterize the reaction gives no extra information. Differentiating eqn (1), we see that the rate of firming, dG/dt, is a compound parameter, depending on both ultimate gel strength G_∞ and the gel time t_g. Its response to compositional or processing changes will thus be a combination of their effect on both parameters. Another way of saying this is that the existence of scaling behaviour reveals the curd elastic modulus and its kinetic development to be largely a function of two independent parameters, the ultimate gel strength or a fixed fraction of it – say shear modulus at a defined multiple of the gel time – and that gel time itself.

The final and most important conclusion to be drawn in this series of observations is the importance of the gel time in defining the rate of development of the mechanical properties of the curd. Generally the process of rennet coagulation is split into a series of distinct, if somewhat overlapping, phases. In the first phase, the actual proteolysis of the casein occurs. Research has shown that some 90% of the κ-casein has to be hydrolysed before the aggregation reaction begins to take place to any significant extent.[14] The distinction between aggregation and the next phase of gel development is by no means clear cut and in many ways it is unhelpful to attempt artificially to make such a division. If the system is left quiescent, as in a cheese vat, the reaction proceeds smoothly through to gelation. If it takes place in a rotating flask, individual clots or large aggregates are formed and no gelation is observed, but it is the difference in the method of observation that alters the outcome and not the reaction itself. Aside from the problems of artificially dividing the later stages of rennet coagulation, nearly all workers regard the proteolysis stage of the reaction to be independent of the second phase, the coagulation of the now destabilized micelles. The gel formation kinetics are timed from what is essentially the end of the first phase, and aggregation is considered to be a completely independent process. The present observation of rheological scaling behaviour, however, clearly demonstrates that this assumption of independence cannot be sustained. The second stage of gel development and its kinetics clearly depend on the rate of the proteolysis reaction, the rate controlling the first stage.

4 Conclusions

Using data on gel formation kinetics obtained by renneting a series of individual cow milks under standard conditions, it has been shown that the individual profiles can be reduced to a single master curve when the time axis is scaled against gelation time for each reaction and the gel strengths for each run are normalized to the value of the elastic modulus at a fixed multiple of the reaction's gel time, typically three times t_g. This scaling behaviour demonstrates that two independent parameters govern the gelation process: the ultimate gel strength G_∞ and the gelation time t_g. The gelation time has also been shown to play a governing role in the kinetics of gel development. The importance of the proteolysis reaction and its rate, which predominantly determines the value of t_g, does not end when gelation begins, but continues on into the aggregation and gel development phase. These are empirical observations, but the quest for a mathematical description and its underlying theoretical basis is well underway and will be presented elsewhere.

Acknowledgements

The author thanks Ms Jo-Ann Smith for expert technical assistance. This research forms part of a LINK project at the Hannah Research Institute funded by the Department of Trade and Industry and the UK Dairy Industry Research Policy Committee. Core funding for the Hannah Research Institute is provided by The Scottish Office Agriculture and Fisheries Department.

References

1. T. A. J. Payens, *J. Dairy Sci.*, 1982, **65**, 1863.
2. D. G. Schmidt, in 'Developments in Dairy Chemistry – 1. Proteins', ed. P. F. Fox, Elsevier Applied Science, London, 1982, p. 61.
3. T. A. J. Payens and J. Brinkhuis, *Colloids Surf.*, 1986, **20**, 37.
4. D. G. Dalgleish, *J. Dairy Res.*, 1988, **55**, 521.
5. D. F. Darling and A. C. M. Van Hooydonk, *J. Dairy Res.*, 1981, **48**, 189.
6. D. B. Hyslop, *J. Dairy Res.*, 1993, **60**, 517.
7. M. Tokita, K. Hikichi, R. Niki, and S. Arima, *Biorheology*, 1982, **19**, 209.
8. L. Bohlin, P. O. Hegg, and H. Ljusberg-Wahren, *J. Dairy Sci.*, 1984, **64**, 729.
9. H. J. M. van Dijk, PhD Thesis, Wageningen Agricultural University, The Netherlands, 1982.
10. P. Dejmek, *J. Dairy Sci.*, 1987, **70**, 1325.
11. P. Zoon, T. van Vliet, and P. Walstra, *Neth. Milk Dairy J.*, 1988, **42**, 249.
12. D. S. Horne and D. G. Dalgleish, *Int. J. Biol. Macromol.*, 1980, **2**, 154.
13. D. G. Dalgleish, E. Patterson, and D. S. Horne, *Biophys. Chem.*, 1981, **13**, 307.
14. D. G. Dalgleish, *J. Dairy Res.*, 1979, **46**, 643.

Sol–Gel Transition of ι-Carrageenan and Gelatin Systems: Dynamic Visco-elastic Characterization

By C. Michon, G. Cuvelier, B. Launay, and A. Parker[1]

LABORATOIRE DE BIOPHYSIQUE DE L'ECOLE NATIONALE SUPERIEURE DES INDUSTRIES AGRICOLES ET ALIMENTAIRES, 91305 MASSY, FRANCE
[1] SANOFI BIO-INDUSTRIE, R&D, FOOD INGREDIENTS DIVISION, BAUPTE, 50500 CARENTAN, FRANCE

1 Introduction

Many food products are structured using hydrocolloids as thickeners or gelling agents. The latter organize themselves in three-dimensional networks throughout the aqueous phase to form reasonably strong gels called 'physical gels'. Contrary to what is found with covalently bonded chemical gels, the cross-links are extended junction zones, consisting of associated segments of two or more flexible chains, and having a finite lifetime.

The fact that gelatin and iota- (ι-)carrageenan form thermo-reversible gels has been known for a long time and is of great importance in industrial applications.[1,2] Even though gelatin is an animal protein, produced from collagen, and ι-carrageenan is a polysaccharide extracted from red seaweed, there are several similarities in their gelling properties. Both biopolymers undergo a conformational disorder–order transition when thermodynamic conditions and solvent quality are changed at sufficiently low temperature. Their extended chains are able to form thermo-reversible networks by associating helices in junction zones stabilized by hydrogen bonds. The mechanisms of their thermo-reversible gelation processes have been extensively investigated.[3-11] Although they are not precisely known, it is generally accepted that the cross-linking junctions in gelatin consist of triple helix structures similar to native collagen.[3,12] For ι-carrageenan, the helicity of the chains and the formation of intertwined double helices are well accepted. But the way they organize themselves in order to create the network is not so well elucidated, although several mechanisms have been successively proposed.[13-15] As the temperatures of conformational change of the two biopolymers are different, the temperature domains of sol–gel transition are quite different also. Both polymers have a

462

broad molecular mass distribution, but the mean molecular mass of the polysaccharide is typically several times higher than that of gelatin. Ranges of biopolymer concentration used are in inverse proportion to molecular mass.

The sol–gel transition is an important phenomenon in the development of material structure. Over the years, several studies have been made of the mechanical properties near this critical point for chemical[16–23] and physical[24–31] gels. Based on their results on plasma desorption mass spectrometry, Winter and Chambon[16,17] have proposed a constitutive equation leading to a power-law variation of the storage modulus G' and the loss modulus G'' with the same exponent Δ as a function of the frequency of oscillation ω. The same type of behaviour has been shown to exist for physical gels:[24,25,27]

$$G'(\omega) \propto G''(\omega) \propto \omega^{\Delta} \qquad (1)$$

Introducing eqn (1) into the Kronig–Kramers relation,[32] irrespective of ω:

$$\tan \delta = \frac{G''}{G'} = \text{constant} = \tan\left(\frac{\Delta\pi}{2}\right) \qquad (2)$$

Using eqn (2), critical parameters (temperature T_c, time t_c, or concentration C_c) corresponding to the formation of an incipient continuous network (at the percolation threshold) can be determined. The functions $\tan \delta = f(T)$, $f(t)$ or $f(C)$ at several frequencies intersect at the critical value of the studied parameter. The additional requirement that $\delta = \Delta\pi/2$ gives a value of the critical exponent Δ. In the absence of any other restrictive hypothesis, Δ values can range theoretically between 0 and 1.[16,17] Experimentally, the value of Δ has been found to range between 0.13 and 0.92 depending on the system (type of polymer, stoichiometry, concentration, etc.) as discussed in a previous paper.[31]

One theoretical approach[33] relates the position of the sol–gel transition to the percolation threshold on a lattice. It corresponds to the critical moment when the largest cluster becomes of infinite size and spans the whole sample volume. The singular mechanical behaviour near the sol–gel transition can be expressed as a power-law function of the distance ε from the critical point [$\varepsilon = (|p - p_c|/p_c)$, with p the reaction extent and p_c its critical value]. The Newtonian viscosity and the equilibrium storage modulus both follow power-laws in ε. Their evolution leads to the determination of exponents s and t, respectively,[34,35] which can be related to Δ by $\Delta = t/(t - s)$. Experimental and theoretical values of the exponents have been compared.[22,26,27,31] In physical thermo-reversible gels, for which structure formation is slow and involves non-permanent junction zones, the determination of ε is difficult and so the exponents t and s cannot be obtained directly. However, as described above [see eqn (2)], estimates of Δ can be obtained from the visco-elastic properties.

This paper is concerned with rheological measurements of the thermo-reversible sol–gel transition of gelatin and ι-carrageenan systems in order to determine the critical parameters of gelation and melting, and to compare the two hydrocolloids in terms of their functionality.

2 Experimental

Acid gelatin and ɩ-carrageenan samples were supplied by Sanofi Bio-Industry. Three gelatin samples (g1, g2 and g3) and two ɩ-carrageenan samples (i1 and i2) were used. Gel permeation chromatography analysis, with PEO calibration, gave the following mean molecular masses in daltons: g1, 182 000; g2, 170 000; g3, 70 000; i1, 1 000 000; i2, 700 000. Aqueous solutions of both polymers were prepared by dispersing the powder at room temperature in distilled water and heating at 60 °C for 45 min (gelatin) or at 75 °C for 30 min and then at 90 °C for 15 min (ɩ-carrageenan). Sodium azide (500 ppm) was added as an antimicrobial agent. The concentration ranges were respectively 1–20 wt% and 0.1–1.6 wt% for the gelatin and ɩ-carrageenan samples.

Dynamic measurements were performed using a Rheometrics Fluids Rheometer (RFR-7800) fitted with coaxial cylinders. The solutions were put in the rheometer at about 60 °C for gelatin and 70 °C for ɩ-carrageenan. A thin layer of paraffin oil was added to protect the samples from dehydration. The linearity of the visco-elastic behaviour was checked over a large strain range (1–30%) at 0.1, 1.0 and 10.0 rad s^{-1} both in the sol and the gel states.

As gelatin and ɩ-carrageenan form thermo-reversible gels, their sol–gel transitions can be studied both from sol to gel (gelation) and from gel to sol (melting). Gelatin systems were thermostated at the required temperature in two different ways: cooling down to a given temperature (T_a) to follow gelation against time or heating step-by-step to follow melting against temperature after an ageing time t_a. For ɩ-carrageenan systems the process of gelation and melting was studied by temperature sweep, cooling or heating step-by-step in order to determine the gelation and melting parameters.

3 Determination of Critical 'Gel' or 'Sol' Points by the Loss-Angle Method

Gelatin and ɩ-carrageenan systems do not appear to organize themselves in the same way. Gelatin systems can be observed evolving slowly before and after the sol–gel transition. Even after 3 weeks of ageing at a given temperature, the storage modulus is still increasing. The formation of triple helices seems to be at the origin of this phenomenon.[5,31] However, ɩ-carrageenan systems reach an apparent equilibrium state much more quickly. The mechanical properties do not evolve significantly after 1 h at a given temperature. The gelation processes of these two types of system cannot therefore be studied in the same way.

Gelation Process *versus* Time for Gelatin Systems

Figure 1 shows an example of gel point determination for a gelatin system at a given temperature ($t = 0$ is arbitrarily defined as the time of starting the cooling-down process from 60 °C). The curves of Figure 1 intersect at the critical value of the studied parameter as given by eqn (2). This enables us to

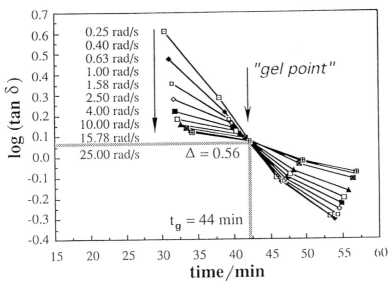

Figure 1 *Gelation of gelatin (g2) (10 wt% in distilled water). Tan δ versus ageing time is plotted for a fast gelatin g2 gelling system for the determination of the critical gel point (gelling time t_g and exponent Δ); $T_a = 28.2$ °C, strain = 5%*

determine a critical gelation time ($t_g = 44$ min) and the corresponding exponent ($\varDelta = 0.56$).

When the system does not evolve significantly during the frequency sweep, it is also possible to calculate $\varDelta$ from the evolution of $\varDelta'$ and $\varDelta''$, which are respectively the slopes $d\log G'/d\log \omega$ and $d\log G''/d\log \omega$ in the vicinity of the percolation threshold: $\varDelta'$, $\varDelta''$ and $\varDelta$ are all equal at the gel point [see eqn (1)].[31] This is not true for the example presented in Figure 1 (the abscissa of each experimental point is the mean time at which a measurement at a given frequency has been done). Here the variation of tan δ at different frequencies using eqn (2) gives a precise determination of the critical conditions and of the value of $\varDelta$. It only supposes that the system remains unchanged during a single measurement at a given frequency.

Equations (1) and (2) do not fix frequency limits. It can be observed that, in the example shown in Figure 1, such power law behaviour is obeyed over two decades (0.25–25 rad s^{-1}). In reality, there are both lower and upper frequency limits. The lower one is an apparent limit due to experimental constraints. It corresponds to a frequency at which moduli are too small to be measured. The upper limit is governed by the organization of the system. The power-law behaviour near the percolation threshold has been theoretically related to the fractal structure of incipient network.[36,37] The upper limit is then linked to the inner fractal structure of the system. At short-length scales, which correspond to very high frequencies, the system is no longer fractal and G' and G'' no longer follow power laws. In addition, this approach does not take into account the contribution of entanglements to the visco-elastic properties of the

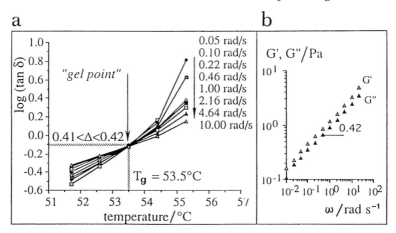

Figure 2   *Gelation of ι-carrageenan (i2) (0.8 wt% in aqueous 0.2 M NaCl). (a) Tan δ*
versus temperature; determination of the critical gel point (T_g *and Δ); step*
time 2 h every 1 °C; strain 5% (b) Storage modulus G' and loss modulus G''
versus frequency ω at $T_g = 53.5$ *°C; strain 5%*

gel. For semi-dilute systems, these contribute significantly to the visco-elastic
behaviour of the system, especially in the high-frequency range. So, in our
gelatin and ι-carrageenan systems, the upper frequency limit is defined in
practice by chain entanglements.

Gelation Process *versus* Temperature for ι-Carrageenan Systems

Gelation has been followed whilst decreasing the temperature step-by-step as
shown in the example of Figure 2(a). The intersection of the tan δ curves
enables the determination of a critical gelation temperature T_g, and of the
corresponding critical exponent Δ as given by eqn (2).

   For the experiment depicted in Figure 2(a), a frequency sweep has been
done at a temperature very close to T_g. Thus, as shown in Figure 2(b), eqn
(1) is verified over a large range of frequencies. The values of Δ' and Δ'', the
slopes on the log scales of $G'(\omega)$ and $G''(\omega)$, are equal at this temperature and
their value (0.42) is the same as that obtained directly from the loss-angle
cross-over method (0.41–0.42). Anyway, as ι-carrageenan systems evolve
quickly, when the frequency sweep is performed at the end of each
temperature step, the system is nearly at equilibrium and does not evolve
during the measurement time. So, in such instances, it is possible to determine
the critical gel point from the evolution of Δ' and Δ''. Both exponents
decrease with temperature as the gelation process occurs and cross at the
critical gel point. This is in good agreement with other published experimental
results.[16,17,23,27,31,32,38] However, this analysis procedure takes much longer
than the tan δ cross-over method, and it often gives less precise values of the
critical parameters.

Melting Process *versus* Temperature for Gelatin and *ι*-Carrageenan Systems

This aspect of the phase transition is interesting because it is specific to a great number of physical gels and has not been previously investigated in detail. For both polymer systems, melting has been followed whilst increasing temperature step-by-step, after an ageing period t_a, at a fixed temperature T_a. The intersection of the $\tan\delta$ curves is called the 'sol point', and it enables us to determine a critical melting temperature T_m and the corresponding critical melting exponent Δ as given by eqn (2). The critical parameters are found to be in good agreement with those obtained from the intersection at the 'sol point' of $\Delta'(T)$ and $\Delta''(T)$.

We note that the values of the critical parameters for gelation (Figure 2) and melting for *ι*-carrageenan at a given concentration are very close ($T_g = 53.5\ °C$, $T_m = 53.6\ °C$, $\Delta_g = \Delta_m = 0.42$). This result seems to confirm that the system is near its equilibrium state at the end of each temperature step, and so we can conclude that there is no significant hysteresis effect.

4 Concentration Effects on the Critical Parameters of the Sol–Gel Transition

ι-Carrageenan systems quickly reach apparent thermodynamic equilibrium. At each concentration, their critical parameters (T_g, T_m, and the corresponding exponents Δ) can be directly obtained as described above. Gelatin systems evolve slowly, however, and they do not reach equilibrium. Their melting parameters (T_m and Δ_m) can also be obtained directly, but with special attention being paid to the thermal history:[31] the critical gelation temperature at each concentration has been obtained by measuring critical gel times at

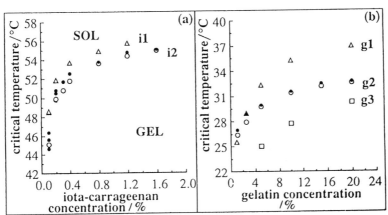

Figure 3   *Phase diagrams showing critical temperature of sol–gel transition versus polymer concentration: (a) ι-carrageenan samples i1 and i2 (0.2 M NaCl); (b) gelatin samples g1, g2 and g3. Open symbols are used to denote gelation and solid ones to denote melting*

different ageing temperatures and by extrapolating T_g to infinite gel time. The corresponding values of Δ during gelation were also extrapolated to the critical gelation temperature.[31]

Thus, critical parameters have been obtained *versus* concentration for all samples (g1, g2, g3, i1, i2). Critical gelation temperatures were used to draw phase diagrams (Figure 3). Only i2 and g2 melting temperatures are reported here in order to clarify the Figures. Only i1 and g2 exponent values *versus* concentration are reported in Figure 4 for the same reason.

Phase Diagrams

For both i2 and g2, T_g and T_m are close to each other. This shows that, if an hysteresis effect exists, it is very small. Therefore, in the chosen experimental conditions, the results are not significantly dependent on kinetic effects.

As expected, the concentration range is much different for ι-carrageenan and for gelatin. The critical temperature range is also not the same: 44–56 °C for ι-carrageenan systems and 24–37 °C for gelatin systems. Critical temperatures increase with concentration and tend towards limiting values depending on polymer type (ι-carrageenan or gelatin) and on sample characteristics. In every case the upper temperature limits seem to have lower values than the conformational transition temperatures, determined by polarimetry measurements, which are approximately 60 °C and 40 °C respectively for our ι-carrageenan and gelatin samples. The coil–helix transition seems to be a condition that is necessary but not sufficient for gel formation. Junction zones have to be present at a sufficiently high concentration and be of sufficient life-times to allow the formation of a permanent network of infinite size.

On the basis of the experimental results for ι-carrageenan, we can conclude that there exists a low concentration limit (about 0.05%) below which no gel could be formed whatever the temperature. This was not found to be the case from the experimental results on gelatin.

The mean molecular mass of the polymer samples has an influence on the phase diagrams. We see in Figure 3 that, at a given concentration, there is a difference of *ca.* 7 °C between the critical temperatures of g1 and g3 and of *ca.* 2 °C between those of i1 and i2. This phenomenon is related to the relative chain lengths of the samples. The mean molecular mass of i1 is clearly higher than that of i2 and, from the molecular mass distribution obtained from chromatography, there is twice the amount of short chains in the g3 sample than in the g1 sample. As proposed by other authors for gelatin,[5,8] it can be hypothesized that the length of stable junction zones depends on temperature. The smallest stable junction zone at a particular temperature is no longer stable at a higher temperature. The chains have to be long enough to be able to participate in several junction zones. So, at a given concentration, it is possible to find a temperature where there are just enough stable junction zones connected to each other to form a network of an infinite size in a given system with long chains but not in a system composed of shorter ones. In the same way, at a given temperature, it is

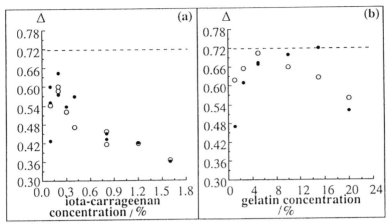

Figure 4   *Critical exponent* Δ *versus concentration for (a)* ι*-carrageenan i2 and (b) gelatin g2. Open symbols are used to denote gelation and solid ones to denote melting*

possible to find a concentration for which a system with long chains is able to percolate but not a system with shorter ones.

Variation of Critical Exponents with Polymer Concentration

All the data for the i2 and g2 samples are collected together in Figure 4. Our results confirm that different values of the critical exponent Δ can be obtained, instead of a unique value[16,18–21,27,28,31] as assumed by the theory of percolation. Figure 4 shows that Δ depends on concentration. For gelatin, at intermediate concentrations, the value of Δ attains, in gelation as well as in melting, a maximum value close to the one predicted by the percolation theory (0.72). The concentration at which a maximum value is observed seems to be lower in gelation (5%) than in melting (10%). This could be due to the fact that, when the 'sol point' is passed, the structure is not the same as at the 'gel point' attained under equilibrium conditions: during melting a structured system in a given state of evolution is broken up. Below the concentration corresponding to the maximum value, it may be assumed that there is no full occupancy of the sample volume by gelatin chains, which then could have some mobility, while the percolation theory assumes that potential junction sites are motionless.[33] At concentrations higher than the concentration corresponding to a maximum value of Δ, gelatin chains are interpenetrating and the entanglements may play a role in determining the visco-elastic properties. In this instance too, the phase angle is lower than predicted by the percolation theory ($\delta < 0.72\pi/2$).

The Δ values of ι-carrageenan systems are all lower than 0.72, and they decrease as concentration increases. In the same way as with gelatin, an effect of entanglements could be suggested. ι-Carrageenan chains are much longer than gelatin ones and, even at the lowest concentrations studied, the chains are presumably interpenetrating. It seems probable that Δ values could tend to 0.72 at concentrations lower than 0.1 wt% if a gel could be obtained. Contrary to gelatin, the Δ values for ι-carrageenan do not seem to be different for gelation and melting. This result again confirms that ι-carrageenan systems are nearly at equilibrium when the sol–gel and gel–sol transitions are being traversed.

References

1. J. Poppe, in 'Thickening and Gelling Agents for Food', ed. A. Imeson, Chapman and Hall, London, 1992, p. 98.
2. W. R. Thomas, in 'Thickening and Gelling Agents for Food', ed. A. Imeson, Chapman and Hall, London, 1992, p. 25.
3. D. Eagland, G. Pilling, and R. G. Wheeler, *Faraday Discuss.*, 1974, **57**, 181.
4. M. Djabourov, J. Maquet, H. Theveneau, J. Leblond, and P. Papon, *Br. Polym. J.*, 1985, **17**, 169.
5. M. Djabourov, *Rev. Gén. Therm. G.*, 1987, **306/307**, 369.
6. M. Djabourov, J. Leblond, and P. Papon, *J. Phys. France*, 1988, **48**, 333.
7. K. Te Nijenhuis, *Colloid Polym. Sci.*, 1981, **259**, 522.
8. K. Te Nijenhuis, *Colloid Polym. Sci.*, 1981, **259**, 1017.
9. C. Rochas, *Food Hydrocolloids*, 1987, **1**, 215.
10. L. Piculell, S. Nilsson, and P. Muhrbeck, *Carbohydr. Polym.*, 1992, **18**, 199.
11. A. Parker, G. Brigand, C. Miniou, A. Trespoey, and P. Vallée, *Carbohydr. Polym.*, 1993, **20**, 253.
12. J. E. Eldridge and J. D. Ferry, *J. Phys. Chem.*, 1954, **58**, 992.
13. D. A. Rees, *Adv. Carbohydr. Biochem.*, 1969, **24**, 267.
14. E. R. Morris, D. A. Rees, I. T. Norton, and D. M. Goodall, *Carbohydr. Res.*, 1980, **80**, 317.
15. O. Smidsrød, in 'IUPAC 27th International Congress of Pure and Applied Chemistry', ed. A. Varmavuori, Pergamon Press, Oxford, 1980, p. 315.
16. H. H. Winter and F. Chambon, *J. Rheol.*, 1986, **30**, 367.
17. F. Chambon and H. H. Winter, *J. Rheol.*, 1987, **31**, 683.
18. J. E. Martin, D. Adolf, and J. P. Wilcoxon, *Phys. Rev. Lett.*, 1988, **61**, 2620.
19. Y. G. Lin, D. T. Mallin, J. C. W. Chien, and H. H. Winter, *Macromolecules*, 1991, **24**, 850.
20. J. C. Scalan and H. H. Winter, *Macromolecules*, 1991, **24**, 47.
21. D. Lairez, M. Adam, J. R. Emery, and D. Durand, *Macromolecules*, 1992, **25**, 286.
22. F. Devreux, J. P. Boilot, F. Chaput, L. Malier, and M. A. V. Axelos, *Phys. Rev. E*, 1993, **47**, 2689.
23. M. In and R. K. Prud'homme, *Rheol. Acta*, 1993, **32**, 556.
24. C. Peigney-Nourry, Doctoral Thesis, Université de Paris VII-XI-ENSIA, 1987.
25. K. Te Nijenhuis and H. H. Winter, *Macromolecules*, 1989, **22**, 411.
26. M. A. V. Axelos and M. Kolb, *Phys. Rev. Lett.*, 1990, **64**, 1457.
27. G. Cuvelier and B. Launay, *Makromol. Chem. Macromolecules Symp.*, 1990, **40**, 23.
28. G. Cuvelier, C. Peigney-Nourry, and B. Launay, in 'Gums and Stabilisers for the Food Industry', ed. G. O. Phillips, D. J. Wedlock, and P. A. Williams, IRL Press, Oxford, 1990, vol. 5, p. 549.

29. T. Ikeda, M. Tokita, A. Tsutsumi, and K. Hikichi, *Jpn. J. Appl. Phys.*, 1990, **29**, 352.
30. J. O. Carnali, *Rheol. Acta*, 1992, **31**, 399.
31. C. Michon, G. Cuvelier, and B. Launay, *Rheol. Acta*, 1993, **32**, 94.
32. R. Muller, E. Gérard, P. Dugand, P. Rempp, and Y. Gnanou, *Macromolecules*, 1991, **24**, 1321.
33. P. G. De Gennes, 'Scaling Concepts in Polymer Physics', Cornell University Press, Ithaca, NY, 1979.
34. B. Derrida, D. Stauffer, H. J. Herrmann, and J. Vannimenus, *J. Phys. Lett.*, 1983, **44**, L701.
35. H. J. Herrmann, B. Derrida, and J. Vannimenus, *Phys. Rev. B*, 1984, **30**, 4080.
36. M. E. Cates, *J. Phys.*, 1985, **46**, 1059.
37. M. Muthukumar, *Macromolecules*, 1989, **22**, 4656.
38. E. E. Holly, S. K. Ventakaran, F. Chambon, and H. H. Winter, *J. Non-Newtonian Fluid Mech.*, 1987, **26**, 17.

Effect of Retrogradation on the Structure and Mechanics of Concentrated Starch Gels

By Christel J. A. M. Keetels, Ton van Vliet, and Hannemieke Luyten

DEPARTMENT OF FOOD SCIENCE, WAGENINGEN AGRICULTURAL UNIVERSITY, PO BOX 8129, 6700 EV WAGENINGEN, THE NETHERLANDS

1 Introduction

Starch is a major structural component in bread and other food products. During bread baking, or in general during heating of starch + water mixtures, several changes occur in the starch structure. These combined changes are usually called 'gelatinization'. During storage of bread, its crumb becomes firmer and more crumbly, resulting in a decrease in its eating quality. This increase in firmness is generally accepted to be a result of recrystallization of amylopectin, loosely called retrogradation.[1-4]

Most literature reports, in which the change in mechanical properties of starch systems due to retrogradation have been studied, refer to rheological measurements at small deformations.[5-8] However, the consumers' perception of product properties is usually related to large deformation properties, including fracture or yielding of the product. Thus, to get more insight into the decrease in eating quality of products containing a high starch content, it is more relevant to study large deformation properties. The aim of this paper is to discuss the relation between the changes in structure of concentrated starch gels during ageing, and the corresponding changes in their rheological and fracture properties.

2 Materials and Methods

Dispersions of 3 wt% potato starch (AVEBE, the Netherlands) or 8 wt% wheat starch (Latenstein BV, the Netherlands) in demineralized water were heated to 65 °C, while being gently stirred. After cooling to room temperature, sufficient starch was added to obtain suspensions with 30 wt% dry matter. PTFE cylindrical moulds with an inner diameter of 15 mm and an inner length of 100 mm were filled with these suspensions. These moulds were heated in an oil bath at 95 °C for 90 min. Then, the gels were cooled to and stored at 7 °C. In this way homogeneous samples without visible defects could be made.

Uniaxial compression tests were performed using a Zwick material testing machine, equipped with a 2000 N load cell. The cylindrical samples with a diameter of 15 mm were cut into test pieces with a height of 20 mm. These were compressed between perspex plates with a diameter larger than the diameter of the test piece, even after compression. The initial strain rate was 1.7×10^{-2} s^{-1}. The compressive stress σ and the Hencky strain ε were determined.[9] From the stress–strain curve, Young's modulus $E[=(\mathrm{d}\sigma/\mathrm{d}\varepsilon)_{\varepsilon \to 0}]$ was calculated. Measurements were performed at 20 °C after various storage times of up to 65 days.

Differential scanning calorimetric (DSC) measurements were performed with a TA Instruments DSC-2910. Just before measuring, 20 mg of starch gel was weighed out in an aluminium-coated pressure cup (25 μl). The gels were heated from 20 to 120 °C at a scanning rate of 5 K min^{-1}. Immediately after heating, the gels were rapidly cooled to 5 °C. These experiments were performed after approximately the same storage times as the compression tests.

3 Results and Discussion

When starch is heated in water, several interrelated processes occur. The granules swell to many times their original size, depending on the amount of water available. Nearly simultaneously, the crystallites melt. These changes are accompanied by a separation of amylose and amylopectin, which is presumably due to amylose and amylopectin being thermodynamically incompatible.[10] Another process during heating of starch suspensions is the breakdown of the amylopectin matrix within the swollen granules, possibly due to loss of entanglements between the starch molecules.[11] These processes together cause a change in the rheological behaviour of starch suspensions. The relation between changes in the structure of starch systems during heating and their behaviour in rheological measurements at small deformations have been discussed before.[11]

After heating suspensions with a rather low starch concentration, the granules become fully swollen. They do not fully occupy the available volume. In such a system, much of the amylose has leached out of the granules during heating; amylose and amylopectin have almost completely phase separated. Consequently, the swollen granules mainly consist of amylopectin. The leached out amylose molecules rearrange and, if their concentration is high enough, form a gel during cooling. Then, an elastic gel develops, consisting of swollen granules in an amylose matrix (Figure 1a). At higher starch concentrations, the granules are only partly swollen after heating and they occupy almost the whole volume. Moreover, amylose and amylopectin have only partly separated, and this results in the swollen granules containing amylose as well as amylopectin. At high concentrations, for instance 30%, gels already form during heating. This can be explained by the assumption that under these

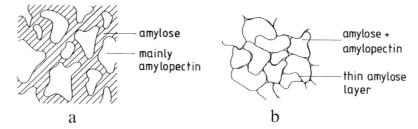

Figure 1   *Schematic presentation of starch gel structures: (a) fully swollen granules acting as a filler in an amylose gel; (b) partly swollen granules that are tightly packed*

conditions, with a high local amylose concentration in between the swollen granules, rearrangements of amylose at high temperatures (approximately 95 °C) are extensive enough to form a thin 'gel' layer between the partly swollen granules. The structure of the resulting concentrated starch gel is illustrated in Figure 1b. In this paper we will discuss only the mechanical properties of the latter type of gel system.

During storage of concentrated starch gels, an increase in the value of Young's modulus E has been observed (Figure 2). The increase in E is mainly due to the swollen granules becoming stiffer, which has been ascribed to reordering of amylopectin molecules.[6,8,11] To learn more about the recrystallization of amylopectin, the increase in stiffness of the gels during ageing was compared with the development of an endothermic transition, as determined by DSC. The melting enthalpy ΔH as measured by DSC is a measure of the amount of (semi-)crystalline amylopectin present in the sample. In Figure 3,

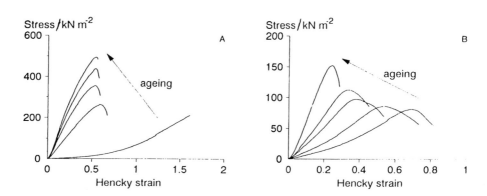

Figure 2   *Stress–strain curves of 30 wt% starch gels: (A) potato starch; (B) wheat starch. Determined after storage times of 0 h and 1, 4, 16 and 65 days at 7 °C (measuring temperature 20 °C, initial strain rate 1.7×10^{-2} s^{-2}). Note that the scales are different*

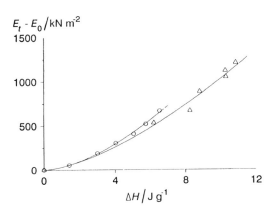

Figure 3   *The increase in stiffness,* $E_t - E_0$, *plotted as a function of the melting enthalpy* ΔH, *as determined by DSC, for 30 wt% potato* ($\triangle$) *and wheat starch* ($\bigcirc$) *gels (storage temperature 7 °C)*

$E_t - E_0$ is shown as a function of ΔH for 30% potato starch and wheat starch gels, where E_0 is the modulus immediately after gel preparation, and E_t is the modulus after storage of the gel at 7 °C for time t. Although 30% potato starch gels show a more rapid increase in modulus and recrystallization of amylopectin (*cf.* Figure 2), the relation between $E_t - E_0$ and ΔH is approximately the same for both gels. The relation between the increase in stiffness of the gels and the recrystallization of amylopectin is not linear, although the deviation from linearity is rather slight.

If large deformations are imposed on concentrated starch gels, fracture occurs around the granules and not through them. This implies that it is the thin amylose layer that is fractured and that the granules stay intact. There is one exception: in *fresh* potato starch gels, fracture proceeds through the granules. Figure 2 shows that fracture stress increases and fracture strain decreases during storage. The change of these properties during storage will be related below to the structure of the concentrated starch gel, and two models of the concentrated starch gel will be considered.

Firstly, we represent a starch gel as a simple system of cubes that are glued together (Figure 4A). In this arrangement the cubes occupy almost the whole volume, with the glue layer being very thin. Moreover, it is assumed that the glue is weaker than the cubes. The Young's modulus of such a system is approximately equal to the Young's modulus of the cubes. Recrystallization of amylopectin causes an increase in the stiffness of the cubes and consequently in the modulus of the whole system. The fracture stress of the system depends on the fracture stress of the glue: the larger the fracture stress of the glue, the larger that of the system as a whole. The strain at fracture is roughly the strain of the cubes at the moment the stress in the glue equals its fracture stress. The strain at fracture is larger for a smaller Young's modulus of the cubes and a larger fracture stress of the glue. The increase in stiffness of the cubes explains

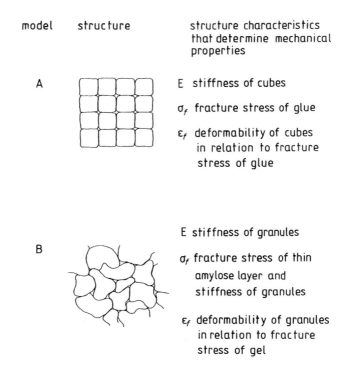

model   structure   structure characteristics
                             that determine mechanical
                             properties

A   E stiffness of cubes

                            σ_f fracture stress of glue

                            ε_f deformability of cubes
                                in relation to fracture
                                stress of glue

                            E stiffness of granules

B σ_f fracture stress of thin
                                amylose layer and
                                stiffness of granules

                            ε_f deformability of granules
                                in relation to fracture
                                stress of gel

Figure 4   *Highly schematic presentations of: (A) cubes that are glued together; (B) partly swollen granules that are tightly packed with thin amylose layers amongst them. Consequences for mechanical properties are briefly indicated (see the text for full description)*

the observed decrease in fracture strain. However, this simple model only explains an increase in the fracture stress if it is also assumed that the amylose 'glue' between the cubes stiffens during storage. Amylose indeed forms a gel during cooling and storage, but the timescale of this gelation process is usually far shorter than a day.[12,13] Consequently, it is unlikely that the slow increase in fracture stress is a result of gelation of amylose molecules. Therefore a more complicated model has to be considered.

Light microscopy has shown that the shape of the granules is far from cubic. They are very irregularly shaped, as has also been observed by others.[14,15] The swollen granules are hooked one on another like pieces of a three-dimensional jigsaw puzzle (Figure 4B). In such a structure, stiffening of the granules would explain the observed effect on not only the modulus and fracture strain, but also the fracture stress. The stiffer the granules, the stronger the 'hooks' and the higher the stress needed to fracture the gels. Consequently, the fracture stress of concentrated starch gels would not only depend on the fracture stress of the thin amylose layer between the swollen granules, but also on the stiffness of the granules; of these two factors, the stiffness of the swollen granules is the more important. This conclusion is supported by the observation that, during storage of 30% potato and wheat starch gels, there is a linear relation between the fracture stress and the modulus (Figure 5). The observed decrease in fracture strain would be due to the granules becoming stiffer, *i.e.* less deformable. However, its decrease with an increase in stiffness of the granules would be less than for model A.

In summary, both models predict a direct relation between the reordering of amylopectin molecules and the Young's modulus. In addition, model B explains why the increase in Young's modulus during storage goes along with an increase in fracture stress.

Besides similarities, differences in the fracture behaviour between potato and wheat starch gels can also be observed (Figures 2 and 5). In fresh potato starch gels the strain at fracture is much larger than in fresh wheat starch gels (Figure 2). This is mainly caused by the lower stiffness of the fresh potato starch granules. However, if the fracture stresses in Figure 5 are extrapolated to zero modulus, the fracture stress of a potato starch gel is still higher than that of a

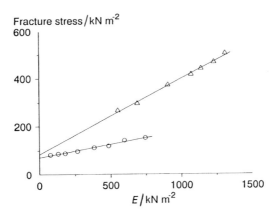

Figure 5   *Fracture stress as a function of Young's modulus for 30 wt% potato (△) and wheat starch (○) gels obtained after various storage times at 7 °C*

wheat starch gel. This may indicate that the fracture stress of the thin amylose layer in potato starch gels is higher, which would then imply that the potato starch granules are more strongly glued to each other. The gelation of amylose may be affected by the average length of the amylose molecules,[12] which is greater for potato starch amylose.[16] Moreover, lipids present in wheat starch possibly affect the gelation of amylose, because it is known that the fracture stress and strain decrease by the addition of amylose-complexing agents like glycerol monostearate.[17] Figure 5 also shows that the increase in fracture stress as a function of the Young's modulus is more pronounced for 30% potato starch gels. There are some indications that swollen potato starch granules are more irregularly shaped than swollen wheat starch granules.[15] Swollen potato starch granules are therefore more hooked one on another than are swollen wheat starch granules; *i.e.* potato starch behaves more like model B. This explains why the fracture stress of potato starch gels increases more rapidly with increasing modulus than that of wheat starch gels.

4 Conclusions

Changes in the rheological and fracture properties of concentrated starch gels during ageing have been studied. Concentrated starch gels consist of partly swollen granules, which occupy the whole available volume. The stiffness of such gels depends on the stiffness of the swollen granules, which increases during storage due to recrystallization of amylopectin.

For relating the structure of a gel to its mechanical properties at large deformations, the irregular shape of the swollen granules has therefore to be considered. Due to their irregular shape, changes in the stiffness of the swollen granules cause changes in the fracture stress and strain of the gels. The presence of a thin layer of amylose gel between the swollen granules affects the initial properties of the gels, but it plays a minor role in the change in mechanical properties of concentrated starch gels during storage.

Acknowledgements

The authors thank J. F. C. van Maanen (TNO Nutrition and Food Research, Wageningen) for performing the DSC experiments. Thanks are also due to Professor P. Walstra and Dr A. H. Bloksma for stimulating and valuable discussions. This work has been supported by the Netherlands Technology Foundation (STW).

References

1. B. L. D'Appolonia and M. M. Morad, *Cereal Chem.*, 1981, **58**, 186.
2. A.-C. Eliasson and K. Larsson, 'Cereals in Breadmaking', Marcel Dekker, New York, 1993.
3. K. Kulp and J. G. Ponte, Jr, *CRC Crit Rev. Food Sci. Nutr.*, 1981, **15**, 1.
4. J. A. Maga, *CRC Crit. Rev. Food Technol.*, 1975, **8**, 443.
5. C. G. Biliaderis and J. Zawiskowski, *Cereal Chem.*, 1990, **67**, 240.
6. M. J. Miles, V. J. Morris, P. D. Orford, and S. G. Ring, *Carbohydr. Res.*, 1985, **135**, 271.
7. P. D. Orford, S. G. Ring, V. Carroll, M. J. Miles, and V. J. Morris, *J. Sci. Food Agric.*, 1987, **39**, 169.
8. Ph. Roulet, M. MacInnes, M. Würsch, R. M. Sanchez, and A. Raemy, *Food Hydrocolloids*, 1988, **2**, 381.
9. H. Luyten, T. van Vliet, and P. Walstra, *Neth. Milk Dairy J.*, 1991, **45**, 33.
10. M. T. Kalichevsky and S. G. Ring, *Carbohydr. Res.*, 1987, **162**, 323.
11. C. J. A. M. Keetels and T. van Vliet, 'Gums and Stabilisers for the Food Industry',

ed. G. O. Phillips, D. J. Wedlock, and P. A. Williams, Oxford University Press, 1994, vol. 7, p. 271.

12. A. H. Clark, M. J. Gidley, R. K. Richardson, and S. B. Ross-Murphy, *Macromolecules*, 1989, **22**, 346.
13. M. J. Miles, V. J. Morris, and S. G. Ring, *Carbohydr. Res.*, 1985, **132**, 257.
14. M. Langton and A.-M. Hermansson, *Food Microstruct.*, 1989, **8**, 29.
15. K. Svegmark and A.-M. Hermansson, *Food Struct.*, 1991, **10**, 117.
16. Y. Takeda, K. Shirasaka, and S. Hizukuri, *Carbohydr. Res.*, 1984, **132**, 83.
17. B. Conde-Petit and F. Escher, *Getreide, Mehl Brot*, 1991, **5**, 131.

Mechanical Properties of Thermo-reversible Gels in Relation to their Structure and the Conformations of their Macromolecules

By Evgeny E. Braudo and Irina G. Plashchina

INSTITUTE OF FOOD SUBSTANCES, ACADEMY OF SCIENCES OF RUSSIA, VAVILOV STR. 28, V-334, MOSCOW, GSP-1, 117813, RUSSIA

Thermo-reversible ('physical') gels are formed under conditions which cause extensive formation of polymer–polymer contacts which involve the less soluble parts of the macromolecules. In other words, the formation of thermo-reversible gels proceeds in thermodynamically poor solvents. The essential contribution of polymer–polymer contacts in the structure of the solution distinguishes thermo-reversible gels from so-called chemical gels which are formed through the cross-linking of macromolecules by covalent bonds. Such cross-linking can proceed also in thermodynamically good solvents.

Extensive polymer–polymer interactions in thermo-reversible gels apparently determine the high local viscosity of structural elements in which the elementary relaxation processes proceed and, thus, the low rate of mechanical relaxation processes proceed and, thus, the low rate of mechanical relaxation as a whole. Although from their elastic moduli[1] and the sign of the Gough–Joule effect,[1-4] thermo-reversible gels like chemical gels fall into the category of highly elastic systems, their mechanical relaxation curves are typical of systems at the beginning or in the middle of the transition zone.[5] As an example, Figure 1 shows concentration-invariant reversible creep curves of some polysaccharide gels.[6] It is seen, from their relaxation behaviour in particular, that pectin gels are more 'glassy' than the gels of furcellaran. Another example – the reversible creep curves of 5 wt% gelatin gels[7] – is shown in Figure 2. In the literature there are a lot of data which give evidence for similar relaxation behaviour of thermo-reversible gels in both creep and stress relaxation experiments. For comparison, Figure 3 shows creep curves of chemical gels, namely gels of casein cross-linked by dialdehyde starch.[8] For these the relaxation behaviour is typical of a highly elastic system. Therefore thermo-reversible gels show ambivalent mechanical behaviour: they are highly elastic according to their elastic moduli and the sign of the Gough–Joule effect, but are more or less glassy according to their relaxation properties.

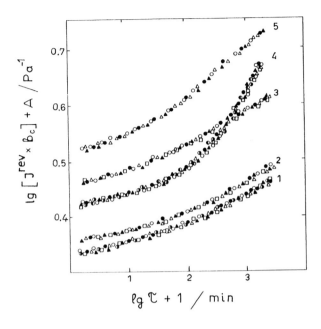

Figure 1 *Concentration-invariant curves of reversible creep compliance J^{rev} of poly-saccharide gels:[6] (1) 2.0–4.4% gels of a high-esterified pectinate (degree of esterification = 58%) in the mixture (1:1) water + glycerine at pH 2.7; (2) 0.5–2.5% gels of a high-esterified pectinate (degree of esterification = 58%) in the mixture (3:7) water + sucrose at pH 3.0; (3) 1.25–2.5% gels of κ-carrageenan in 0.11 M KCl at pH 7.6; (4) 0.5–2.0% gels of a calcium salt of furcellaran in the mixture (3:7) water + sucrose at pH 7.6; (5) 1.5–2.5% gels of κ-furcellaran in 0.15 M KCl at pH 7.6*

The low rate of mechanical relaxation makes unlikely the attainment of a steady state and, therefore, the determination of an equilibrium elasticity modulus in creep or stress relaxation experiments, and especially not in dynamic experiments. However, it is well known that a frequency-invariant value of G' is typical of thermo-reversible gels. Apparently, there is a gap in the time relaxation spectra of thermo-reversible gels which manifests itself in the lack of mechanical relaxation at frequencies of *ca.* 1 Hz or higher. Thus, Ferry[9] has not observed frequency dependence of elasticity moduli of gelatin gels in the ranges 1250–2500 Hz or 320–630 Hz, depending on the rigidity of the gel. At the same time, gels of gelatin have shown rather slow relaxation behaviour in creep experiments.[7]

So, we think that the equilibrium elastic modulus of thermo-reversible gels cannot usually be determined in standard experiments, such as dynamic mechanical measurements, creep, or stress relaxation. Therefore, it is very dangerous to interpret values of elastic moduli obtained in such experiments

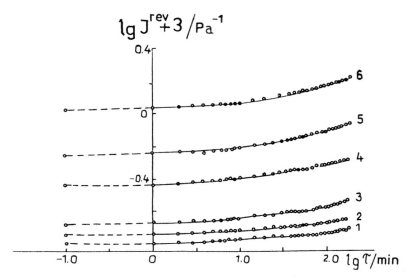

Figure 2   *Reversible creep compliance J^{rev} of a 5% gelatin gel:[7] (1) 13.4 °C; (2)*
14.0 °C; (3) 16.0 °C; (4) 20.0 °C; (5) 22.0 °C; (6) 24.0 °C

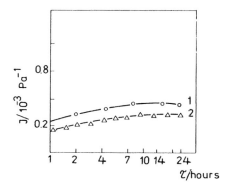

Figure 3   *Creep compliance J as a function of time t for a gel of casein (10%) cross-*
linked by dialdehyde starch (2.5%) at pH 6.25.[8] Time of structure formation
at 3 °C: (1) 8 days; (2) 15 days

from the viewpoint of theories developed for equilibrium systems. At the same
time, it is possible to determine reliable scaling parameters from temperature- or
concentration-invariant plots.[5] In particular, in creep experiments such plots for
thermo-reversible gels can be built up by shifting the curves of reversible creep
along the axes of time and compliance.[10,11] The shift factors determined from
the shift along the time axis characterize the effect of temperature and
concentration on the rate of relaxation processes; the shift factors determined
from the shift along the compliance axis characterize effects of these factors on
the equilibrium elastic modulus and, therefore, on the number of cross-links.

The low strength of cross-links in thermo-reversible gels allows the flow of gels even at strains small enough to correspond to the region of linear visco-elasticity. In this region reversible and irreversible strains are independent of each other. Therefore, it is possible to characterize thermo-reversible gels both from the elastic modulus and the limiting low-stress Newtonian viscosity. Taking into account the impossibility of attaining the regime of steady-state flow, due to the low rate of relaxation, the correct separation of both components of strain in creep experiments is possible only after determination of the irreversible strain by a relaxation experiment.[12] This is illustrated in Figure 4.

It has been shown[6] that both the limiting low-stress Newtonian viscosity and the moduli of thermo-reversible gels are exponential functions of the reciprocal value of temperature and power functions of the polymer concentration (more exactly, on the reduced concentration $(c/c^* - 1)$, where c^* is the gelation threshold). Thus, we can postulate a set of six scaling parameters, namely:

E_η^G — activation energy of viscous flow;

E_{a_T} — activation energy of mechanical relaxation;

$\Delta H^o_{b_T}$ — breakdown enthalpy of cross-links (an exponent of the van't Hoff isobar);

r — an exponent in the dependence of the macroscopic viscosity on the polymer concentration;

s — an exponent in the dependence of the mechanical relaxation rate or the local viscosity on the polymer concentration;

t — an exponent in the dependence of the equilibrium elastic modulus or the number of cross-links on the polymer concentration (generally it is equal to 1.8–2.5).

Whereas absolute values of viscosities and moduli of thermo-reversible gels are sensitive to individual features of the polymer sample, such as average molecular mass and degree of polydispersity, the scaling parameters are

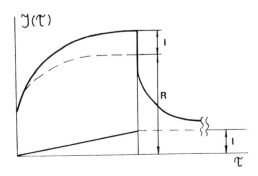

Figure 4   *Scheme of the separation of reversible (R) and irreversible (I) components of strain in a creep experiment, according to Kargin and Sogolova.[12] Creep compliance J is plotted against time τ*

representative of inherent structural and conformational features of a given type of polymer. Thus, based on scaling parameters, it is possible to determine such characteristics as the enthalpy of cross-links and the extent of homogeneity of a gel.

The enthalpy of cross-links can be determined by two independent methods.[6] According to Schultz and Myers,[13] the enthalpy of cross-link breakdown is equal to $\Delta H^{\circ}_{b_T}$. In the second method the enthalpy of cross-link breakdown is taken to be equal to the structural activation enthalpy of the viscous flow[14] which, in its turn, is taken to be equal to the difference between the activation energies of viscous flow of a gel and a solvent. Both methods of calculating the enthalpy of cross-links give comparable results: for 10 systems studied the ratio is 1.3 ± 0.26. Higher values of the enthalpy calculated from the activation energies of viscous flow appear to be the results of neglect of the contributions of polymer–solvent interactions.[15] The coincidence of the values obtained by the two methods supports the validity of the model of viscous flow of thermo-reversible gels in the linear viscoelasticity region, which forms the basis of the second method. According to this model, proposed for gelatin gels by Miller *et al.*,[16] an elementary act of flow includes the breakdown of a cross-link and its re-establishment in a new equilibrium state.

The comparison of the values of the enthalpy of cross-links, obtained from rheological experiments, with the values calculated on the basis of the concentration dependence of the melting temperature according to the Eldridge–Ferry equation[17] (these values characterize the enthalpy of the most stable cross-links surviving up to the melting point) allows us to evaluate the energetic heterogeneity of cross-links. For nine systems studied the ratio of Eldridge–Ferry values of enthalpy to those calculated from rheological experiments has varied from 1.1 (hydrogels of maltodextrin) to 16.7 (gels of agarose in the mixture, 1:1, of 0.75 M KCl and dimethyl sulfoxide). Cross-links of gels obtained in water are more homogeneous than those of gels obtained in mixtures of water with an organic solvent.

If both the melting enthalpy of a gel and the enthalpy of cross-links are known, the concentration of cross-links at a given temperature can be calculated. This allows us to calculate the entropy of cross-links.[6]

In Table 1 are shown the enthalpies and entropies of cross-links of a number of biopolymer gels. It is seen that the values can be divided into two groups. For the most part ΔH does not exceed 40 kJ mol^{-1}. Respective values of ΔS are no more than 135 J mol^{-1} K^{-1}. These values are characteristic of (a) gels of a semi-rigid chain polymer (highly esterified pectinate) which does not undergo co-operative conformational transitions and (b) gels of flexible-chain polymers at temperatures either below or above temperature ranges of helix–coil transitions. Values of ΔH and ΔS nearly an order of magnitude higher characterize flexible-chain polymers (agarose, κ-carrageenan, and gelatin) in the temperature ranges in which a helix–coil transition occurs. At these temperatures the helical conformation of macromolecules can exist only by being stabilized by intermolecular interactions in junction zones. Therefore,

Table 1 *Thermodynamic parameters of cross-link breakdown of thermo-reversible gels*

Composition of gel		Concentration	Temperature	ΔH^a	ΔS
Polymer	Solvent	wt%	°C	kJ mol^{-1}	J mol K^{-1}
1. High-esterified pectinate (sodium salt)	Water + sucrose, 3:7 (pH 3.0)	0.5 1.0 1.5 2.0 2.5	25.0–55.0	7±1 19±1 24±1 28±1 30±1	
2. High-esterified pectinate (sodium salt)	Water + glycerin, 1:1 (pH 2.7)	2.0	11.5–48.5	39±2	
3. Agarose	0.75 M KCl + dimethyl sulfoxide, 1:1	1.0	1.3–20.0 20.0–50.3 50.3–70.0 70.0–82.0	16±1 ~0 103±2 6±2	
4. κ-Carrageenan (sodium salt)	0.11 M KCl (pH 7.6)	1.5	9.0–41.0 41.0–54.0 54.0–66.0	36±2 235±2 11±2	105 (at 10 °C) 135 (at 25 °C) 780 (at 41 °C)
5. κ-Furcellaran (sodium salt)	0.15 M KCl (pH 7.6)	2.0	15.0–50.0	42±1	
6. Furcellaran (calcium salt)	Water + sucrose, 3:7 (pH 7.6)	1.25	25.0–50.0	19±2	
7. Maltodextrin	Water	20	25.6–45.0	25±2	25 (at 25 °C)
8. Gelatin	Water	5.0	18.0–24.0	295±8	

^{a} Determined from data derived from measurements of irreversible strain.

the breakdown of a cross-link is accompanied by the unwinding of adjacent helical segments, and *vice versa*.

The values of E_{aT} (50–160 kJ mol^{-1}),[6] show that elementary relaxation processes are co-operative and include the breakdown of several non-covalent bonds. The fact that temperature dependencies of the rate of mechanical relaxation follow the Arrhenius equation rather than the equation of Williams–Landel–Ferry[5] means that, despite the low rate of mechanical relaxation, the process is not limited by the free volume.

A great discrepancy is observed between the concentration dependencies of the macroscopic viscosity of thermo-reversible gels of polysaccharides and their local viscosity which manifests itself in the rate of mechanical relaxation. Thus, for five polysaccharide gels the parameter r – an exponent of the concentration dependence of the limiting low-stress Newtonian viscosity – is equal to 3.3 ± 0.1;[6] this does not differ from similar characteristics of semi-dilute liquid solutions of polysaccharides.[18] It can be shown that this value is typical of polymer solutions in good solvents. However, the parameter s, which characterizes the concentration dependence of the rate of mechanical relaxation, for the same gels is equal to zero. In other words, the rate of mechanical relaxation in these gels does not depend on the concentration. Similar regularity was found for gels of a heat-denatured protein.[19] This regularity was explained by the microheterogeneity of gels, whose network is built up from aggregates of macromolecules, relaxation processes being localized in these aggregates.

Generally, the extent of homogeneity of a gel (q) can be characterized by the ratio $s/3.3$. The value $q = 1$ corresponds to a fully homogeneous gel. In this case the local viscosity would depend on the concentration in the same manner as the macroscopic viscosity. The value $q = 0$ corresponds to a microheterogeneous gel. It holds for the majority of investigated polysaccharide gels. Only in the case of agarose gels in the 1:1 mixture of 0.75 M KCl and dimethyl sulfoxide is $q = 0.40-0.55 \pm 0.01$. It can be supposed that dimethyl sulfoxide has a solubilizing action on agarose.

Thus, the approaches proposed here to the interpretation of the results of creep experiments allows us to deduce information about the structure of gels, as well as the conformational state of macromolecules, without using models. The use of the methods of temperature–time and concentration–time superposition permits us to compensate in part for limitations associated with the low rate of mechanical relaxation in thermo-reversible gels.

References

1. N. Hirai, *Bull. Inst. Chem. Res., Kyoto Univ.*, 1955, **33**, 21.
2. N. Hirai and S. Seno, *Nippon Kagaku Zasshi*, 1954, **75**, 695 (in Japanese).
3. Yu. K. Godovsky, I. I. Maltseva, and G. L. Slonimsky, *Vysokomol. Soedin., Ser. A*, 1971, **13**, 2768 (in Russian).
4. E. Niwa, E.-S. Chen, S. Kano, and T. Nakayama, *Nippon Suisan Gakkaiski*, 1988, **54**, 249 (in Japanese).
5. J. D. Ferry, 'Viscoelastic Properties of Polymers', Wiley, New York, 3rd edn, 1980.
6. E. E. Braudo, I. G. Plashchina, and V. B. Tolstoguzov, *Carbohydr. Polym.*, 1984, **4**, 23.
7. E. E. Braudo, I. G. Plashchina, N. S. Kuzmina, and V. B. Tolstoguzov, *Vysokomol. Soedin. Ser. A*, 1974, **16**, 2240 (in Russian).
8. K. D. Schwenke, L. Prahl, E. E. Braudo, and V. B. Tolstoguzov, *Nahrung*, 1978, **22**, 915.
9. J. D. Ferry, *J. Am. Chem. Soc.*, 1948, **70**, 2244.
10. K. Arakawa, *Bull. Chem. Soc. Jpn*, 1962, **35**, 309.
11. K. Ninomiya and J. D. Ferry, *J. Polym. Sci., Part A-2*, 1967, **5**, 195.
12. V. A. Kargin and T. I. Sogolova, *J. Fiz. Khim.*, 1949, **23**, 540 (in Russian).
13. R. K. Schultz and R. R. Myers, *Macromolecules*, 1969, **2**, 281.
14. S. Glasstone, K. J. Laidler, and G. Eyring, 'The Theory of Rate Processes', McGraw-Hill, New York, 1941.
15. Y. Freckel, 'Kinetic Theory of Liquids', Dover Publications, New York, 1955.
16. M. Miller, J. D. Ferry, E. W. Schremp, and J. E. Eldridge, *J. Phys. Colloid Chem.*, 1951, **55**, 1387.
17. J. E. Eldridge and J. D. Ferry, *J. Phys. Chem.*, 1954, **58**, 992.

18. E. R. Morris, A. N. Culter, S. B. Ross-Murphy, and D. A. Rees, *Carbohydr. Polym.*, 1981, **1**, 5.
19. T. M. Bikbov, V. Ya. Grinberg, H. Schmandke, T. S. Chaika, I. A. Vaintraub, and V. B. Tolstoguzov, *Colloid Polym. Sci.*, 1981, **25**, 675.

Effect of Hydrocolloid Concentration on Mechanical Behaviour of Orange Gels

By S. M. Fiszman, M. C. Trujillo, and L. Durán

INSTITUTO DE AGROQUÍMICA Y TECNOLOGÍA DE ALIMENTOS (CSIC), JAIME ROIG, 11, 46010 VALENCIA, SPAIN

1 Introduction

Gelled quince pulp with a high level of sucrose is a traditional Spanish confection. In the last decade some commercial gelled products made with other fruit pulps have been produced with carrageenan + locust bean gum (LBG) as the gelling agent. Effects of some ingredients on the mechanical and textural characteristics of various polysaccharide gels have been studied by several authors. The effects of gum concentration,[1,2] addition of fruit pulp,[3,4] and of sugars[5-7] have been considered. In this paper, the effects of hydrocolloid type and concentration on several mechanical parameters in orange gels are reported.

2 Materials and Methods

κ-Carrageenan (Genugel UPC type, Copenhagen Pectin Factory, Denmark), alginate (Satialgine S 550), xanthan gum (Satiaxane CX91) (Sanofi Bio-Industries, France), gellan gum (Kelcogel, Kelco, Merck and Co., Chicago, IL) and LBG (Ceratonia S.A., Hercules, Spain) were used. Other chemicals were analytical grade: KCl, $CaHPO_4.2H_2O$, $(NaPO_3)_6$, and glucono-δ-lactone. Commercial sucrose and canned orange pulp (pH 3.51 and 11.7° Brix) were used.

Some 25 experimental sweet orange gels samples were made up with different amounts of gelling agents and 15 wt% orange pulp, adjusted to final 55° Brix with sucrose. Concentrations of each hydrocolloid were selected from preliminary experiments to cover the commercial range of consistencies.

Carrageenan (0.5, 0.7, 0.95, 1.2 or 1.4 wt%) and carrageenan + LBG (1:1) (0.4, 0.6, 0.75, 0.9 or 1.1 wt%) samples were prepared by dispersing dry hydrocolloid powders, with 0.5 wt% KCl and sugar, in a mixture of deionized water and orange pulp and heating to boiling. Water lost by evaporation was replaced, and solutions were left to set. Gellan (0.25, 0.4, 0.55, 0.7 or 0.8 wt%)

and gellan + xanthan + LBG (2 : 0.5 : 0.5) (0.35, 0.5, 0.7, 0.9 or 1.05 wt%) samples were prepared by dispersing dry hydrocolloid powders with a small amount of sugar in preheated deionized water (≈80 °C) and agitating until total dissolution. This solution was then added to preheated (≈90 °C) orange pulp with the rest of the sugar and was heated (80–90 °C) for 10 min. Water lost by evaporation was replaced, and the solution was left to set. Alginate (0.25, 0.4, 0.6, 0.8 or 0.95 wt%) samples were prepared by dispersing the hydrocolloid blended with sugar in deionized water under vigorous agitation. A slurry containing $CaHPO_4.2H_2O$ and $(NaPO_3)_6$ (0.7 and 0.5 wt% in final product, respectively) in deionized water was added to the alginate solution. Separately, glucono-δ-lactone (1.25 wt% in final product) was dissolved in the orange pulp and immediately added to the other components while agitating. The mixture was left to gel. All orange gel samples were stored at 4–6 °C and 100% relative humidity for 24 h prior to measurement and testing.

Stabilized orange gels were conditioned at room temperature (20 ± 1 °C). Samples of 17 × 17 mm cylindrical probes were cut as previously reported.[8] Three replicates of each sample were prepared on different days. Five probes from each sub-sample were measured. Compression tests were performed with an Instron model 6021 using a 50 mm diameter plunger and a cross-head speed of 30 mm min^{-1}. Wet filter papers were placed between the sample and instrument surfaces to avoid slipping. Tests were conducted up to rupture of the gels. Four parameters were recorded: maximum rupture force at the break point, percentage deformation at the break point, energy dissipated as the area of the force–compression curve up to the break point, and the modulus between 5 and 15% deformation. Double-cycle compression tests (instrumental texture profile analysis[9]) were also performed up to 30% compression of the gels. Five parameters were recorded: hardness, springiness, cohesiveness, gumminess and chewiness. Penetration with a $\frac{1}{2}$ inch cylinder (TA 10) and with a 45° cone (TA 15), and cutting tests with a knife (TA 7), were performed with a Stevens apparatus at a speed of 30 mm min^{-1} and 15 mm displacement. From each penetration test, two parameters were recorded: the resistance to penetration, and the area under the penetration curve up to maximum force. From the cutting test, five parameters were recorded: the maximum force, the deformation up to the maximum force, the total area under the curve, the partial area up to the maximum force, and the partial remaining area. Absolute values of the correlation coefficients were used for clustering the parameters into groups. Two-way analysis of variance was applied to analyse the effects of the gelling system and of concentration on parameter values.

3 Results

Five mechanical parameters, one from each group obtained by clustering, were selected: (i) force, (ii) deformation and (iii) deformability modulus at the break point from the compression up to rupture test, (iv) resistance to cone penetration, and (v) cohesiveness from instrumental texture profile analysis.

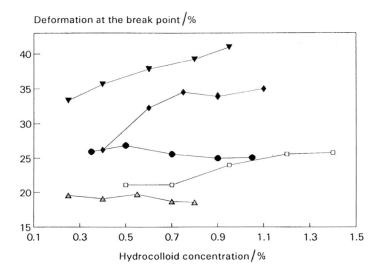

Figure 1   *Percentage deformation at break point values for orange gels at different concentrations of the following gelling systems:* ▼, *alginate;* □, *carrageenan;* ◆, *carrageenan + LBG;* △, *gellan;* ●, *gellan + xanthan + LBG*

As expected, 'force at break point' values from the compression test showed a general increase with concentration for all systems. Higher absolute values corresponded to alginate and lower ones to carrageenan and carrageenan + LBG systems. 'Deformation at break point' values from the compression test increased with concentration in carrageenan, carrageenan + LBG and alginate systems, absolute values increasing in this order. No changes in deformation were detected with increasing gellan and gellan + xanthan + LBG concentrations, the values for gellan being the lowest ones (Figure 1). 'Deformability modulus' values from the compression test increased with concentration in all systems; both absolute values and rate of increase were higher for gellan and alginate, confirming the high firmness of these gelling systems. 'Resistance to cone penetration' values increased with concentration in all cases, the highest absolute values corresponding to gellan gels. 'Cohesiveness' values from instrumental texture profile analysis decreased slightly with concentration in gellan and carrageenan + LBG gels and did not vary in other systems. Gels prepared with gellan exhibited the highest values of cohesiveness and alginate the lowest ones.

4 Conclusion

Results show that, when the hydrocolloid concentration is increased to enhance the gel strength of fruit gels, other textural parameters may change differently depending on the gelling agent used.

Acknowledgements

This work has been financed by Comisión Interministerial de Ciencia y Tecnología (Project ALI 91–0395). Hydrocolloid samples were kindly supplied by the Spanish agents of Hercules and Sanofi Bio-Industries, and by Vedeqsa (Kelco, Spain).

References

1. A. Nussinovitch, M. M. Ak, M. D. Normand, and M. Peleg, *J. Texture Stud.*, 1990, **21**, 37.
2. A. Nussinovitch, I. J. Kopelman, and S. Mizrahi, *Lebensm.-Wiss. Technol.*, 1991, **24**, 513.
3. S. M. Fiszman and L. Durán, 'Gums and Stabilisers for the Food Industry', ed. G. O. Phillips, D. J. Wedlock, and P. A. Williams, IRL Press, Oxford, 1990, vol. 5, p. 545.
4. G. Kalentunc, A. Nussinovitch, and M. Peleg, *J. Food Sci.*, 1990, **55**, 1759.
5. S. M. Fiszman, E. Costell, and L. Durán, *Food Hydrocolloids*, 1986, **1**, 113.
6. D. I. Gerdes, E. E. Burns, and L. S. Harrow, *Lebensm.-Wiss. Technol.*, 1987, **20**, 282.
7. D. Oakenfull, A. Scott, and E. Chai, 'Gums and Stabilisers for the Food Industry', ed. G. O. Phillips, D. J. Wedlock, and P. A. Williams, IRL Press, Oxford, 1990, vol. 5, p. 243.
8. L. Durán, E. Costell, and S. M. Fiszman, 'Physical Properties of Foods – 2', Elsevier Applied Science, London, 1987, p. 429.
9. M. Bourne, 'Food Texture and Viscosity', Academic Press, London, 1982, p. 114.

Effect of Starter Culture on Rheology of Yoghurt

By H. Rohm

DEPARTMENT OF DAIRY SCIENCE AND BACTERIOLOGY, UNIVERSITY OF AGRICULTURE, A-1180 VIENNA, AUSTRIA

1 Introduction

The textural and rheological properties of yoghurt represent an essential feature in consumer acceptance. Whereas protein enrichment and preheat treatment have been studied extensively,[1] only limited data are available on the effects of starter properties on yoghurt characteristics. These data mainly cover effects of specific starter strains on steady shear viscosity of stirred products.[2,3] Dynamic rheological measurements, which allow interpretation in terms of product structure, have been performed on model gels coagulated by acid precursors as well as on commercial stirred yoghurt.[4,5]

The paper presents findings on gel characteristics and texture properties of yoghurt acidified by starters with different properties, and it gives some implications for the structure of the resulting products.

2 Materials and Methods

Yoghurts were prepared from enriched UHT milk (2.5 wt% fat, 2 wt% skim milk powder, preheat treatment 90 °C/20 min, 10 starter cultures) in beakers and, simultaneously, in centrifugation tubes and in the Couette system of an RFS II Rheometer until pH 4.6. Dynamic gel measurements were carried out at 15 °C. Gel firmness was measured with an Instron 1011 (35 mm plunger, speed 25 mm min^{-1}, 15 °C) from the initial slope of the force–deformation curve. Drained whey was measured after centrifugation of yoghurt tubes (600 g, 10 min, 15 °C). Dynamic measurements on the stirred yoghurt was m with the RFS II using the cone/plate system at 15 °C.

3 Results and Discussion

Physical analyses resulted in marked differences between the products (Table 1). Linear visco-elastic behaviour was observed up to a critical strain

Table 1   *Rheological properties of yoghurt as affected by starter culture*

Starter code	Description	Yogurt gel				Stirred yoghurt	
		G'/Pa^b	$\tan \delta^b$	*Firmness*^{c} N mm^{-1}	*Whey drainage*/%	G'/Pa^b	$\tan \delta^b$
1	Classical	1120	0.235	0.75	3.0	79.5	0.248
2	Viscous	938	0.251	0.67	2.1	63.2	0.263
3	Classical	954	0.233	0.68	3.8	102	0.259
4	Classical	988	0.234	0.68	3.5	95.1	0.253
5	Viscous	928	0.249	0.65	3.0	79.6	0.273
6	Classical	882	0.234	0.64	3.4	104	0.258
7	Classical	1010	0.228	0.65	3.4	66.2	0.270
8	Highly viscous	621	0.262	0.56	2.1	42.4	0.284
9	Classical	1510	0.221	0.80	3.5	126	0.241
10	Viscous	778	0.249	0.61	2.7	54.2	0.273

[a] Description as outlined by culture suppliers.
[b] G' and $\tan \delta$ refer to the linear visco-elastic region after performing dynamic experiments at 15 °C by sweeping strain from minimum to maximum.
[c] Gel firmness from initial slope of force–deformation curve.

amplitude $\gamma_0 \approx 0.02$. Compared with yoghurt gels, the storage moduli G' of stirred products were about 15 times lower, whereas higher $\tan \delta$ values pointed to a greater relative importance of viscous contributions. Both G' and $\tan \delta$ of set and stirred yoghurt were, however, significantly interrelated ($p < 0.01$), which indicated that basic properties were independent of the product state. There also existed a direct proportionality between G' and gel firmness as measured by plunger penetration ($R^2 = 0.916$).

Evaluation of results with respect to descriptions provided showed that higher G' and lower $\tan \delta$ values are characteristics of classical starters, whereas viscous starters produce yoghurt with lower G' and a somewhat higher phase shift. Regardless of product state there is an inverse exponential relationship between G' and $\tan \delta$ ($p < 0.01$). Analysis of variance and Duncan grouping confirmed the starter description provided by suppliers.[6]

It has been pointed out for casein gels that dynamic moduli primarily depend on the number and strength of bonds between casein particles and on the strand distribution, whereas $\tan \delta$ is mainly a function of the relaxation behaviour.[7] It is evident that more viscous yoghurt gels are less susceptible to whey leakage forced by external pressure ($p < 0.01$). There are, however, no significant correlations between firmness and susceptability to syneresis.

In Figure 1 the complex viscosity η^* and the value of $\tan \delta$ of selected yoghurts are plotted *versus* the angular frequency ω. Shapes of $\tan \delta$–ω functions imply relatively more elastic contributions to visco-elastic behaviour at shorter timescales, or at higher ω values. Power law fits resulted in a more pronounced decay of η^* with increasing ω for classical starters than for viscous ones, and for stirred yoghurt than for set yoghurt gels. As ω is directly related to a shear strain-rate amplitude $\dot{\gamma}_0$ by $\dot{\gamma}_0 = \omega \, \gamma_0$, this viscosity decay can be interpreted in terms of shear-thinning behaviour.

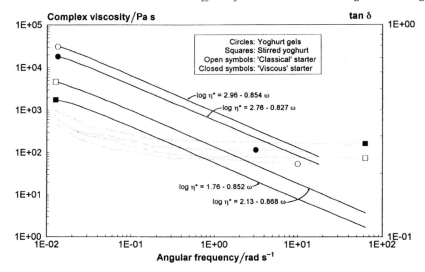

Figure 1   *Complex viscosity (——) and tan δ (....) of yoghurt gels and stirred yoghurt as a function of angular frequency ω (γ₀=0.005). Power law fits were calculated for 0.1 < ω/rad s⁻¹ < 60*

The relaxation modulus G_t of yoghurt gels was measured directly and, additionally, calculated from the dynamic data. Apart from varying magnitude, log G_t *versus* log time curves also showed different slopes but, in all cases, asymptotic behaviour and an approach to nearly constant values at $t > 10^3$ s.[6] Such a behaviour points to the existence of a true equilibrium modulus, which, in line with the shape of dynamic functions, may serve as an indicator of a permanent network system. Relaxation spectra calculated from dynamic and transient data imply that interactions with long relaxation times contribute less to the modulus in viscous products, and that the stress relaxes in a shorter time. Structural phenomena such as the formation of extraneous polysaccharides by mucigenic starters and formation of polysaccharide–protein cross-links are obviously responsible for the differences in the rheological properties.

References

1.   A. Y. Tamime and R .K. Robinson, 'Yoghurt Science and Technology', Pergamon Press, Oxford, 1985.
2.   S. M. Schellhaass and S. M. Morris, *Food Microstruct.*, 1984, **4**, 279.
3.   J. A. Teggatz and S. M. Morris, *Food Struct.*, 1990, **9**, 133.
4.   P. Zoon, S. P. F. M. Roefs, B. deCindio, and T. van Vliet, *Rheol. Acta*, 1990, **29**, 212.
5.   A.J. Steventon, C. J. Parkinson, P. J. Fryer, and R. C. Bottomley, in 'Rheology of Food, Pharmaceutical and Biological Materials with General Rheology', ed. R. A. Carter, Elsevier Applied Science, London, 1990, p. 196.
6.   H. Rohm and A. Kovac, *J. Texture Stud.*, 1994, **25**, 311.
7.   S. P. F. M. Roefs, T. van Vliet, H. J. C. M. Bijgaart, A. E. A. deGroot-Mostert, and P. Walstra, *Neth. Milk Dairy J.*, 1990, **44**, 159.

Rheology of Mixed Carrageenan Gels: Opposing Effects of Potassium and Iodide Ions

By Alan Parker

SANOFI BIO-INDUSTRIES, R&D, FOOD INGREDIENTS DIVISION, BAUPTE, 50500 CARENTAN, FRANCE

1 Introduction

Iota (ι)- and kappa (κ)-carrageenan are two important gel-forming food polysaccharides.[1] Despite the similarity of their molecular structures, their gels have very different properties: those of the κ form are hard, strong, and brittle, whereas those of the ι form are soft, weak, and fracture at large strains. It is known[2-6] that samples of both κ- and ι-carrageenan are usually contaminated by more or less significant quantities of the other form.

Determination of the purity of carrageenan samples is essential for fundamental studies. This is especially true for ι-carrageenan, as the presence of very small amounts of the κ-form can greatly modify its rheology[5,6] and the apparent ion selectivity.[4] ^{13}C NMR is certainly the method of choice,[7] but it is expensive and demands long measurement times.

In a previous publication[6] it was shown that a mass fraction of the κ form as low as 2.5% in impure samples of ι-carrageenan can lead to the appearance of 'two step gelation', when the elasticity is followed during cooling. The previous method of determining the amount of κ-carrageenan in a sample of the ι form had the disadvantage of requiring an expensive rheometer. Here, a simple, sensitive, rheological method is described, which requires only two room-temperature gel-strength measurements. The method depends on the fact that potassium and iodide ions have large and opposing effects on the gel strength of κ-carrageenan,[8] whilst scarcely modifying the behaviour of the ι form.[9] The method allows easy detection of mass fractions as low as 1% of κ-carrageenan in a sample of the ι form, if both components are of typical food grade.

2 Materials and Methods

The carrageenan samples were typical industrial products manufactured by Sanofi Bio-Industries (Carentan, France). The sample of ι-carrageenan was

produced from cultivated *Eucheuma denticulatum*. It had been studied
previously (sample $\iota 3$ in reference 6) and was shown to be pure (κ fraction $\leqslant 2$
mol%) by ^{13}C NMR. The sample of κ-carrageenan was produced from
cultivated *Kappaphycus alvarezii*. It contained 11% of the ι form, a typical
value for industrial samples (A. Parker, unpublished results). Both samples
were in the mixed potassium/sodium form ($\approx 70\%$ K$^+$/30% Na$^+$). To
determine the behaviour of completely pure ι-carrageenan, the sample in the ι
form was treated with pure κ-carrageenanase (kindly supplied by B. Kloareg,
University of Brest, France) to remove any traces of the κ form.[2]

Cylindrical gel samples (2 cm × 2 cm) were prepared from 15 g l^{-1}
carrageenan solutions which had been heated in a water bath at 95 °C for
30 min. After 17 h at room temperature, the sample rheology was measured
by unlubricated uniaxial compression at a rate of 20 mm min^{-1}. The force
at the maximum in the force–distance curve, F_{max}, was used as a measure of
the gel strength. The variance was typically 7% for the average of four
samples. For each carrageenan composition, the F_{max} values of gels prepared
in 0.2 M KCl and in a solution containing 0.1 M NaCl + 0.1 M NaI were
compared.

3 Results and Discussion

Figure 1 shows the effect on F_{max} of replacing from 0 to 40% of ι-carrageenan
by the κ form. There is a significant difference between the values of F_{max} in

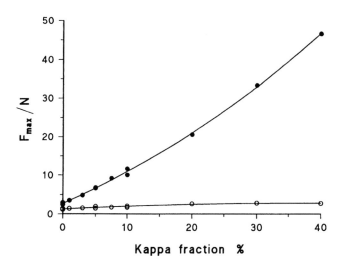

Figure 1   *Effect of substituting κ-carrageenan for ι-carageenan on the maximum force*
F$_{max}$ measured during uniaxial compression of gels prepared in 0.2 M KCl
(upper line) and 0.1 M NaI + 0.1 M NaCl (lower line). Lines are drawn to
guide the eye

potassium-rich and iodide-rich solutions at 0% substitution, *i.e.* 2.5 N and 1.2 N, respectively. On replacing ι-carrageenan by the κ form, the value of F_{max} for the potassium gels increases rapidly and almost linearly, its extrapolated value passing through the value for the unsubstituted ι-carrageenan. However, F_{max} for the iodide gels remains practically constant. The most surprising aspect of these results is the fact that no lower limit is found for the reinforcing effect of substitution. The smallest substitution of the κ form (1%) corresponds to a solution concentration of only 150 mg l^{-1}, yet this is sufficient to increase the maximum force from 2.5 to 3.5 N.

Previous studies[5,6] have concluded that mixed ι/κ-carrageenan gels must be both phase separated and bi-continuous, without mixed junction zones. On diluting such a system sufficiently, it must pass from a two-phase region into a single-phase one. In the single-phase region, the total gel strength is given by the sum of each component's individual rheological contribution as Zhang and Rochas[10] observed for κ-carrageenan + agarose mixtures. Since it is easily demonstrated that a 150 mg l^{-1} solution of our κ-carrageenan is not gelled, and so would not reinforce a one-phase system, we have to conclude that the 1% substituted system is still in the two-phase region of the phase diagram. Considering the similar chemical structures of the two carrageenans, it is remarkable that no behaviour characteristic of the one-phase system is observed, even when the κ-carrageenan concentration is so very low.

We suggest that the existence of equal gel strengths in iodide-rich and potassium-rich solvents provides a rheological definition for pure ι-carrageenan. To see whether a completely pure ι-carrageenan gives a ratio of exactly unity for F_{max} in iodide-rich and potassium-rich solvents, gels were prepared from the ι-carrageenan after treatment with κ-carrageenanase[2] to eliminate any undetected κ-carrageenan. The values of F_{max} for the enzyme-treated sample were indeed equal (1.1 N) and so our proposed definition seems to be valid. This result was confirmed by low deformation oscillatory measurements of the same samples, using the same protocol as previously.[6] At 20 °C, the value of G' was 140 Pa in the iodide-rich solvent and 130 Pa in the potassium-rich solution.

This method for determining the purity of ι-carrageenan is simpler, cheaper and more sensitive than ^{13}C NMR. In practice, given the extreme sensitivity of the method, a ratio of even a factor of two in the maximum forces corresponds to a very high degree of purity.

Acknowledgement

Thanks are due to Patrick Vallée for technical assistance and to Lennart Piculell for his critical reading of the draft manuscript.

References

1. N. F. Stanley, in 'Food Gels', ed. P. Harris, Elsevier Applied Science, London, 1990, p. 79.

2. C. Bellion, G. K. Hamer, and W. Yaphe, *Proc. Int. Seaweed Symp.*, 1981, **10**, 379.
3. C. Rochas, M. Rinaudo, and S. Landry, *Carbohydr. Polym.*, 1989, **10**, 115.
4. L. Piculell and C. Rochas, *Carbohydr. Res.*, 1990, **208**, 127.
5. L. Piculell, S. Nilsson, and P. Muhrbeck, *Carbohydr. Polym.*, 1992, **18**, 199.
6. A. Parker, G. Brigand, C. Miniou, A. Trespoey, and P. Vallée, *Carbohydr. Polym.*, 1993, **20**, 253.
7. A. I. Usov, *Botanica Marina*, 1984, **27**, 189.
8. H. Grasdalen and O. Smidsrød, *Macromolecules*, 1981, **14**, 1842.
9. O. Smidsrød, I.-L. Andersen, H. Grasdalen, B. Larsen, and T. Painter, *Carbohydr. Res.*, 1984, **80**, C11.
10. J. Zhang and C. Rochas, *Carbohydr. Polym.*, 1990, **13**, 257.

Rheology of Semi-sweet Biscuit Doughs

By Ged Oliver and Sarabjit S. Sahi

FLOUR MILLING AND BAKING RESEARCH ASSOCIATION, CHORLEYWOOD, HERTFORDSHIRE WD3 5SH, UK

1 Introduction

Semi-sweet biscuit doughs resemble bread doughs in that they contain a developed gluten network. However, the elastic properties of the gluten may lead to shrinkage of the biscuit dough during processing.[1] This can lead to variability in product quality. Sodium metabisulfite (SMS) may be used in the recipe formulation to counteract dough contraction.[2] SMS acts as a source of sulfur dioxide which is able to break some of the disulfide bridges in the gluten proteins. This weakens the elastic properties of the gluten and reduces the likelihood of dough shrinkage. SMS is also responsible for a reduction in dough mixing time.[2] However, there is a move away from the use of chemical additives, and the present work considers the effect of eliminating SMS addition to semi-sweet biscuit doughs prepared from a number of single-variety wheat cultivars.

2 Materials and Methods

The wheat cultivars studied were Tara, Hunter, Riband, Galahad, Galahad7 and Galahad77. Biscuit doughs were produced with and without SMS at a water level optimized for the inclusion of SMS. The effect of dough contraction was measured from the change in the eccentricity of the biscuit (biscuit length/biscuit width): dough pieces were cut to an eccentricity of 1.0625, and so the deviation from this could be determined and used as a measure of the ability of the dough to relax. Dynamic oscillatory measurements were performed using a Bohlin VOR rheometer on dough samples immediately after mixing at an applied strain of 4×10^{-4}.

3 Results

For all the varieties investigated, the use of SMS in the recipe formulation leads to a reduction in dough mixing time and dough shrinkage. This is in agreement with the known effects of SMS on dough behaviour.[2] Addition of

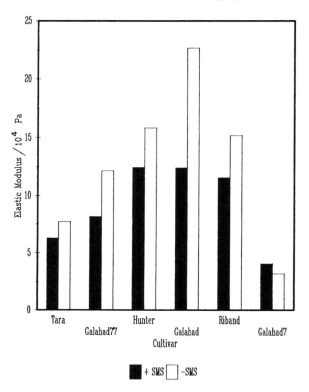

Figure 1   *Influence of SMS addition on elastic modulus of semi-sweet biscuit doughs produced from single wheat cultivars*

SMS leads to a reduction in the magnitude of the elastic modulus of doughs made from all varieties except Galahad7 (Figure 1). This is consistent with the use of SMS to produce a more uniform product, and it confirms that for most varieties SMS has the effect of decreasing the extent of cross-linking in the dough.

Galahad7 produces a phase angle of approximately 50° for doughs with and without SMS, whereas all other doughs give phase angles in the region 24–30° for both recipe formulations (Figure 2). A phase angle of 50° indicates that the factors contributing to the viscous nature of the dough are predominant with Galahad7, whereas the elastic properties dominate the rheological behaviour with the other varieties studied. The data in Figure 1 show that Galahad7 produces doughs with a smaller elastic modulus than the other varieties, suggesting that the interactions which cause elasticity are less abundant in Galahad7 doughs. SMS appears to have no consistent effect on the phase angle. This property is determined primarily by variety characteristics.

Dough viscosity seems to be related to biscuit eccentricity (Figure 3). Doughs with high viscosity give the lowest values of biscuit eccentricity

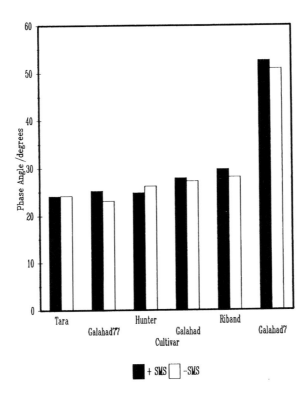

Figure 2 *Influence of SMS addition on phase angle of semi-sweet biscuit doughs produced from single wheat cultivars*

suggesting increased dough shrinkage. However, it is clear that there are two distinct populations in the data. One comprises doughs with SMS and the second comprises doughs without SMS. Linear regression on the two separate populations produces correlation coefficients of 0.77 and 0.82, respectively. This may indicate that the relationship is non-linear (*e.g.* hyperbolic).

4 Conclusions

The presence of SMS reduces the rheological parameters in doughs from all varieties except Galahad7. This is due to the well known ability of SMS to break disulfide linkages between protein molecules which weakens the cohesive and elastic structure of the dough. Galahad7 doughs exhibit rheology with phase angles of approximately 50° indicating that elastic properties contribute less to the dough rheology. This may account for the inability of SMS to weaken the dough structure in this variety.

Dough viscosity shows a relationship with biscuit eccentricity. Two populations of data are observed, one comprising doughs containing SMS and a second grouping of doughs without SMS. Doughs of high viscosity undergo greater shrinkage during processing, leading to low values of biscuit eccentricity.

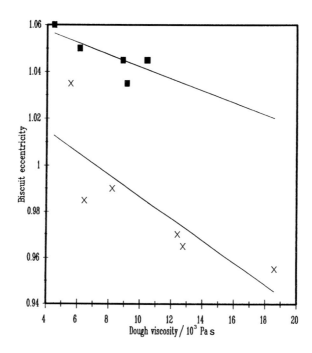

Figure 3   *Influence of dough viscosity on biscuit eccentricity for systems with (■) and without (×) SMS*

References

1. D. Thacker, in 'Food Technology International Europe 1993', ed. A. Turner, Sterling Publications, London, 1993, p. 154.
2. P. Wade, 'Biscuits, Cookies and Crackers: The Principles of the Craft', Elsevier Applied Science, London, 1988, vol. 1, p. 94.

Bulk and Surface Rheological Properties of Wafer Batters

By Ged Oliver and Sarabjit S. Sahi

FLOUR MILLING AND BAKING RESEARCH ASSOCIATION, CHORLEYWOOD, HERTFORDSHIRE WD3 5SH, UK

1 Introduction

For optimum processing performance wafer batters should be homogeneous and free of gluten aggregates. Little is known about the mechanisms by which gluten aggregation occurs in wafer batters, or about the factors which can control it. Standard flour quality tests cannot predict gluten development in wafer batters, but tests to measure gluten hydration and dispersion in these systems have been developed.[1] However, these tests are empirical in nature, and they do not provide information about the interactions present in the batter. In this study, the tests have been carried out alongside rheological measurements in an attempt to provide an improved understanding of the physical characteristics of the batter.

2 Experimental

The wheat cultivars studied here were Haven, a hard milling variety producing high starch damage, and Galahad, Beaver, Riband and Galahad7, which are soft milling varieties giving low starch damage. Wafer batters were produced to a constant solids:water ratio (1:1.2). Bulk rheological properties of the batters were measured by dynamic oscillatory experiments in a Bohlin VOR rheometer at an applied strain of 1.8×10^{-3}. Surface rheological measurements[2] were carried out on batter liquor produced by centrifugation of the batter at 62 000 g force.

3 Results and Discussion

Batters produced from Haven were found to have the highest values of storage and loss moduli (Table 1). Haven contained the greatest amount of damaged starch which would tend to absorb a greater proportion of the available water compared with the soft milling varieties.[3] Less water would be available to

Table 1   *Measured rheological parameters at 1 Hz of batters prepared from different flours*

	Galahad	Beaver	Haven	Galahad7	Riband
Phase angle/deg	27.8	29.7	35.5	35.7	23.3
Storage modulus/Pa	152.5	145.0	205.5	108.0	94.9
Loss modulus/Pa	80.2	82.1	146.5	77.4	40.6

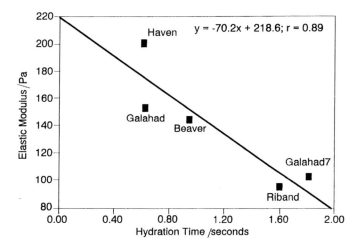

Figure 1   *Plot of the elastic (storage) modulus of wafer batters* versus *gluten hydration time*

dilute the Haven gluten which may account for the higher moduli of batters prepared from Haven compared with the other varieties. The elastic (storage) modulus of the batters showed a linear relationship with measured gluten hydration times, $r = 0.89$ (Figure 1), suggesting that gluten proteins in flours producing weak batters require longer hydration times.

The lowest phase angle was produced by the Riband variety, which suggests that its elastic properties play a greater role in determining overall rheological behaviour than for the other varieties (Table 1). This variety also produced the longest gluten dispersion time and there was a linear relationship between the phase angle of the batters and gluten dispersion time, $r = 0.91$ (Figure 2). These findings suggest that batters with dominant elastic properties, *i.e.* low phase angles, may produce long gluten dispersion times and hence potential production difficulties.

Batter liquor appears in a batter when the water absorption capacity of a flour has been exceeded. Values of batter liquor yields (Table 2) appear to be related to the level of damaged starch in the flours. Haven, with the highest level of damaged starch, produced the lowest amount of batter liquor. Surface tension values (Table 2) indicate that there are differences in the composition of the batter liquors with respect to the surface-active materials. Flour polar

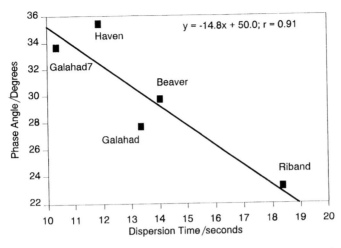

Figure 2   *Plot of the phase angle of wafer batters* versus *gluten dispersion time*

lipids are the surface-active materials in flour most likely to lower the surface tension to the value observed for Riband.[4]

Galahad batter liquor produced the strongest surface elasticity whereas Riband displayed the lowest value (Table 2). This behavior is opposite to the trend observed with the surface tension values. This type of behaviour has been observed with bread dough liquors[2] and was shown to be highly sensitive to the level of lipid in the dough liquor phase. There was found to be no correlation between the surface rheology of the batter liquor and the gluten hydration and dispersion measurements.

Table 2   *Batter liquor yields and surface properties of batter liquors*

	Galahad	Beaver	Haven	Galahad7	Riband
Batter liquor yield/%[a]	35.5	32.6	26.1	27.8	34.6
Surface tension/mN m^{-1}	45.3	42.6	40.9	42.9	34.9
Surface elasticity/mN m^{-1}	12.5	2.1	2.0	5.1	0.8

[a] Expressed as a fraction of the batter weight.

4 Conclusions

Surface properties do not correlate directly with measures of batter quality, but they can help to identify lipid and protein interactions within the batter. These properties appear to have greater application in the study of bread doughs where the rheology of the dough liquor material influences product quality.[2] Bulk rheological measurements have been able to detect differences in the properties of wafer batters prepared to a constant solids content. Gluten hydration times of the batters were found to be related to the bulk elastic moduli. Gluten dispersion time was found to correlate with the phase angle of the batter. Haven, a hard milling variety with high starch damage, produced stronger batters than soft milling varieties. These findings suggested that a soft milling wheat variety, producing flour with dominant viscous properties, is the ideal material for wafer production.

References

1. D. Thacker, in 'Food Technology International Europe 1993', ed. A. Turner, Sterling Publications, London, 1993, p. 154.
2. S. S. Sahi, 'Food Colloids and Polymers: Stability and Mechanical Properties' ed. E. Dickinson and P. Walstra, Special Publication No. 113, The Royal Society of Chemistry, Cambridge, UK, 1993, p. 410.
3. A. H. Bloksma and W. Bushuk, in 'Wheat: Chemistry and Technology', ed. Y. Pomeranz, American Association of Cereal Chemists, St. Paul, USA, 3rd edn, 1988, vol. 2, p. 131.
4. T. L.-G. Carlson, PhD Thesis, University of Lund, Sweden, 1981.

Effect of Dry Ultra-fine Size Reduction on Physico-chemical Properties of Pea Starch

By Sophie Jacqmin and Michel Paquot

FACULTÉ DES SCIENCES AGRONOMIQUES, UER DE TECHNOLOGIE DES INDUSTRIES AGRO-ALIMENTAIRES, 5030 GEMBLOUX, BELGIUM

1 Introduction

Pea starch is now considered as a high-quality natural food ingredient. Compared with other native starches, it is shown to be more resistant to heat and to mechanical treatment. The 'micronization', or ultra-fine size reduction, of pea starch causes a degree of damage. By X-ray diffraction, the damage to barley starch by milling may be attributed partly to the conversion of starch into a more amorphous form.[1] In addition, the water absorption, the water solubility, and the swelling are known to increase with starch damage.[1-3] This paper examines the effect of damage generated by the 'micronization' and evaluates the resulting physico-chemical properties. To characterize 'micronized' pea starch, the paste viscosity, the gelatinization, and the water absorption are studied.

2 Material and Methods

Native pea starch Nastar was provided by Provital (Warcoing, Belgium). The 'micronization' of starch was performed using an Alpin Multi-processing system with a circoplex classifier mill 50 ZPS. The degree of starch damage was measured with an SD4 Chopin according to the principles described by Medcalf and Gilles.[4] The results are expressed in Chopin–Dubois units (CDU) and corresponding values for the Farrand,[5] Audidier et al.,[6] and American Association of Cereal Chemists' (AACC) methods can be obtained. The Brabender viscograph was used to develop viscosity curves of the 'micronized' starch (bowl speed = 75 rpm; torsion spring = 700 cm g; concentration of starch = 8.5 wt%). The temperature was increased from 25 °C to 97 °C at a rate of 1.5 °C min^{-1}; it was maintained at 97 °C for 20 min and then decreased at a rate of 1.5 °C min^{-1} until 25 °C. Differential scanning calorimetry (DSC) was carried out on a Perkin Elmer machine (DSC-7, TAC 7-3). Samples were

Table 1   *Effect of turbine speed V_t on degree of starch damage (CDU = - Chopin–Dubois units; % AACC = percentage from American Association of Cereal Chemists; s = standard deviation)*

V_t (r.p.m.)	CDU	%AACC	Farrand units	Audidier
0	19.4 (s = 0.2)	7.5	27.6	16.0
2500	20.5 (s = 0.1)	9.0	32.2	17.2
3750	21.1 (s = 0.0)	9.9	35.0	17.9
5000	21.7 (s = 0.1)	10.7	37.3	18.4
6250	24.3 (s = 0.1)	14.4	48.1	21.1
7500	25.1 (s = 0.0)	15.6	52.2	22.1

prepared in pans of 25 μl; 2 μg were weighed and 8 μl of distilled water were added. An empty pan was used as the reference. Analyses were performed in a nitrogen atmosphere. Sorption moisture isotherms were executed with a Novasina aw-centre at 25 °C. The curves were established following manufacturer's recommendations.

3 Results and Discussion

Two parameters were varied in the 'micronization' of pea starch: the rotation speed of the classification turbine (V_t) and the rotation speed of the impact mill (V_m). The 'micronization' of pea starch led to damage of starch granules. The increase of V_t (for a constant V_m) increased the degree of damage (Table 1) whereas the variation of V_m (for constant V_t) had little effect (Table 2). The degree of starch damage of the untreated starch was already at 19.4 CDU (*i.e.* 7.5% damaged starch content with the AACC method); this might be caused by the shearing action of the separation procedure of the starch from the pea flour. For the effect of physical damage on barley starch granules, Stark and Yin[7] have also observed an increase in iodine affinity after damage as shown by potentiometric iodine titration.

Brabender curves have been established for each 'micronized' starch sample and for the untreated starch. Because of the high amylose content of pea starch (about 30%), a high peak viscosity did not appear after the gelatinization point. During the holding period at 97 °C, the consistency of all samples increased slightly and all starches displayed a large increase in paste consistency upon cooling (from 97 to 25 °C). To characterize the curves, three viscosity values were highlighted: at 48 min (97 °C), at 68 min (97 °C) and at

Table 2   *Effect of mill speed V_m on degree of starch damage*

V_m (r.p.m.)	CDU	%AACC	Farrand units	Audidier
0	19.4 (s = 0.2)	7.5	27.6	16.0
16000	20.7 (s = 0.1)	10.7	37.3	18.4
18000	22.1 (s = 0.0)	11.3	39.1	18.6
20000	23.3 (S = 0.1)	12.9	44.0	20.1
22000	23.4 (S = 0.1)	13.2	44.8	20.3

Table 3   *Effect of* V_t *on hot paste viscosity, expressed in Brabender units, of 'micronized' starch measured with a Brabender Amylograph*

V_t (r.p.m.)	97 °C (48 min)	97 °C (68 min)	57 °C (98 min)
0	450	515	750
2500	450	500	750
3750	445	500	720
5000	420	480	700
6250	405	455	665
7500	405	455	655

Table 4   *Effect of* V_m *on hot paste viscosity, expressed in Brabender units, of 'micronized' starch measured with a Brabender Amylograph*

V_m (r.p.m.)	97 °C (48 min)	97 °C (68 min)	57 °C (98 min)
0	450	515	750
16000	420	480	700
18000	420	475	680
20000	435	485	695
22000	430	485	685

88 min (57 °C) (Tables 3 and 4). For the low V_t values, the 'micronization' did not affect paste viscosity, but as V_t increased the viscosity decreased (Table 3). For increased V_m values, the viscosity did not change significantly, but decreased with respect to the untreated starch. The damaged starch level of 'micronized' starch could explain this rheological behaviour.

Gelatinization involves swelling, loss of birefringence, release of exudate, and viscosity changes. In the gelatinization of wheat starch, the increase in viscosity is due mainly to release of exudate from the granule rather than to the swelling of the granule.[8] An increase in the swelling caused by the starch damage has also been observed.[1-3]

In pea starch, the swelling is quite limited since the amylopectin content is lower than in other starches; therefore, the increase in viscosity is due rather to the dispersible components. Evers and Stevens[9] suggest that declining molecular integrity is associated with declining hot paste viscosity and a higher degree of damage, with greater fragmentation of the polymers amylose and amylopectin.

For each sample of 'micronized' starch and for the untreated starch, the gelatinization peak was found using DSC. The endothermic temperature, the onset temperature, and the enthalpy were calculated. The DSC analyses were performed with a large amount of water (80 wt%) so that only a single-peak transition was observed. Neither the variation in V_t nor the variation in V_m affected significantly the peak of gelatinization, its enthalpy, or its onset.

Amylose is the component of starch that contributes the characteristic of gelling. Gelling is the precipitation or crystallization of the amylose fraction. Amylopectin is the non-gelling portion of the starch.[10] No effect of 'micronization' on gelatinization parameters would seem to mean that the 'micronization'

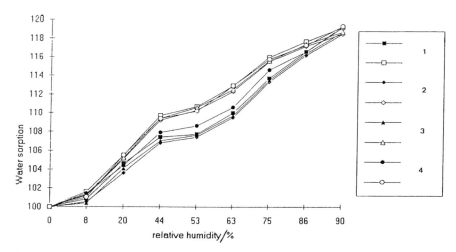

Figure 1   *Moisture absorption isotherms of pea starch: (1) untreated starch; 'micronized' starch with $V_m = 1600$ r.p.m.; (2) micronized starch with $V_t = 2500$ r.p.m.; (3) micronized starch with $V_t = 5000$ r.p.m.; (4) micronized starch with $V_t = 7500$ r.p.m. Water sorption is expressed as mass of water per unit mass of dry matter*

affects amylopectin molecules rather than amylose molecules. Amylopectin molecules are large enough to be cleaved during granule damage. Some amylose will also be released, but the molecule is much smaller and therefore less likely to be cleaved.[7]

Moisture absorption isotherm determination allows the observation of how much water a product can fix or loose while it is exposed to atmospheres of different relative moistures. For 'micronized' and untreated pea starch, the absorption isotherms (25 °C) have been similar and quasi-superimposable (Figure 1). This implies that 'micronization' has no effect on moisture absorption.

4 Conclusion

'Micronization' of pea starch was found to lead to granule damage. Starch damage increased with the rotation speed of the classification turbine (V_t). Starch damage led to a decrease of the hot paste viscosity. This was associated mainly with a degradation of the integrity of the molecules of the exudate. Amylopectin molecules were preferentially cleaved rather than amylose molecules which are smaller. 'Micronization' had no effect on water absorption of pea starch.

Acknowledgement

This work was carried out with the help of Walloon Region (Belgium) as part of the programme FIRST (Formation et Impulsion à la Recherche Scientifique et Technologique).

References

1. J. Lelièvre, *Die Stärke*, 1974, **3**, 85.
2. O. Paredes-Lopez, E. C. Maza-Calvino, and R. Montes-Rivera, *Starch*, 1988, **40**, 205.
3. C. Mok and J. W. Dick, *Cereal Chem.*, 1991, **68**, 409.
4. D. G. Medcalf and K. A. Gilles, *Cereal Chem.*, 1965, **42**, 546.
5. E. A. Farrand, *Cereal Chem.*, 1964, **41**, 98.
6. Y. Audidier, J. F. de la Guerivière, Y. Seince, and K. Benoualid, *Ind. Aliment. Agric.*, 1966, **83**, 1597.
7. J. R. Stark and X. S. Yin, *Starch*, 1986, **38**, 369.
8. B. S. Miller, R. I. Derby, and H. B. Trimbo, *Cereal Chem.*, 1973, **50**, 271.
9. A. D. Evers and D. J. Stevens, *Starch*, 1988, **40**, 297.
10. T. E. Luallen, *Food Technol.*, 1985, **39**(1), 59.
11. M. Ladoudaki, P. G. Demertlis, and M. G. Kontominas, *Food Sci. Technol.*, 1993, **26**, 512.

Influence of Fat Globule Size on the Rheological Properties of a Model Acid Fresh Cheese

By C. Sanchez, K. Maurer, and J. Hardy

LABORATOIRE DE PHYSICOCHIMIE ET GÉNIE ALIMENTAIRES, ENSAIA, 2 AV. DE LA FORÊT-DE-HAYE, 54500 VANDŒUVRE-LÈS-NANCY, FRANCE

1 Introduction

High-fat fresh cheeses are complex dairy emulsions since they contain three phases, emulsified milk fat, aggregated caseins, and a serum phase with essentially all the whey proteins and minerals. The textural properties of these food materials are mainly governed by the functional properties of the individual components as well as their three-dimensional arrangement and interactions. These relationships need to be understood in order to produce high- and constant-quality fresh cheeses and to develop new products.

As far as the emulsified milk fat is concerned, it is generally found that a fat content increase of soft cheeses results in a firmness and elasticity decrease. However, milk fat globules (MFG) can positively influence the rheological properties of milk products such as acid-induced composite milk gels. The most relevant parameters affecting this influence are the MFG size and number, and the membrane composition and deformability.[1-3] Since high-fat fresh cheeses are based on acid-induced composite gels, it was expected that their textural properties could also be improved by modifying the MFG characteristics. It is shown in this paper how MFG size affects firmness, flow and dynamic visco-elastic properties of a high-fat model fresh cheese.

2 Experimental

Six recombined creams were made by homogenizing at 65 °C a pre-emulsion of 20 wt% anhydrous milk fat and 18.6 wt% low-heat skim milk powder dispersed in distilled water. An ALM homogenizer type 2 was used at different homogenization pressures and passes in order to obtain different MFG sizes [4 MPa/1 pass (cream 1), 10 MPa/2 passes (2), 15 MPa/3 passes (3), 15 MPa/4 passes (4), 20 MPa/4 passes (5), 24 MPa/4 passes (6)]. Reconstituted skimmed

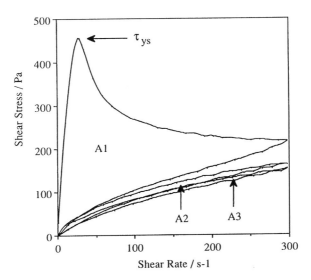

Figure 1 *Typical flow curves obtained at 10 °C for the model fresh cheeses. The parameters indicated are: A1–A3, areas of the loops 1–3; τ_{ys}, 'static yield value'*

milk (12 wt% total solids) was mixed with the creams and 2.5 wt% glucono-δ-lactone, and 0.02 wt% NaN_3 was added to avoid microbial spoilage. The mixture was left for 18 h at 30 °C. The acid curd was drained by pressing (3 wt%), packed into cylinders of varying sizes in order to perform the rheological measurements, and stored for 3 days at 5 °C. The model fresh cheeses 1–6 had about 63.2% moisture, 40.3% fat in dry matter, 6.1% total proteins, and a pH of 4.6.

The volume–surface mean diameter d_{vs} of the MFG was measured with an image analysis system calibrated with latex beads of 1.02 μm diameter (Sigma). Ten images per cream were analysed (not less than 500 particles per image).

Firmness was estimated with a capillary extrusion apparatus fitted to an Instron Universal Testing Machine type 1122 as previously described.[4] Flow curves were measured with a Rheomat 120 rheometer equipped with concentric cylinders in the shear rate range 0–300 s^{-1} with a sweep time of 144 s. Three successive shearing cycles were recorded. Dynamic visco-elastic parameters were measured on the Rheomat 120 rheometer operating in oscillatory mode with a frequency sweep of 0.062–31.2 rad s^{-1} (strain amplitude 0.017). All measurements were performed at 10 °C.

3 Results and Discussion

All the model fresh cheeses exhibited hysteresis loops similar to those drawn in Figure 1. Such time-dependent flow behaviour was also shown by commercial

Table 1   *Rheological parameters of model fresh cheeses determined at 10 °C from flow curves and capillary extrusion profiles*

Model fresh cheese number	Rheological parameters					
	$d_{vs}/\mu m$	τ_{ys}/Pa	$A1/Pa\ s^{-1}$	$A2/A1/\%$	$A3/A1/\%$	F_e/N
1	0.76	298	26071	17	10	24.3
2	0.65	391	34455	16	10	36.4
3	0.61	443	39994	13	10	46.7
4	0.59	519	45836	13	9	54.5
5	0.56	721	70308	13	7	86.0
6	0.53	852	81385	12	7	136.0

high-fat fresh cheeses.[5-7] The area of the hysteresis loop represented the extent of structural breakdown during a shearing cycle. It may be estimated that 83–88% of the structure of model fresh cheeses was broken after the first cycle and only 5–7% after the second one (Table 1). The loop 1 area increased with the d_{vs} decrease, indicating that cheeses with smaller MFG were initially more strongly structured (Table 1). These systems were seemingly more strongly destructured after the first shearing cycle as suggested by the smaller A2/A1 ratio.

The peak observed on the 'up' curve of the first cycle is the 'static yield value' τ_{ys} which should first be overcome in order that flow can occur in the food material (Figure 1). This parameter characterizes a three-dimensional, gel-like structure[8] and is an estimation of its firmness. Firmness estimated by both τ_{ys} and from the capillary extrusion force F_e were found to increase with decrease in d_{vs} (Table 1). Similar behaviour was observed for acid-induced milk gels.[2,3]

The limited frequency dependence of the storage and loss moduli, G' and G'' respectively, was indicative of a three-dimensional network, and it may have been related to the presence of τ_{ys} on the flow curves (Figure 2). The irregularities in the G'' plots were not so unusual since they have already been found with cream cheese and Mozzarella.[9,10] The loss angle (G''/G') increase at high frequencies was certainly induced by irreversible bond breakage (linear visco-elastic behaviour could not be longer sustained). There was not a clear relationship between the MFG size and G'' (Figure 2). In contrast, the G' value of model fresh cheeses was clearly improved by reducing d_{vs}. In this case, more numerous elastic bonds were created.

The three-dimensional structure of high-fat fresh cheeses, which brings about the observed rheological behaviour, is microscopically defined as a network of casein–MFG aggregates immobilizing a dilute oil-in-water emulsion.[11] Since the recombined fat globules are assimilated as pseudo-proteins, smaller and more numerous globules increase the protein interactions in and amongst the aggregates. The developing proteinaceous phase is more continuous and the individual aggregates are firmer. The resulting model fresh cheeses are consequently more elastic and firmer.

It can be concluded that high-fat fresh cheese texture may be improved by

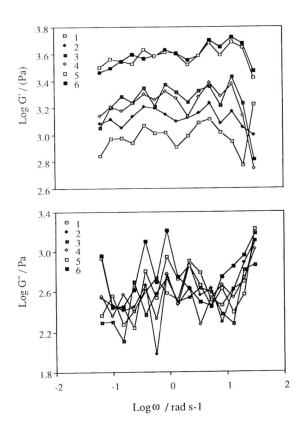

Figure 2   *Frequency dependence at 10 °C of storage modulus G' and loss modulus G"
of the model fresh cheeses 1–6 based on the six recombined creams: (1) 4
MPa/1 pass; (2) 10 MPa/2 passes; (3) 15 MPa/3 passes; (4) 15 MPa/4
passes; (5) 20 MPa/4 passes; (6) 24 MPa/4 passes. Logarithm of modulus
is plotted against logarithm of frequency ω*

decreasing the milk fat globule size in the initial mixture. We suggest that other
features like fat globule rigidity could also be involved. Study in this last area
is in progress in our laboratory; it involves fractionating milk fat to make
recombined MFG with different solid fat contents at a given temperature.

References

1. T. van Vliet and A. Denterer-Kikkert, *Neth. Milk Dairy J.*, 1982, **36**, 261.
2. Y. L. Xiong and J. E. Kinsella, *Milchwissenschaft*, 1991, **46**, 207.
3. Y. L. Xiong, J. M. Aguilera, and J. E. Kinsella, *J. Food Sci.*, 1991, **56**, 920.
4. C. Sanchez and J. Hardy, *Sci. Aliments*, 1993, **13**, 611.

5. S. Massaguer-Roig, S. S. H. Rizvi, and F. V. Kosikowski, *J. Food Sci.*, 1984, **49**, 668.
6. C. Sanchez, J. L. Beauregard, M. H. Chassagne, A. Duquenoy, and J. Hardy, *J. Food Eng.*, 1994, **23**, 595.
7. J. Korolczuk and M. Mahaut, *J. Texture Stud.*, 1989, **20**, 167.
8. A. N. Martin, G. S. Banker, and A. H. C. Chun, 'Rheology', Academic Press, New York, 1964, vol. 1, p. 20.
9. C. Sanchez, J. L. Beauregard, M. H. Chassagne, J. J. Bimbenet, and J. Hardy, *J. Food Eng.*, 1994, **23**, 579.
10. H. A. Diefes, S. S. H. Rizvi, and J. A. Bartsch, *J. Food Sci.*, 1993, **58**, 764.
11. C. Sanchez, INPL Thesis, Nancy, France, 1994.

Glasses

Influence of Macromolecules on the Glass Transition in Frozen Systems

By D. Simatos, G. Blond, and F. Martin

LABORATOIRE DE PHYSICO-CHIMIE ET PROPRIÉTÉS SENSORIELLES, ENSBANA, UNIVERSITÉ DE BOURGOGNE, 1 ESPLANADE ERASME, 21000 DIJON, FRANCE

1 Introduction

What is the influence of macromolecules on the glass transition in frozen aqueous systems? There are two reasons for raising this question. Firstly, there is the issue of the stability of frozen foods. More precisely, is it possible to increase the shelf-life of frozen foods by increasing the proportion of polymeric ingredients? Secondly, there is the issue of the role of polymers which are used as stabilizers in the manufacture of ice-cream.

Before looking particularly at the influence of macromolecules, some basic facts concerning frozen aqueous systems have to be reviewed. The state diagram of sucrose + water systems may be considered as representative of the physical state of most food products at low temperature, the numerical values of temperature and concentration only varying with the product composition (Figure 1). At the usual temperatures of storage a frozen food product is composed of ice crystals embedded in a freeze-concentrated phase. Under the practical conditions used for the freezing of foods, we may assume that the formation of ice progresses with temperature according to the cryoscopic (T_m) curve. The freezing of water is arrested when the temperature and concentration of the freeze-concentrated phase arrive at or close to the values T_g' and C_g' which correspond to the intersection of the T_m curve and the glass transition curve.

According to the theoretical argument developed by Franks,[1,2] and much experimental evidence, the freezing of water stops as a result of the strong increase in viscosity which occurs when the temperature approaches the temperature of glass transition. The temperature T_g' is then the temperature of the glass transition for the maximally freeze-concentrated phase. Over recent years, the importance of the glass transition for the stability of food systems has been highlighted in many papers.[3-10] If the temperature of storage is below T_g', the freeze-concentrated phase is a glassy solid; the evolution of the product

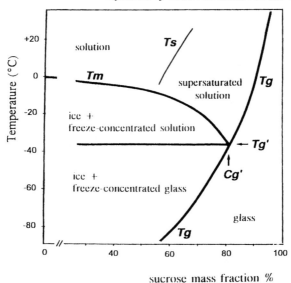

sucrose mass fraction %

Figure 1   *State diagram of the sucrose + water system: T_m = equilibrium ice formation and melting curve; T_s = solubility curve; T_g = glass transition curve; C_g' = sucrose fraction of maximally freeze-concentrated solution; T_g' = glass transition temperature of maximally freeze-concentrated solution*

is then expected to be very slow. In contrast, if the storage is carried out at a temperature higher than T_g', the freeze-concentrated phase is a supercooled melt, which may be either a viscous liquid, or a rubbery material, depending on its composition. When the temperature is increased above T_g', a tremendous reduction in viscosity may be expected, and is actually observed,[10,11] because of (i) the decrease in viscosity associated with the glass transition and (ii) the dilution of the freeze-concentrated phase due to the melting of ice.

This decrease in viscosity may result in an acceleration of all diffusion-controlled reactions. It is very well known, indeed, that the stability of frozen foods is strongly dependent upon temperature, with a rate of evolution which may be multiplied by a factor ranging between 2 and 30 for a 10 °C increase in the temperature of storage.[12] From well established knowledge in the field of polymer science, we know that the temperature of the glass transition increases with the average molecular mass of the system. It has thus been claimed that the addition of polymers could increase the T_g' of a product and hence its stability at a given temperature T above T_g'.[3-5] This possibility is discussed in the present paper. With the support of experimental data obtained for model systems, two aspects are considered: is the T_g' of sugar + water solutions raised by the addition of macromolecules, and what is the effect of this addition on the molecular mobility around T_g'?

Macromolecules (especially polysaccharides) are commonly used in ice-cream formulations because they are recognized as being very effective at

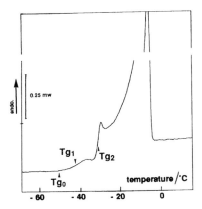

Figure 2   *DSC thermogram of a sucrose + water solution (50 wt%)*

improving the texture of the product. Their mode of action, however, is not well understood. Although numerous mechanisms have been suggested, only two are explored in the present paper: (i) increase in T_g', resulting in a slowing down of ice crystal growth during freezing and storage, and (ii) the influence on the mechanical properties of the freeze-concentrated phase at temperatures above T_g'.

2 The Glass Transition in Frozen Sugar + Water Systems

The differential scanning calorimetry (DSC) thermograms of frozen sugar solutions show two jumps in heat capacity, indicated as T_{g_1} and T_{g_2} in Figure 2. Following the interpretation given by Levine and Slade,[3-5] the values reported in the literature as T_g' data most often are the T_{g_2} values. The interpretation of the T_{g_1}/T_{g_2} features, however, is still a matter of controversy. Consequently, the determination of T_g' is still uncertain.[10,13] This point is not only of purely academic interest. The question is not so much what is the true value of T_g', but more about the molecular mobility in that temperature range.

Based on DSC studies, including annealing experiments and measurements of heat capacity increments, our interpretation is that the two jumps in heat capacity (T_{g_1} and T_{g_2}) are part of a single glass transition with enthalpy relaxation.[9,14] The measurement of visco-elasticity by dynamic mechanical thermal analysis (DMTA) confirms that some mobility appears at the temperature T_{g_0} (Figure 2) during rewarming.[10] The high value of the heat capacity increment, corresponding to both jumps T_{g_1} and T_{g_2}, which is much larger than the value to be expected for the glass transition of the freeze-concentrated phase, is explained by the fact that the beginning of ice melting is superimposed on this glass transition.[15] Since the determination of T_g' from

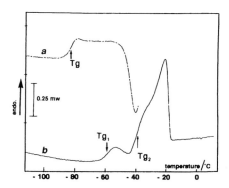

Figure 3   *DSC thermogram of a galactose + water solution (60 wt%). After a first rewarming to −40 °C (curve* a*) the sample is cooled again and then rewarmed (curve* b*)*

DSC thermograms is ambiguous, it is suggested that the intersection of the glass transition curve (T_g), and the ice formation curve (T_m) (Figure 1) could provide a better estimation.[16] It is generally considered, however, that the experimental values of T_m as obtained by DSC for concentrated solutions (beyond 40% for sucrose) are not reliable. The T_m curve represented in Figure 1 is therefore derived from the values of water activity predicted using the Universal Quasi-chemical (UNIQUAC) model, as modified by Larsen.[10] The T_g' value thus obtained for sucrose solutions is −37 °C.[10] Using the method of optimal annealing temperature, Ablett *et al.*[13] have arrived at a value of $T_g' = -40$ °C.

It must be stressed that the two-step increase in heat capacity is observed only when the samples contain ice.[9] When a concentrated sugar solution is quench-frozen, the sample is completely vitrified; the rewarming thermogram exhibits an increase in heat capacity with a small overshoot, which is typical of a glass transition (Figure 3a). After some ice has been formed during rewarming, the two-step increase is shown (Figure 3b). This kind of feature has been demonstrated by polymer chemists to result from enthalpy relaxation when the material displays a broad distribution of relaxation times, as a consequence of microstructural defects having various energy barriers. In the glass, microstructural defects, or regions of anomalous density, are 'frozen', *i.e.* immobilized. When the temperature is raised, approaching the temperature of the glass transition, these defects regain mobility at a temperature depending on their energy level. Microstructural heterogeneity in the glass appears to result from the interface of ice crystals with the freeze-concentrated phase. These features of two-step increase in heat capacity have been described not only for sugars and polyols, but also for frozen aqueous solutions of electrolytes, and other small molecular mass solutes (Table 1). With a series of

Table 1 *Temperatures (mid-point) for the two steps of the heat capacity increase determined by DSC with frozen aqueous solutions*

Solute	T_{g_1}	T_{g_2}	$T_{g_2} - T_{g_1}$
	°C	°C	°C
Sorbitol	− 60	−47	13
Galactose	− 56	−38	18
Sucrose	− 40	−32	8
β-Alanine[a]	− 89	−65	24
CaCl$_2$[a]	−109	−95	14
Polyvinylpyrrolidone	− 33	−21	12

[a] Data from reference 18

Table 2 *Temperatures for the two steps of the heat capacity increase determined by DSC for frozen aqueous solutions (50% dry solid content)*

Solute	wt%	T_{g_0}	T_{g_1}	T_{g_2}	$T_{g_2} - T_{g_0}$
		°C	°C	°C	°C
Sucrose		− 50	− 40	− 32	18
Dextran/sucrose	2:48	− 50	− 40	− 32	18
	10:40	− 50	− 40	− 28	22
	25:25	− 50	− 38	− 25	25
Gelatin/sucrose	2:48	− 56	− 48	− 31	25
	10:40	− 56	− 50	− 26	30
	25:25	− 58	− 51	− 20	38
Polyvinylpyrrolidone	0:50	− 65	− 60	− 47	18
/sorbitol	25:25	− 69	− 65	− 39	30
	50:0	− 37	− 30	− 20	17

malto-oligomers, Ablett *et al.*[17] have reported that when the molecular mass increases, there is an increase in the temperatures of both transitions (T_{g_1}, T_{g_2}), a decrease in the temperature differential between them, and a decrease in the change in heat capacity associated with the initial one.

3 Influence of Macromolecules Added at a High Concentration

When added at a rather high concentration to solutions of low molecular mass solutes, polymers induce an increase in the value of T_{g_2}. Figure 4 demonstrates this effect for various proteins added to lactose, sorbitol, glycerol and Tris. The same trend is shown in Table 2. A significant increase of T_{g_2} is observed when dextran or gelatin are added to a sucrose solution at a concentration of 20% of the dry matter. These results show, however, that the values of T_{g_0} and T_{g_1} do not increase, or even decrease, which may indicate that the temperature at which some molecular mobility appears is not raised on addition of the

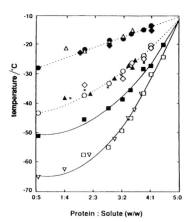

Figure 4   *Effect of proteins on the* T_{g_2} *values of other solutes:[18]* ●, *lactose + bovine serum albumin (BSA);* △, *lactose + lysozyme;* ◆, *lactose + ribonuclease A;* ○, *sorbitol + BSA;* ▲, *sorbitol + lysozyme;* ◇, *sorbitol + ribonuclease A;* ×, *sorbitol + elastase;* ■, *Tris + BSA;* □, *glycerol + BSA;* ▽, *glycerol + lysozyme*

polymer. It is a common observation that, in mixed materials, the glass transition broadens to a wide range of temperature.[19] For instance, with mixtures of salts, Senapati and Angell[20] have reported a broadening of the glass transition and a spreading out to a very wide range of temperature of the enthalpy relaxation processes which are enhanced by the annealing. This behaviour was interpreted to be the result of microstructural heterogeneity. We then suggest that the broad distribution of relaxation times which may be promoted by the complex chemical composition is here amplified by the presence of ice.

If the enthalpy relaxation process can give some information on the molecular mobility, the visco-elastic properties are certainly more representative of the macroscopic mobility. In order to monitor the large change of mechanical behaviour of frozen materials as a function of temperature, we have carried out our mechanical spectroscopy studies using different measuring devices (Metravib Viscoanalyzer). A compression device gives a better sensitivity to study the behaviour of completely frozen samples, and to detect the beginning of softening which indicates the glass transition. However, an annular shear cell is more appropriate to monitor the tremendous reduction in the rigidity and viscosity above T_g'.[21,22]

For a sucrose solution (50 wt%), the compression test gives evidence of a decrease in the storage modulus E' beginning at a temperature close to $T_{g_0} = -50$ °C, and of a maximum in the loss modulus E'' at -37 °C, *i.e.* at a temperature close to T_g'.[10] The classical frequency effect associated with the glass transition can be observed, *i.e.* a shift in the above temperatures to higher

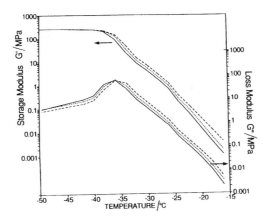

Figure 5   *Evolution of the storage and loss moduli (annular shear test) during the rewarming of a 50% sucrose + water solution, for three measurement frequencies: ———, 5 Hz; . . . ., 10 Hz; - - -, 20 Hz*

Table 3   *Temperatures T_{g_1} and T_{g_2} measured on DSC thermograms and the temperature corresponding to the maximum of the loss modulus (from DMTA) for aqueous solutions (50% dry solids)*

		DSC		DMTA
Solute	*wt%*	T_{g_1}	T_{g_2}	$T_{(G''max)}$
Sucrose		−40	−32	−36
Dextran/sucrose	2:48	−40	−32	−36
	25:25	−38	−25	−29
Gelatin/sucrose	2:48	−48	−31	−36
	25:25	−51	−20	−30

values when the frequency of measurement is increased. With the shear test, the same effects are observed; the measurements are, however, less reliable at the lower temperatures, but they can be carried out in the higher temperature range, despite a tendency of the sample to flow (Figure 5). The temperature corresponding to the maximum in the loss modulus G'' (Table 3) shows a significant increase on addition of a polymer, dextran or gelatin, at least when added at a high concentration (*e.g.* 50 wt% of the solid content).

We may conclude that, when added at a rather high concentration to a sucrose solution, a polymeric material induces an increase in T_g'. From a practical point of view this raised T_g' is certainly significant if we consider, for instance, the resistance to structure collapse during a freeze-drying operation. However, the observed enthalpy relaxation below T_g' should remind us that molecular mobility exists in the temperature range below T_g' and that other kinds of instability may also exist.

4 Influence of Macromolecules Added at a Low Concentration

Macromolecules (various polysaccharides or gelatin) are commonly added to ice-cream mixes as 'stabilizers', as they are known to exert a beneficial effect on the texture of the final product. The investigations performed in support of the various suggested mechanisms, however, still give rise to contradictory reports.[23,24] In a recent study Goff *et al.*[25] have demonstrated that the addition of 'stabilizers' (0.15 wt% locust bean gum + 0.02 wt% carrageenan) to an ice-cream mix results in a smaller size of the ice crystals immediately after freezing, and in a smaller rate of growth after 24 weeks of storage at abusive temperatures. The same authors, however, have been unable to detect any change in the T_g' (or T_{g_2}) values in the DSC thermograms obtained with 20 wt% sucrose solutions to which 0.6% 'stabilizer' (locust bean gum, carrageenan, guar, or xanthan) had been added.[24]

Our DSC data lead to the same conclusion. When added to a sucrose solution, dextran or gelatin (4 wt% of the dry matter content) does not induce any significant change in the T_{g_2} value (see Table 2). The same lack of an effect is observed with maltodextrin at 4 wt% or guar gum at 1 wt%.[21] This absence of an effect on the value of T_g' can be explained on the basis of the modification of the state diagram (Figure 1) to be anticipated upon addition of a small proportion of polymer.[21] A dry mixture of sucrose and 4 wt% dextran shows a well defined glass transition (indicative of the good miscibility of the two materials) with a 4 °C shift above the temperature of the glass transition of pure sucrose. This increment is quite consistent with the one that is predicted from the Couchman expression,[26] taking for sucrose a value of $T_g = 67$ °C, and for dextran the estimated value of $T_g = 225$ °C. With 20 wt% water, however, the sucrose + dextran (4 wt%) mixture has the same T_g as sucrose with 20 wt% water (a water content very close to that of the maximally freeze-concentrated phase[21]). Should the dextran solution be exhibiting ideal behaviour, the ice formation curve of sucrose would be shifted slightly upwards upon addition of 4 wt% dextran, resulting in a T_g' value at the intersection with the glass transition curve raised by *ca.* 5 °C. Most probably, however, the sucrose + dextran solutions do not behave ideally, and a steepening of the ice curve is to be expected, suggesting that the addition of small amounts of polymer does not significantly change the T_g' value of sucrose solutions.[21]

If T_g' does not appear to be raised by the addition of macromolecules at a low concentration, interesting effects can still be observed concerning the molecular mobility. In the first place, DSC studies show the influence of polymer addition on the enthalpy relaxation processes. The overshoot which is

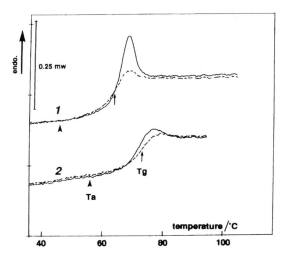

Figure 6   *DSC thermograms for sucrose (curve 1) and a mixture of sucrose + dextran (90:10) (curve 2). Solid lines are the curves obtained after annealing for 2 h at T_a; broken lines are the second scans*

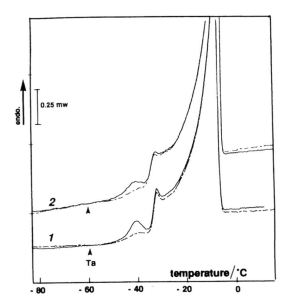

Figure 7   *DSC thermograms for a 50% sucrose + water solution (curve 1) and a 50% sucrose/dextran (48:2) solution (curve 2). Solid and broken lines, as in Figure 6*

Table 4 *Values for the storage and loss moduli, and of the loss factor, at temperature* $T = T_{G''max} + 10\ °C$ *(annular shear test frequency of measurement = 5 Hz; aqueous solutions 50% dry matter)*

Solute	G'/MPa	G''/MPa	tan δ
Sucrose	1.84	1.64	0.89
Sucrose/maltodextrin (48:2)	3.28	2.6	0.79
Sucrose/dextran (48:2)	5.1	3.44	0.67
Sucrose/guar (49.5:0.5)	9.2	5.48	0.59
Sucrose/gelatin (48:2)	0.167	0.262	1.57

observed at the end of the heat capacity increment can be increased by the process of annealing, *i.e.* holding the sample for some time in the temperature range below the glass transition; it is representative of the structural relaxation which takes place during the annealing treatment. The amplitude of this overshoot is much reduced in the dry sucrose + dextran mixture, indicating that the rate of relaxation is slowed down by the presence of the polymer (Figure 6). In sucrose + dextran solutions (water content 50 wt%, dextran 4 wt% of the dry matter content), the 'pre-peak' is much reduced after annealing, as well as the overshoot, as a result of the same effect (Figure 7).[21]

The addition of macromolecules also influences the visco-elastic properties of the solution in the temperature range above T_g'. When maltodextrin or dextran are added to a sucrose solution (50 wt% water content) as 4 wt% of the dry matter content, or guar gum as 1 wt%, the temperature corresponding to the maximum of the loss modulus G'' is not significantly changed, consistent with the fact that no change in T_{g1} and T_{g2} is observed on the DSC thermograms. But, above these temperatures, the actual values of the loss modulus G'' or storage modulus G', however, are higher than for the sucrose solution. Simultaneously, the loss factor (tan δ) decreases, which means that the G' increase is more pronounced than the increase in G'' (Table 4), and the material becomes more solid-like.

It should be recalled that the change in the visco-elastic properties recorded for frozen materials depends on the behaviour of both co-existing phases: the ice crystals and the freeze-concentrated fraction. After the first decrease in the storage moduli (E', G') corresponding to the glass transition (beginning at a temperature close to T_{g0}), the freeze-concentrated fraction undergoes a continuing decrease in G' and a strong decrease in G''. This evolution is a consequence of the incipient melting of ice which is superimposed on the direct effect of the rising temperature on the viscosity, *i.e.* viscosity change with temperature above T_g according to the Vogel–Tammann–Fulcher (VTF) or Williams–Landel–Ferry (WLF) expressions. The G', G'' and tan δ values recorded for an 80 wt% sucrose solution at a temperature 10 °C above the maximum of G'' denote highly viscous behaviour, with a rather low rigidity (no 'rubbery plateau' is to be expected in the temperature range above T_g for a low molecular mass material[21]). With a 50 wt% sucrose solution, G'' is much smaller, due to the dilution of the freeze-concentrated fraction, resulting from the melting of ice; but G' is higher (and tan δ much lower) than with the 80 wt% sample. This high apparent rigidity is attributed to the ice phase.[21] These last characteristics are even more visible with less concentrated sucrose solutions.[21] Although the addition of macromolecules may be able to change the texture of the ice fraction, as will be shown later, the increase in G' and G'' at a temperature 10 °C above the maximum of G'', which is observed when maltodextrin, dextran or guar gum are added to the 50 wt% sucrose solution, is probably due to modification of the visco-elastic behaviour of the freeze-concentrated fraction. The decrease in tan δ gives evidence for an enhanced rigidity, with a tendency towards the development of a rubbery plateau. These effects increase in order of increasing shear-thinning behaviour of the polysaccharides, *i.e.* maltodextrin < dextran < guar gum, although guar gum is used at a lower concentration. As is well known, guar gum has a particularly low critical concentration for entanglement, and exhibits a steep increase in viscosity above that concentration.[27] In the freeze-concentrated phase, the polysaccharide concentrations are well above their critical concentrations for entanglement.

When gelatin is added to the 50 wt% sucrose solution, what is observed at a temperature 10 °C above the maximum of G'' is not an increase in the G' value but a decrease, despite the elevated shear-thinning character of the polymer.[28] As has been demonstrated previously[29] the higher the rigidity of a polymer + - water system, the more effective it is at decreasing the ice propagation rate, resulting in smaller ice crystals. The higher rigidity of the sucrose + gelatin system is probably responsible for a thinner ice structure, leading to lower G' values in the partly frozen material.

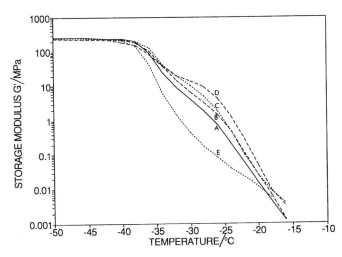

Figure 8 *Evolution of the storage modulus during the rewarming of 50% (dry matter) solutions of sucrose (A), sucrose/maltodextrin (48:2) (B), sucrose/dextran (48:2) (C), sucrose/guar gum (49.5/0.5) (D), sucrose/gelatin (48:2) (E). Annular shear test; measurement frequency 5 Hz*

5 Discussion

This discussion proposes a combined view of the effects to be observed, or to be expected, following the addition of macromolecules to a frozen sugar solution. For each of the discussed items, we distinguish between the effects that can be anticipated on the basis of the polymer science concepts and the extra effects that have to be added because we are dealing with frozen aqueous systems.

Temperature of the Glass Transition

For a series of homologous, linear polymers, it is well known that the temperature of the glass transition is inversely proportional to the molecular mass. The same is shown to be true also for sugars.[30] Other molecular properties, such as branching, may play a secondary role. For a mixture, the temperature of the glass transition can be predicted from the characteristics of the individual components by means of expressions such as

$$T_g = (\sum_{i=1}^{n} X_i \; \Delta C_{pi} \; T_{gi}) \, / \, (\sum_{i=1}^{n} X_i \; \Delta C_{pi})$$

where X_i is the mole fraction, T_{gi} is the glass transition temperature, and ΔC_{pi} is the incremental change in heat capacity at T_{gi} for each pure component i. This expression has been shown to satisfactorily describe the compositional variation of T_g for compatible polymer mixtures[26] and also for anhydrous, binary, sugar mixtures.[31]

The temperature of the glass transition for the maximally freeze-concentrated fraction (T_g') corresponds to the intersection of the glass transition curve and the ice curve. It is then also under the control of the water activity in the concentrated solution. The primary effect of this parameter is also an increase in T_g' with an increase in molecular mass (influence of the molecular mass of the solute on the water mole fraction); a secondary effect is the consequence of molecular interactions (influence on the activity coefficient).

Data on a 'true value' for T_g' is almost non-existent, and so discussion is based at present on numbers which are actually values of T_{g2}. In the case of binary solutions, these values have been shown to increase with the molecular mass of the solute.[3,17] The addition of macromolecules to an aqueous solution of a low molecular mass solute induces an increase in T_{g2} when this addition represents a rather high proportion of the solid content.

Molecular Mobility in the Temperature Range around T_g'

The evolution of molecular mobility above the temperature of the glass transition may be very different depending on the material, as is shown by the strong/fragile classification of Angell and co-workers.[32,33] 'Strong' liquids are those with Arrhenius or near-Arrhenius behaviour above T_g'; in contrast 'fragile' liquids show VTF behaviour, *i.e.*, an important increase of mobility with increasing temperature. According to data very recently reported by Angell *et al.*[34] aqueous solutions of sugars and polyalcohols may be classified as intermediate between strong and fragile materials; poly-L-asparagine seems to be as 'strong' as or 'stronger' than polyisobutylene, which is the 'strongest' chain polymer.

In frozen systems, the presence of ice appears to add a further complication: ice seems to promote heterogeneity in the microstructure, which could be favourable to structural relaxations in the glass, over a broad temperature range below T_g'. The rate of these structural rearrangements may be slowed down by the added polymer,[21] but the temperature range may be especially large in the case of polymer-containing systems.

Consequence for Stability of Frozen Food Products

The curve representing the temperature of the glass transition *versus* water content has been proposed as a frontier in a 'stability map' between a domain (temperature and water content below the glass transition curve) where diffusion is strongly limited and stability is expected, and a domain where the kinetics of evolution are 'dramatically' increased with temperature and/or water content.[3–5] Experimental evidence is being accumulated demonstrating the close relationship between the glass transition and the kinetics of various phenomena (caking,[36] crystallization,[37] non-enzymic browning,[38] *etc.*), at least under conditions where the kinetics are truly diffusion-controlled.

The temperature of the glass transition of the freeze-concentrated fraction has been claimed to play the same role as the stability limit.[3–5] It is probably right to say that, as a general rule, T_g' corresponds to a temperature range corresponding more or less to a threshold of instability, and that increasing T_g' by the addition of a polymer may result in increased stability. It may be safer, however, to discuss this statement more precisely, according to the nature of the process, or the material. For instance, the value of T_g' may be a useful characteristic for predicting the resistance to structure collapse of a product to be freeze-dried.[39] This may be particularly the case when the material is homogeneously composed of low molecular mass solutes, or when it comprises a proportion of polymer large enough to affect the T_g' value. It has been well demonstrated, however, that small amounts of polymers (polysaccharides or proteins), insufficient to induce a change in the transition temperatures (T_{g_1}, T_{g_2}) in a DSC thermogram, are truly able to increase the resistance to collapse during freeze-drying, probably because structure collapse depends more on the rheological properties of the freeze-concentrated phase than on the mobility of small solutes.[40] As regards chemical/biochemical changes, it appears that the stability of frozen products is partly controlled by the difference between the storage temperature and their individual T_g' values, but only partly so.[41] Other phenomena suggested to play a part are also, for instance, the specific cryoprotective effect of some solutes and the volume of the freeze-concentrated phase.[42]

It should also be emphasized that the tremendous reduction in viscosity which takes place at temperatures above T_g' depends only partly upon the glass transition (the VTF or WLF effect); an important contribution results from the melting of ice.[9] Again, the water activity of concentrated solutions is to be considered. When food products are stored at temperatures above their T_g' values, it is very generally observed that the kinetics of physico-chemical or biochemical evolutions are much slower than could be expected on the basis of the viscosity decay above T_g'.[9,10]

The Texture of Ice-Cream

As mentioned before, the polysaccharides used as stabilizers do not induce any visible effect on the temperatures of the transitions, under circumstances where they favourably affect the texture. On the basis of the observations reported in this paper, the following explanations can be suggested.

> They could be able to restrict the molecular mobility around T_g' (and above), as suggested by their influence on the enthalpy relaxation process, and thus slow down the ice crystal growth.
> They limit the ice crystal growth as a consequence of the rigidity they impart to the freeze-concentrated phase at temperatures above T_g'.
> They are directly responsible for the smoothness of the texture.

Finally, therefore, it may be concluded that polymers can significantly modify the glass transition behaviour of frozen aqueous systems. From a practical point of view, however, their effect on the glass transition is only one aspect of how they work.

References

1. F. Franks, in 'Water: A Comprehensive Treatise', ed. F. Franks, Plenum, New York, vol. 7, p. 215.
2. F. Franks, in 'Properties of Water in Foods', ed. D. Simatos and J. L. Multon, Nijhoff, Dordrecht, 1985, p. 497.
3. H. Levine and L. Slade, *Cryoletters*, 1988, **9**, 21.
4. H. Levine and L. Slade, *Commun. Agric. Food Chem.*, 1989, **1**, 315.
5. H. Levine and L. Slade, in 'Thermal Analysis of Foods', ed. V. R. Harwalkar and C. Y. Ma, Elsevier Applied Science, London, 1990, p. 221.
6. M. Karel and I. Saguy, in 'Water Relationships in Foods', ed. H. Levine and L. Slade, Plenum, New York, p. 157.
7. Y. Roos and M. Karel, *Biotechnol. Progr.*, 1990, **6**, 159.
8. M. Karel, in 'Drying 91', ed. A. S. Meyumdar and I. Filkova, Elsevier, Amsterdam, 1991, p. 26.
9. D. Simatos and G. Blond, in 'Water Relationships in Food', ed. H. Levine and L. Slade, Plenum, New York, 1991, p. 139.
10. D. Simatos and G. Blond, in 'The Glassy State in Foods', in J. M. V. Blanshard and P. J. Lillford, Nottingham University Press, 1993, p. 395.
11. R. J. Bellows and C. J. King, *AIChE Symp. Ser.*, 1973, **69**, 33.

12. M. Jul, 'The Quality of Frozen Food', Academic Press, London, 1984.
13. S. Ablett, A. H. Clark, M. J. Izzard, and P. J. Lillford, *J. Chem. Soc., Faraday Trans.*, 1992, **88**, 789.
14. G. Blond, *Cryoletters*, 1989, **10**, 299.
15. G. Blond and D. Simatos, *Thermochim. Acta*, 1991, **175**, 239.
16. R. H. M. Hatley, C. Van den Berg, and F. Franks, *Cryoletters*, 1991, **12**, 13.
17. S. Ablett, A. H. Darke, M. J. Izzard, and P. J. Lillford, in 'The Glassy State in Foods', ed. J. M. V. Blanchard and P. J. Lillford, Nottingham University Press, 1993, p. 189.
18. B. S. Chang and C. S. Randall, *Cryobiology*, 1992, **29**, 632.
19. B. Wunderlich, 'Thermal Analysis', Academic Press, New York, 1990.
20. H. Senapati and C. A. Angell, *J. Non-Cryst. Solids*, 1991, **130**, 58.
21. G. Blond, *J. Food Eng.*, 1994, **22**, 253.
22. G. Blond, *J. Rheol.*, in the press.
23. J. M. V. Blanshard and F. Franks, in 'Food Structure and Behaviour', ed. J. M. V. Blanshard and P. Lillford, Academic Press, London, 1987, p. 51.
24. H. D. Goff, K. B. Caldwell, and D. W. Stanley, *J. Dairy Sci.*, 1993, **76**, 1268.
25. H. D. Goff, K. B. Caldwell, and D. W. Stanley, *Food Struct.*, 1992, **11**, 11.
26. P. R. Couchman, *Macromolecules*, 1978, **11**, 1156.
27. E. R. Morris, in 'Frontiers in Carbohydrate Research', ed. R. P. Millaire, J. N. Bemitter, and R. Chandrasekarian, Elsevier Applied Science, London, 1989, p. 132.
28. G. Blond and F. Martin, Comptes rendus des 6èmes Rencontres Agoral., ENSIA Massy, 1993, p. 115.
29. G. Blond, in 'Properties of Water in Foods', ed. D. Simatos and J. L. Multon, Nijhoff, Dordrecht, 1985, p. 531.
30. P. D. Orford, R. Parker, S. G. Ring, and A. C. Smith, *Int. J. Biol. Macromol.*, 1989, **11**, 91.
31. L. Finegold, F. Franks, and R. H. M. Hatley, *J. Chem. Soc., Faraday Trans. 1*, 1989, **85**, 2945.
32. C. A. Angell, A. Dworkin, P. Figuière, A. Fuchs, and H. Szwarc, *J. Chim. Phys.*, 1985, **82**, 773.
33. C. A. Angell, *J. Phys. Chem. Solids*, 1988, **49**, 863.
34. C. A. Angell, R. D. Bressel, J. L. Green, H. Kanno, M. Oguni, and E. J. Sarre, *J. Food Eng.*, 1994, **22**, 115.
35. G. E. Downton, J. L. Flores Luna, and C. J. King, *Ind. Eng. Chem. Fundam.*, 1982, **21**, 447.
36. Y. Roos and M. Karel, *J. Food Sci.*, 1991, **56**, 38.
37. Y. Roos and M. Karel, in 'The Glassy State in Foods', ed. J. M. V. Blanshard and P. J. Lillford, Nottingham University Press, 1993, p. 207.
38. R. Karmas, M. del Pilar Buera, and M. Karel, *J. Agric. Food Chem.*, 1992, **40**, 873.
39. F. Franks, R. H. M. Hatley, and S. F. Mathias, *Pharm. Technol. Int.*, 1991, Oct., 24.
40. M. le Meste, F. Diallo, and D. Simatos, *Proc. Int. Congress on Refrigeration*, 1979, p. 261.
41. W. L. Ken, M. H. Lim, D. S. Reid, and H. Chen, *J. Sci. Food Agric.*, 1993, **61**, 51.
42. M. H. Lim and D. S. Reid, in 'Water Relationships in Foods', ed. H. Levine and L. Slade, Plenum, New York, 1991, p. 103.

Phenomenon of Enthalpy Relaxation at the Glass Transition in Granular Starches

By Chee Choon Seow

FOOD TECHNOLOGY DIVISION, SCHOOL OF INDUSTRIAL TECHNOLOGY,
UNIVERSITI SAINS MALAYSIA, 11800 PENANG, MALAYSIA

1 Introduction

The gelatinization of aqueous starch systems has been extensively studied using differential scanning calorimetry (DSC). A common feature observed for most starches is that the typical monophasic endotherm obtained in systems with excess water is transformed into biphasic endotherms (usually labelled as the G and M_1 endotherms[1]) in limited water systems. Interpretation of these thermal events is still very much a subject of controversy. The various hypotheses advanced have been thoroughly reviewed by several authors.[1-3]

The most attractive hypothesis, which is fast gaining universal acceptance, views granular starches as partially crystalline polymer systems and gelatinization as a non-equilibrium process involving a prerequisite glass transition or softening of the amorphous phase prior to irreversible melting of crystallites, with water acting as a plasticizer.[2,4-6] Based on the expected characteristic shift in heat capacity on passage through a glass transition, the operative or effective glass transition temperature T_g associated with gelatinization is believed to be located at the leading edge of the DSC mono- or biphasic endotherms.[6,7]

Although the bulk of evidence indicates that the G and M_1 transitions represent distinct thermal events in the amorphous and crystalline regions of the starch granules, respectively,[1,4,5] the exact nature and origin of the lower temperature (or G) transition remain unresolved. Recent studies[8] suggest that it may, in reality, be an apparent endotherm arising from a glassy-to-rubbery state transformation associated with enthalpy relaxation superimposed on the crystallite melting (M_1) endotherm. In this paper, current knowledge concerning the non-equilibrium thermal behaviour of granular starches is examined in the light of such an interpretation.

2 Sub-T_g Annealing of Granular Starches

The fact that the G 'endotherm' in the DSC thermograms of most starches at intermediate moisture contents is typically the dominant component of the biphasic endotherms has probably given the impression that it is an actual endotherm derived from melting of crystallites, and it indeed has been so misinterpreted by several researchers.[9,10] However, there are atypical cases where it is represented by a small shoulder, which appears to be a clearly discernible glass transition,[5,8] at the leading edge of a dominant M_1 peak. DSC studies carried out by Seow and Teo[8] have shown that the effects of isothermal annealing are qualitatively similar for rice starches exhibiting either type of thermal profile. What is obviously more important is the annealing temperature T_a relative to the effective T_g associated with gelatinization. Different and, at times, rather misleading interpretations have previously arisen because of the difficulty in determining the exact temperature location of the glass transition in granular starches. Annealing, defined as the holding of a starch + water mixture at intermediate to high water contents for limited periods of time at temperatures either slightly above or below its effective onset T_g (which, in the present context, may be taken as the onset gelatinization temperature T_o[6-8]), has been shown to produce different structural changes in the starch granules.[8]

Effects on DSC Thermal Transitions

The main concern of this paper is with sub-T_g annealing or physical aging and its effects on the thermal behaviour of granular starches as determined by DSC. In this respect, the variables of interest include the annealing temperature, the moisture content, and time. The important changes associated with sub-T_g annealing of granular starches may be summarized as indicated below.

1. There is a narrowing of the gelatinization range $(T_c - T_o)$[8,11-13] which is primarily due to a shift in T_o to a higher temperature with little change in the crystallite melting (T_m) and conclusion (T_c) temperatures (Figure 1). The G transition becomes narrower and longer, but the M_1 endotherm is not much affected.

2. As T_a approaches T_o, the G endotherm becomes progressively narrower and longer and the increase in T_o becomes more pronounced, as shown in Figure 2.

3. At a particular T_a and moisture content, T_o increases relatively sharply over the first few hours before levelling off[8,11,12] (Figure 3). Limited water systems reach a higher T_o plateau than corresponding systems with excess water.

4. An increase in the overall enthalpy of gelatinization, ΔH_G, determined from the area of the monophasic or biphasic endotherms, may or may not be obtained depending on the thermal history of the sample tested.[12]

5. No alterations in X-ray crystallinity occurs on annealing at $T_a < T_o$,[1,13] thus indicating little or no annealing of crystallites, which is consistent with the fact that the M_1 endotherm also remains unaffected.

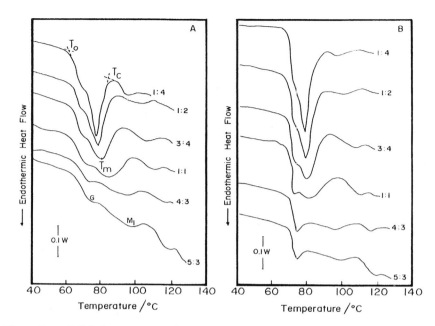

Figure 1   *DSC thermograms of an atypical rice starch at various starch/water ratios:*
A, untreated; B, annealed (55 °C for 4 h).
(Reproduced, by permission of VCH Verlagsgesellschaft, from reference 8)

Interpretation of the Effects Observed

The changes in the G transition described above may be ascribed to free
volume and enthalpy relaxation (*i.e.*, to the decrease in excess volume and
enthalpy) in the amorphous regions of the starch granules on sub-T_g annealing.
Such non-equilibrium behaviour of the glassy state is commonly encountered
in amorphous or partially crystalline synthetic polymers,[14–18] biopolymers,[19,20]
and low molecular mass glassy systems.[21–23]

Enthalpy and volume relaxations are known to occur in materials where the
polymer molecules have not been able to attain a more stable conformation
and packing in the glassy state. Although polymer molecular or segmental
mobility is hindered to a large extent by the extremely high viscosity in the
glass, conformational rearrangement and/or packing can occur if a driving
force exists to push the system towards a state of lower energy and volume.[17]
Thus, excess volume and enthalpy quenched into the system would decrease
during sub-T_g annealing, while the enthalpy recovered during subsequent
heating to temperatures above T_g would increase with increasing annealing
time. This gives rise to correspondingly larger and sharper apparent
endothermic peaks (as shown in Figures 1 and 2) which reflect enhanced
homogeneity and/or metastability in the glassy regions of the starch granules.[8,12,24,25]

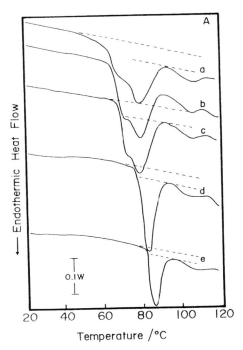

Figure 2 *DSC thermograms of an atypical rice starch at a starch/water ratio of 3:4 annealed under different temperature–time conditions: (a) untreated; (b) 45 °C for 4 h; (c) 55 °C for 4 h; (d) 65 °C for 4 h; (e) 70 °C for 1 h.* (Reproduced, by permission of VCH Verlagsgesellschaft, from reference 8)

Isothermal annealing of granular starches have been variously reported to increase, decrease, or have no significant effects on ΔH_G. Such conflicting observations may be resolved by taking into consideration the following factors. (a) The annealing temperature T_a: is it above or below the T_g associated with gelatinization? (b) The thermal history of the starch sample which has an important bearing on the condition of the amorphous regions in the starch granules: has the sample been subjected to any previous thermal treatment, intentionally or accidentally?

At T_a less than but close to T_g, ΔH_G would be expected to remain practically unchanged over realistic annealing times, or to increase, depending on the condition of the glassy state in the original sample. If a large excess enthalpy had been trapped in the amorphous regions of the starch granules, then sub-T_g annealing would result in enthalpy recovery during subsequent passage through the glass transition and, therefore, an increase in ΔH_G, as has been observed by most workers.[12,26–28] An annealing time as short as 15 min at 50 °C has been found to be sufficient to cause this approach of enthalpy to an apparent 'equilibrium' value.[12] It is worthwhile emphasizing that this apparent equilibrium condition does not imply that physical ageing has stopped

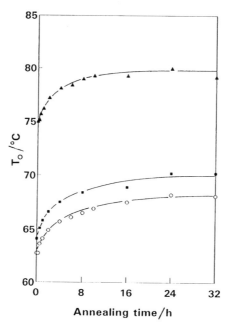

Figure 3   *Changes in onset temperature of gelatinization, T_o, of granular rice starch with time of annealing at 50 °C:* ■, *4:3 starch + water system;* ▨, *1:2 starch + water system;* ▲, *1:2:1 starch + water + sucrose system*

completely, since the metastable glassy state can never reach a true equilibrium state.[29] However, if for some reason the starch sample has already attained the quasi-equilibrium ΔH_G value with reference to any particular T_a and moisture content, then there would obviously be no further significant enthalpy relaxation on sub-T_g annealing over realistic times and no change in ΔH_G would be obtained.[8,30]

Decreases in ΔH_G resulting from annealing[4,8,26] should logically occur only when T_a has exceeded T_g. Under such circumstances, the glassy-to-rubbery state transformation would have already taken place during annealing, and the apparent enthalpy arising therefrom would no longer contribute to ΔH_G on subsequent heating of the annealed starch; hence the decrease in ΔH_G observed. Furthermore, since the glass transition is overlayed at the leading edge of the melting endotherm, it is likely that some melting of less thermostable (deformed or imperfect) crystallites would have occurred during above-T_g annealing, thereby leading to a further decrease in ΔH_G. The higher T_a is above T_o, the greater is the decrease in ΔH_G.[8] Some recrystallization and perfection of crystallites (which would contribute positively to ΔH_G) may, however, occur during annealing at $T_g < T_a < T_m$, leading to narrower endotherms.[4,26]

Changes in ΔH_G, if any, do not appear to parallel changes in T_o on sub-T_g annealing. It is an interesting feature of the non-equilibrium behaviour of the glassy state in granular starch that T_o can continue to increase long after ΔH_G has attained its apparent 'equilibrium' value.[12] This may be interpreted to mean that any excess volume originally quenched into the system can continue to be redistributed (which, by implication, suggests that conformational rearrangement and packing of polymer molecules can go on) without any substantial change in internal energy. Thus, an increase in T_o on sub-T_g annealing need not necessarily be accompanied by an increase in ΔH_G as this would depend on the magnitude of the excess thermodynamic properties in the amorphous regions of the starch granules under study, which, in turn, would be affected by the thermal history of the material such as the conditions prevailing during flour milling or starch isolation.

3 Starch–Sucrose Interactions

It is a well-known fact that the presence of sucrose, which is a common ingredient in most starch-based products, elevates the gelatinization temperature of a starch + water system. This has been attributed to the lower plasticizing power of the water + sugar mixture (which possesses a higher average molecular mass) compared with water alone in mobilizing and increasing the free volume of the amorphous fringes in the 'fringed micelle' structure of a starch granule, thereby resulting in a smaller depression of T_g.[2,29]

Since sub-T_g annealing induces a relaxation of the amorphous regions of the starch granules to lower energy states, a pertinent question that arises is:

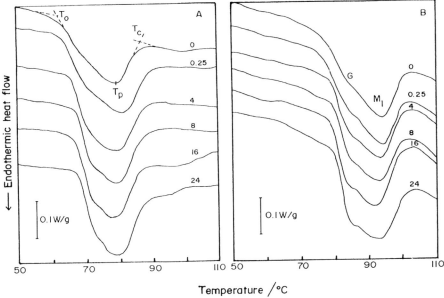

Figure 4	*DSC thermograms of (A) a binary 1:2 rice starch + water system and (B) a ternary 1:2:1 starch + water + sucrose system, labelled with time of annealing (in hours) at 50 °C[12]*

would such time-dependent changes have any effects on starch–sucrose interactions? We have attempted to answer this question by studying the effects of sucrose addition to granular rice starch, physically aged at 50 °C (which is below T_g) for varying time periods.[12] Figure 4 compares the DSC thermograms of (A) a binary (1:2) rice starch + water system and (B) a ternary (1:2:1) starch + water + sucrose system in which the sugar was added after annealing of the binary system for specified periods of time. As expected, the presence of sucrose increased all transition temperatures associated with gelatinization. Both the binary and ternary systems exhibited similar trends where changes in T_o with time of annealing are concerned (Figure 3).

Table 1   *Effects of sucrose on ΔH_G of a 1:2 rice starch + water system after sub-T_g annealing at 50 °C for different periods of time[12]*

System	ΔH_G/J g^{-1}		
	0 h	*0.25 h*	*24 h*
Binary system	12.2 ± 0.5^a	15.3 ± 0.7^a	15.2 ± 0.1^a
Ternary system	13.7 ± 0.8^b	14.7 ± 0.6^a	15.5 ± 0.5^a

[a,b] Means within a column with same letter (a or b) are not significantly different at $P < 0.05$ (each mean value quoted is the average of 5 determinations).

The effects of sucrose on ΔH_G are shown in Table 1. Addition of sucrose to the original (unannealed) starch was found significantly ($P < 0.05$) to increase ΔH_G due possibly to stabilization of the glassy regions of the starch granules as a result of starch–sugar interactions, as suggested by several workers[31,32] and experimentally observed through the use of techniques such as ESR[33] and NMR.[34] In contrast, the addition of sucrose to physically aged starch in which the glassy regions had attained a quasi-equilibrium energy state did not significantly ($P < 0.05$) affect ΔH_G. It is possible that the high degree of packing and low segmental mobility of starch chains in the stabilized amorphous regions of the annealed starch granules do not permit any significant development of starch–sucrose interactions. It is apparent that the effect of sucrose addition on ΔH_G would depend largely on the physical condition of the amorphous regions of the starch granules which can be altered, intentionally or adventitiously, by sub-T_g annealing. Contradictory observations concerning the effect of sucrose on ΔH_G reported in the literature, as highlighted by Eliasson,[35] are probably a consequence of the different thermal histories of the granular starches studied.

4 Concluding Remark

Enthalpy relaxation at the glass transition, induced by sub-T_g annealing, provides a cogent explanation for the existence of the G 'endotherm' in DSC thermal profiles associated with starch gelatinization in limited-water systems.

References

1. J. M. V. Blanshard, in 'Starch: Properties and Potential', ed. T. Galliard, John Wiley, New York, 1987, p. 16.
2. L. Slade and H. Levine, in 'Frontiers in Carbohydrate Research', ed. R. P. Millane, J. N. BeMiller, and R. Chandrasekaran, Elsevier Applied Science, New York, 1989, p. 215.
3. C. G. Biliaderis, in 'Water Relationships in Foods', ed. H. Levine and L. Slade, Plenum, New York, 1991, p. 251.
4. T. J. Maurice, L. Slade, R. R. Sirett, and C. M. Page, in 'Properties of Water in Foods', ed. D. Simatos and J. L. Multon, Martinus-Nijhoff, Dordrecht, 1985, p. 211.
5. C. G. Biliaderis, M. C. Page, T. J. Maurice, and B. O. Juliano, *J. Agric. Food Chem.*, 1986, **34**, 6.
6. L. Slade and H. Levine, in 'Industrial Polysaccharides', ed. S. S. Stivala, V. Crescenzi, and I. C. M. Dea, Gordon and Breach, New York, 1987, p. 387.
7. L. Slade and H. Levine, *Carbohydr. Polym.*, 1988, **8**, 183.
8. C. C. Seow and C. H. Teo, *Starch/Stärke*, 1993, **45**, 345.
9. I. D. Evans and D. R. Haisman, *Starch/Stärke*, 1982, **34**, 224.
10. P. Colonna and C. Mercier, *Phytochemistry*, 1985, **24**, 1667.
11. I. Larsson and A.-C. Eliasson, *Starch/Stärke*, 1991, **43**, 227.
12. C. C. Seow and C. K. Vasanti Nair, *Carbohydr. Res.*, 1994, **261**, 307.
13. R. Stute, *Starch/Stärke*, 1992, **44**, 205.
14. S. E. B. Petrie, *J. Polym. Sci. Part A-2*, 1972, **10**, 1255.
15. S. E. B. Petrie, in 'Physical Structure of the Amorphous State', ed. G. Allen and S. E. B. Petrie, Marcel Dekker, New York, 1977, p. 225.
16. L. C. E. Struik, 'Physical Aging in Amorphous Polymers and Other Materials', Elsevier, Amsterdam, 1978.
17. M. R. Tant and G. L. Wilkes, *Polym. Eng. Sci.*, 1981, **21**, 874.
18. A. R. Berens and I. M. Hodge, *Macromolecules*, 1982, **15**, 756.
19. I. A. M. Appleqvist, D. Cooke, M. J. Gidley, and S. J. Lane, *Carbohydr. Polym.*, 1993, **20**, 291.
20. J. W. Lawton and Y. V. Wu, *Cereal Chem.*, 1993, **70**, 471.
21. Z. H. Chang and J. G. Baust, *Cryobiology*, 1991, **28**, 87.
22. T. R. Noel, S. G. Ring, and M. A. Whittam, *Carbohydr. Res.*, 1991, **212**, 109.
23. D. Simatos and G. Blond, in 'Water Relationships in Foods', ed. H. Levine and L. Slade, Plenum, New York, 1991, p. 139.
24. F. Nakazawa, S. Noguchi, J. Takashi, and M. Takada, *Agric. Biol. Chem.*, 1984, **48**, 2647.
25. H. F. Zobel, S. N. Young, and L. A. Rocca, *Cereal Chem.*, 1988, **65**, 443.
26. D. A. Yost and R. C. Hoseney, *Starch/Stärke*, 1986, **38**, 289.
27. B. R. Krueger, C. A. Knutson, G. E. Inglett, and C. E. Walker, *J. Food Sci.*, 1987, **52**, 715.
28. B. R. Krueger, C. E. Walker, C. A. Knutson, and G. E. Inglett, *Cereal Chem.*, 1987, **64**, 187.
29. L. Slade and H. Levine, *CRC Crit. Rev. Food Sci. Nutr.*, 1991, **30**, 115.
30. C. A. Knutson, *Cereal Chem.*, 1990, **67**, 376.
31. R. D. Spies and R. C. Hoseney, *Cereal Chem.*, 1982, **59**, 128.

32. S. S. Kim and C. S. Setser, *Cereal Chem.*, 1992, **69**, 447.
33. J. M. Johnson, E. A. Davis, and J. Gordon, *Cereal Chem.*, 1990, **67**, 286.
34. L. M. Hansen, C. S. Setser, and J. V. Paukstelis, *Cereal Chem.*, 1989, **66**, 411.
35. A.-C. Eliasson, *Carbohydr. Polym.*, 1992, **18**, 131.

Kinetic Processes in Highly Viscous, Aqueous Carbohydrate Liquids

By Timothy R. Noel, Roger Parker, and Stephen G. Ring

INSTITUTE OF FOOD RESEARCH, NORWICH RESEARCH PARK, COLNEY LANE, NORWICH NR4 7UA, UK

1 Introduction

Although dried and frozen foods have long been known to act as encapsulation and preservation systems, the physico-chemical basis for this action is not well understood. Information on the mobility of molecules dissolved in these highly viscous matrices is fundamental to understanding kinetic processes in these systems.

For the case of diffusion, some pointers to the overall picture can be obtained from the synthetic polymer literature.[1,2] There is a progressive slowing of the diffusion of small molecules in polymers as the material is cooled from the 'rubber' to the glass,[1] but, quantitatively, the rate of diffusion is much higher than would be expected on the basis of the 'viscosity'.[2] Ehlich and Sillescu,[2] summarizing the current state of knowledge, expect that the molecular size of the diffusant will control the extent of the coupling between diffusion and viscosity. Physical chemists have thoroughly tested the Stokes–Einstein relationship for solutes in low-viscosity liquids.[3] A qualitatively similar view emerges, i.e. hydrodynamic theories of diffusion breakdown when the size of the diffusing molecule approaches that of the medium in which it is moving and when the viscosity is increased. Measurements on food systems, however, are rare. The spray-drying literature[4] provides some measurements on the diffusion of water and low molecular mass volatiles in biopolymers in the vicinity of their glass transition which indicate that diffusion is strongly retarded as the water content falls below 20 wt%.

In the present work the link between bulk viscosity and ionic and molecular mobility in carbohydrate liquids is examined from both a theoretical and an experimental viewpoint. The theory of diffusion-controlled chemical reactions is briefly reviewed and model calculations are presented for a partially frozen sucrose + water mixture. This illustrates the basic structure of a quantitative theory which could potentially be tested by experiment. On the experimental side, translational motions have been studied in conductivity and drying experiments and rotational motions in dielectric relaxation studies. In this work, the three sugars glucose, maltose and xylose in mixtures with water have been chosen for several reasons: (1) they are the monomer units of polysaccharides and plasticizers of polysaccharides; (2) they have intrinsic interest as encapsulation and preservation systems; (3) their 'rubbery' state can readily be studied in the absence of water; and (4) importantly, for dielectric and conductivity measurements, they are readily available in forms which are relatively free from ionic contaminants.

2 Theory of Diffusion-Controlled Reactions

It has been proposed[5] that chemical reactions approach the diffusion-controlled limit in systems close to the glass transition temperature T_g. This type of behaviour is well known in low-viscosity liquids.[6-9] The conditions under which diffusion control occurs can be demonstrated for a two-stage reaction scheme as follows:[6,8]

$$A + B \underset{k_{-d}}{\overset{k_d}{\rightleftharpoons}} (AB) \overset{k_1}{\rightarrow} P$$

The reactants A and B diffuse until sufficiently close to form an 'encounter pair' (AB) at a rate described by a second-order rate law with rate constant k_d. The encounter pair can either react to give a product P or follow the reverse reaction back to the freely diffusing reactants at rates described by first-order rate constants, k_1 and k_{-d}, respectively. Under the condition that $k_1 \gg k_{-d}$, the overall rate of product formation is

$$\frac{d[P]}{dt} = k_d [A][B] \tag{1}$$

and the reaction is said to be 'diffusion controlled'. In low-viscosity liquids it is reactions with little activation energy that are diffusion controlled.[6-8] By solving the diffusion equation with the appropriate boundary conditions, it can be shown that

$$k_d = \frac{8RT}{3\eta} \tag{2}$$

where η is the viscosity, T is the absolute temperature, and R is the gas constant. In deriving this equation use has been made of the Stokes–Einstein relation,

$$D = \frac{kT}{6\pi\eta a} \tag{3}$$

to relate the diffusion coefficient D to the viscosity, the hydrodynamic radius a, and the Boltzmann constant k. This assumption is, however, not an essential part of the theory; diffusion coefficients could be determined by independent experiment or from alternative theoretical approaches, *e.g.* the free volume theory.[3]

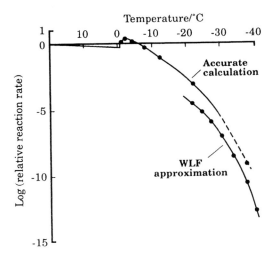

Figure 1   *Theoretical prediction of the rate of a diffusion-controlled reaction for reactants in a sucrose + water mixture at temperatures between 20 and −40 °C*

Equation (2) shows that there is a simple relationship between the rate constant, the viscosity and the temperature. A half-life $t_{\frac{1}{2}}$ for the reaction can be defined in terms of the initial concentration of one reactant, $[A]_0$, *i.e.*

$$t_{\frac{1}{2}} = \frac{1}{k_d\,[A]_0} \tag{4}$$

Taking $[A]_0 = 1$ mM as a typical concentration for a reactant, we get $t_{\frac{1}{2}} \approx 1\ \mu s$ for aqueous systems. However, as the viscosity increases, eqn (3) predicts a retardation of diffusion and a corresponding decrease in the rate of diffusion-controlled reactions [from eqn (2)]. At the glass transition the viscosity is $\eta \approx 10^{12}$ Pa s,[10] *i.e.* the system is 10^{15} times more viscous than water; and this gives $t_{\frac{1}{2}} \approx 10^9$ s, *i.e.* tens of years.

Comprehensive studies[11–14] on the state diagram, viscosity and density of sucrose + water mixtures enable predictions to be made for the rate of diffusion-controlled reactions in partially frozen systems. Figure 1 shows the effect of temperature on the reaction rate for reactants dissolved in a 10 wt% sucrose solution. The overall rate of reaction includes effects of (1) freeze concentration on the concentrations of reactants, (2) freeze concentration and temperature on the diffusion (via the viscosity), and (3) temperature on the diffusion. Unit reaction rate has been taken to be for 20 °C in a 10 wt% sucrose solution. As soon as ice freezing starts, the slope of the melting curve is such that a small decrease in temperature results in a large decrease in the volume occupied by the freeze-concentrated solutes. Freeze concentration is the dominant effect and this leads to an increase in the reaction rate. However, as the temperature is lowered further, the freeze concentration effect weakens and the retardation of diffusion effect dominates. Only approximate calculations are possible at viscosities above 10^4 Pa s (Figure 1). At each water content the T_g value is estimated from the Gordon–Taylor equation,[14] and then the viscosity is calculated using Williams–Landel–Ferry (WLF) theory[15] assuming that the viscosity at T_g is 10^{12} Pa s.[10]

It is only recently that experiments designed to explore the relationship between reaction kinetics and the glass transition have been reported for low water[16] and partially frozen[17] model food systems. Whilst some reactions are strongly retarded on passing into the glassy state, the question of how closely they approach to the diffusion-controlled limit still remains open. Several points are worth bearing in mind when designing experiments to test for diffusion control. (i) A prerequisite is a knowledge of the diffusive behaviour of the reactants since, in general, the Stokes–Einstein relationship cannot be relied upon.[2,3] (ii) For the scheme described previously involving reactants A and B, the reaction is second order. (iii) Absolute rate studies are required. (iv) For most reactions, as diffusion is retarded by varying temperature or water content, a cross-over between reaction control and diffusion control should be observed.

In the next section some experiments to probe molecular mobility are described. While translational mobility is important to reactants which are dilute, at higher concentrations rotational motions and motions associated with internal degrees of freedom may become rate limiting. The mobility of small molecules is examined since it is for this case that the hydrodynamic model is most severely tested.

3 Experimental

Aqueous carbohydrate liquids were prepared by melting the crystalline material with the appropriate amount of water. For the conductivity measurements potassium chloride was added as an aqueous solution prior to melting the sugar crystals. The conductivity cell had circular parallel-plate electrodes with a 20 mm diameter inner electrode, a 4 mm wide guard electrode, and an electrode separation of 3 mm. The d.c. current was measured using a Keithley 617 programmable electrometer with a 10 V applied voltage.

For the drying experiments the liquids were contained in tubes (5 mm internal diameter) open at both ends. The mass was measured by a Cahn 2000 recording balance during drying over phosphorus pentoxide for up to 18 000 s. In this initial study the data were analysed assuming that the diffusion coefficient was independent of concentration and that the concentration of water at the surface was zero. Under these conditions the mass of water lost from the surface $M(t)$ was related to the diffusion coefficient by[18]

$$\frac{M(t)}{M(\infty)} = 2\left(\frac{Dt}{\pi l^2}\right)^{\frac{1}{2}} \tag{5}$$

where $M(\infty)$ is the total amount of water, t is the time elapsed after the creation of the concentration step, and $2l$ is the sample thickness.

Dielectric measurements were made with a Polymer Laboratories dielectric thermal analysis system as described previously.[19]

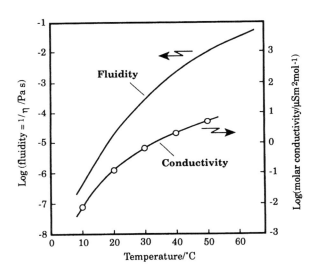

Figure 2   *Effect of temperature on the molar conductivity and fluidity of a maltose + - water + KCl mixture above its glass transition temperature. Composition: water, 12.1 wt%; KCl, 0.05 wt% (9.9 mM)*

4 Results and Discussion

The variation of the molar conductivity of a maltose + water + KCl mixture with temperature is shown in Figure 2. For comparison the fluidity,[20] the reciprocal of viscosity, is shown over the same temperature range. As the temperature is reduced from 50 to 10 °C the molar conductivity Λ_m decreases by 2.7 decades whilst the fluidity decreases by 4.4 decades. The application of Stokes' formula to an ion moving in an electric field[21] suggests that its velocity should be inversely proportional to viscosity. This is in agreement with the empirically based Walden rule[21] which proposes that the product $\Lambda_m \eta$ is constant. The differing temperature dependences of the molar conductivity and the fluidity show that there is a breakdown of this relationship. Quantitatively, the Walden product, $\Lambda_m \eta \approx 1.5 \times 10^{-5}$ N S s mol^{-1} for KCl in water at 25 °C, increases by over two orders of magnitude to $\approx 2.4 \times 10^{-3}$ N S s mol^{-1} in 12.1 wt% maltose–water. The value of T_g for maltose + water with a water content of 12.1 wt% is -5 °C, and so at 25 °C the mixture is

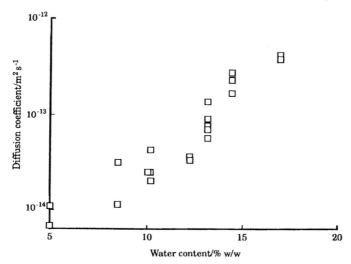

Figure 3   *Influence of water content on the mutual diffusion coefficient of maltose + -water mixtures at 300 K*

30 °C above T_g. It can be concluded that ionic mobility is to some extent uncoupled from viscosity at temperatures above T_g.

The drying data were analysed by plotting the reduced mass loss, $M(t)/M(\infty)$, against $t^{\frac{1}{2}}/l$. These plots were found to be linear. The apparent diffusion coefficients for maltose + water over a range of water contents are shown in Figure 3. At 300 K the glass transition of maltose + water is at a water content of 5.5 wt%. The diffusion coefficient decreases by 1.7 decades as the water content is reduced from 17 to 5 wt%. The concentration dependence of the diffusion coefficient shows that our initial analysis could be improved by taking this into account. However, it is the order of magnitude of the diffusion coefficient which is important here. A typical value of 10^{-14} m^2 s^{-1} can be compared with the mutual diffusion coefficient for maltose + water at infinite dilution of 5.0×10^{-10} m^2 s^{-1}. Despite the viscosity approaching a value of 10^{12} Pa s at T_g, the diffusion coefficient in maltose + water mixtures is only a factor of 5×10^4 below that at infinite dilution. It can be concluded that diffusion is extensively decoupled from viscosity in these systems.

While the above techniques allow translational motions to be studied, complementary studies of the rotation of whole molecules or parts of molecules can be made using dielectric techniques. The variation of dielectric loss with temperature for a series of maltose + water mixtures is shown in Figure 4. In the temperature range studied, there are two loss peaks on each scan: the one at higher temperatures is known as the α-loss peak and the other lower peak is the β-loss peak. It is the α-loss peak which is associated with the glass transition and studies at lower frequencies find that this relaxation slows to ≈ 1 mHz at the calorimetric glass transition. The temperature of the α-peak

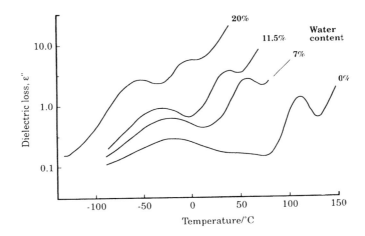

Figure 4   *Effect of temperature on the dielectric loss ε″ of maltose and maltose + water mixtures with various water contents at 10 kHz*

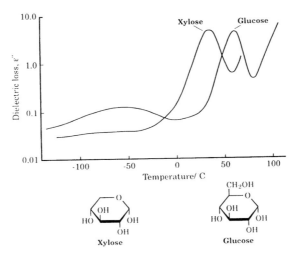

Figure 5   *Effect of temperature on the dielectric loss ε″ of xylose and glucose at 1 kHz, and their structures in pyranose form*

decreases with increasing water content. This is associated with the depression of T_g and the reduction of viscosity which occurs on addition of water. The β-loss peak occurs at temperatures below T_g (T_g for dry maltose is 95 °C), *i.e.* within the glassy solid, and increases in strength with addition of water. At present the physical origin of the β-process cannot be unambiguously identified.

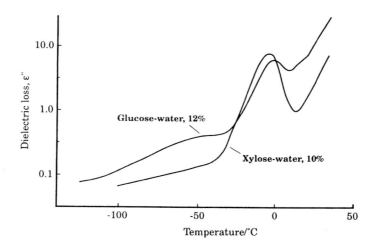

Figure 6   *Effect of temperature on the dielectric loss ε″ of xylose + water and glucose + water mixtures at 1 kHz*

Figure 5 shows a comparison of the dielectric loss of dry glucose and dry xylose together with the structural diagrams of these sugars in their pyranose forms. It is tempting to interpret the β-relaxation in glucose as being a direct consequence of the motion of the pendant hydroxymethyl group. This effect has been observed in polysaccharides.[22] Recently, using ^{13}C NMR, Girlich and Ludemann[23] have shown that, in concentrated aqueous solution, each of the three hydroxymethyl groups of sucrose relax independently from the tumbling of the molecule as a whole. Figure 6 shows a comparison of the dielectric loss of a glucose + water mixture and a xylose + water mixture. The dielectric loss of both glasses is enhanced by the addition of water although the differences between the two sugars in the dry state remains. It appears that the increased strength of the β-relaxation on addition of water can be linked with the presence of pendant hydroxymethyl groups.

In summary, the overall picture that is emerging is one in which there is significantly more molecular mobility in highly viscous and glassy food materials than might previously have been expected.[5]

Acknowledgement

This work was supported by the Agricultural and Food Research Council.

References

1. 'Diffusion in Polymers', ed. J. Crank and G. S. Parks, Academic Press, New York, 1968.
2. D. Ehlich and H. Sillescu, *Macromolecules*, 1990, **23**, 1600.

3. H. J. V. Tyrrell and K. R. Harris, 'Diffusion in Liquids: A Theoretical and Experimental Study', Butterworths, London, 1984.

4. S. Bruin and K. Ch. A. M. Luyben, in 'Advances in Drying', ed. A. S. Majumdar, McGraw-Hill, New York, 1980, vol. 1, p. 155.

5. H. Levine and L. Slade, *Carbohydr. Polym.*, 1986, **6**, 213.

6. P. W. Atkins, 'Physical Chemistry', Oxford University Press, 1978, p. 908.

7. L. Stryer, 'Biochemistry', Freeman, New York, 3rd edn, 1988, p. 191.

8. S. A. Rice, 'Comprehensive Chemical Kinetics, Vol. 25, Diffusion Limited Reactions', ed. C. H. Bamford, C. F. H. Tipper, and R. G. Compton, Elsevier, Amsterdam, 1985.

9. M. E. Davis, J. D. Madura, J. Sines, B. Luty, S. A. Allison, and J. A. McCammon, *Methods Enzymol.*, 1991, **202**, 473.

10. G. Tammann, 'Der Glaszustand', Leopold Voss, Leipzig, 1933.

11. R. J. Bellows and C. J. King, *AIChE Symp. Ser.*, 1973, **69**, 33.

12. M. P. Mageean, J. U. Kristott, and S. A. Jones, 'Physical Properties of Sugars and their Solutions', BFMIRA Scientific and Technical Survey, no. 172, BFMIRA, Leatherhead, 1991.

13. S. Ablett, M. J. Izzard, and P. J. Lillford, *J. Chem. Soc., Faraday Trans.*, 1992, **88**, 789.

14. Y. Roos, *Carbohydr. Res.*, 1993, **238**, 39.

15. J. D. Ferry, 'Viscoelastic Properties of Polymers', Wiley, New York, 3rd edn, 1980.

16. R. Karmas, M. P. Buera, and M. Karel, *J. Agric. Food Chem.*, 1992, **40**, 873.

17. W. L. Kerr, M. H. Lim, D. S. Reid, and H. Chen, *J. Sci. Food Agric.*, 1993, **61**, 51.

18. J. Crank, 'The Mathematics of Diffusion', Oxford University Press, 2nd edn, 1975.

19. T. R. Noel, S. G. Ring, and M. A. Whittam, *J. Phys. Chem.*, 1992, **96**, 5662.

20. T. R. Noel, S. G. Ring, and M. A. Whittam, *Carbohydr. Res.*, 1991, **212**, 109.

21. S. I. Smedley, 'The Interpretation of Ionic Conductivity in Liquids', Plenum, New York, 1980.

22. S. A. Bradley and S. H. Carr, *J. Polym. Sci., Polym. Phys. Ed.*, 1976, **14**, 111.

23. D. Girlich and H.-D. Ludemann, *Z. Naturforsch., Teil C*, 1993, **48**, 407.

Calculation of Glass Transition Temperatures of Food Proteins and Plasticizer Effects of Different Ingredients

By Yu.I. Matveev

INSTITUTE OF FOOD SUBSTANCES, RUSSIAN ACADEMY OF SCIENCE, VAVILOV STR. 28, 117813 MOSCOW GSP-1, RUSSIA

1 Introduction

Processing of food proteins as a rule is associated with heating of the proteins to take them to a highly viscous state. In this context it is necessary to know important thermal characteristics of food proteins such as glass transition temperature T_g, and the temperatures of denaturation and degradation. In polymer physics, various additive schemes have been successfully used to calculate glass and degradation temperatures of homopolymers, copolymers and polymer networks taking into account contributions of weak (dipole–dipole) and strong (hydrogen bonding) intermolecular interactions.[1] Here we show that additive schemes can also be applied to extimate T_g values of dry proteins, and to analyse the contribution of different additional factors (disulfide bonds, degree of polymerization) on T_g, so indicating the plasticizer effect of other components (water and lipids).

2 Additive Scheme for Calculation of T_g

Considering protein molecules as complicated copolymers, we can obtain the following additive scheme for the protein glass temperature:

$$T_g^{-1} = \sum \phi_i T_{gi}^{-1} \tag{1}$$

where $\phi_i = n_i \Delta V_i / \Sigma n_i \Delta V_i$, and ΔV_i is the van der Waals volume of the ith amino acid residue, n_i is the number of amino acid residues of the ith type per molecule, and T_{gi} is the partial contribution of the ith residue to the overall protein glass temperature. The values T_{gi} and ΔV_i are determined as described elsewhere.[1] In Table 1 the values of ΔV_i and T_{gi} calculated for each type of amino acid residue are given. Determined in this way, the protein T_g can be arbitrarily named as 'polymeric', *i.e.* it is suggested that the polymerization degree is sufficiently high that it does not depend on the molecular mass.

Table 1 *Partial glass transition temperature contribution* T_{gi} *and van der Waals volume* ΔV_i *of individual amino acids (i = 1–20)*

i	Amino acid	T_{gi}/K	$\Delta V_i/Å^3$
1	Gly	504	47.6
2	Ala	544	64.7
3	Val	442	98.8
4	Leu	382	115.9
5	Ile	382	115.9
6	Phe	504	140.4
7	Pro	681	87.2
8	Trp	574	170.1
9	Ser	563	72.1
10	Thr	428	89.1
11	Met	463	117.1
12	Asn	406	94.8
13	Gln	357	111.9
14	Cys-SH	645	82.2
15	Asp	536	90.7
16	Glu	433	107.8
17	Tyr	511	147.0
18	His	555	119.5
19	Lys	295	127.4
20	Arg	326	146.6

By means of eqn (1), the values of T_g for various food proteins have been calculated: proteins from milk, eggs, meat, seeds (cereals and beans), and blood. We consider 32 proteins in all. The calculations show that the proteins can be divided into three classes: low, middle, and high glass transition temperature proteins corresponding to values of $135 \,°C < T_g < 165 \,°C$, $T_g \approx 176 \pm 10 \,°C$ and $187 \,°C \leqslant T_g \leqslant 217 \,°C$. Most proteins fall into the first (12 proteins) or the second (16 proteins) classes. There are three amino acid residues that appear to have a remarkable effect on the T_g value: glycine, alanine, and leucine. The analyses show that for the cysteine content that is typically met in proteins, including food proteins, the contribution of disulfide bridges to the computed T_g value can be neglected, and so the value of the protein T_g is defined by the chemical structure of the linear polypeptide chain.

3 Influence of Plasticizer on the Protein T_g

We estimate the plasticizer effect by means of the function $Pl(C)$ which has the form

$$Pl(C) = T_g(C)/T_{gp}$$

where T_g (C) is the protein T_g value at a given concentration C of plasticizer, and T_{gp} is the dry protein T_g value at $C=0$. Using the additive approach an explicit form of the function Pl(C) can be obtained. The relation found for Pl(C) permits us to go to the limit of dilute solutions as $C \rightarrow 1$. In this case, we have

$$\text{Pl}(C \rightarrow 1) = \frac{\omega}{1 + \tilde{n} T_{gp} < T_{gh} >} \cdot \frac{\varepsilon_p}{\varepsilon_{po}} \tag{2}$$

where $\omega = \Delta V_{pl}/ < \Delta V >$, $< \Delta V > = \Sigma n_i \Delta V_i / \Sigma n_i$, $< T_{gh} > = -b_h/ < \Delta V >$, ΔV_{pl} is the van der Waals volume of the plasticizer, ε_{po} and ε_p are the values of the protein dielectric constant with 'blocked' hydrogen connections and in the absence of plasticizer, respectively, and $\tilde{n}$ is the mean number of hydrogen connections accessible for plasticization.

4 Comparison of Experimental and Calculated Values of T_g

Experimental[2-4] values of T_g for (i) collagen, (ii) elastin and (iii) gelatin, and values calculated from the amino acid composition by means of eqn (1) and the data of Table 1, are respectively (i) 192 °C and 208 °C, (ii) 200 ± 20 °C and 213 °C, (iii) 217 °C and 194 °C. The calculated T_g values agree rather well with the experimental values (the error being less than 5%). So it seems that proteins can be considered as polymers. However, many proteins have a polymerization degree N below N_c (the critical value) and therefore their T_g values depend on N. This effect can be estimated as described previously.[5] At $N < N_c$, we have $T_g^* = T_f$, where T_g^* is the glass transition temperature of the polypeptide chain in terms of $N < N_c$, and T_f is the temperature of the transition of the polymer into its viscous flow state. For myoglobin, calculation of T_g by means of eqn (1) gives $T_g = 151$ °C. The recalculated T_g value is equal to $T_g^* = 122$ °C. Comparison of T_g with the experimental data[3] makes use of eqn (2). In an aqueous solution of glycerol, containing 75 vol% glycerol, myoglobin has a $T_g = 175$ K.[3] The value of T_{gp} recalculated by means of eqn (2) gives $T_{gp} = 124$ °C which is near to the calculated value of T_g.

We note that the number of proteins belonging to the different classes changes when we take into account the effect of the polymerization degree: 21 proteins belong to the first class, eight proteins to the second class, and just three proteins to the third class.

References

1.   A. A. Askadskii and Yu. I. Matveev, 'Chemical Structure and Physical Properties of Polymers', Khimiya, Moscow, 1983 (in Russian).
2.   H. Batzer and U. Kreiblich, *Polym. Bull.*, 1981, **5**, 585.
3.   I. E. T. Iben, D. Braunstein, W. Doster, H. Frauenfelder, M. K. Hong, J. B.

Johnson, S. Luck, P. Ormos, A. Schulte, P. J. Steinbach, A. H. Xie, and R. D. Young, *Phys. Rev. Lett.*, 1989, **62**, 1916.
4.   A. S. Marshall and S. E. B. Petrie, *J. Photograph. Sci.*, 1980, **28**, 128.
5.   Yu. I. Matveev and A. A. Askadskii, *Polym. Sci.*, 1993, **35**, 50.

Water Adsorption and Plasticization of Amylopectin Glasses

By Kirsi Jouppila, Tarja Ahonen, and Yrjö Roos

DEPARTMENT OF FOOD TECHNOLOGY, PO BOX 27 (VIIKKI B), FIN-00014, UNIVERSITY OF HELSINKI, HELSINKI, FINLAND

1 Introduction

Various food polymers exist in an amorphous, metastable state, which in foods has an effect on stability and rates of deteriorative changes.[1,2] Such materials (*e.g.* starch in cereal products such as pasta and snacks) show thermoplastic properties, and they are significantly plasticized by water.[3] Water plasticization causes a drastic decrease of the glass transition temperature T_g.[3,4] Various structural changes and crystallization are kinetically delayed in the vicinity of T_g, but they may proceed at temperatures above T_g.

The objective of the present study is to obtain the water adsorption properties of amylopectin and the effect of water plasticization on the T_g of amylopectin. The state diagram is established by modelling T_g and water sorption data; this allows prediction of the critical water content and water activity a_w for amylopectin.

2 Materials and Methods

Amylopectin from waxy corn starch (Sigma) was gelatinized by heating in a beaker with excess water (5 wt% amylopectin) in a boiling water bath. A clear, transparent paste was obtained and the solubilization of amylopectin was confirmed by observing the loss of crystallinity using polarized light microscopy. The gelatinized material was immediately transferred to glass vials (5 ml aliquots in 20 ml vials), frozen at -40 °C (for 12 h), and then freeze-dried (Lyovac GT2 freeze dryer, Amsco Finn-Aqua; $p < 0.1$ mbar) for 48 h. The freeze-dried material was stored over P_2O_5 for 1 week in a vacuum desiccator.

Samples were stored in duplicate (at least) at constant relative humidity (RH) at 24 °C for 2 weeks over saturated salt solutions in vacuum desiccators. The water sorption was determined gravimetrically. The salts used were LiCl, CH_3COOK, $MgCl_2$, K_2CO_3, $Mg(NO_3)_2$, $NaNO_2$, NaCl, KCl, and KNO_3 (pro analysis, E. Merck), with respective RH values of 11.5, 23.9, 33.0, 44.4, 53.8, 66.2, 76.4, 85.8, and 93.6%, giving a_w values of 0.01 × the percentage RH at equilibrium.[5] The steady-state water content was taken as the average of nine determinations using samples stored for 24 to 336 h at each RH value.

Triplicate samples (approximately 3 mg) equilibrated for 24 h under various RH conditions were hermetically sealed in differential scanning calorimeter (DSC) pans and used for the determination of the effect of water on T_g using a Mettler TA 4000 DSC analysis system with a DSC-30 low-temperature cell, a TC10A TA processor, and GraphWare TA 72AT.2 thermal analysis software. The samples were scanned at 5 °C min^{-1} from $T_g - 30$ °C to $T_g + 30$ °C, and the T_g value was determined from the onset of the endothermal step change occurring over the T_g.

The BET and GAB models were used to model the water sorption data.[2] The effect of water on T_g was modelled with the Gordon–Taylor equation using the value of T_g of -135 °C for water and an estimated T_g value of 243 °C for amylopectin.[3]

3 Results and Discussion

Water adsorption of amylopectin was found to level off to constant values within 24 and 48 h at RH $\leqslant 85.8\%$ and 93.6% RH, respectively. Steady-state water contents are shown in Figure 1. Volman *et al.*[6] reported almost equal water sorption for amylopectin to that obtained in the present study. The

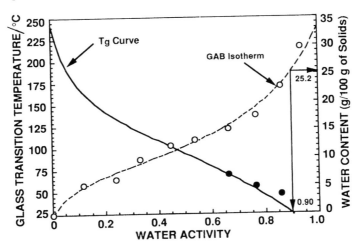

Figure 1   *A modified state diagram that shows relationships between glass transition temperature, water content, and water activity for amylopectin. The sorption isotherm is predicted with the GAB model. The T_g curve is estimated using water contents obtained with the GAB model and the Gordon–Taylor equation (k = 5.4). The values of critical water activity and critical water content (indicated with the arrows) are the limits that depress the T_g value to below 25 °C*

GAB model is valid over the whole experimental a_w range. The BET model was found to be applicable for $a_w < 0.4$, as also reported by Volman *et al.*[6]

The T_g value was observed for amylopectin stored over the RH range 66.2–93.6% with corresponding water contents of 14.9–25.3 g per 100 g of solids. Determination of T_g values for samples with lower water contents was found not to be feasible due to flattening and broadening of the change in the specific heat over the T_g range. Corresponding water contents for the determination of T_g for starch or starch components have also been reported in other studies.[4,7] The T_g decreased with increasing water content, and this could be predicted using the Gordon–Taylor equation ($k = 5.4$). Zeleznak and Hoseney[4] and Kalichevsky *et al.*[7] have reported T_g (mid-point) values for native wheat starch and amylopectin from waxy maize starch, respectively, which agree with those obtained in the present study. Our reported T_g values for amylopectin at various water contents are above those reported for maltodextrins;[3] and this gives further evidence that the T_g value of anhydrous starch is close to 250 °C.

The decrease of T_g with increasing a_w has the typical sigmoid relationship of food materials, as previously shown by Roos.[2] The estimated T_g curve as a function of a_w and the GAB isotherm can be used to predict the critical water content and a_w values, *i.e.* water content and a_w corresponding to $T_g = 25$ °C, as shown in the modified state diagram (Figure 1). The critical water content at 25 °C is 25.2 g per 100 g of solids with the corresponding a_w value of 0.90. These values agree with the finding of Zeleznak and Hoseney[4] that the T_g value for wheat starch with a water content of 28 g per 100 g of solids is below 25 °C. State diagrams with critical water content and a_w values for starch or starch components have not been reported in previous studies.

4 Conclusions

The physical state of amorphous amylopectin is related to the T_g value and the water content. The modified state diagram including the critical water content and a_w values is important in the characterization of the physical state of amylopectin and starch-containing foods. Changes in the physical state that occur above T_g may affect storage stability, quality, and textural properties (*e.g.* crispness) of such low-moisture foods as flour, pasta, cookies, and extruded products. In bread and other bakery products, the value of T_g may control amylopectin crystallization, which is related to staling.

References

1.   L. Slade and H. Levine, *Crit. Rev. Food Sci. Nutr.*, 1991, **30**, 115.
2.   Y. H. Roos, *J. Food Process. Preserv.*, 1993, **16**, 433.
3.   Y. H. Roos and M. Karel, *J. Food Sci.*, 1991, **56**, 1676.
4.   K. J. Zeleznak and R. C. Hoseney, *Cereal Chem.*, 1987, **64**, 121.

5. T. P. Labuza, A. Kaanane, and J. Y. Chen, *J. Food Sci.*, 1985, **50**, 385.
6. D. H. Volman, J. W. Simons, J. R. Seed, and C. Sterling, *J. Polym. Sci.*, 1960, **46**, 355.
7. M. T. Kalichevsky, E. M. Jaroskiewics, S. Ablett, J. M. V. Blanshard, and P. J. Lillford, *Carbohydr. Polym.*, 1992, **18**, 77.

Influence of Moisture Content on Glass Transition Temperature of the Amorphous Matrix in 'Xixona Turrón'

By Nuria Martínez, M. Pilar Betrán, and Amparo Chiralt

DEPARTMENT OF FOOD TECHNOLOGY, UNIVERSIDAD POLITÉCNICA DE VALENCIA, 46022 VALENCIA, SPAIN

1 Introduction

'Xixona turrón' is a typical Spanish confectionery product, whose microstructure can be described as an amorphous and porous matrix of a concentrated sugar syrup (85–90 wt%) with 1 wt% of reconstituted ovalbumin, infiltrated with almond oil, which traps almond globular protein aggregates and other almond cell fragments.[1,2] The usual sugar composition is a ternary mixture of sucrose + glucose + fructose. Water content and sugar composition are relevant to the prediction of the stability and textural properties of the product. The moisture of Xixona turrón is not well controlled in the manufacturing process, its value lying in a range where small changes in composition could lead to a glassy or rubbery product, with very different physical properties, at the storage and normal eating temperature.

The aim of the present work is to analyse the influence of water activity a_w and composition (sugar and other components) on the glass transition temperature T_g of the amorphous matrix of Xixona turrón.

2 Experimental

Over-saturated syrups of binary mixtures (1:1 by weight) of fructose + glucose (sample 1), fructose + sucrose (sample 2) and glucose + sucrose (sample 3), and a ternary mixture (1:1:1) of the three sugars (sample 4) were prepared by vacuum concentration. Samples 5 and 6 were prepared by adding 1 wt% (calculated on the basis of a syrup with 5 wt% moisture content) of ovalbumin reconstituted with distilled water (ratio 1:7) to samples 1 and 4, respectively. One part of samples 5 and 6 was homogenized with 64 wt% (calculated on the basis of a syrup with 5 wt% moisture content) of toasted and ground almonds to obtain samples 7 and 8, respectively. Ovalbumin and almonds were mixed in this ratio to model Xixona turrón *suprema* quality.

The water content of all samples was measured by Karl Fischer titration on a reflux-extracted sample with methanol[3] in a titrator (Mettler DL-18). Afterwards they were conditioned at different water activity and moisture levels. Samples were prepared by weighing 10–20 mg in 40 μl aluminium differential scanning calorimetry (DSC) pans, which were equilibrated at 25 °C in vacuum desiccators with P_2O_5 ($a_w = 0$) and over-saturated salt solutions of CH_3COOK ($a_w = 0.23$), K_2CO_3 ($a_w = 0.44$) and $CuCl_2$ ($a_w = 0.68$). The final water content was determined by mass change, taking into account the initial moisture values.

The DSC used was a Mettler TA 4000 system with a DSC 30 measuring cell. The apparatus was calibrated for temperature and heat flow with distilled water, gallium and indium. An empty aluminium pan was used as the reference sample and the heating rate was 5 °C min^{-1} in all experiments. A nitrogen gas flow of 20–30 ml min^{-1} was used to avoid water condensation in the measuring cell.

3 Results and Discussion

Table 1 shows the final moisture content achieved for each sample at different a_w levels. Syrups with lower mole fractions of sugars (those containing sucrose) presented lower moisture values, probably due to solute–solvent interactions, despite the low water contents in the samples. Ovalbumin addition did not change significantly the syrup a_w–moisture content relations. Nevertheless the almond mixtures (12 wt% protein, 31 wt% fat) showed a strong decrease in water uptake capacity because of a decrease in the weight fraction of polar compounds. All samples had unfreezable water.

The glass transition temperature (onset and mid-point) of the samples and the change of specific heat (Δc_p) for the anhydrous ones are shown in Table 2. The T_g values of carbohydrate mixtures had a negative deviation from linearity with mole fraction, especially for the ternary ($\Delta T = 18$ °C) and glucose fructose mixture ($\Delta T = 10$ °C). Also, from T_g and the heat capacity change ΔC_p of pure sugars, the equation of Couchman and Karasz[4] gave higher predicted values.[5] The addition of ovalbumin to samples 1 and 4 did not affect significantly the T_g, but the almond mixture increased this value strongly ($\Delta T \approx 29$ °C).

The variation of T_g with a_w and the water fraction (w) of the samples are plotted in Figure 1. Because the glass transition of samples containing almond with $a_w = 0.44$ overlapped with the melting endotherm of almond fat (between -20 and 0 °C), a defatted sample was prepared to obtain the T_g data, but the possible plasticizing effect of oil was not analysed. The T_g of all studied samples varied linearly with a_w (Table 3) as was observed for other carbohydrates in the range 0–0.6 of a_w.[5,6] For the evolution of T_g with w, a linear model (Table 3) and the Gordon–Taylor equation[7] were fitted to the experimental results. The last equation gave better predictions than the linear model. Table 3 shows the empirical k values obtained for each case which increased almost linearly as the T_g value of the anhydrous samples increased.

Table 1 Moisture content (wt%) reached by the studied samples over different saturated salt solutions

a_w	Sample number							
	1	2	3	4	5	6	7	8
0.68	24.60±0.17	22.1±0.4	14.5±0.7	24.1±0.9	24.7±1.3	24.60±0.17	13.1±0.5	12.2±0.8
0.44	12.3±0.2	9.1±0.8	9.0±0.7	12.97±0.12	13.5±0.3	12.93±0.06	5.4±1.6	4.7±0.8
0.23	7.7±1.4	4.5±0.5	5.0±0.5	9.33±0.15	9.0±0.5	9.4±0.4	2.6±0.0	2.8±0.2
0	0	0	0	0	0	0	0	0

Table 2 T_g values (°C) for the studied samples (upper value corresponds to the onset and lower value to the mid-point of the glass transition)

a_w	Sample number							
	1	2	3	4	5	6	7	8
0.68	−62.1±0.11	−57.3±0.7	−53.1±1.3	−60.9±0.1	−61.2±1.3	−62.0±0.3	−44.5±0.6	−47.7±1.0
	−58.7±1.3	−53.97±0.15	−50.2±1.2	−58.0±0.3	−58.1±1.4	−58.50±0.10	−42.9±0.2	−45.7±0.2
0.44	−33.8±0.2	−22.1±1.1	−19.17±0.06	−29.0±0.9	−31.7±0.5	−29.6±0.4	−28.1±1.2	−21.5±1.3
	−31.3±0.4	−19.2±1.3	−15.47±0.15	−26.1±0.7	−28.6±0.7	−26.5±0.6	−16.3±1.5	−16.0±1.2
0.23	−16.6±1.8	−6.3±0.8	12.3±1.2	−14.9±1.5	−15.9±1.1	−14.5±1.6	8.6±1.3	9.6±1.1
	−13.7±1.1	−4.3±0.2	15.1±1.0	−12.2±1.0	−12.5±1.0	−11.7±1.4	10.5±0.9	11.1±0.8
0	9.2±1.7	22.9±1.1	46.2±1.1	10.4±0.4	10.3±0.9	12.1±1.4	41.1±1.2	41.3±1.5
	13.0±1.2	25.1±1.2	47.0±1.9	14.1±0.7	14.7±1.5	17.1±1.3	42.5±1.3	43.4±1.4
$\Delta c_p{}^a$ ($J\,g^{-1}\,K^{-1}$)	0.74	0.58	0.31	0.80	0.78	0.71	0.15	0.13

[a] Experimental values for anhydrous samples.

Table 3   Coefficients of different models fitted to variation of T_g with a_w and w

| | Sample number | | | | | | | |
Coefficient	1	2	3	4	5	6	7	8
α_1 [a]	12.275	24.834	47.432	14.079	14.118	16.433	40.874	42.281
β_1 [a]	−103.55	−112.37	−143.18	−102.61	−104.43	−107.65	−125.70	−130.69
α_2 [b]	9.649	16.982	47.338	14.371	13.806	16.339	30.793	32.257
β_2 [b]	−289.90	−336.97	−676.92	−301.04	−296.03	−308.88	−613.14	−692.02
k [c]	2.89	3.7	5.6	2.60	2.66	2.75	7.2	8.1
ΔC_p [d] (J mol^{-1} K^{-1})	158	123	63	159	170	167	92	75
M_e [e]	214	212	203	199	218	235	613	577

[a] Coefficients of linear model $T_g = \alpha_1 + \beta_1 a_w$ (for all regressions, $r^2 > 0.992$).
[b] Coefficients of linear model $T_g = \alpha_2 + \beta_2 w$ ($r^2 > 0.901$).
[c] Empirical k values for Gordon and Taylor equation[7] ($r^2 > 0.991$).
[d] Empirical heat capacity change of anhydrous samples obtained from Couchman and Karasz equation[4] ($r^2 > 0.991$).
[e] Effective relative molecular mass of solids.

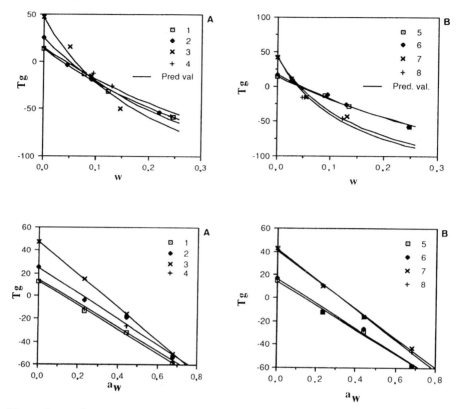

Figure 1   *Plot of* T_g *values versus water weight fraction* w *and* a_w *for (A) samples 1 to 4 and (B) samples 5 to 8*

The theoretical equation of Couchman and Karasz[4] was fitted to the experimental T_g values when water mole fractions x_w were considered. For samples 5 and 6, ovalbumin content was not taken into account to calculate the total number of moles of solutes per gram of sample, because of its low mass fraction and high molecular mass. The same hypothesis was considered for almond components in samples 7 and 8. In this case, the x_w calculation was made from the mass fraction of water referred to the defatted sample weight. T_g and ΔC_p of pure water were taken as -135 °C and 34.9 J mol^{-1} K^{-1}.[5] Values of ΔC_p for the anhydrous mixture were empirically determined (Table 3) and compared with the experimental Δc_p, obtaining an effective molecular mass M_e for each mixture (Table 3). It can be seen that the mixtures of sugars seemed to act with an effective molecular mass similar to that expected from their individual values. The error associated with this parameter did not allow us to analyse the possibly small differences between the syrups. Nevertheless, it allowed us to see, for samples 7 and 8, a strong increase in the effective molecular mass of the carbohydrate because of almond component interactions. The plasticizing action of water attenuated this effect and, for small increases in *w*, a higher T_g decay than in the syrups was observed, both reaching similar values for moisture contents between 10 and 15 wt%. This behaviour was also observed in high molecular mass maltodextrins.[6]

From the results obtained here, we can control the physical state (glassy or rubbery) for Xixona turrón at storage and normal eating temperature by varying the sugar ratio and water content. This is important for the expected texture of the product which is the main parameter for defining its quality.

References

1. A. Chiralt, M. J. Galotto, and P. Fito, *J. Food Eng.*, 1991, **14**, 117.
2. M. A. Lluch, M. J. Galotto, and A. Chiralt, *Food Struct.*, 1992, **11**, 181.
3. H. Scholz, *Dtsch. Lebensm.-Rundsch.*, 1988, **84**, 1.
4. P. R. Couchman and F. E. Karasz, *Macromolecules*, 1978, **11**, 117.
5. P. D. Orford, R. Parker, and S. G. Ring, *Carbohydr. Res.*, 1990, **196**, 11.
6. Y. Roos and M. Karel, *J. Food Sci.*, 1991, **56**, 1676.
7. M. Gordon and J. S. Taylor, *J. Appl. Chem.*, 1952, **2**, 493.

Phase Transitions of Tapioca Starch

By Veronique Garcia, Alain Buleon, Paul Colonna, Guy Della Valle, and Denis Lourdin

INRA, BP 527, 44026 NANTES CEDEX 03, FRANCE

1 Introduction

Gelatinization of starch is an order–disorder transition during which granules, heated in excess water, loose their crystallinity and swell. Under limited-water conditions (volume fraction of water $0.45 < \phi_1 < 0.65$), endothermic transitions observed by differential scanning calorimetry (DSC) have been interpreted differently. Some authors[1-4] have proposed that starch gelatinization is a solvent–facilitated melting process involving crystal melting following a Flory–Huggins thermodynamic treatment. They have suggested that the endothermic peaks are melting transitions of crystalline material exhibiting a large range of stability at various diluent levels. However, this biphasic behaviour of starch has also been interpreted[5] as a non-equilibrium process of partial melting, followed by recrystallization and final melting during DSC measurements. No studies have apparently been reported on tapioca starch transitions. The purpose of the present study is therefore to identify these thermal transitions in the case of tapioca starch granules, when heated above glass transition temperatures and at different moisture contents, by combining X-ray diffraction, light microscopy and DSC techniques. In this work no attention is given to the glass transition.

2 Experimental

Tapioca starch was obtained from Tipiak (Nantes, France).

DSC experiments were performed on a DSC 111 Setaram (France). The required amounts of water and starch, either native or thermally modified, were weighed into stainless steel pans, hermetically sealed, and allowed to equilibrate overnight at 4 °C before measurement. Samples were then heated from 25 to 180 °C at a rate of 3 °C min^{-1} with a sensitivity of 0.1 mV. A known weight of water was used in the reference pan to determine the heat capacity of the sample pan. A blank sample was run against the reference pan and the resulting data substracted from all the curves.

Starch samples, equilibrated at different moisture contents from 20 to

80 wt% water, were placed in flat, sealed discs of aluminium (diameter 100 mm, height 3 mm) and immersed in a hot oil bath for 1 h at a constant temperature (± 1 °C). The products were then either immediately examined by light microscopy or quenched in liquid nitrogen, freeze-dried and stored before use in the DSC and X-ray diffraction experiments.

In light microscopy experiments, sample aliquots were dispersed in water and observed in brightfield or polarization modes. The state of the starch granules (*i.e.* native size or swollen; intact or disrupted; birefringence) was noted.

For X-ray diffractometry, heated samples were moistened to 20–25 wt% water. Diffraction patterns were recorded in transmission mode with an Inel generator working at 40 kV and 30 mA. The CuKα_1 ($\lambda = 0.15405$ nm) ray was selected. X-diffraction photons were counted by a curve position sensitive (CPS 120) detector at 120° (Θ). The crystalline index (%) was determined using Wackelin's method.[6] The crystallinity level was estimated taking native starch as 100% and amorphous starch as 0% crystallinity.

3 Results and Discussion

DSC Thermal Profiles of Starch + Water Mixtures

With water volume fractions higher than 0.7, a single endotherm (P1) was observed at 68 °C, with initial and final temperatures of 57 °C and 81 °C, respectively. The reproducibility of the temperature was ± 0.3 °C. For volume fractions of water, ϕ, between 0.65 and 0.45, two endothermic peaks were observed (Figure 1). As the water content was reduced from $\phi_1 = 0.65$ to $\phi_1 = 0.45$, the size of this P1 endotherm was found to be progressively reduced while a second endotherm P2 started to develop at higher temperatures and became predominant at low water contents ($\phi_1 = 0.45$). Whereas the P1 endotherm remained essentially at the same temperature, P2 shifted to higher temperatures when ϕ_1 was decreased. The multiple transitions have been characterized by their characteristic temperatures: the maximum peak temperatures T1 and T2, the temperature T3 between endotherms P1 and P2, and the completion temperature T4.

The Flory–Huggins equation, which describes the phase behaviour of polymer + diluent mixtures, has been applied to evaluate thermodynamic parameters of melting of starch crystallites. According to this theory, based on assumptions of equilibrium thermodynamics, the equilibrium melting point of a polymer, T_m^o, in the presence of a diluent is depressed to a new value, T_m:

$$1/T_m - 1/T_m^o = (R/\Delta H_u)\,(V_u/V_1)\,(\phi_1 - \chi_1\phi_1^2) \qquad (1)$$

where R is the gas constant, ΔH_u is the enthalpy of fusion per repeating glucose unit, V_u/V_1 is the ratio of the molar volumes of the repeating unit and the diluent, and χ_1 is the Flory–Huggins polymer–diluent interaction parameter. The melting-point of the diluent–polymer mixtures, T_m, is taken as the upper temperature limit of the endotherm P2. The classical use of eqn (1) is based upon the assumption that the melting of the second endotherm is

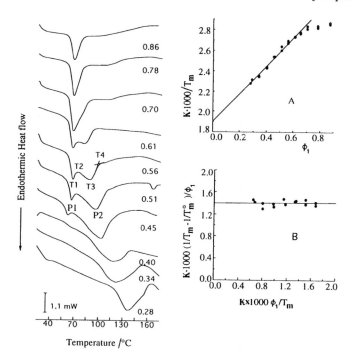

Figure 1 *DSC thermograms of tapioca starch + water mixtures labelled with their volume fraction ϕ_1 of water. T1, T2, T3, T4 are the characteristic temperatures of each thermogram. Insets A and B show plots of melting data according to the Flory–Huggins theory for the determination of T_m^o, ΔH_u and B*

determined by the average water content in the cell. For the lowest ϕ_1 values, the product $\chi_1\,\phi_1^2$ is negligible and a straight line can be drawn (Figure 1A). The intercept at $\phi_1 = 0$ gives the reciprocal of the melting-point of the most perfect crystallites of the undiluted polymer, $1/T_m^o$. For tapioca, T_m^o has been determined to be 253 °C which is far higher than corresponding values of 168 °C for potato[2] and 210 °C for wheat[4]. These melting temperatures are far lower than the temperature range (260–330 °C) where thermal decomposition of starch occurs. By introducing $B = \chi_1 RT/V_1$, the energy of polymer–diluent interaction, an alternative form of eqn (1)

$$(1/T_m - 1/T_m^o)\,/\,\phi_1 = (R/\Delta H_u)\,(V_u/V_1\,\phi_1/R\,T_m) \qquad (2)$$

gives the ΔH_u value, which has been calculated to be 34.7 ± 1.4 kJ mol^{-1}, and B which was computed from the slope to be 1.25 ± 0.48 J ml^{-1} (Figure 1B). This positive value of B suggests that water is a poor solvent for starch; a liquid–liquid phase separation could then occur above the melting temperature (253 °C).[1] This thermodynamic treatment is based only on the second melting transition and therefore thermodynamic parameters should be reconsidered

V. Garcia et al.

with regard to the existing first transition. No studies have been reported on the starch state in the temperature range T1–T3. The originality of the second part of this work has concerned a study of structural changes of starch at intermediate temperatures T1 (67 °C) and T2 (73 °C).

Characterization of Endothermic Transitions at Intermediate Water Content

In order to know if multiple endotherms could be interpreted by a successive disorder–order–disorder transition, starch samples were heated at the different characteristic temperatures T1, T2, T3, T4 observed in the DSC curve at $\phi_1 = 0.56$.

When observed by light microscopy, native starch granules heated between 67 °C (T1) and 73 °C (T2) at $\phi_1 = 0.56$ underwent progressive swelling (from an average size of 10 ± 4 μm to an average of 19 ± 12 μm for 73 °C heated starch). A progressive disruption of granules was also noticed, concomitant with the loss of birefringence. The total disappearance of polarizing pattern was associated with the completion of the second endotherm in agreement

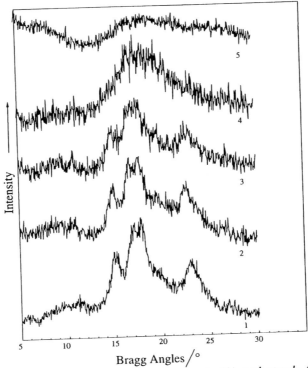

Figure 2 *X-ray diffraction patterns of native starch (1) and starch (45% water content) heated at 67 °C (2), 73 °C (3), 88 °C (4), and 99 °C (5) for 1 h with crystallinity levels of (1) 100%, (2) 74%, (3) 75 °C (4) 55%, and (5) 0%*

with previous studies on other granular starches.[7] However, the size distribution of native and treated granules displayed a population of small size (3–5 μm) which was more resistant to swelling and remained birefringent at T3 (88 °C). It may be assumed that these granules were more stable, their crystallites melting at higher temperatures.

X-ray diffraction of heated starches revealed the progressive disappearance of the initial A pattern between T1 and T3, with the broadening of peaks and the completion of crystallinity loss at T4 (Figure 2). These changes were in agreement with crystalline indices which were found to be 74%, 75%, 55% for 67 °C, 73 °C, 88 °C heat-treated starches, respectively. However, the accuracy of these crystalline levels depended on the value for the amorphous state of starch taken as 0% for the reference. It must be emphasized that our X-ray diffraction technique had not provided any evidence of annealing during heating, thus confirming the heat-programmed X-ray diffraction experiments of Zobel et al.[8]

This transition from an ordered to a disorganized state of the granule had been checked by DSC. Table 1 lists the characteristic values for the gelatinization of heat-treated starches when analysed by DSC in an excess of water. Damage state of starch (%), defined as the ratio of treated starch enthalpy to native starch enthalpy, decreased from 62 to 5% when starch samples were heated between 67 °C and 88 °C. This shift showed that the fraction of ungelatinized granules after hot bath treatment was reduced as a function of temperature. This quantitative estimation assumed the additive character of the melting reaction. The narrowing of residual gelatinization peaks as a function of treatment temperature demonstrated that a higher fraction of granules was transformed when starch was heated under the same conditions as described previously. However, no recrystallization occurred during heating since the final temperature remained constant (81 °C). Moreover, the weak shift of peaks to higher temperatures after the heat treatment at T1 and T2 could not be attributed to a retention of the crystallites in the starch granules.

Table 1   *Peak temperatures and residual enthalpy δH (45% water content) measured by DSC after thermal treatment. Crystallinity levels (%) are also shown*

Temperature of heat treatment [a]/ °C		Gelatinization temperature/ °C			δH/J g^{-1}	Crystallinity level/%
		Initial	*Maximum*	*Final*		
No treatment		57	68	81.4	16	100
T1	68	64	71.5	81.7	10	74
T2	73	64.3	73	81.4	7	75
T3	88	71	77.6	81.4	0.8	55
T4	99		No peak		0	0

[a] For 1 h.

The combination of structural analysis with DSC confirms the interpretation of Evans and Haisman[3]. The melting of a fraction of the granules during the first transition induces the disordering of polysaccharide chains and probably an increase in water absorption by the amorphous phase. This makes water unavailable for the remaining ungelatinized granules which will therefore melt at higher temperatures. Estimation of the volume fraction of water involved in each phenomenon can demonstrate this heterogeneous repartitioning of water in starch heated at intermediate water contents.

References

1. P. J. Flory, 'Principles of Polymer Chemistry', Cornell University Press, Ithaca, NY, 1953, ch. 13.
2. J. W. Donovan, *Biopolymers*, 1979, **18**, 263.
3. I. D. Evans and D. R. Haisman, *Starch/Stärke*, 1982, **34**, 224.
4. J. Lelievre, *J. Appl. Polym. Sci.*, 1973, **18**, 293.
5. C. G. Biliaderis, C. M. Page, T. J. Maurice, and B. O. Juliano, *J. Agric. Food Chem.*, 1986, **34**, 6.
6. J. H. Wakelin, H. S. Virgin, and E. Crystal, *J. Appl. Phys.*, 1959, **30**, 1654.
7. D. J. Burt and P. L. Russell, *Starch/Stärke*, 1983, **35**, 354.
8. H. F. Zobel, S. N. Young, and L. A. Rocca, *Cereal Chem.*, 1988, **65**, 443.

Concluding Remarks

By Denis Lorient

ENSBANA, CAMPUS UNIVERSITAIRE, 1 ESPLANADE ERASME, 21000 DIJON, FRANCE

Despite the great number and extreme diversity of topics discussed in this volume, most of the lectures and posters presented at the symposium can be classified under the following headings: surface properties of food macromolecules, rheological properties of macromolecular systems, and organization of dispersed food systems. Substantial progress has been made in these fields over the past 5 years or so, and it appears that some of the results are now becoming very useful for the technology of food products.

Surface Properties of Food Macromolecules

New aspects of surface properties have been demonstrated in solid dispersed systems such as bread and ice-cream (solid foams). It is becoming possible to explain the mechanisms of formation and stabilization of foams in terms of components such as hydrocolloids or proteins. The effect of interfacial properties of dispersed particles has been related to the microstructure in systems of varying degrees of complexity, including multiple emulsions.

Considerable advances in methodology have been described. The gap between our knowledge of the structure of proteins and polysaccharides and their interfacial behaviour is being successfully bridged by the application of a number of new physical and microscopic techniques: light scattering studies of dispersions and emulsions, ESR spectroscopy studies of lipid–protein interactions, and electron microscopy and image analysis studies of microstructure. Techniques such as immunochemical and biochemical methods have been used with success for the characterization of interfacial protein layers.

The influence of enzymic activity on the structure of the interface and the stability of emulsions has been investigated. Three examples of surface enzymology were presented: (i) the relationship between the activity and the competitive adsorption of proteins onto fat globules, (ii) the effect of proteolytic activity on the structure of a protein layer adsorbed on a lipid surface, and (iii) the effect of transglutaminase on proteins adsorbed on emulsion droplets. In each of these cases, the enzyme acts as a controller of the surface properties.

Rheological Properties of Macromolecular Systems

The second major theme of the conference was concerned with rheological aspects of food macromolecules. This is because, as the molecular structures of pure food macromolecules become better known, it is easier to establish reliable relationships between structure and rheological properties. Significant progress has been made in understanding factors controlling the mechanical properties of concentrated gels made from starch, pectin and milk proteins.

The complexity of real food systems makes it necessary to study the rheology of model mixed systems as a way of identifying the strength of the interactions between different components. Several studies of this type have been reported here: (i) synergistic effects in model polysaccharide mixtures (xanthan + galactomannan, κ-carrageenan + galactomannan, mixed carrageenan systems), and (ii) compatibility of macromolecular components in mixtures (milk protein + pectin, gelatinized starch + caseinate, mixed milk protein systems). A better knowledge of the macromolecular interactions in these model systems allows better control of the sensory properties in real foods. By using multiparametric techniques, it becomes possible to estimate the effect of various factors on these interactions.

Research on gels and glasses has led to new insights. Specific cases of interest are mixed polysaccharide gels, the influence of polymers on the glass transition of the freeze-concentrated phase, and gelation of proteins induced by high pressure. In this last case it is clear that for whey proteins the gelation mechanism is quite different from that induced by heat.

Organization of Dispersed Food Systems

The third topic of the conference was concerned with how the properties of small dispersed molecules are affected by the structure of dispersed systems and by the different phases in which they are solubilized or upon which they are adsorbed. This topic includes dispersions of liquid crystals in oils as well as volatile compounds in emulsions and/or protein systems. The activity of these small dispersed molecules depends on the partition coefficients between the different phases and on diffusion coefficients through interfaces during system formation or storage.

Final Conclusions

Several main trends can be identified from this conference. It is clear that, in order to make progress, scientists from various fields must work together, *i.e.* biochemists, organic chemists, physical chemists, biologists, as well as technologists. This is the reason why it may be difficult in the future to restrict the scope of topics for such a meeting. For example, the range of experimental techniques being used to study the properties of surface layers is now very broad (scattering techniques, immunochemical, and enzymic methods). However, the list of studied macromolecules is still rather narrow. Proteins studied are very often just dairy proteins: in the future we need to encourage the use of other well defined proteins from vegetables, cereals, meat, and eggs.

Kinetic aspects of dispersed systems are attracting increased attention. In the future, microstructures must be considered as moving or dynamic systems which are continuously modified by changes in the environmental conditions. These kinetic aspects must be taken even more into account, including molecular transport between phases, if we are to obtain an improved understanding of the functional behaviour of food macromolecules.

The new trends in research presented in this volume may be helpful in defining new programmes for the progress of food macromolecular science and for developing new products and ingredients with improved functionality.

Subject Index

Absorption isotherm, 508, 510
Acid milk drinks, 349–55
Acoustic properties of food, 453–4 (*see also* ultrasonics)
Activation energy, rheological, 483, 484
Activation volume of reaction, 134
Activity coefficients, infinite dilution, 128
Actomyosin, 125, 192
Adhesion
    of fat crystals, 418–19, 420, 425
    work of, 82
Adhesiveness of particles, 352
Admul Wol, 236–9
Adsorbed protein layers
    composition of, 25–7
    diffusion in, 293–4
    diffusion through, 154–63
    dynamics of, 2, 43–9, 115–9, 293–4
    effect on emulsion stability, 255, 272–4
    effect of ethanol on, 104–7
    effect of ionic strength on, 92–3
    effect of pH, 27, 77–80, 206–7
    effect of sucrose on, 104–7
    effect of temperature on, 100–2, 107, 270–2
    interactions in, 4, 27–8, 30, 103–8, 211–13, 272, 294
    modelling of, 4–7, 71–6, 79–80, 114–19
    overview of, 572
    relaxation processes in, 47–9
    structure of, 2–7, 23–42, 77–80, 90–4
Adsorption
    onto amylopectin, 556–9
    associative, 201, 204, 246
    competitive, *see* competitive adsorption

of emulsifiers, 201–2
of fat crystals, 195–6, 365
onto fat crystals, 196
of fat droplets, 307, 366
onto ice crystals, 141
isotherms, 217–21, 352
of hydrocolloids, 145, 208–13, 351–5
kinetics of, 86, 101–2, 105, 115–19
Langmuir-type, 53
of Lennard–Jones spheres, 114–19
of polymers, 71–6, 202
of polysaccharide, 206–7, 351–5
of protein, 4–5, 24–5, 77–80, 86, 88, 90–4, 99–102, 161, 202–5, 210–12, 217–18
onto protein particles, 351–5
of water, 556–9
Aeration (*see also* foam formation, whipping)
    of ice-cream, 303–5
Aerosol whipped cream, 309–11
Agarose, 379, 484, 486, 497
Aggregation (*see also* flocculation)
    of emulsion droplets, 61, 63
    of fat crystals, 418–19, 422–3
    kinetics of, 171–7, 438–43, 458–9
    of polysaccharide, 387
    of protein, 171–7, 349–55, 390, 393, 394, 395, 408, 428, 437–44, 456, 458–60
    simulation of, 398
Alanine, 523, 553
Alginate, 211, 356–9, 488, 489, 490
Amphiphiles, *see* surfactants
Amylase, 34, 35
Amylopectin, 10, 472, 473–8, 509, 510, 556–9

Amylose, 344, 346, 347, 473–8, 509, 510
8-Anilino-1-naphthalenesulfonic acid
    (ANS), *see* surface hydrophobicity
Annealing, near glass transition, 524, 528,
    534–40, 570
Antibody activity, effect of adsorption
    on, 34–5
Antibody interaction with adsorbed
    proteins, 38–9
Antifreeze biopolymers, 141, 144
Apo-α-lactalbumin, 168, 169, 411–12
Apple pectin, 431
Arabinogalactan, 329, 331
Aroma compounds, 123–33, 154–63
Arrhenius activation energy, 173, 174
Arrhenius behaviour, near glass
    transition, 530

Baking, 282–3, 472
Ballotini, 298–305
Batter, 503–6
Beer foam, 285, 296
Biaxial extension, 278, 280, 281
Biaxial stress, 280–1, 282
Bile salts, 59, 68, 206–7, 210, 244
Binding isotherms, 290, 433
Bingham yield stress, 420, 423 (*see also*
    yield stress)
Binodal, 328, 330, 331
Biphasic endotherm, 534, 535, 566
Biscuit dough, 499
Blood serum proteins, 202 (*see also*
    bovine serum albumin *et al.*)
Bohlin Rheometer, 420, 427, 456, 499,
    503
Bovine serum albumin (BSA), 7, 10–11,
    40, 45–8, 103–7, 125, 126, 129, 144,
    164–6, 168–70, 236, 237, 239, 241,
    242, 312–15, 329, 334, 410, 524
Boyle's law, as applied to gas bubble,
    316
Brabender viscograph (Amylograph),
    341, 507, 509
Breadmaking, 277, 282, 283, 472
Bread dough, 277–84, 285, 499, 505
Bread staling, 558 (*see also*
    retrogradation)
Bridging flocculation, 92, 94, 212, 422
Bromelain, 356–9
Brownian dynamics simulation, 5, 12–14
Brookfield viscometer, 135, 249
Bubbles
    coalescence of, 279, 282, 283, 312
    deformation of, 281
    detachment of, 317, 318

disproportionation of, 279–81, 283,
    309–11, 312
formation of, 278–9, 304
in gels, 406, 408
growth of, 279, 281, 282, 316–18
rupture of, 278
shrinkage of, 280, 310
size distribution of, 277, 278–9, 281,
    283, 304
stability of, 277–84
Butter oil, 81, 84
Buttermilk, 349, 364

Cabbage fibre, 246
Calcium caseinate, 184
Calcium ions, 9, 93–4, 136, 168–70, 368–72,
    406, 408, 410, 411, 431–6, 523
Calcium phosphate, 31, 32
Capillary extrusion, 513, 514
Capillary tube, bubble detachment from,
    318
Capillary viscometer, 350
Carbohydrate liquids, kinetic processes
    in, 543–51
Carbon dioxide, 279, 283
Carboxymethylcellulose (CMC), 144,
    211, 329, 331
Carob, 376, 380–3
Carob flour, 322
Carrageenan, 211, 356–9, 488–90, 495–9,
    526
    ι-carrageenan, 376–9, 462–71, 495–9
    κ-carrageenan, 321–7, 329, 331, 335–6,
        340, 367–72, 376–9, 481, 484, 485,
        488, 490, 495–9
κ-Carrageenanase, 496, 497
Carri-med rheometer, 322, 330, 342, 391,
    432
Casein(ate)(s), 23–32, 50–7, 90, 95–102,
    128–9, 136, 146–62, 182–8, 193–5,
    203, 204, 215–22, 236, 238, 239–41,
    299, 341–55, 364–7, 371, 372, 450,
    480, 482, 514
    α_s-(α_{s1}-)casein, 9, 26, 32, 36–7, 38,
        39–40, 51–2, 94, 99–101, 221, 458–9
    β-casein, 2–4, 7–9, 25, 26, 30, 37, 45–7,
        51–2, 77–80, 91–4, 99–102, 161, 221,
        256–8
    κ-casein, 31, 32, 51–6, 90, 92–4, 99–102,
        456, 460
    para-κ-casein, 52, 53, 54
Casein gels, 138–40, 349–55, 512–15 (*see
    also* milk gels, yoghurt)
Casein micelles, 14, 24, 31–2, 50, 58, 136,
    140, 217, 365, 456

Casein particles, 349–55, 493
Caseinomacropeptide, *see*
    glycomacropeptide
Catalase, 34, 35
Cavity, bubble growth at, 316–18
Centrifugation, 51, 61, 136, 216, 249, 329,
    330, 351, 352, 353, 354, 439, 457, 492,
    503
Cheese
    making, 50, 57, 89, 456, 460
    rheology, 512–16
Chitosan, 206–7, 210
Chlorogenic acid, 426, 427
Cholesterol, 205
Churning, 216
Chymosin, 37, 50–7, 457
Chymotrypsin, 36
Circular dichroism, 288, 289, 294
Citrate, effect on whey protein
    adsorption, 27
Citrus pectin, 431
Cluster–cluster aggregation, 398
Clustering of fat droplets, 217
Coacervation, 329
Coagulation, *see* aggregation
Coalescence, 10–11, 96, 215, 217, 245–6,
    282, 312
    orthokinetic, 11, 216, 253–5, 272–4
Coconut oil, 366
Cohesiveness, measurement of, 489, 490
Coil–helix transition, 368, 379, 380, 383,
    468, 484
Collagen, 462, 554
Colloidal stability, acid milk drinks, 349–
    55
Colour characteristics, fish protein gels,
    401–4, 406
Competitive adsorption
    between emulsifiers, 201–2
    between Lennard–Jones spheres, 114–19
    between proteins, 9, 25–7, 41, 64, 95–8,
        99–102, 202
    between proteins and emulsifiers, 7–9,
        26, 27–8, 60, 66–8, 202, 256–60,
        270–2, 285–96
    simulation of, 8, 114–19
    during spray drying, 202–4
Composite gel, 512
Compression test, *see* uniaxial
    compression
Computer simulation, 4–5, 8, 12–14,
    114–19, 398
Concentrated starch gels, 472–9
Conductivity
    aqueous solution, 544, 546–7

foam, 287, 290
Contrast matching, 3, 77, 391
Contraves Rheomat, 341
Controlled release, 235
Co-operative adsorption, *see* adsorption,
    associative
Co-operative binding, 129, 287, 290, 432,
    434-5
Co-operative relaxation, 484, 485
Couchman expression, 526, 561, 563, 564
Coulter counter, 298, 300
Crab analogues, 189–93
Crack propagation, 447, 450, 454
Cream cheese, 514
Creaming, 52, 212, 219–34, 261, 266, 267
    (*see also* sedimentation)
Creep compliance, 481–3
Crispness, 448, 453–4
Critical coalescence time, 241
Critical micelle concentration (CMC), 8,
    287–8, 292
Critical point, 331, 334
Cross-linking of proteins, enzymic,
    410–17
Cross-links in thermo-reversible gels,
    480–6
Cryoprotection, 531
Cryoscopic curve, 519
Crystallization
    in emulsions, 9, 223
    of fat, 418–25
Crystals
    ice, 141, 211, 297, 302, 303, 519, 521,
        522, 529
    triglyceride, 418–25
Curd, 456, 458, 460, 513
Cyclodextrin, 130

Dairy cream, 301
Debye scattering region, 392
Deformable particles, 4–5, 346, 448, 474,
    476
Denaturation (*see* proteins, denaturation
    of)
Depletion flocculation, 1, 12, 212, 221,
    224, 266, 267
Dephosphorylated β-casein, 92–4
Detergent, *see* surfactants
Dextran, 10, 209, 210, 265–7, 328–37,
    523, 525–9
Dextran sulfate, 10–11, 312–15
Dextrin, 131
Diacetyl tartaric acid ester of
    monoglyceride (DATEM), 271–4
Dialysis, 60, 123, 126, 287, 357, 428

Dielectric relaxation, 544, 547, 548–50
Differential scanning calorimetry (DSC),
    28, 29, 167–77, 377–9, 408, 427, 429,
    473, 474, 507, 509, 521–3, 525–8,
    531–42, 557, 561, 566–8, 570
Diffusion coefficients, 147, 156–61, 544–8
Diffusion-controlled reactions, 543–6
Diffusion layer, 157
Diffusion-limited aggregation, 398
Digestion, 69, 244, 356, 359
Diglycerides, 85, 89, 196, 211, 212
Dimethylpyrazine, 155, 157, 158, 159
Dimethyl sulfoxide, 484, 485, 486
Dispersed food systems, overview of, 573
Disproportionation, 14, 279–81, 283,
    309–11, 312
Disruption of flocs, 11–12
Disulfide bonds, 4, 11, 23, 28, 32, 71, 93,
    100, 136–7, 191, 285, 406, 411, 437,
    438, 499, 501, 552, 553
n-Dodecane, 58, 59, 62–9
Dough, 71, 277–84, 285, 449, 499–502
Dried foods, 543
Droplet size
    effect of protein concentration on,
        24–5, 253
    effect of shear flow on, 240–1, 252–5
    effect on emulsion gel rheology, 512–16
    in ice-cream, 302
    time-dependent changes in, 9–12, 237,
        245–6, 252–5
Droplet-size distribution, *see* particle-size
    distribution
Drug delivery, 205, 236
Drying experiments, 546, 548
Dry heating, complexes from, 10
Dynamic coupled column liquid
    chromatography (DCCLC), 129–32
Dynamic light scattering, 438, 439 (*see
    also* photon correlation spectroscopy)
Dynamic mechanical thermal analysis
    (DMTA), 521, 525
Dynamic surface properties, 43–9 (*see
    also* surface diffusion, surface
    dilational rheology, surface shear
    viscosity)

Egg, effect of high pressure on, 134
Egg white, 190, 286, 294, 295
Egg yolk, 204
Elastase, 524
Elastin, 554
Eldridge–Ferry equation, 484
Electron microscopy, 15, 136, 139, 146,
    147–8, 150–2, 182–3, 188, 204, 222,
349, 365–7, 369, 370, 372–3, 393, 395,
    414–15
Electron spectroscopy for chemical
    analysis (ESCA), 203
Electron spin resonance (ESR), 36, 81–4,
    540
Electrophoretic mobility, 206, 237, 241,
    242, 245, 256–60
Electrostatic stabilization, 250, 351
Ellipsometry, 205
Emulsification, *see* homogenization
Emulsifier(s), *see* surfactants
Emulsifying activity index (EAI), 164, 166
Emulsifying capacity of protein, 86
Emulsion rheology, 12, 249–50, 261, 264,
    267, 416
Emulsion stability
    effect of heat treatment on, 269–74
    effect of hydrocolloids on, 207–13,
        244–7, 261, 266
    effect of pH on, 207
    effect of protein content on, 215–22,
        249–50, 364–5
    effects on salt on, 249–50
    prediction of, 9–12, 230–2
    in shear flow, 216–22, 239, 240–1,
        252–5, 272–4
Encapsulation, 235, 242, 543, 544
Engineered food structure, *see* structure
    engineering
Entanglements, 466, 469, 473, 529
Enthalpy of cross-link breakdown, 483,
    484
Enthalpy of denaturation, 167–72
Enthalpy relaxation, 521, 522, 524, 526,
    534–42
Entropy of cross-links, 484
Enzymic activity, 51–2, 61–2, 65–6
    effect of adsorption on, 34–5
Enzymic coagulation, *see* renneting
Enzymic cross-linking of protein, 410–17
Enzymic hydrolysis, *see* lipolysis,
    proteolysis
Enzymology, surface, 572
Esterasic activity in emulsions, 65–6
Ethanol, effect on adsorbed layers, 104–7,
    110–13
Ethyl acetate, 155, 157, 158, 159
Ethyl butanoate, 155, 157, 158, 159
Ethyl hydroxycellulose, 209
Ethylenediaminetetraacetic acid (EDTA),
    401, 406–8
Excluded volume, 332–3
Experimental techniques, new, 572
Exponential dilution, 127–9

Extruded caseinate, 184
Extrusion cooking, 189–93

Faba bean protein, 178–81
Falling ball method, 381
Fast protein liquid chromatography
   (FPLC), 257
Fat crystals
   aggregation of, 418–19, 422, 425
   effect of milk proteins on, 365
   growth of, 425
   networks of, 418–25
   at the oil–water interface, 194–7, 365
Fat extraction technique, 217
Fat globules
   in cheese, 512–16
   in ice-cream, 302–3
Fat removal from whey, 85–9
Fatty acids, *see* free fatty acids
Fermentation, 279–82
Films
   drainage of, 288, 292–3
   elasticity of, 109–13
   extension of, 281, 282
   stabilization of, 109
   thinning of, 288
Fish protein, 189–93, 400–9, 414
Flavour release, 126–7, 154, 387
Flavour removal, 123
Flocculation (*see also* aggregation)
   of casein dispersions, 350–2
   effect on creaming, 11–12, 52, 221, 224,
     266, 267
   effect of pH on, 207, 212
   of fat crystals, 422, 425
   by hydrocolloids, 208–13, 266, 267
   kinetics of, 208
   orthokinetic, 11
   by protein, 92
   by salt, 208, 250
Flory–Huggins theory, 566, 567
Flour, 277, 282, 504, 506, 558
Flour milling, 539
Fluoresence measurements, 59, 287, 288,
   290, 292, 413
Foam formation, 104, 281, 303–4, 313–14
   (*see also* aeration, whipping)
Foamability, 287
Foaming properties of lipid-binding
   protein, 285–96
Foam stability, 14–15, 86, 179, 180, 287,
   288–91, 294, 295, 309–15
Folding test, 401
Food colloids, interdisciplinary approach
   to, 573

Form factor, 391
Formol tritration, 182
Fractal aggregates, 253, 393, 395, 398–9
Fractal dimension(ality), 1, 12, 253, 352,
   391, 393, 395, 397, 398–9
Fractal structure, 5, 12, 465
Fracture properties, 374, 447–55, 472–9
Fracture energy, 447, 453, 454
Fracture stress (strain), 374, 384, 385,
   387, 447, 448, 449, 451, 475
Free fatty acids, 59, 89, 196, 296
Freeze-drying, 525, 531
Freezing, 141–5, 298, 519–33, 545
Frozen food products, 519, 520, 531–2, 543
Fructose, 560
Fruit pulp, 488
Functionality, biopolymer, 363–75
Furcellaran, 480, 481, 485
Future trends in food colloids research,
   573–4

Galactomannan mixed systems, 321–7,
   340, 371, 380–3
Galactose, 522, 523
Galacturonic acid, 353
Gas cells
   in bread, 281–3
   in ice-cream, 302–4
Gas bubbles, *see* bubbles
Gas diffusion to a cavity, 316–18
Gas–liquid chromatography (GLC), 86,
   87, 128, 155
Gas–liquid transition, 14
Gas microcells, 14–15
Gel(s)
   chemical, 462, 463, 480
   filled, 336–8, 452, 453, 474
   fracture of, 387, 447–55
   particle, 12–14, 139, 349, 372–4, 395,
     415–16
   physical, 462, 463, 480
   polymeric, 12, 321–7, 366–72, 377–89,
     426–30, 480–7, 495–9
   weak, 346, 401
Gel filtration, 123, 129
Gel permeation chromatography, 53, 173,
   179, 180, 439, 464
Gel(ation) time/point, 458–61, 465, 466,
   467, 469
Gelatin, 99–102, 144, 236, 237, 239–40,
   241, 364, 365, 366, 448, 449, 462–71,
   480, 482, 484, 485, 523, 525, 526, 528,
   529, 554
Gelatinization, 277, 282, 340–8, 472, 507,
   508, 509, 534–6, 537, 566, 570

Gelation
    amylose, 344, 473, 509
    bean globulins, 413–16
    ι-carrageenan, 462–71, 495–9
    κ-carrageenan, 495–9
    egg, 134
    emulsions, 416
    fish myofibrillar protein, 400–9
    gelatin, 448, 449, 462–71, 484, 485
    gellan gum, 383–7
    kinetics of, 344, 345, 347, 429, 456–61
    β-lactoglobulin, 390–9, 429, 437
    milk, 135, 138–40, 456–61, 492–4, 512
    mixed biopolymers, 12, 321–7, 335–8,
        371–2, 376–89
    pectin, 431–6
    pressure-induced, 400–9
    starch, 340–8
    sunflower protein, 426–30
    theory of, 463
    types of, 12, 448
Gellan, 383–7, 488, 489, 490
Gibbs surface, 4, 6
Glass transition curve, 519, 520, 531
Glass transition (temperature, T_g)
    of amylopectin, 556–8
    of food proteins, 552–5
    in frozen systems, 519–33
    in granular starches, 534–42
    in Xixona turrón, 560–6
Glassy state, properties of, 536, 537, 539,
    543, 546, 550
Glassy-to-rubbery state transformation,
    534, 538
Gliadins, 71–6, 277, 278, 280
Globular protein(s)
    aggregation of, 390, 437–44
    conformational stability of, 167–70,
        390
    denaturation of, 137, 167–70, 189–93,
        335–8, 372–4, 390–8, 402–8, 429,
        437–44, 486
Glucono-*δ*-lactone, 488, 489, 513
Glucose, 544, 549, 550, 560
    effect on adsorbed layers, 110–13
Glutamyl residues, reactivity of, 410,
    412–13
Gluten, 71, 277–84, 499, 504, 505
Glycerol, 523, 524, 554
Glycerol lacto-palmitate gel, 450
Glycerol monostearate (GMS), 9, 81–3,
    477 (*see also* monoglycerides)
Glycine, 203–4, 553
Glycinin, 413, 414, 416
Glycomacropeptide, 52, 456

Glycosylation of *κ*-casein, 92-3
Gordon–Taylor equation, 546, 557, 558,
    561, 563
Gouch–Joule effect, 480
Guar gum, 144, 244, 246, 247, 321–6, 329,
    331, 336, 340, 526, 528, 529
Guinier plateau, 392
Guinier plot, 3
Gum tragacanth, 209, 210, 353

Haake viscometer, 183, 356
Hard-disc model, 45
Hard-sphere model, 5, 353
Head-space analysis, 127, 155
Heat-induced denaturation, *see* thermal
    denaturation
α-Helix(-ices), 39–40, 165–6, 289
Helix–coil transition, *see* coil–helix
    transition
Hencky strain, 450, 452, 473, 474
2-Heptanone, 125
Herschel–Bulkeley equation, 343
Heterogeneous bubble growth, 316
Heterogeneous nucleation, 145
n-Hexadecane, 9
High pressure liquid chromatography
    (HPLC), 53, 129–32, 178–9, 248–9,
    330, 413
Hill equation, 287, 290, 435
High pressure processing, 134–40, 400–9,
    573
Holo-*α*-lactalbumin, 168, 169, 411–12
Homogenization, 25, 26, 31–2, 34, 51,
    95–8, 146, 147, 148, 152, 208, 215,
    236–7, 249, 257, 298, 349, 350, 512
Homogenization clusters, 217
Human serum albumin, 165, 167
Hydration time, gluten, 504, 505
Hydrocarbon–water interface, 8–9,
    58–70, 194–7, 270–2
Hydrocolloids
    effect on crystalline ice, 144, 145, 526
    effect on protein hydrolysis, 356–9
    effect on starch, 340
    gelling properties, 463, 488–91
    stabilizing properties, 207–13, 236, 526
Hydrodynamics
    during creaming, 230
    of rotating diffusion cell, 156–7, 160
Hydrolysate of BSA, emulsifying
    properties of, 164–6
Hydrophile–lipophile balance (HLB),
    204
Hydrophobic (interaction)
    chromatography, 60, 165

Hydrophobic interaction, 55, 59, 112, 123, 136, 152, 191, 201–2, 406, 427
Hydrophobicity, protein, 8, 35–6, 39–40, 59, 99, 125, 179, 180, 294

Ice crystals, 141, 211, 297, 302, 303, 519, 521, 522, 529
Ice dispersions, 298, 301–3
Ice-cream, 89, 211, 212, 297–308, 519, 520, 526, 532
Image analysis, 143, 374, 513
Imidazole, 27, 44, 147, 237, 241, 242
Immunochemical analysis of adsorbed proteins, 38–9
Immunoglobulin, 34, 35
Incompatibility, *see* thermodynamic incompatibility
Instron, 378, 384, 489, 492, 513
Interactions between surfaces, 207–8
Interfacial complexation, 236, 239, 270, 272, 292, 294, 295, 312
Interfacial energy, *see* surface energy
Interfacial resistance, 155, 161–2
Interfacial shear viscosity, *see* surface shear viscosity
Interfacial tension, *see* surface tension
Intrinsic viscosity, 332, 353, 431
Iodide ions, effect on carrageenan gels, 495–8
β-Ionone, 126, 128, 129, 131
Isoelectric point, 3, 9, 77, 79, 80, 242, 330, 331, 333, 338, 349
Isoelectric precipitation, 60, 427
Isostress/isostrain boundaries, 378, 379

Jet homogenizer, 60, 236, 253
Junction zones in gelation, 468, 469, 484

Kefir, 349
Kerosene oil, 236–42
Kinetics in highly viscous liquids, 543–51
Kjeldahl analysis, 330, 357
Klotz plot, 124
Krieger–Dougherty equation, 353, 354
Kronig–Kramers relation, 463
Kubelka–Munk scattering model, 299–305

α-Lactalbumin, 24, 28, 168–70, 329, 333–4, 338, 410–13
β-Lactoglobulin, 7–9, 14, 24, 28, 38–9, 40, 44, 45–7, 67, 68, 94, 125, 131, 168–77, 252–60, 269–74, 312–15, 329, 333–4, 338, 373–4, 390–9, 429, 437–44
Lactose, 203–4, 523, 524

Lamellar phase, 204
Langmuir film balance, 96, 109
Laplace equation, 312, 316
Laplace pressure, 14, 278, 281, 309, 316
Lard, 83
Laser diffraction, 51
Latex(-ices), 28, 30, 90–4, 146, 353, 513
Le Chatelier's principle, 134
Lecithin(s), 9, 27, 30, 146, 209, 210, 419
Legumin, 411, 413–15
Lennard–Jones potential, 115
Leucine, 553
Levich equation, 156
Levich plots, 157–61
Ligand binding, 123–4, 287, 290–2, 295
Light absorption, theory of, 300
Light microscopy, use of, 142, 245, 288, 330, 332, 335, 337, 350–1, 364–6, 373, 424, 425, 476, 556, 566, 567
Light scattering, 9, 135, 147, 297–308, 438, 439
Limonene, 126, 128–31
Linalool, 128, 129
Linear visco-elastic region, 396, 483, 492–3
Lipid analysis, 86, 87
Lipid-binding protein, wheat, 285–96
Lipid
  in flour, 504–5
  foam destabilization by, 285, 294, 296
Lipid digestion, 244
Lipid–protein interactions, *see* protein-lipid interactions
Lipolysis, 58–70, 85–9
Lipoprotein particles, 205
Liposomes, 90, 92, 94
Liquid crystals, 245, 246
Liquid–liquid partition, 123, 126–7
Locust bean gum (LBG), 211, 321–6, 340, 371–2, 488, 489, 490, 526
Low-fat food products, 2, 14, 363, 364, 372, 418, 449
Lysophosphatidylcholine (LPC), 285–95
Lysozyme, 35, 45, 236, 241, 242, 524
Lysyl residues, reactivity of, 410, 412

Macromolecular interactions, overview of, 573
Macromolecules, effect on glass transition, 519–33
Maltodextrin(s), 15, 298–303, 329–36, 484, 526, 528, 529, 558, 565
Maltose, 544, 547, 549
Margarine, 364, 448, 449
Mastersizer, 10, 11, 61, 183, 216, 237, 245, 252, 257, 350

Mechanical properties, gels, 447–55 (*see also* rheology, viscosity)
Mechanical relaxation of gels, 480–7
Mechanical spectroscopy, *see* oscillatory rheology
Melting temperature, gel, 323–6, 378–9, 381
Membrane, mass transport through, 126, 155–62, 262
Mercaptoethanol, 61, 401, 406, 411, 428
Methionine, 356, 359
Methoxyl pectin, 349, 355, 431
Methylcellulose, 209, 329–38
Methyl ketones, 123, 125, 131
Microcrystalline cellulose, 236
Microfluidizer, 25, 95, 97, 147, 207, 208
Microgel particles, 386, 387
Micronization, 507–10
Microstructure, *see* electron microscopy, light microscopy, structure
Mie scattering, 300
Milk component 3 protein, 58–70
Milk fat globule membrane, 69, 95–8, 512
Milk gels, 136, 138–40, 448, 456–61, 512, 514 (*see also* casein gels, yoghurt)
Milk processing, 31, 134–40
Milling of starch, 503, 506, 507–11
Mixed biopolymer systems, 2, 321–59
Mobility near glass transition, 530–1, 532, 543–51
Model systems, relevance of, 16
Modelling
  of adsorption, 4–7, 71–6, 79–80, 114–19
  of coalescence, 253
  of competitive displacement, 114–19
  of creaming, 230–2
  of gelation, 12–14
  of networks, 395
Modification of proteins
  chemical, 2, 410, 411
  physical, 2, 178–81, 372, 411
Molecular dynamics simulation, 114–19
Molten globule state, 4, 411–13
Monoclonal antibodies, 38–9
Monoglyceride(s), 9, 85, 89, 103–13, 196, 209, 210, 211, 212, 419 (*see also* glycerol monostearate)
Monolayer stability, 110–12
Monophasic endotherm, 534, 535
Monte Carlo simulation, 4
Mozzarella cheese, 514
Multilayers, 77, 79–80, 201–13
Multiple emulsions, 235–43
Myofibrillar proteins, 400–9
Myoglobin, 554

Myosin, 192, 193, 401, 408

Networks
  fat crystals, 418–25
  interpenetrating, 326, 379
Neutron reflectivity, 3–4, 8, 36, 77–80
Neutron scattering, 390–9
Nitrous oxide, 309
2-Nonanone, 124, 125
Non-starch polysaccharide, 244–7
Notch sensitivity, 447, 451, 453
Nuclear magnetic resonance (NMR), 216, 495, 497, 540, 550
Nucleation
  fat crystal, 422
  ice crystal, 145

2-Octanone, 125
Oil-in-water emulsion, *see* emulsion
Olive oil, 244
Optical microscopy, *see* light microscopy
Optical rotation, 379, 381
Orange gels, 488–91
Orthokinetic stability, 11, 216–22, 252–5, 269, 272–4
Oscillatory rheology
  bulk, 322, 330, 342, 378, 381, 384, 386, 391–2, 396, 397, 401, 408, 427, 432, 448–9, 458, 481, 493, 497, 499, 503, 513
  surface, 44, 288, 310
Osmotic pressure of emulsions, 261–8
Ovalbumin, 410, 560, 561, 564
Overlap concentration, 6–7, 72, 75
Overrun, 309

Palm (kernel) oil, 299, 366, 419, 420–3
Palm stearin, 420–3
Papain, 356–9
Partial coalescence, 216
Particle packing, 352, 386, 387
Particle-size distribution, 8, 9–12, 228, 231, 252–5, 350, 364–6
Particulate gels, *see* gels, particle
Partition coefficients, 157–60
Pea flour, 508
Pectin(ate), 208, 209, 244, 246, 247, 329, 331, 349–55, 431–6, 481, 485
Pepsin, 37, 164, 356–9
Peptides as emulsifiers, 40, 164–6
Percolation theory, 436, 463, 469
Phase diagrams, 328–9, 331–4, 432, 433, 467, 468
Phase separation, 14, 266, 328–39, 340, 347, 348, 379, 473, 497, 568

in droplets, 365
Phase transitions, tapioca starch, 566–71
Phenol–sulfuric acid method, 330, 342
Phosphatidic acid, 205, 210
Phosphatidylcholine, 31, 94, 146–53, 205, 244–7
Phosphatidylglycerol, 94
Phospholipids, 27–8, 30, 31, 94, 146–53, 210, 285–96
Phosphoserines, casein, 92, 94
Photon correlation spectroscopy (PCS), 92, 146, 147, 148 (see also dynamic light scattering)
Physiological emulsions, 244–7
Phytate, 426, 427
Plant proteins, 178–81
Plasticizers, 534, 539, 544, 552–8
Polarimetry, 468
Polyacrylamide gel electrophoresis (PAGE), 61, 179, 189, 411, 412
Polyalcohols, 531
Polydispersity, effect on light scattering, 300
Polymer–polymer interactions in gels, 480
Polymorphism, relation to fat crystal sintering, 423, 425
Polyoxyethylene emulsifier, 256, 258–60 (see also Tweens)
Polysaccharides, see carrageenan, hydrocolloids, pectin, starch, etc.
Polysorbates, see Tweens
Polystyrene particles, see latex
Polyvinyl alcohol, 299
Polyvinylpyrrolidone, 523
Porcine pancreatic lipase, 58–70
Pore-size distribution, 12, 374
Potassium ions, effect on carrageenan gels, 495–8
Potentiometry, 431, 432, 434
Pressure transducer, 261–3, 312, 314
Protective colloid, pectin as, 351, 352
Protein(s)
    chemical modification of, 2, 126, 410, 411
    denaturation of, 14, 27, 28, 125, 137, 167–81, 269–74, 329, 331, 335–8, 400, 411–13, 429, 437–44, 552
    displacement of, 7–9, 50, 66–8, 81, 99, 107, 256–7, 270–2, 295
    foaming of, 179–81
    gelation of, 14, 138–40, 189–93, 349–55, 426–30
    glass transition temperatures of, 552–5
    physical modification of, 2, 178–81, 372, 411 (see also high pressure

processing, thermal denaturation, etc.)
Protein adsorbed layer
    ageing of, 9, 28, 99, 256, 258–60
    density profile, 3, 77–80
    dynamic behaviour, 2, 43–9, 115–9, 293–4
    effect of charge on, 90–4
    effect of pH on, 27, 77–80, 206–7
    effect of solvent quality on, 73, 75, 104–7
    effect of temperature on, 100–2, 107, 270–2
    mass transport through, 154–63
    multilayers, 24, 27, 79–80, 149
    overview of, 572
    theory of, 4–7, 71–6
    thickness, 2–4, 6–7, 28–31, 36, 77–80, 92–4, 149
Protein–aroma interactions, 123–33, 154
Protein–ligand interactions, 123–4, 129–30, 290–2
Protein–lipid interactions, 36, 39, 81–4, 103–8, 146–53, 205, 287, 290–6
Protein–polysaccharide interactions, 10–11, 312–15, 328–59, 371–2, 494
Protein–protein interactions, 50–7, 99–102, 137, 138, 189, 329, 331
Protein–surfactant interactions, 27, 30, 81–4, 272, 274
Proteolysis, 29, 36–7, 92, 147–53, 164, 356–9, 427, 456–61
Proteose-peptone, 58, 60, 68
Puroindolines, 285–96

Rapeseed oil, 419, 423, 424
Reaction kinetics, second-order, 544
Reaction-limited aggregation, 398
Recrystallization, amylopectin, 472–9
Reflectance studies, 297–308, 401–3
Refractive index ratio, 298, 300, 301
Relaxation modulus, 494
Renneting, 32, 456–61
Retinol, 131
Retrogradation, 340, 472–9
Rheology (see also viscosity)
    acid milk gels, 512, 514
    batter, 503–6
    cheese, 512–16
    dough, 277–84, 449, 499–502
    emulsions, 12, 249–50, 261, 264, 267, 416
    foam, 366
    fractal networks, 395
    geletin gels, 464–7, 480–2

Rheology (*cont.*)
  globular protein gels, 12, 374, 390–2,
    395–9, 414, 427–8
  mixed biopolymer gels, 321–7, 336,
    341–8, 371, 378, 381–3, 495–8
  myofibrillar protein gel, 400–9
  orange pulp gels, 488–91
  overview of, 573
  rennet milk gels, 456–61
  single polysaccharide gels, 368, 370,
    383–7, 466–7, 480–1, 485–6
  starch gels, 341–8, 472–9
  yoghurt, 492–4
Rheomat rheometer, 513
Rheometrics rheometer, 464
Ribonuclease A, 524
Rigid-sphere model, *see* hard-sphere
    model
Rotating diffusion cell, 155–62

Salting-out effect, 128, 370, 441
Scaling behaviour, rheological, 459–61,
    463, 483
Scaling theory of adsorbed polymers, 6–7,
    71–6
Scatchard plot, 124, 126, 287, 434, 435
Scattering phenomena, 226–7, 233,
    297–308, 391, 441
Second-order kinetics, 544
Sediment volume, 350, 352–3, 420–3
Sedimentation, 349–55, 420 (*see also*
    creaming)
Semi-dilute solution, 7, 71–6, 466, 485
Sensory properties, 373, 374, 453–4
Serum proteins, *see* whey proteins
β-Sheet, 289
Shear flow
  coalescence in, 11, 216, 253–5, 272–4
  effect on gelation, 336–8
Shock waves, 454
Sintering of fat crystals, 418–25
Silica, adsorption on, 205
Skim milk powder, 81, 211, 215–22, 364,
    366, 492, 513
Small-angle neutron scattering (SANS),
    390–9
Smoluchowski theory, 253
Sodium caprylate, 208, 209, 210
Sodium caseinate, *see* casein(ate)(s)
Sodium chloride, effect on β-
    lactoglobulin aggregation, 437–44
Sodium dodecyl sulfate (SDS), 27, 61,
    190–2, 221, 261, 263–7, 286, 401,
    405–6, 411, 412
Sodium metabisulfite (SMS), 499–502

Sodium stearoyl-2-lactylate (SSL), 278,
    280, 281, 283
Sodium taurocholate, 244–7
Sol–gel transition (*see also* gelation)
  ι-carrageenan, 462–71
  gelatin, 462–71
  β-lactoglobulin, 390–9
  pectin, 432, 435–6
Solubility
  of aroma compounds, 157
  of caseinates, 183, 186–8
  of emulsifiers, 202
  gas, 309
Solubility index, 342, 345–7
Solubilization
  of amylopectin, 556
  of fish protein, 190
  of wheat starch, 345–6, 347
Sorbitan esters, *see* Span(s), Tween(s)
Sorbitol, 523, 524
Sorption isotherm, 508, 510, 557
Soybean lecithin, *see* lecithin(s)
Soya (soybean) oil, 25, 26, 29, 35, 50, 51,
    207, 208, 236–42, 286, 287, 294, 295,
    416, 419–24
Soybean flour, 356–9
Soy(bean) protein, 45, 124, 125, 356–9, 413
Soybean trypsin inhibitor, 130–1
Span(s), 9, 204, 209, 210, 236–41, 366, 367
Spin probes, 82, 83–4
Spray drying, 202–4, 543
Spray freezing, 147
Stability map for frozen foods, 531
Stabilization of acid milk drinks, 349–55
Starch
  barley, 508
  damaged, 503, 506, 507–10, 570
  dialdehyde, 480, 482
  milling of, 507–11
  pasting of, 340–8
  pea, 507–11
  potato, 451, 472–8, 568
  rice, 209, 535–40
  swelling of, 342, 345, 347
  tapioca, 566–71
  waxy corn, 556
  waxy maize, 558
  wheat, 282, 340–8, 452, 453, 472–8,
    503, 509, 558, 568
Starch gels, 341–8, 451, 452, 453, 472–9
Starch granules, 346, 348, 473–8, 508,
    510, 534–42, 566, 567
Starch–sucrose interactions, 539–40
Steric stabilization, 7, 90, 250, 257, 352,
    355

Stevens apparatus, 489
Sticky particles, 352, 353–4
Stokes–Einstein relation, 147, 439, 543, 544, 546
Storage/loss moduli, *see* oscillatory rheology
Strain hardening, 281, 282, 283
Stress relaxation of adsorbed layers, 44–5, 48
Structure
    of adsorbed layer, 2–7, 23–42, 77–80, 90–4
    of aggregates, 393, 394, 395
    of caseinate powder, 182–3, 188
    of gels, 12–14, 390, 393, 406, 408, 414–16, 465, 472–9, 480–7
    of high-fat cheese, 514
Structure engineering, 329, 336, 338, 363–74
Structure factor, 391
Sucrose
    effect on adsorbed layers, 104–7, 110–13
    effect on starch gelatinization, 538, 539–40
Sucrose esters, 8, 9, 14, 15, 209
Sucrose + water systems, 519–29, 539, 543, 545, 560, 561
Sugar solutions, freezing of, 519–33, 545–6
Sulfur dioxide, 499
Sunflower oil, 194–6, 223, 230, 231, 233
Sunflower proteins, 426–30
Supercooling, 520
Supersaturation, 316, 317
Superstrands, polysaccharide, 368, 369, 372
Supramolecular structure, κ-carrageenan, 368–71
Surface activity, *see* adsorption, surface tension
Surface diffusion, 293–4
Surface dilational rheology, 43–9, 109–13, 277–84, 288, 310, 311, 503–6
Surface energy of ice crystals, 141–5
Surface enzymology, 572
Surface equation of state, 45–6, 72–3, 109–13
Surface hydrophobicity, 35, 135, 137–8, 140, 270
Surface pressure, 6–7, 43–9, 71–3, 86, 88, 96, 97, 109–10
Surface relaxation, 109–13
Surface shear viscosity, 2, 28, 194–7, 270, 271, 272
Surface tension, 8–9, 59, 60, 62, 81–3, 86,

88, 99–108, 204, 278, 310, 311, 312, 316, 505
Surfactants
    effect on emulsion gels, 416
    as emulsifiers, 261–8
    interaction with fat crystals, 418–19
    interaction with proteins, 27, 30, 81–4, 272, 274
    monolayers of, 109–13
Surimi, 189, 190, 400, 401, 408
Swelling, starch granules, 473, 509, 569, 570
Swelling index, 342, 345–7
Syneresis, 432, 493

Tail–train–loop theory, 4, 6, 72, 195
Tara gum, 321
Terpenyl acetate, 128, 129
n-Tetradecane, 8, 67, 104, 107, 194, 195, 236, 252–4, 269–72
Texture analysis, fish protein gels, 401–4, 406
Texture of ice-cream, 520–1, 526, 532
Thermal denaturation, 11, 14, 28–9, 125, 137, 167–77, 189–93, 335–8, 372–4, 390–8, 402–8, 426–30, 437–44, 485
Thermal hysteresis in biopolymer gelation, 324
Thermodynamic incompatibility, 1, 328–39, 340, 473
Thermodynamics of binding, 123–5
Thermodynamics of denaturation, 167–70
Thermo-reversible gelation, 338, 378, 380–3, 462, 464, 480–7
Thin film, *see* films
Thin layer chromatography, 86
Thixotropy, 343, 346, 351
Train–loop–tail model, 4, 6, 72, 195
Transglutaminase (TGase), 410–17
Transmembrane protein, 39, 152, 153, 285
Triglycerides, properties of, 69, 158, 419
Triglyceride–water interface, 9, 81–4, 194–7
Trypsin, 36, 37, 147–53, 164, 356–9, 427
Turbidimetry, 208
Turbulent flow, 253, 270
Tween(s), 9, 26, 27, 60, 66–8, 81–3, 204, 224, 236, 242, 253, 256–61, 263–7, 270–2, 286, 366, 367

Ultrafiltration, 123
Ultra-fine size reduction of starch, 507–11
Ultrasonics, 12, 223–34
Ultrasound scattering, 226–7, 232

Ultrasound velocity scanning, 223–34
Uniaxial compression, 278, 384, 385, 414,
    452, 473, 489, 496
Universal Quasi-chemical (UNIQUAC)
    model, 522
Urick equation, 223, 225–9, 232, 233

Van der Waals attraction, 349, 355
Van der Waals volume, 552–4
Vanillin, 125
Vesicles, 146–53
Vibration milling, 178, 180
Visco-elasticity, *see* rheology
Viscosity (*see also* rheology)
    casein(ate) dispersion, 186–7, 351,
        353–4
    dough, 278, 283, 502
    emulsion, 249–50, 264
    ice-cream, 303–4
    intrinsic, 332, 353, 431
    measurement of, 135, 139, 183, 356
    near glass transition, 543–8
    soybean flour, 356, 357
    starch paste, 507, 509
    sucrose solution, 520
Vogel–Tammann–Fulcher (VTF)
    behaviour, 528, 531, 532
Volume change on reaction, 134, 136
Voluminosity, 352
Vroman series, 202

Wafer batters, 503–6
Walden rule, 547
Water, physical properties of, 223
Water absorption capacity
    caseinate, 183, 186–8
    flour, 504
    starch, 507, 510
Water activity, 530, 532, 557, 558, 560,
    561
Water bridges between fat crystals, 419

Water-in-oil (W/O) emulsions, 364–5
Water-in-oil-in-water (W/O/W)
    emulsions, 235–43
Water-in-water emulsion, 329
Weber number, 278
Wheat cultivars, 499–506
Wheat lipids, 277, 278, 280
Wheat proteins, 277–96
Wheat starch, 282, 340–8, 452, 453,
    472–8, 503, 509, 558, 568
Whey, enzymic treatment of, 85–9
Whey proteins, 23–30, 167–70, 248–51,
    328–39 (*see also* α-lactalbumin, β-
    lactoglobulin)
Wheying off, 349
Whipped cream(s), 309–11, 365
Whipped emulsions, 364, 365–6
Whipping, 15, 95, 97, 211, 215, 216, 303
    (*see also* aeration, foam formation)
Wilhelmy plate, 60, 72, 81, 86, 100, 103
Williams–Landel–Ferry equation, 485,
    528, 532, 545, 546

Xixona turrón, 560–5
X-ray scattering (diffraction), 36, 507,
    566, 567, 569–70
Xanthan, 209, 224, 299, 329, 331, 340,
    376, 380–3, 488, 489, 490, 526
Xylose, 544, 549, 550

Yield analysis, multiple emulsion, 236–40
Yield force, 401, 405, 406–8
Yield stress, 343, 344, 346, 418, 419, 420,
    423, 425, 448, 513, 514
Yielding behaviour, 447, 450, 472
Yoghurt, 135, 139, 349, 350, 353, 492–4
    (*see also* casein gels, milk gels)
Young's modulus, 384, 385, 448, 451,
    473, 475

Zeta potential, 241, 242, 258–60, 351

New and Recent books on
FOOD SCIENCE AND NUTRITION

McCance and Widdowson's The Composition of Foods
5th Edition 3rd Reprint 1994
Edited by B. Holland and A.A. Welch, The Royal Society of Chemistry
I.D. Unwin, Information Consultant; D.H. Buss, Ministry of Agriculture, Fisheries and Food
A.A. Paul, MRC Dunn Nutritional Laboratory; D.A.T. Southgate, AFRC Institute of Food Research, Norwich

Softcover   xiii + 462 pages   ISBN 0 85186 391 4   1991   Price £35.00

Supplements to McCance and Widdowson's The Composition of Foods

Miscellaneous Foods
Edited by W. Chan and J. Brown, The Royal Society of Chemistry
D. H. Buss, Ministry of Agriculture, Fisheries and Food

Softcover   Approx 160 pages   ISBN 0 85186 360 4   1994   Price £27.50

Fish and Fish Products
Edited by B. Holland and J. Brown, The Royal Society of Chemistry
D.H. Buss, Ministry of Agriculture, Fisheries and Food

Softcover   vii + 135 pages   ISBN 0 85186 421 X   1993   Price £24.50

Vegetable Dishes
Edited by B. Holland and A.A. Welch, The Royal Society of Chemistry
D.H. Buss, Ministry of Agriculture, Fisheries and Food

Softcover   viii + 242 pages   ISBN 0 85186 396 5   1992   Price £24.50

Fruit and Nuts
Edited by B. Holland, The Royal Society of Chemistry
I.D. Unwin, Information Consultant; D.H. Buss, Ministry of Agriculture, Fisheries and Food

Softcover   vii + 136 pages   ISBN 0 85186 386 8   1992   Price £24.50

Vegetables, Herbs and Spices
Edited by B. Holland, The Royal Society of Chemistry
I.D. Unwin, Information Consultant; D.H. Buss, Ministry of Agriculture, Fisheries and Food

Softcover   163 pages   ISBN 0 85186 376 0   1991   Price £24.50

Milk Products and Eggs
Edited by B. Holland, Royal Society of Chemistry
I.D. Unwin, Information Consultant; D.H. Buss, Ministry of Agriculture, Fisheries and Food

Softcover   146 pages   ISBN 0 85186 366 3   1989   Price £20.50

Prices subject to change without notice.

To order please contact:
Turpin Distribution Services Ltd., Blackhorse Road, Letchworth, Herts SG6 1HN, UK.
Tel: +44 (0) 1462 672555. Fax: +44 (0) 1462 480947.

For further information please contact:
Sales and Promotion Department, Royal Society of Chemistry,
Thomas Graham House, Science Park, Milton Road, Cambridge CB4 4WF, UK.
Tel: +44 (0) 1223 420066. Fax: +44 (0) 1223 423623. E-mail: (Internet) RSC@RSC.ORG.

RSC members are entitled to a discount on most RSC products, and should contact
Membership Administration at our Cambridge address.

THE ROYAL
SOCIETY OF
CHEMISTRY

Information
Services